An Introduction to Planetary Atmospheres

An Introduction to Planetary Atmospheres

Agustín Sánchez-Lavega
University of the Basque Country
Bilbao, Spain

CRC Press
Taylor & Francis Group
Boca Raton   London   New York

CRC Press is an imprint of the
Taylor & Francis Group, an **informa** business

A TAYLOR & FRANCIS BOOK

Taylor & Francis
6000 Broken Sound Parkway NW, Suite 300
Boca Raton, FL 33487-2742

© 2011 by Taylor and Francis Group, LLC
Taylor & Francis is an Informa business

No claim to original U.S. Government works

Printed in the United States of America on acid-free paper
10 9 8 7 6 5 4 3 2 1

International Standard Book Number: 978-1-4200-6732-3 (Hardback)

Library of Congress Cataloging-in-Publication Data

Sanchez-Lavega, Agustin.
    An introduction to planetary atmospheres / by Agustin Sanchez-Lavega.
        p. cm.
    Includes bibliographical references and index.
    ISBN 978-1-4200-6732-3
    1. Planets--Atmospheres--Textbooks. I. Title.

QB603.A85S36 2011
551.50999'2--dc22   2010010718

Visit the Taylor & Francis Web site at
http://www.taylorandfrancis.com

and the CRC Press Web site at
http://www.crcpress.com

To my wife María Jesús who has greatly enriched my life.

To the memory of my parents.

Contents

Preface

"Planetary atmospheres" is a relatively new, interdisciplinary subject that embraces different areas of the physical and chemical sciences, including geophysical fluid dynamics, atomic and molecular spectroscopy, chemical reactions, interaction of radiation with matter (light absorption and scattering by gases and particles), thermodynamics, microphysics, turbulence, chaotic motions, etc. It is therefore a subject that straddles geophysics and geophysical fluid dynamics, atmospheric sciences, and astronomy and astrophysics. Planetary scientists derive atmospheric properties from in situ measurements obtained by probes and "landers," and also perform remote-sensing observations from orbiting satellites around planets and from telescopes placed on Earth and in space. Therefore, a scientific expert in planetary atmospheres is difficult to ascribe to a well-defined scientific discipline. Most planetary scientists come from an astronomy background, followed by those coming from geophysical fluid dynamics and atmospheric disciplines. In fact, research on the subject is published in a number of journals pertaining to all these areas. This book introduces current knowledge on atmospheres and the fundamental mechanisms operating on them to students and researchers working in this broad array of fields.

This book has been written mainly for undergraduate students in these areas. It may also be helpful for scientists specialized in a particular type of atmosphere but unfamiliar with other types, and planetary scientists who are looking for an overview of the subject.

The subjects covered are treated in a comparative manner among the different solar system bodies—what we call "comparative planetology." Comparative planetology developed rapidly with the advent of the space exploration era, and made enormous strides when the first spacecraft visited Venus and Mars in the 1960s. The comparative vision of the physical and chemical processes that occur in planetary atmospheres represents an important step in the knowledge of Earth's atmosphere, since this is best understood in the broad context of planetary atmospheres. The variety of properties and circumstances in planetary atmospheres is so large that they are in fact natural laboratories where we can test the theories developed to explain the mechanisms operating in the terrestrial atmosphere. In planets and satellites with substantial atmospheres, we find large differences in size (a factor of 62 between Jupiter and Pluto), gravitational force (a factor of 32 between the same extreme cases), in rotation angular speed (a factor of 583 between Venus and Jupiter), in the presence or not of a surface as a boundary condition (between terrestrial and gaseous planets), in the existence or not of an internal energy source, in the strength of the solar radiation (a factor of 2900 between Venus and Pluto), in the duration and importance of the annual insolation cycle (differences in the tilt of the rotation axis angle relative to the orbital plane), in chemical composition (between the heavy atmosphere of Venus and the much lighter ones of giant planets), and in many other properties that influence the state of atmospheres (planetary magnetic field, surface nature, topography, etc.). The common property that is relevant in all cases is that all phenomena in planetary atmospheres are subjected to the spherical geometry constraint.

From this point of view, the Earth's atmosphere is just one example among the families of atmospheres we find in the solar system. It is certainly fundamental for our lives, and both its day-to-day behavior and long-term evolution, including man-made and naturally induced variability, represent a major sociological concern. Natural disasters caused by atmospheric activity (e.g., hurricanes), polar ozone hole depletion, and the greenhouse effect and climatic change are fresh in everyone's mind. In other planets near the Earth, we find extreme versions of the phenomena found on Earth. For instance, there are huge rotating eddies in the major planets, some of which have survived for decades, and gigantic convective storms that ridicule those seen on Earth. The runaway carbon dioxide greenhouse effect in Venus produces temperatures of the order of a hundred Celsius. Saturn's satellite, Titan, exhibits a rich hydrocarbon chemistry, where a methane cycle probably plays a role similar to the hydrology cycle on Earth. Finally, a great variety of "rare clouds" (sulfuric acid, ammonia, methane, and others) are found in these atmospheres. Their properties serve to test and modify the models we have to explain similar phenomena on Earth's atmosphere, adding or deleting terms in the differential equations that describe them, modifying appropriately their boundary conditions, or introducing new ideas and their mathematical modeling for their interpretation. In other words, by experimenting with state-of-the-art models for Earth's atmosphere, we can reproduce what we see in other planetary atmospheres on the computer, thus increasing our learning and allowing us to improve the models.

In addition, the study of planetary atmospheres is fundamental to understand the origin of the solar system and the formation mechanisms of planets and satellites, closely related with the elemental and isotopic abundances of the different chemical compounds present in the atmospheres and their deep structure in the interiors of the giant planets, for example. In the long term, Earth's paleoclimate can be best addressed by comparatively studying the circumstances that have led to the "extreme climates" (at least in the case of life-forms) we currently find on Venus and Mars, and the cyclical variations produced by astronomical mechanisms in all of them. The details of the evolution of Earth's atmosphere in its composition and mass, for example, is best understood when we compare it with our two neighboring planets. The study of atmospheres is also important in astrobiology to understand the origins of life on Earth and its development from simple to complex organisms. The concept of "habitable zone" is directly related to the presence or absence of an atmosphere and to its properties. The rapidly increasing discovery of extrasolar planets and the first measurements of their "exotic" atmospheric properties (for solar system standards) has expanded even more the comparative view described above. Finally, the future manned exploration of nearby planets, such as Mars, and the exploration of distant planets employing robotic spacecrafts make "meteorological forecasting" a vital tool for the safety of these missions.

Each chapter in the book is dedicated to a selected atmospheric topic in the way classically done for the Earth's atmosphere, and summarizes the most important aspects in this field. Every chapter concludes with a set of proposed exercises along with their solutions in order to help students. Some of these problems can also be seen as additional examples that complete the content of the chapters. The observational methods and the current facilities used to explore planetary atmospheres on Earth and in space are summarized in an appendix at the end of the book. Since a large number of symbols have been used in the text, these have been listed along with their meaning at the end of the front matter of the book. A selected bibliography, classified by subjects, has also been provided. More specialized references can also be found in many of the figure captions. I hope the reader will find in this book a useful reference and guide for study, teaching, and research in planetary atmospheres.

Acknowledgments

This book would not have been possible without the support of several people. I would first like to thank my family for patiently standing by me when I was working on this project during weekends and holidays. From a scientific point of view, the support and motivation I have received from my students and colleagues of the "Grupo de Ciencias Planetarias" at the Universidad del País Vasco (UPV-EHU) have been invaluable. I would like to express my gratitude to the current members at UPV-EHU: Ricardo Hueso, Santiago Pérez-Hoyos, José Félix Rojas, Jon Legarreta, Teresa del Río, Javier Peralta, Naiara Barrado, and Jesús Arregi; my former student, Juan Ramón Acarreta; and my Catalonian colleagues, Enrique García-Melendo and Josep M. Gómez. All of them gave me invaluable support and advice. I would also like to acknowledge my department colleagues, in particular, Agustín Salazar at the Escuela Técnica Superior de Ingeniería, for the enriched collaboration in the studies of heat propagation and for reviewing Chapter 1. Apart from the above-mentioned collaborators, the following colleagues helped in revising and improving the other chapters: J. Sáenz, J. L. Ortiz, M. A. López-Valverde, M. López-Puertas, E. Karkoschka, P. Read, and T. Dowling. A long-term research collaboration with Glenn Orton and his team at the Jet Propulsion Laboratory, and with the Pic-du-Midi Observatory and their French team, Jean Lecacheux, François Colas, Pierre Drossart, and Pierre Laques (deceased), has also been fundamental for my research. I am also very grateful to John Navas at Taylor & Francis for encouraging me to write this book and to the editorial staff for their advice on editorial issues. I would also like to thank the staff vice-rector at UPV-EHU, Prof. J. J. Unzilla, for granting me a one-year first course teaching leave between September 2008 and July 2009 to write a large part of this book. This work would never have been completed without the research grants given to my group by the Spanish Ministry of Science and Innovation (currently MICINN) and by the Basque Country Government that supported our research in planetary atmospheres. Finally, I would like to thank the directorial staff at the School of Engineers and to the Diputación Foral de Bizkaia for their support in sustaining the master's course on space sciences and the Aula EspaZio Gela.

Author

Agustín Sánchez Lavega is a full professor of physics (Catedrático) at the Engineering School of the Universidad del País Vasco (UPV-EHU) in Bilbao (Spain) since 1994. After graduation at UPV-EHU, he worked at the Centro Astronómico Hispano-Alemán (Calar Alto Observatory) in Almería (Spain). Following his PhD on Saturn's atmosphere, he returned to UPV-EHU, where he currently serves as the head of the Grupo de Ciencias Planetarias (Planetary Sciences Group). He is also the director of the Aula EspaZio Gela, a facility at the School of Engineers at UPV-EHU, where a master's course on space science and technology is being developed. He served as a member of the Solar System Working Group of the European Space Agency between 2004 and 2006. Currently, he is also head of the Department of Applied Physics I at UPV-EHU. He has coauthored book chapters on giant and extrasolar planets and has authored more than 150 papers on the subject in refereed journals.

List of Tables

List of Symbols

For some common and usual cases, the same symbol is used to represent different magnitudes or parameters. However, its meaning is easily interpreted in the context of the corresponding chapter. Units are indicated for selected magnitudes to help their identification.

a, a_i	Semimajor orbital axis (for a body or mass i)
a	Particle radius (spherical)
a_0	Bohr atomic radius
a_ν	Opacity factor ($cm^5\ g^{-2}$)
$a_{V\nu}$	Volumetric gas opacity ($cm^{-1}\ amagat^{-2}$)
$\vec{a}$	Acceleration vector
A_B	Bond albedo
A	Atomic mass number
A_k	Chemical activity coefficient
A_D	Diffusion coefficient
A_ν	Absorptance
$\langle A \rangle$	Bond absorptance
A_R	Rayleigh coefficient
A_L	Albedo of a Lambert surface
A_{cl}	Albedo of a cloud
A_T	Gas conductivity coefficient
A_H	Horizontal (plane) area
B	Semiminor orbital axis
b_c	Binary collision parameter for diffusion
$\vec{B}$	Magnetic field vector (B_r, B_r, B_ϕ)
B_E	Magnetic field intensity at distance R_E
$\vec{B}_{ext}$	External magnetic field vector
$B_\lambda(T), B_\nu(T)$	Planck function
B_R	Rayleigh coefficient
$B_\oplus$	Sub-observer planetocentric latitude
$B_\odot$	Subsolar planetocentric latitude
c	Speed of light
c_s	Speed of sound
c_w	Phase speed of a wave
c_i	Phase speed of a wave in direction $i\ (= x, y, z)$
c_i	Complex phase speed
c_r	Real phase speed
c_g	Group velocity for a wave
c_{gi}	Group velocity of a wave in direction $i\ (= x, y, z)$
C_D	Drag coefficient (nondimensional)
C_{Radfor}	Cloud radioactive forcing

C_i	Concentration of a radiogenic element i
C_α	Photometric phase function constant
C_p	Specific heat at constant pressure
C_V	Specific heat at constant volume
c_p	$= C_p/\mu$, C_p/n (specific heat at constant pressure per unit mass, per unit mole)
c_v	$= C_V/\mu$, C_V/n (specific heat at constant volume per unit mass, per unit mole)
C_{vm}	Heat capacity of moist air at constant volume
C_{pm}	Heat capacity of moist air at constant pressure
C_{vd}	Heat capacity of dry air at constant volume
C_{pd}	Heat capacity of dry air at constant pressure
C_{vV}	Heat capacity of vapor at constant volume
C_{pV}	Heat capacity of vapor at constant pressure (indexes: v = volume; V = vapor; m = moist; d = dry)
C_K	Kolmogorov constant (turbulence)
C_ξ	Entropy constant (turbulence)
d	Atmospheric layer thickness
d_E	Ekman layer thickness
D_T	Thermal diffusivity
D_i	Diffusion coefficient for species i
D_v	Diffusion coefficient for a vapor
D_{eddy}	Turbulent diffusion coefficient
$D(K, P)$	Energy dissipation function
e	Orbital eccentricity
E_D	Dissociation energy
E_i	Ionization energy
E_p	Potential energy
E	Total energy
$E_{p,eff}$	Effective potential energy
E_l	Rotational energy levels
ΔE_e	Energy gap between electronic levels
ΔE_v	Energy gap between vibrational levels
E_K	Ekman number
$\vec{E}$	Electric field vector
EW	Equivalent width of a spectroscopic line
$f(y)$	Coriolis parameter function of the meridional direction
f_0	Coriolis parameter at latitude φ_0
f_{EP}	Oblateness
$f(k)$	k-Distribution probability function
f_S	Scattering parameter
f_a	Fraction of aerosol heating rate
f_{ce}	Cyclotron frequency
f_e	Fraction of electronic to atomic energy conversion
$\vec{F}$	Generic force vector
$\vec{F}_G$	Gravitational force vector

$\vec{F}_{ext}$	External force to a system
F_{BV}, F_B	Buoyancy force
F	Helmholtz function (J)
$\vec{F}_E$	Electrostatic Coulomb force
F_{int}	Internal heat flux (W m^{-2})
$F_{\odot p}$	Energy flux from the Sun (W m^{-2})
F_{*p}	Energy flux from a star (W m^{-2})
F_{obs}	Reflected energy flux from a planet (W m^{-2})
F_{abs}	Absorbed energy flux (W m^{-2})
$\vec{F}_Q$	Heat flux vector (W m^{-2})
F_{xuv}	Stellar flux in XUV
F_ν, F_λ	Flux density (frequency, wavelength) (W m^{-2} Hz^{-1}, W m^{-2} µm^{-1})
$\vec{g}$	Gravitational field intensity or gravity acceleration vector
g_r	Radial component of the gravitational field
g_φ	Perpendicular to the radial component
g_0	Reference acceleration of gravity
g_n^m	Magnetic Gauss coefficients
g_n, g_i	Degeneracy of an atomic/molecular energy level
g_J	Degeneracy level for a rotational molecular state
$g(k)$	Cumulative R-distribution function
g_s	Asymmetry parameter for scattering
g_{ph}	Photon scattering coefficient
G	Gravitational constant
G	Gibbs function
G_p	Geometrical cross section of particles per unit volume
$G(K, P)$	Energy generation function (kinetic, potential)
h	Planck constant
h	Height (altitude) variable
$h_\odot$	Solar hour angle
h_n^m	Magnetic Gauss coefficients
H	Enthalpy (J)
H	Isothermal scale height
H_p	Particle scale height
H_{cl}	Cloud scale height
H_i	Radiogenic power per unit mass (W kg^{-1})
H_{int}	Internal heat power per unit mass (W kg^{-1})
$H(\varpi, \mu_\varphi)$	Chandrashekar H-functions
i	Planetary spin to orbit tilt angle
i_z	Solar zenith angle
I	Moment of inertia
I_E	Electric current intensity
I_λ, I_ν	Intensity of radiation (W m^{-2} sr^{-1} µm^{-1}, W m^{-2} sr^{-1} Hz^{-1})
I_s	Sound intensity
$\vec{j}$	Electrical current density vector
j_ν	Mass emission coefficient (W kg^{-1} sr^{-1} Hz^{-1})
J	Total angular momentum number (atomic)

J	Photochemical production rate (photons cm^{-1} s^{-1} Hz^{-1})
J	Molecular rotational quantum number
J_n	Gravitational spherical moments
J_ν	Mean intensity
k	Wavenumber (x-direction)
$\hat{k}$	Cartesian and natural coordinate unit vector
k_B	Boltzmann constant
k_ν	Mass absorption coefficient (cm^2 kg^{-1})
k_{ice}, k'_{ice}	Particle shape factor
k_{xy}, k_{ij}	Reaction rate for a two-body reaction (cm^3 s^{-1})
k_{xym}	Reaction rate for a three-body reaction (cm^6 s^{-1})
k_K	von Kármán constant
k_M	Mass entrainment rate parameter
k_I	Energy injection wavenumber in a turbulent motion
k_f	Energy dissipation wavenumber in a turbulent motion
k_T	Tidal constant
$\vec{K}$	Wavenumber vector
K	Chemical reaction constant
$\bar{K}, K'$	Kinetic energy (mean, turbulent)
K_T	Thermal conductivity
K_R	Radiative thermal conductivity
K_B	Bulk modulus
K_P	Gas polytropic constant
K_i	Chemical equilibrium constant for compound i
Kn	Knudsen number
K_f	Friction coefficient for a falling particle
K_M	Minnaert coefficient
K_m	Eddy viscosity
K_θ	Eddy heat coefficient
ℓ	Quantum number
l	Wavenumber (y-direction)
ℓ	Path length
l_{fp}	Mean free path for a particle
l_{eddy}, l'	Eddy size (free turbulent path length)
$\vec{\ell}(s)$	Position rector for a fluid element
l_c	Cloud condensate mixing ratio
$\vec{L}_0$	Angular momentum vector (generic)
L	Angular momentum (atomic and molecular)
L_z	z-Component of angular momentum
L_{atm}	Angular momentum of the atmosphere
L_{hur}	Angular momentum in a hurricane
$L_1 - L_5$	Lagrange orbital points
L_{rad}	Radiogenic energy power (W)
L_{int}	Internal energy power (W)
L_*	Star luminosity (W)
$L_\odot$	Sun's luminosity (W)

L_p	Total power received on a planet (W)
L_{abs}	Absorbed power by a planet
L_{out}	Emitted power by a planet
L_m	Magnetic length scale
L_{ev}	Evaporation parameter (atmosphere loss by impact)
L_H	Latent heat (general)
L_i	Latent heat for process i
	$i = v$ (vaporization)
	$i = f$ (fusion)
	$i = s$ (sublimation)
L_e	Electrons loss rate
L_{Ni}	Particles loss rate
L_β	Rhines scale
L_D	Rossby deformation radius
L_x, L_y, L_z	Characteristic length scale (x, y, z direction)
m	Wavenumber (z-direction)
m, M	Mass
m_i	Generic mass of element-i
m_i	Magnetic angular momentum number
m_j	Magnetic total angular momentum (atomic)
m_s	Magnetic quantum spin number
m_{rock}	Mass of rocky material
dm	Differential mass element
m_p	Particle mass
m_e	Mass of the electron
m_n	Mass of the atomic nucleus
m_v	Mass of a vapor
$(1/m)(dm/dz)$	Entrainment mass rate
m_w	Precipitation mass rate
$\vec{M}$	Force torque vector
M_D	Dipolar magnetic moment
M_P	Mass of a planet or satellite
M_*	Mass of a star
M_C	Mass of an impacting body
M_S	Mass released by sublimation per unit area
$\vec{n}$	Normal vector, unit vector in natural coordinates
n	Distance in direction along the normal to a parcel trajectory
$\hat{n}$	Unit vector normal to a parcel trajectory
n	Mean orbital motion
$n = 1, 2, \ldots$	Natural number series
n	Principal quantum number
n	Refraction index
n_g	Refraction index for a gas
n_0	Mole number
n_i	Number of moles of compound i
N	Number density of molecules (cm^{-3})

Nu	Nusselt number
N_p	Number of particles
N_{ex}	Number density of particles at exobase
N_i	Number of atoms/molecules in a given state (every level)
N_i	Number density of an ionizing compound
N_i	Total number of molecules of a gas
N_{iE}	Number density of particles under diffusive equilibrium
N_l	Number of molecules at level l
N_T	Total number of molecules
N_A	Avogadro number
N_0	Loschmidt number
N_B	Brunt–Väisälä frequency
$\vec{p}$	Linear momentum
P_n	Legendre polynomials
p	Generic parameter
$p(\varphi)$	Optical phase function
p_R	Rayleigh scattering phase function
p_{HG}, p_{2HG}	Henyey–Greenstein phase function
p_0	Geometric albedo
P	Pressure
P_C	Pressure at center of a planet
P_s	Surface pressure
P_i	Partial pressure of compound i
P_0	Reference pressure
P_{STP}	STP pressure
P_v	Vapor partial pressure
P_{vs}	Saturation vapor pressure
P_{vso}	Saturation vapor pressure at T_0
P_{cl}	Pressure for the base of cloud level formation
P_e	Electron production rate
P_{Ni}	Production rate of particles i (cm^{-3} s^{-1})
$\bar{P}, P'$	Available potential energy (mean, eddy)
Pr	Prandtl number
q	Specific humidity
q_α	Phase integral
q_R	Planetary rotation parameter
q_{Ti}	Planetary tidal parameter
$\vec{q}_T$	Thermal wave vector
q_e	Electron charge
q_g	Quasigeostrophic potential vorticity
Q	Heat power per unit volume (W m^{-3})
Q_{sol}	Solar radiation heating per unit volume
Q_{IR}	Thermal infrared cooling per unit volume
δQ	Infinitesimal thermal energy (heat)
$Qxuv$	Heat by XUV radiation absorption (W m^{-3})
Q_H	Column integrated heating rate (W m^{-2})

Q_R	Rotational potential in the gravitational field
Q_e, Q_S, Q_a	Efficiency scattering factors (extinction, scattering, absorption)
Q_T	Tidal quality factor
r	Distance, polar radial coordinate (r_i)
r	Generic particle size
$\vec{r}, \vec{r}'$	Position vector
r_Z	Zonal radius of curvature
r_M	Meridional radius of curvature
$\langle r \rangle$	Mean orbital distance
r_{L_1}	Position of the Lagrange point L_1
$\vec{R}$	Vector position for the center of mass of a system
R	Local curvature radius
R_p	Planetary (or satellite) radius
R_E	Planetary equatorial radius
R_P	Planetary polar radius
R_L	Roche limit
R_g	Universal gas constant
R_g^*	Molar gas constant
R_d^*	Specific dry atmosphere constant
Ra	Rayleigh number
Re	Reynolds number
Re_m	Magnetic Reynolds number
R_s, R_r	Isotopic ratios (sample and reference)
R_{xuv}	Distance to the center of a planet (altitude) where F_{xuv} is deposited
R_c	Radius of an impacting object
R_R	Rydberg constant
R_R^*	Modified Rydberg constant
RH	Relative humidity
R_V^*	Specific vapor constant
Ri	Richardson number
R_t	Radii of curvature of a parcel trajectory
R_s	Radii of curvature of a streamline
s	Distance along a parcel trajectory
s	Spin quantum number
$\hat{s}$	Unit vector along a parcel trajectory
s	Temperature exponent for diffusion
s_ℓ	Line strength intensity (cm kg^{-1})
S	Entropy (J K^{-1})
S	Spin angular momentum
S_z	Spin angular momentum (z-component)
$\vec{S}, d\vec{S}$	Vector area (infinitesimal area)
S	Area modulus
St	Strouhal number
$S_{\odot p}$	Solar constant in a planet (W m^{-2})
S_F	Strength force
S_B	Absorption band strength

S_ν	Source function for radiative transfer (W m^{-2} sr^{-1} µm^{-1})
S_T	Static stability (K km^{-1})
t	Time
t_0	Reference time
t_{LT}	Local time (Sun-synchronous reference frame)
T	Temperature
T_{eq}	Equilibrium temperature
T_{eff}	Effective temperature
T_c	Curie temperature
T_s	Surface temperature
T_0	Reference temperature
T_{STP}	Temperature at STP
T_V	Virtual temperature
T_{cl}	Cloud temperature
T_ν	Line transmittance
$\langle T \rangle$	Averaged band transmittance
T_T	Tidal torque
u	x-Component of the velocity (zonal direction)
u'	x-Component of the turbulent velocity (zonal direction)
$\bar{u}, \langle u \rangle$	Mean zonal velocity
u_g	x-Component of the geostrophic velocity (zonal direction)
$\vec{u}_I$	Velocity vector in an inertial frame
$\vec{u}_R$	Velocity vector in an rotating frame
u_E	Equilibrium zonal velocity
u_M	Absorber mass (kg m^{-2})
U	Internal energy
U_G	Total gravitational potential
U_b	Binding energy of surface atoms
U_{VK}	von Kármán vortex speed
v	y-Component of the velocity (meridional direction)
v_i	i-Component of the velocity
v_r	Radial component of the velocity
v_θ	Orthogonal (to the radial) component of the velocity
v_{esc}	Escape velocity
v_{th}	Thermal velocity
v_{eff}	Effusion velocity
v_{dif}	Diffusion velocity
v_{sw}	Solar wind velocity
v_g	y-Component of the geostrophic velocity (meridional direction)
$\vec{V}$	Generic velocity vector
V	Velocity modulus
V_T	Tangential velocity component (in vortices)
V_R	Radial velocity component (in vortices)
$\vec{V}_g$	Geostrophic wind velocity
V_G	Gravitational potential
V, V_0, Vol	Volume

V_S	Gravitational potential at the surface of a body
V_M	Magnetic potential
V_v	Specific volume ($1/\rho_v$)
w	Vertical component of velocity
w_L	Limiting velocity for a falling particle
w_i	Vertical velocity for a species i
dW	Infinitesimal work
x	Parameter, Cartesian coordinate (zonal direction)
x_{vi}	Mole fraction or volume mixing ratio
x_{Mi}	Mass mixing ratio
x_{MV}	Mass mixing ratio of a vapor
x_S	Saturated mass mixing ratio
x_{VV}	Vapor volume mixing ratio
X^*	Excited state of species X
X	Neutral species
X^+	Ionized species
y	Parameter, Cartesian coordinate (meridional direction)
y_ν	Pressure coefficient parameter
Y_{sput}	Sputtering yield
Y_n	Sputtering yield by neutrons
Y_e	Sputtering field by electrons
Y_H	Meridional extent of a Hadley cell
z	Cartesian coordinate (vertical direction)
z_{ex}	Altitude level of the exobase
z_{cl}	Altitude level of cloud formation
Z	Atomic number
Z_A	Column abundance of a gas (cm-amagat)
α_I	Inertial factor
α	Phase angle
α_T	Thermal expansion coefficient
$\alpha_R, \tilde{\alpha}$	Rosseland mean opacity
α_r	Radiative recombination coefficient ($cm^{-3}\ s^{-1}$)
α_{TD}	Thermal diffusion factor
$\alpha_B, \alpha_{BC}, \alpha_{Bm}$	Baroclinic parameter
β	Meridional gradient of the Coriolis parameter ($= df/dy$)
β_{nm}	Gravitational density–pressure coefficient
β_a	Volume absorption coefficient
β_s	Volume scattering coefficient
β_e	Volume extinction coefficient
β_R	Rayleigh scattering coefficient (in volume)
γ	Adiabatic coefficient
γ_D, γ_L	Half width of an absorption line (Doppler, Lorentz)
γ_E	Ekman layer parameter
Γ	Temperature lapse rate ($K\ km^{-1}$)
Γ_d	Dry adiabatic lapse rate
Γ_s	Wet adiabatic lapse rate

Γ_c	Circulation of the velocity field
$\Gamma_{CI}, \Gamma'_{CR}$	Circulation of the velocity field (relative and absolute)
δ	Symbol for infinitesimal processes
δ	Stretching quantum constant
δ	Separation between spectral lines
δ_M	Magnetic dipole tilt
δ_i	Differential isotope composition
Δ_{EP}	Earth–planet distance
ε	Emissivity from a body
ε_s	Emissivity from a surface
ε_p	Ratio of planetary equatorial to polar radius
ε_0	Dielectric permittivity
ε_m	Efficiency factor for the mass loss rate
ε_c	Droplet collision efficiency
$\varepsilon_\mu = \mu_v/\mu$	Ratio between the vapor molecular weight and the mean molecular weight
ε_j	Photoionization heating efficiency
ε_I	Turbulent kinetic power per unit mass
ε_ε	Turbulent entropy supply rate
$\varepsilon_V = L_y/L_x$	Vortex aspect ratio (meridional/zonal scales)
χ	Angular direction of the wind
χ_F	Angle in a baroclinic front
η_D	Viscosity coefficient (dynamic)
$\eta_p = r/R_p$	Normalized distance to the planet radius
$\eta(r)$	Particle size distribution
$\eta(\vec{v})$	Particle velocity distribution
π_D	Dynamic pressure
κ	Adiabatic index
υ	Vibrational quantum number
υ	Polytropic index in the equation of state
υ_v	Viscosity coefficient (molecular or kinematic)
υ_{eff}	Variance of the effective particle distribution
σ	Vertical "sigma" coordinate
σ_B	Stephan–Boltzmann constant
σ_C	Electric conductivity
σ_ℓ	Line intensity distribution
σ_T	Tension force at surface (modulus)
σ_R	Rayleigh scattering cross section
σ_ν	Scattering coefficient
σ_P	Particle scattering cross section
σ_x	Collision cross section
σ_i	Ionization cross section
σ_a	Absorption cross section
ν	Frequency
ν_e	Electronic frequency

ν_v	Vibrational frequency
ν_l	Rotational frequency
$\tilde{\nu}$	Wavenumber
λ	Wavelength
λ_m	Magnetic diffusivity
λ_{ex}	Atmosphere escape parameter
Λ	Elsasser number
Λ_0	Wind vertical shear
ϑ	Longitude of the perihelion
ξ	Entropy
ψ	Inclination of the orbital plane relative to the ecliptic
ψ	Stream function
ψ_i	Phase of a wave
φ	Planetary latitude coordinate (generic)
φ_g	Planetographic latitude
φ_c	Planetocentric latitude
$\varphi_\odot$	Subsolar latitude
ϕ	Planetary longitude coordinate (generic)
$\Phi(\alpha)$	Photometric phase function
Φ_{th}	Jeans flux (atoms cm^{-2} s^{-1})
Φ_i	Particle-i flux (cm^{-2} s^{-1})
Φ_L	Limiting diffusion flux
Φ_G	Geopotential
θ	Potential temperature
θ	Angular coordinate (polar angle)
θ_{an}	Orbital true anomaly
$\theta_\odot$	Orbital longitude of the Sun
θ_p	Longitude of the orbit planet perihelion
Θ	Scattering angle
Θ_n	Hough functions
τ	Optical depth, timescale
τ_c	Orbital circularization timescale
τ_R	Rayleigh scattering optical depth
$\tau_{a,g}$	Optical depth for gas absorption
$\tau_{s,g}$	Optical depth for scattering by a gas
$\tau_{a,p}$	Optical depth for particle absorption
$\tau_{s,p}(\tau_p)$	Optical depth for scattering by particles (particle optical depth)
τ_{cl}	Cloud optical depth
τ_{orb}	Planet orbital period
τ_{syn}	Planet synodic rotation period
$\tau_{sid}, (\tau_d, \tau_{rot})$	Planet sideral period (length of the planetary day, rotation period)
τ_{in}	Half-lifetime of radioactive nuclei
τ_m	Magnetic diffusion time constant
τ_{esc}	Mean loss time

τ_{ch}	Photodissociation lifetime
τ_D	Molecular diffusion timescale
τ_{eddy}	Eddy diffusion timescale
τ_f	Viscous dissipation timescale (turbulence)
τ_k	Eddy timescale (turbulence)
τ_E	Ekman pumping timescale
τ_ξ	Entropy timescale
τ_F	Rayleigh friction (damping) timescale
$\vec{\tau}_F$	Stress force vector
$\bar{\mu}$	Mean molecular weight
μ_I	Molecular weight of compound i
μ_M	Reduced mass (orbital)
μ_T	Thermal diffusion length
μ_0	Magnetic permeability
$\mu_\varphi, \mu'_\varphi$	Cosine of a light beam incident and emission angles
μ_H	Vertical scale wavenumber
ρ, ρ_i	Density (density for component-i)
ρ_P	Planet density
ρ_S	Satellite density
ρ_E	Electrostatic charge density
ρ_{SW}	Density of particles in the solar wind
ρ_{STP}	Gas density at STP
ρ_V	Vapor density or specific humidity
ρ_{cl}	Cloud density
ω	Angular frequency
ω_{orb}	Angular orbital velocity
$\tilde{\omega}_0$	Single scattering albedo
ω_θ	Vertical velocity in isentropic coordinates
ω_p	Vertical velocity in pressure coordinates
$\vec{\omega}$	Vorticity
$\vec{\omega}_I$	Vorticity in an inertial frame
$\vec{\omega}_R$	Vorticity in a non-inertial frame
Ω	Rotation angular velocity
Ω_0	Reference rotation angular velocity
$\delta\Omega$	Infinitesimal rotation angular velocity
Ω_s	Solid angle

1 Introduction to Planets and Planetary Systems

1.1 PLANETARY SYSTEMS

The bodies of the solar system formed 4650 million years ago (4.65 Gy [gigayears]) together with the Sun from a mass of gas, mainly hydrogen, mixed with small quantities of heavier elements formed in older stars, and some additional dust. Gravitational contraction of this mass and the formation of the protoplanetary disk due to angular momentum conservation were the first steps in the formation of the system. Whereas most of the mass was concentrated at the center of gravity of the system, forming the Sun, about 1% remained in the disk, becoming the prime material for forming the planets, satellites, and other minor bodies. The action of gravity combined with magnetic forces and viscous effects redistributed the mass and angular momentum within the disk to reach a final distribution in which about 99.9% of the mass and 2% of the total angular momentum of the system reside in the Sun.

Whereas there is a consensus about how the terrestrial planets formed from the gravitational accretion of small growing clumps of solid matter, from grains to kilometer-sized "planetesimals" to finally the planet body, there are two alternative possibilities that have been proposed so far to explain the formation of giant gas planets. The first is based on the formation of an embryo protoplanet, a giant planet core (containing few Earth planet masses) that grew by a similar accretion mechanism and was followed by the capture of surrounding gas material from the nebula that finally formed a large shell, which is the main gaseous body of the planet. The gas capture (basically hydrogen) took place at sufficiently large distances from the Sun where the sunlight heat and the solar wind cannot impede its aggregation into the core. The second hypothesis is based on the so-called gravitational disk instability, a mechanism acting on dense disks that fragments the ring-like structure of disks into massive gaseous clumps that form the massive protoplanet.

These processes are fast compared with the current age of the planetary system and planet formation that reached its final basic structure in about 5–10 million years (My). The major satellites of the planets formed around them in a somewhat similar accretion mechanism, whereas other satellites such as the Moon formed following large impacts with the rest of the sparse material within the nebula. Minor satellites were incorporated into planetary orbits from the capture of the rest of the disk debris. During the following 2000 My, a large quantity of solar system debris left an impact on the planets and satellites, forming surface craters and eroding planetary gaseous envelopes (the atmospheres) that formed in the planets and major satellites.

A major advance in the study of planetary systems occurred in the early 1990s with the first detection of extrasolar planets. It took place in two very different places: around a dead star (a pulsar, in 1992) and around a solar-type star (in 1995). These discoveries broke the paradigm of the solar system as a unique archetype of the formation and structure of a planetary system. In main sequence stars (like the Sun), the basic mechanism previously described is expected to operate in a regular way, but the processes acting in systems with multiple stars or in evolved (dead) stars that contain planets opened new possibilities. New discoveries, with the presence of giant massive planets very close to the stars, required the action of new mechanisms that could redistribute planets into very different orbits from where they had been formed, or could create isolated ("floating") planets not gravitationally bounded to a star. The prediction that the basic architecture of a planetary system would consist of terrestrial planets close to the star and large gaseous giant and icy planets distant to it was rapped as a result of these findings. We can therefore expect planetary diversity in an orbital configuration.

Figure 1.1 shows a scheme of the basic types of planetary configurations so far observed in extra solar systems. In cases A, B, C, D, E, F, G, and H, the bodies are gravitationally trapped to a normal star (a main sequence star) as occurs in our own solar system. Some cases include double stars (cases E, F, and H) and even a triple stellar system (case G). The most complicated situation for classification purposes corresponds to the recognition of an object as a planet when it is embedded in a population of bodies with similar physical properties, that is, when it is located within an orbital "belt of objects" (cases B and C). Case B, in particular, parallels the solar system family of trans-Neptunian objects (TNOs), which includes the new class of "minor planets" (e.g., Pluto and Eris) and objects such as Ceres, previously classified as an asteroid. In addition, there is a battery of peculiar orbital cases, unpredicted on theoretical grounds. Case I corresponds to "isolated planets" not bounded

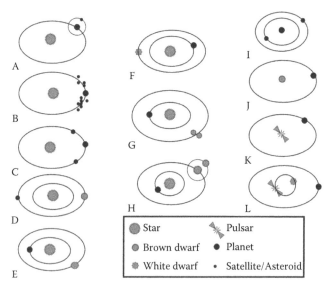

FIGURE 1.1   Observed orbital configurations of planetary systems. (From Sánchez-Lavega, A., *Cont. Phys.*, 47, 157, 2006.)

gravitationally to any massive stellar body. Case J has a brown dwarf (an object without the mass required to ignite thermonuclear reactions in its core) as a central mass body with a planet orbiting about it. This was the orbital configuration of the first direct detection of an extrasolar, young, and intrinsically hot planet. The low brightness of the brown dwarf and the wide separation of the hot (and therefore bright and young) planet permitted its direct detection. "Pulsar planets" orbiting around a neutron star correspond to case K and case L is really singular with a planet orbiting a pulsar—white dwarf binary. It is not yet clear if these "pulsar planets" formed before the supernova explosion that led to the pulsar, or if they formed from the material ejected during the explosion. There is also indirect evidence of planetary systems around stars surrounded by a dust belt.

The orbital configuration of a planetary system basically depends on the formation mechanisms of the protoplanetary system (i.e., the initial conditions) on the planet/disk tidal interactions and on the subsequent orbital evolution due to the mutual gravitational interaction between their components (migration mechanisms and gravitational scattering). Models of the gravitational interactions between growing planets and between the planets and the disk can explain, for example, the actual configuration of the external parts of our solar system and the existence of the "hot" and "very hot Jupiter" families of extrasolar planets due to the migration mechanism (the change of a planet from an outward to an inward orbit). The characteristics of this disk (metallicity, i.e., the proportion of elements heavier than hydrogen and helium, mass and mass distribution) will determine the planetary-type system and its long-term evolution and stability.

Probable protoplanetary dust has been detected around the small "pseudo-stars" called brown dwarfs, demonstrating that material exists around these low-mass objects to form planets.

1.1.1 SOLAR SYSTEM PLANETS

During its General Assembly in Prague (Czech Republic), the International Astronomical Union (IAU) approved on August 24, 2006, the resolutions 5A and 6A on the definition of a planet and other bodies except satellites in the solar system.

Resolution 5A

1. A "planet"* is a celestial body that (a) is in orbit around the Sun, (b) has sufficient mass for its self-gravity to overcome rigid body forces so that it assumes a hydrostatic equilibrium (nearly round) shape, and (c) has cleared the neighborhood around its orbit.
2. A "dwarf planet" is a celestial body that (a) is in orbit around the Sun, (b) has sufficient mass for its self-gravity to overcome rigid body forces so that it assumes a hydrostatic equilibrium (nearly round) shape,[†] (c) has not cleared the neighborhood around its orbit, and (d) is not a satellite.

---

* The eight "planets" are Mercury, Venus, Earth, Mars, Jupiter, Saturn, Uranus, and Neptune.
[†] An IAU process will be established to assign borderline objects into either dwarf planet or other categories.

FIGURE 1.2   The solar system planets according to the classification of the International Astronomical Union (IAU) of August 2006. (Courtesy of IAU.)

3. All other objects* except satellites orbiting the Sun shall be referred to collectively as "small solar system bodies."

Resolution 6A

Pluto is a "dwarf planet" by the above definition and is recognized as the prototype of a new category of TNOs.

As a consequence, the largest asteroid Ceres has been reclassified as a dwarf planet together with Pluto and Eris, the largest known object of the TNO family. Two more previous TNOs were added in 2008: "Makemake" (discovered on March 31, 2005) and "Haumea" (discovered on December 28, 2004) (Figures 1.2 through 1.4).

1.2   ORBITAL MOTION

This first section is devoted to the study of the basics of orbital motion under the action of gravity. The formulation we follow is that of classical "Newtonian" mechanics. The general theory of relativity, which must be used to study the motion of bodies in the proximity of very large masses, is important in some extreme situations but is not required to understand the basics of atmosphere physics, which is the aim of this book.

1.2.1   BASICS OF ORBITAL MECHANICS

For our purposes, the basic elements describing the orbit followed by a planet or satellite are the orbit semimajor axis a, orbital eccentricity e, and the tilt of the orbit relative to the equatorial plane of the Sun (or central star or planet).

---

* These currently include most of the solar system asteroids, most TNOs, comets, and other small bodies.

FIGURE 1.3 Terrestrial planets, dwarf planets, and major satellites (top to bottom, left to right): Earth, Mars, Mercury, Moon; Io, Europa, Ganymede, and Callixto; Venus (atmosphere and surface), Titan (atmosphere and surface), Triton, and Pluto. (Image mosaic from different spacecrafts courtesy of NASA.)

1.2.1.1 Newton's Law of Universal Gravitation

In 1687, Isaac Newton introduced a law that describes the gravitational force between two bodies with masses m and M separated by a distance r

$$\vec{F}_G = -G\frac{mM}{r^2}\hat{r} \tag{1.1}$$

where $G = 6.67 \times 10^{-11}$ N m^2 kg^{-2} is the universal gravitational constant. This kind of force is known as "central" since the vector $\hat{r}$ always points in the radial direction joining the centers of the two bodies.

Newton also introduced the three general laws of motion. The second law establishes that if a body of mass m is subjected to a force $\vec{F}$, its linear momentum $\vec{p} = m\vec{v}$ changes according to the equation

$$\vec{F} = \frac{d\vec{p}}{dt} \tag{1.2}$$

where $\vec{v}$ is its velocity.

FIGURE 1.4 Giant planets: Uranus, Jupiter, Saturn, Neptune (Earth at the center for comparison). (Image mosaic from NASA Voyager missions and Hubble Space Telescope courtesy of NASA/ESA.)

1.2.1.2 Motion under a Central Conservative Force

The gravitational force is conservative and therefore has an associated potential energy $E_p(r)$ that obeys the relationship

$$\vec{F}_G(\vec{r}) = -\nabla E_p(r) = -\frac{\partial E_p}{\partial r} \tag{1.3}$$

Integrating (1.3) from infinity, where the force is zero according to (1.1) and the potential energy is taken to be 0, we have

$$\int_0^{E_p} dE_p = GMm \int_\infty^r \frac{dr}{r^2} \tag{1.4}$$

Therefore, the potential energy associated with the two masses separated by a distance r is

$$E_p(r) = -G\frac{Mm}{r} \tag{1.5}$$

If a body of mass m moves with velocity $\vec{v}$ under the gravitational attraction of a much larger mass M, its total mechanical energy is given by

$$E = \frac{1}{2}mv^2 + E_p(r) \tag{1.6}$$

In the absence of other forces, E remains constant for this body. It can be positive, zero, or negative since the potential energy is negative according to (1.5).

On the other hand, if we take the origin of the coordinate system O at the position of the massive object M, the angular momentum for the body of mass m with respect to O, given by $\vec{L}_0 = m\vec{r} \times \vec{v}$, also remains constant since the torque of a central force is null

$$\vec{M}_0 = \frac{d\vec{L}_0}{dt} = \hat{r} \times \vec{F}(r) = 0 \tag{1.7}$$

Constant angular momentum implies that the motion of the point mass m is confined to a plane that contains O and $(\vec{r}, \vec{v})$. It is therefore convenient to use polar coordinates in the plane of motion (r radial and θ orthoradial). In these coordinates, velocity has the two components $v_r = dr/dt$ and $v_\theta = r\, d\theta/dt$ and the angular momentum becomes

$$\vec{L}_0 = mr^2 \frac{d\theta}{dt} \hat{k} = \text{Const} \tag{1.8}$$

where $\hat{k}$ is the unit vector perpendicular to the plane of motion (Figure 1.5). Note that this argument is valid for any central force, conservative or not.

Equations 1.6 and 1.8 are two constants of motion of the mass m under the gravitational action of the mass $M \gg m$.

With the aid of Equation 1.8, the total velocity of the mass m can be written as (Figure 1.6)

$$v^2 = v_r^2 + v_\theta^2 = v_r^2 + \frac{L_0^2}{m^2 r^2}$$

and the total energy of the body can be rewritten as

$$E = \frac{1}{2} mv_r^2 + \frac{L_0}{2mr^2} + E_p(r) \tag{1.9}$$

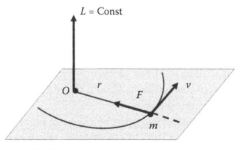

FIGURE 1.5 Orbital path in the plane of a body of mass m subjected to a central force F, and its angular momentum (L).

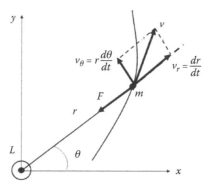

FIGURE 1.6 Motion of a body of mass m subjected to a central force and its velocity components.

The sum of the last two terms is usually referred to as the "effective potential energy" $E_{p,eff}(r)$,

$$E_{p,eff} = \frac{L_0}{2mr^2} + E_p(r) \tag{1.10}$$

Once the constants E and L_0 are established from the initial conditions of the motion, the trajectory followed by the mass m in polar coordinates $r(t)$ and $\theta(t)$ can be obtained upon integration of (1.8) and (1.9):

$$\frac{dr}{dt} = \left\{ \frac{2}{m} \left[E - E_{p,eff}(r) \right] \right\}^{1/2} \tag{1.11a}$$

$$\frac{d\theta}{dt} = \frac{L_0}{mr^2} \tag{1.11b}$$

By combining these two equations, we can obtain the equation for the body trajectory $r(\theta)$:

$$\left(\frac{dr}{d\theta} \right)^2 = \frac{2mr^4}{L_0^2} \left[E - E_{p,eff}(r) \right] \tag{1.12}$$

This equation allows us to determine the geometry of the orbit described by the mass m about M once the effective potential is known. Conversely, $E_{p,eff}(r)$ can be determined if $dr/d\theta$ is obtained from the orbit.

In the case of the central force described by (1.1) and potential energy described by (1.5) (see Figure 1.7), there are three types of orbits that can be classified according to E, all of them conic sections with one of their foci in the origin of coordinates. For $E>0$, the orbit is open (unbounded) and is a hyperbole. For $E=0$, the orbit is

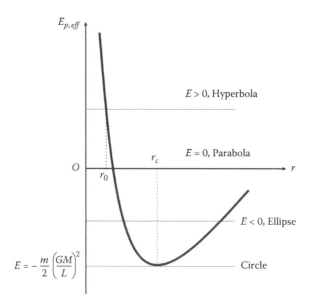

FIGURE 1.7 Effective potential $E_{p,eff}$ for a central force that varies as the inverse square law with distance. The energy levels for the different type of orbits are shown. The minimum energy value corresponds to the case of a circular orbit.

parabolic. For $E<0$, the orbit is closed (bounded) and is an ellipse with extreme distances to the focus r_a and r_b. For the particular case $r_a=r_b=r_0$ (corresponding to a minimum in the effective potential energy), the elliptic motion reduces to a simple circle. In summary, according to the values of the total energy E, the following types of orbits for the mass m about M ($\gg m$) result in the following:

$$E > 0, \quad \text{Hyperbolic}$$

$$E = 0, \quad \text{Parabolic}$$

$$E < 0, \quad \text{Ellipse}$$

$$E = -\frac{m}{2}\left(\frac{GM}{L}\right)^2, \quad \text{Circular}$$

To the first order, planets and satellites follow an elliptical motion, described by the equation $r(\theta)$

$$r = a\frac{1-e^2}{1+e\cos\theta} \tag{1.13}$$

where
    a is the semimajor axis of the ellipse
    e is the eccentricity (defined as the ratio of the distance between the two foci of
       the ellipse and the distance $2a$)

For an ellipse, $0 < e < 1$ ($e = 0$ for circular motion). The total energy and angular momentum for mass m can be written as

$$E = -\frac{GmM}{2a} \tag{1.14}$$

$$L_0 = m\sqrt{GMa(1 - e^2)} \tag{1.15}$$

Combining (1.14) and (1.15), we get the relationship between the eccentricity and the two integrals of motion E and L_0:

$$e^2 = 1 + \frac{2E}{m}\left(\frac{L_0}{GMm}\right)^2 \tag{1.16}$$

According to the previous energy signs and orbital classification, the eccentricity is $e < 1$ for a parabolic orbit and $e > 1$ for a hyperbolic orbit.

1.2.1.3   Kepler's Laws of Planetary Motion

In our solar system, the mass of the Sun ($M_\odot$) dominates that of any other body m ($M_\odot \gg m$) (see Tables 1.1 and 1.2) and the above equations can be used. They lead the laws of planetary motion introduced by Johannes Kepler in 1609. The same analysis is valid in general for the satellites moving around the planets with the exception of the case of the minor planet Pluto and its moon Charon.

First law: The planets follow elliptical paths with the center of the Sun at one of the foci of the ellipse.

Using Figure 1.8, we can determine the semi-axes and eccentricity of the orbit from the knowledge of the closest position to the Sun or perihelion P (periapsis for a satellite), and the most distant point in the orbit or aphelion A (apoapsis for the case of a satellite). Since the eccentricity $e = OS/OP$, the perihelion distance is $PS = a (1 - e)$ and the aphelion distance is $AS = a(1 + e)$. AP, the apsidal line, corresponds to the major axis $2a$ and a relation between the eccentricity and mayor and minor semi-axes is easily deduced:

$$e^2 = 1 - \frac{b^2}{a^2} \tag{1.17}$$

Second law: The radius vector joining the center of the Sun to the planet sweeps out equal areas in equal periods of time.

This law results from the conservation of angular momentum (1.8). The area S swept by the radius vector is given by

$$\frac{dS}{dt} = \frac{1}{2}r^2\frac{d\theta}{dt} = \frac{L_0}{2m} = \text{Const} \tag{1.18}$$

TABLE 1.1
Orbital Characteristics of the Planets

Sun

$1M_\odot = 1.989 \times 10^{30}$ kg

$1R_\odot = 6.96 \times 10^8$ m

$T_{eff}(\odot) = 5785$ K

$1L_\odot = 3.9 \times 10^{26}$ W

I. Planets

Planet	Mean Distance to Sun ($\times 10^8$ km)	Orbital Eccentricity	Orbital Tilt (Degrees)	Insolation (W m^{-2})	Orbital Period (Years)	Number of Satellites	Rings
Mercury	0.58	0.2056	7	9040	0.24	0	—
Venus	1.08	0.0068	177.4	2620	0.61	0	—
Earth	1.49	0.0167	23.4	1370	1.00	1	—
Mars	2.28	0.0934	24	590	1.88	2	—
Jupiter	7.78	0.0483	3.08	50.6	11.86	60	(1)
Saturn	14.27	0.0560	26.7	15.1	29.5	31	(2)
Uranus	28.69	0.0461	97.9	3.72	84.01	22	(3)
Neptune	44.96	0.0097	28.8	1.52	164.79	11	(4)

Notes for rings: D (distance from planet's center); τ (maximum optical depth); *m* (estimated mass in grams).
(1) $D = 1.4$–$3.2R_J$ (Jupiter's radius); $\tau = 3 \times 10^{-6}$; $m = 10^{16}$ g.
(2) $D = 1.4$–$2.2R_S$ (Saturn's radius); $\tau = 2.5$; $m = 2.8 \times 10^{25}$ g.
(3) $D = 1.5$–$1.95R_U$ (Uranus's radius); $\tau = 2.3$.
(4) $D = 1.68$–$2.53R_N$ (Neptune's radius); $\tau = 0.15$.

II. Dwarf Planets

Planet	Mean Distance to Sun ($\times 10^8$ km)	Orbital Eccentricity	Orbital Tilt (Degrees)	Insolation (W m^{-2})	Orbital Period (Years)	Number of Satellites
Ceres	4.14	0.077	10.58	180	4.60	0
Pluto	59.0	0.2482	17.15	0.9	248.54	3
Eris	67.6	0.44	44.2	0.7	557	1
Haumea	65.01	0.189	28.19	0.74	285.4	2
Makemake	68.68	0.159	28.96	0.66	309.9	0

and according to (1.15),

$$\frac{dS}{dt} = \frac{1}{2}\sqrt{GMa(1-e^2)} \qquad (1.19)$$

Third law: The ratio of the cube of the semimajor axis of the orbit to the square of its period is constant.

The equation describing this law is

$$\frac{a^3}{\tau_{orb}^2} = \text{Const}$$

TABLE 1.2

Orbital Properties of Satellites with Atmospheres

Satellite (Planet)	a (km)	τ_{orb} (Days)	e	i (°)
Moon (Earth)	384,400	27.32	0.05	5.16
Io (Jupiter)	421,600	1.769	0.04	0.04
Europa (Jupiter)	670,900	3.552	0.01	0.47
Ganymede (Jupiter)	1,070,000	7.155	0.00	0.21
Callisto (Jupiter)	1,883,000	16.689	0.01	0.28
Titan (Saturn)	1,221,850	15.945	0.03	0.33
Enceladus (Saturn)	238,020	1.37	0.004	0.02
Triton (Neptune)	354,760	5.877	0.000	158.8

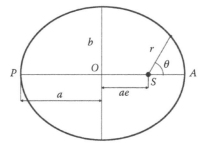

FIGURE 1.8   Elements of an elliptical orbit.

where τ_{orb} is the planet orbital period. The value of the constant can be obtained upon integration of (1.18):

$$\int_0^{\tau_{orb}} \frac{dS}{dt}dt = S = \frac{L_0}{2m}\tau_{orb} \qquad (1.20)$$

By using (1.15) for the angular momentum and the area of the ellipse $S = \pi ab$, we get

$$\frac{a^3}{\tau_{orb}^2} = \frac{GM}{4\pi^2} \qquad (1.21)$$

1.2.2 MOTION OF TWO BODIES UNDER THEIR MUTUAL GRAVITATIONAL ATTRACTION

Next, we consider the motion of two bodies with masses m_1 and m_2 moving under the action of their mutual gravitational attraction and an external force $\vec{F}_{ext} = \vec{F}_1 + \vec{F}_2$ ($\vec{F}_1$ acting on m_1 and $\vec{F}_2$ acting on m_2), a study known as a "two-body problem." For most cases of interest, $\vec{F}_1/m_1 - \vec{F}_2/m_2 = 0$ and to a first approach, the motion of these masses is described by the following two equations:

$$\vec{F}_{ext} = M\frac{d^2\vec{R}}{dt^2} \tag{1.22a}$$

$$Gm_1m_2\frac{\vec{r}}{r^3} = -\mu_m\left(\frac{d^2\vec{r}}{dt^2}\right) \tag{1.22b}$$

Here, $M = m_1 + m_2$ is the total mass of the system and $\mu_m = (m_1m_2)/(m_1 + m_2)$ is the *reduced mass* of the system. The position of the *center of mass* of the system (CM) is given by the vector

$$\vec{R} = \frac{m_1\vec{r}_1 + m_2\vec{r}_2}{M} \tag{1.23}$$

and $\vec{r} = \vec{r}_1 - \vec{r}_2$ is the position vector between the masses (see Figure 1.9). According to (1.22a), the CM of the system moves under the action of the external force, whereas the motion of the two masses takes place around the CM under their mutual gravitational interaction.

From (1.22), Kepler's laws are followed (elliptic orbit) and (1.21) can be rewritten as

$$\frac{a^3}{\tau_{orb}^2} = \frac{G(m_1 + m_2)}{4\pi^2} \tag{1.24}$$

From (1.23), $m_1/m_2 = a_2/a_1$ and $a = a_1 + a_2$, which is the semimajor axis of the relative orbit of the two masses (a_1 and a_2 are the semimajor axis for bodies 1 and 2 relative to the CM). Equation 1.22a applies, for example, directly to the motion of the CM of a system formed by a planet and their satellites around the Sun, and Equation 1.22b

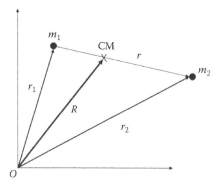

FIGURE 1.9   Coordinates employed to study the motions of two bodies under their mutual interaction and subjected to any external force. CM is the center of mass.

applies to the relative motion of a satellite around a planet. Equation 1.24 reduces to (1.21) when one mass is much greater than the other, as occurs on planets and most of their satellites, as previously described.

1.2.3 THREE-BODY PROBLEM

Whereas the equation of motion of two bodies due to their mutual gravitational interaction can be integrated and their relative trajectories can be described by simple conic sections, adding a third body makes the problem difficult and the motion equations not directly integral. In such cases, analytic solutions cannot be found except for limiting situations. The problem is quite complex and the numerical resolution of the motion equations is needed. A full treatment is out of the scope of this book, so we will only comment on the orbital situation corresponding to the so-called *restricted three-body problem*.

We consider the case of two bodies with masses m_1 and m_2 moving around their common CM and a third body (a test particle) with negligible mass $m \ll m_1, m_2$ (so that it does not affect the motion of the two bodies or that of the CM) that moves in the plane determined by the motion of m_1 and m_2. In the reference frame of the CM (i.e., with origin at the CM and moving with the masses, i.e., m_1 and m_2 are stationary), and assuming that the two masses describe a circular orbit with angular velocity Ω, Lagrange found that there are five points of equilibrium for the mass m in the plane of motion (Figure 1.10). A test particle placed at rest would feel no net force in the rotating frame when located in one of these points. The three *Lagrangian points* L_1, L_2, and L_3 are aligned with the masses m_1 and m_2. The other two (L_4 and L_5) form equilateral triangles with m_1 and m_2. When m_1 and m_2 have at least 96% of the total mass of the system, points L_4 and L_5 are stable (i.e., a test particle slightly displaced from the point will oscillate back and forth around it). However, points L_1, L_2, and L_3 are unstable and if the test particle is slightly displaced from this point, it will move away.

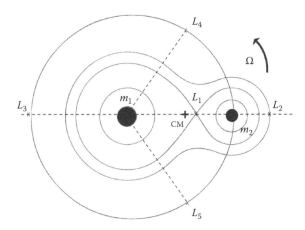

FIGURE 1.10   Equipotential gravitational curves, Lagrange points (L_i) and Roche lobe for two masses m_1 and m_2.

The Lagrangian point L_1 is important in some cases of study of planetary atmospheres. Point L_1 exists, for example, if a planet has a close moon or if the planet is placed close to its parent star as occurs in observed cases in extrasolar giant planets. Through L_1, the planet's atmosphere can escape, for example, when the two masses are at a close distance. The situation is well visualized when drawing the equipotential gravitational curves in the plane of the masses (Figure 1.10). When the atmosphere of one or both masses fills the limiting surfaces surrounding them (those intersecting in the common point L_1), mass transfer can occur.

The region enclosed by this surface is called the *Roche lobe*. The distance r between m_2 and the point L_1 can be calculated using Newton's second law, since a mass at point L_1 rotates about the CM with angular velocity $\Omega = 2\pi/\tau_{orb}$ under the action of the gravitational pull of m_1 and m_2

$$\frac{Gm_1 m}{(a-r)^2} - \frac{Gm_2 m}{r^2} = m\Omega^2 (a_2 - r) \tag{1.25}$$

where a_2 is the orbital radius for m_2 and $a = a_1 + a_2$ (the distance between m_1 and m_2). In the limit $m_2 \ll m_1$, the Roche lobe reaches a radius that is essentially equivalent to the *Hill sphere*, defined as the region of gravitational dominance of the mass m_1 (usually considered the primary, the Sun or a star, or a planet) against m_2 (a planet or a moon, respectively) and given by

$$R_H \approx a \sqrt[3]{\frac{m_2}{3m_1}} \tag{1.26}$$

1.2.4 ORBITAL PERTURBATIONS AND RESONANCES

The motion of the bodies in a planetary system is in fact much more complex than described in previous sections since we are dealing with the mutual gravitational interaction between N-bodies. As indicated before, in the case of the solar system, the Sun dominates the planetary motion due to its large mass and planet–Sun separation. The same is valid for most of the planets and their satellites (one exception is that of minor planets Pluto and Charon). Keplerian motions can therefore reasonably be assumed to represent in first order the motions of planets about the Sun and those of satellites around a planet. However, the gravitational action of other bodies on the Keplerian motions, although low, perturb the orbits producing changes in the planet's orbits with important effects on the long-term evolution of planetary atmospheres. The perturbed orbits undergo the following effects: (a) perihelion (or periapsis) changes by rotation of the major axis of the orbit, (b) changes in the semimajor axis length ("short-period" variations), and (c) a change of orbital eccentricity and inclination (secular variations). These changes can be periodic (most long-term perturbations are) or chaotic depending on the initial conditions.

When two (or more bodies) have an orbital configuration such that their periods are commensurable, they are said to be in *mean motion resonance*. Resonances are important in the study of planetary atmospheres as they can be involved in the production of a strong tidal heating within the bodies (see Sections 1.3.2 and 1.4.2.3). There are numerous examples of mean motion resonance in the solar system, an important one being that occurring between the Galilean satellites of Jupiter Io, Europa, and Ganymede. These satellites obey the following relationship:

$$n_I - 3n_E + 2n_G = 0 \qquad (1.27)$$

where n_I, n_E, and n_G are their mean motions, defined as $2\pi/n_i$ ($i = I,E,G$). Their tenuous atmospheres owe their existence to and are controlled by the heat generated in the satellites' interiors by this resonance.

Another important kind of resonance from the point of view of planetary atmospheres occurs when the mean motion and the body's rotation rate are equal, a phenomenon known as *synchronous rotation* (spin and orbit are coupled so their periods are in a 1:1 ratio). The origin of this coupling resides in the mutual tidal forces between the bodies that we shall discuss later and has profound influences in the rotation rate of a planet or satellite, a fundamental parameter in the study of planetary atmospheres. All the satellites with regular orbits are in a 1/1 spin–orbit coupling with its parent planet, and it is also expected that "hot giant extrasolar planets" (those close to their parent star) are also in such states.

1.3   GRAVITATIONAL FIELD AND THE SHAPE OF PLANETS AND MAJOR SATELLITES

Up until now, we have considered the planets and satellites to be point masses, as it is appropriate to study their mutual gravitational interaction. In fact, the gravitational force in the exterior of a spherically symmetric body is the same as that of a point mass having the total mass of the body, and the planets and major satellites are all nearly spherical due to the symmetry of the gravitational force.

Nevertheless, the actual planet shape is further constrained by its rotation angular velocity (Ω) and by the gravitational tidal effects exerted on the planet by the star or by close massive satellites. In the case of a deformable fluid body, the equilibrium figure resulting from the action of these forces is an ellipsoid. Rapidly spinning planets can widen along the equatorial axis due to the centrifugal force. In the solar system fluid planets, tidal deformations can also be produced by close satellites, although their effect on the planet shape is negligible when compared with that produced by rotation. However, in the extrasolar giant planets of "hot" or "very hot" Jupiter types, the tidal effect raised by the star on the planet can be important when compared with the effect of rotation for two reasons. First, they can be located very close to the star (typically ~0.01 AU; 1 AU [astronomical unit] = 1.49×10^8 km) and second, these planets are supposed to be slow rotators due to the tidally developed spin–orbit synchronism (rotation periods ~1–4 days). Therefore, we can expect these planets to be elongated in one axis becoming "spheroidal" in shape. In Tables 1.3 and 1.4, we summarize the most important properties of the planets and the major satellites (with atmosphere).

TABLE 1.3
Physical Characteristics of Planets, Dwarf Planets, and Major Satellites

Object	R_p (km)	M_p (kg)	$\langle\rho\rangle$ (g cm^{-3})	g (cm s^{-2})	Rotation Period	Rotation Parameter (q_r)	Oblateness
Mercury	2,439	3.30×10^{23}	5.43	370	58.65 days	9.5×10^{-7}	0
Venus	6,050	4.86×10^{24}	5.20	884	-243.01 days	6.1×10^{-8}	0
Earth	6,378	5.98×10^{24}	5.51	981	23.93 h	0.0035	0.0033
Moon (S)	1,738	7.35×10^{22}	3.34	162	27.32 days (s)	7.6×10^{-6}	
Mars	3,398	6.42×10^{23}	3.91	376	24.62 h	0.0046	0.005
Ceres (dP)	467	9.46×10^{20}	2.08	27	9.08 h	0.06	0
Jupiter	71,300	1.90×10^{27}	1.33	2288	9.84 h	0.089	0.061
Io (S)	1,822	8.93×10^{22}	3.53	181	1.77 days (s)	0.0017	0
Europa (S)	1,565	4.80×10^{22}	2.99	131	3.55 days (s)	5×10^{-4}	0
Ganymede (S)	2,631	1.48×10^{23}	1.94	142	7.15 days (s)	1.8×10^{-4}	0
Callisto (S)	2,410	1.08×10^{23}	1.83	125	16.7 days (s)	3.6×10^{-5}	0
Saturn	60,100	5.68×10^{26}	0.69	950	10.23 h[a]	0.16	0.09
Titan (S)	2,575	1.34×10^{23}	1.88	135	15.95 days	4×10^{-5}	0
Enceladus (S)	249	6.5×10^{19}	1.13	7	1.37 days (s)	0.01	0
Uranus	25,500	8.68×10^{25}	1.32	869	17.9 h	0.027	0.03
Neptune	24,800	1.02×10^{26}	1.64	1100	19.2 h	0.018	0.03
Triton (S)	1,355	2.15×10^{22}	2.05	77	5.88 days (s)	2.6×10^{-4}	0
Pluto (dP)	1,150	1.32×10^{22}	2.0	72	6.38 days	2.2×10^{-4}	0
Eris (dP)	1,200	1.66×10^{22}	2.3	75	~0.3 days	—	0
Haumea (dP)	575	4.2×10^{21}	2.6–3.3	~40	—	—	—
Makemake (dP)	750	~4×10^{21}	~2	~50	—	—	—

Note: (s) indicates spin–orbit synchronization. dP, dwarf planet; S, satellite.

[a] In the case of Saturn, the rotation axis and magnetic field are closely aligned (long-term variability in System III has been found) making it difficult to establish the radio rotation period. Until a definitive internal rotation period is adopted for Saturn, the radio rotation System III determined by the Voyager missions is recommended to make previous atmospheric dynamic studies coherent.

1.3.1 GRAVITATIONAL FIELD OF A PLANET

The gravitational field produced by a body is determined by its total mass and the density distribution $\rho(\vec{r})$ within it. We shall assume first that the density distribution depends mainly on the radial direction (it has spherical symmetry). Density and mass distribution are then related through the equation

$$\frac{\partial M}{\partial r} = 4\pi r^2 \rho(r) \qquad (1.28)$$

TABLE 1.4

Albedo, Internal Energy, and Temperature of Planets, Dwarf Planets, and Major Satellites

Object	Geometric Albedo	Internal Energy $(W\ m^{-2})$	T_{eff} (K)	T_{eq} (K)
Mercury	0.106	—	Eq	446
Venus	0.65	0	Eq	238
Earth	0.367	0.075[a]	Eq	263
Moon (S)	0.12	0.026	Eq	277
Mars	0.15	0.04	Eq	222
Ceres (dP)	0.11	—	Eq	167
Jupiter	0.52	5.44	124.4	113
Io (S)	0.61	2.5	⟨Eq⟩	130
Europa (S)	0.64	0.02	Eq	102
Ganymede (S)	0.42	0.003	Eq	110
Callisto (S)	0.20	0.003	Eq	134
Saturn	0.47	2.01	95	83
Enceladus (S)	0.99	~0.005–0.01	Eq	75
Titan (S)	0.21	—	Eq	94
Uranus	0.51	0.042	59.1	60
Neptune	0.41	0.433	59.3	48
Triton (S)	0.76	—	Eq	38
Pluto (dP)	0.6	—	Eq	44
Eris (dP)	0.86	—	Eq	42
Haumea (dP)	—	—	Eq	32
Makemake (dP)	—	—	Eq	30

Note: Eq means that the equilibrium and effective temperatures are the same. For Io, we adopt the equilibrium and effective temperatures to be the same outside the hot volcanic and vent areas. dP, dwarf planet; S, satellite.

[a] Averaged value over the whole planet. In geothermal areas the heat power per unit area reaches ~1.7 W m⁻².

The total mass of the body can be obtained upon the integration of (1.28) from the center to its outer total radius R_p

$$M = 4\pi \int_0^{R_p} \rho(r)r^2\, dr \qquad (1.29)$$

The conservative nature of the gravitational force allows us to calculate the gravitational field $\vec{g}$ produced by a mass distribution in terms of gravitational potential $V_G(\vec{r})$ (which is the potential energy per unit mass produced by this mass distribution) through an equation equivalent to (1.3)

$$\vec{g} = -\nabla V_G(\vec{r})\tag{1.30}$$

A test particle with mass m will suffer a force $m\vec{g}$.

The structure of the gravitational force and the use of vector relationships imply that the gravitational potential obeys Poisson's equation:

$$\nabla^2 V_G(\vec{r}) = -4\pi G\rho(r)\tag{1.31}$$

The gravitational potential can be obtained by integration, leading to

$$V_G(\vec{r}) = -G\iiint\limits_{Vol} \frac{\rho(r')}{|\vec{r} - \vec{r}'|}\, d^3\vec{r}'\tag{1.32}$$

where

$d^3\vec{r}'$ is an infinitesimal volume element of the body
$\vec{r}'$ is its position
$\vec{r}$ is the position where the potential is to be evaluated

For a homogeneous sphere of mass M, the gravitational potential outside the sphere at a distance r is simply given by

$$V_G(r) = -\frac{GM}{r}\tag{1.33}$$

equivalent to that produced by a point mass. At the surface of the body, the gravitational potential is $V_S = -GM/R_p$. Inside the sphere, assuming a spherically symmetric density distribution, the gravitational potential is obtained from the contribution from the mass shells outside and inside a distance r and then from (1.32), we have

$$V_G(r) = -G\left[\frac{M(r)}{r} + 4\pi\int_r^{R_p}\rho(r')r'\,dr'\right]\tag{1.34}$$

For a spherical body with uniform mass distribution, the gravitational potential inside the sphere is

$$V_G(r) = -\frac{GM}{2R_p}\left(3 - \frac{r^2}{R_p^2}\right)\tag{1.35}$$

and from (1.30), the gravitational field is

$$\vec{g}(r) = -\frac{\partial V(r)}{\partial r}\hat{r} = -\frac{GM}{R_p^3}r\hat{r}\tag{1.36}$$

In general, planets and major satellites have a nearly spherical shape and the above expressions for the gravitational field (excluding rotation) are appropriate to first order. However, a complete description of the planet shape must include deviations from sphericity. In this case, the solution of Equation 1.31 for the external gravitational potential of a nonrotating body, can be fully represented using the harmonic expansion

$$V_G(r,\varphi,\phi) = \frac{GM}{r}\left\{1 - \sum_{n=2}^{\infty}\left(\frac{R_E}{r}\right)^n J_n P_n(\sin\varphi)\right.$$

$$\left. + \sum_{n=2}^{\infty}\sum_{m=1}^{n}\left(\frac{R_E}{r}\right)^n P_n^m(\sin\varphi)\left[C_n^m \cos m\phi + S_n^m \sin m\phi\right]\right\} \quad (1.37)$$

where
  (r,φ,ϕ) are the spherical coordinates (radial, latitude, and longitude)
  R_E is the equatorial radius
  P_n, P_n^m are the Legendre and associate Legendre polynomials

The gravitational shape of the planet is characterized by the dimensionless numbers J_n (the zonal harmonics or gravitational moments) and C_n^m, S_n^m (the *tesseral harmonics*), which depend on the mass density distribution within the body. The gravitational coefficients in (1.37) represent the contributions to the planet gravity potential of the perturbations due to tidal forces, rotation, and departures intrinsic to the planet's shape from hydrostatic equilibrium.

In the case of solar system giant planets, Earth and Mars, the most important perturbation to the gravity potential is due to rotation Q_R. Its inclusion requires us to rewrite Equation 1.31 as

$$\nabla^2 U_G(\vec{r}) = \nabla^2 Q_R(\vec{r}) - 4\pi G\rho(r) \quad (1.38)$$

where $U_G = V_G + Q_R$ and the rotational potential for uniform rotation is written as

$$Q_R(r,\varphi) = \frac{1}{3}\Omega^2 r^2\left[1 - P_2(\sin\varphi)\right] \quad (1.39)$$

where Ω is the angular velocity about the axis perpendicular to the equatorial plane (in the z-direction for the Cartesian reference frame centered on the body). At the equator, $Q_R(r,\varphi = 0°) = (1/2)\Omega^2 R_E^2$.

The internal structure of the planets is of interest in the context of this book in the case of the giant planets, whose fluid atmospheres are deep enough to leave their signatures in U_G. Let us assume that hydrostatic conditions prevail (i.e., that surfaces of constant gravitational potential and constant density coincide, see Section 1.5.2) and that the potential V_G in Equation 1.37 is symmetric about the rotation axis. Then the third term in brackets in (1.37) can be neglected, that is the effect in longitude is null or the tesseral harmonics are zero. Moreover, a planet with its mass symmetrically distributed about the planet's equator has $J_n = 0$ for odd n. In these conditions, Equation 1.37 involves only the even coefficients J_{2n} and polynomials P_{2n} and the total potential, including rotation, is given by

$$U_G(r,\varphi) = -\frac{GM}{r}\left[1 - \sum_{n=1}^{\infty}\left(\frac{R_E}{r}\right)^{2n} J_{2n}P_{2n}(\sin\varphi)\right] - \frac{1}{2}\Omega^2 r^2 \cos^2\varphi \qquad (1.40)$$

the second term being the contribution to the potential of planetary spin (also called *centrifugal potential*). The coefficients J_{2n} are known as the *gravitational moments* and are determined by the planet's mass distribution:

$$J_{2n} = \frac{1}{MR_E^{2n}}\iiint\limits_{Vol}\rho(r)r^{2n}P_{2n}(\sin\varphi)d^3\vec{r} \qquad (1.41)$$

Note that the total mass is obtained from a generalized version of (1.29):

$$M = \iiint\limits_{Vol}\rho(r,\varphi)d^3\vec{r} \qquad (1.42)$$

The representation of the gravitational potential given in Equation 1.40 has been found to be very good for the giant planets that are mainly fluid and highly deformable. For the terrestrial planets and major satellites that are approximately spherical, the "J-coefficients" are small and can be neglected. However, the oblateness of the giant planets makes the J-coefficients substantial, giving important information on their internal structure and deep atmospheres. They are measured by observing the gravitational perturbations acting on satellites, rings, and spacecrafts in orbit around the planet or during a fly-by. As discussed in Chapter 7, a family of models for the atmospheric circulation in the giant planets proposes that motions extend deep in the atmosphere, occupying a significant fraction of the planetary radius. If this is the case, the planetary rotation vector should incorporate a term due to atmospheric motions. Under the assumption that the deep circulation is steady zonal (along longitude), and occurring in cylinders parallel to the rotation axis (see Chapter 7), the planet rotation vector can be written as $\vec{\Omega} = \vec{\Omega}_0 + \delta\vec{\Omega}$, where $\vec{\Omega}_0$ is the planetary rotation and $\delta\vec{\Omega}$ is the contribution due to the zonal circulation (derived from motions observed at the top of the atmosphere). By introducing this modification in (1.39), we can expect that the gravitational potential U_G will be affected if a deep zonal variable rotation is present. Thus, precise measurements of the gravitational field from satellite motions can give information on the deep motions in a giant planet. This is the purpose of the *Juno mission* to Jupiter.

From Equations 1.30 and 1.40, the component of the gravitational acceleration in the radial direction $g_r = \partial U_G/\partial r$ becomes

$$g_r(r,\varphi) = -\frac{GM}{r^2}\left[1 - \sum_{n=1}^{\infty}(2n+1)J_{2n}\left(\frac{R_F}{r}\right)^{2n}P_{2n}(\sin\psi)\right] + \frac{2}{3}\Omega^2 r\left(1 - P_2(\sin\varphi)\right)$$

(1.43a)

and the latitudinal component $g_\varphi = r^{-1}\partial U_G/\partial\varphi$

$$g_\varphi(r,\varphi) = -\frac{GM}{r^2}\left[\sum_{n=1}^{\infty}\left(\frac{R_E}{r}\right)^{2n}J_{2n}\frac{dP_{2n}(\sin\varphi)}{d\varphi}\right] - \frac{1}{3}\Omega^2 r\left[1 - \frac{dP_2(\sin\varphi)}{d\varphi}\right]$$ (1.43b)

The total intensity of the acceleration of gravity is given by

$$g = \sqrt{(g_r^2 + g_\varphi^2)}$$

(1.44)

As stated above, rotational effects are important in determining the shape of rapidly rotating fluid giant planets. Under hydrostatic equilibrium (see Section 1.5.2), the shape of a deformable body under fast rotation is that of a biaxial ellipsoid (Maclaurin spheroid) with an equatorial radius R_E and polar radius R_P. For such a figure, a plane cutting the ellipsoid containing the rotation axis (and center of the body) is an ellipse with semiminor axis R_P and semimajor axis R_E. The cross-section perpendicular to the rotational axis is then a circle. We define the oblateness f_{EP} of this body as

$$f_{EP} = \frac{R_E - R_P}{R_E}$$

(1.45)

The intensity of the rotational effects on the shape of a planet can be characterized by a parameter that gives the ratio between the rotational energy and the gravitational binding energy of the body (see Section 4.2.2):

$$q_R = \frac{\Omega^2 R_E^3}{GM}$$

(1.46)

Under a hydrostatic equilibrium, it is possible to express the even zonal harmonics as a power series of q_R

$$J_{2n} = \sum_{n=0}^{\infty}\beta_{nm}q_R^{n+m}$$

(1.47)

where β_{nm} are coefficients that depend on the pressure–density equation of the planet's interior. For a uniform incompressible fluid sphere, we have

$$J_2 = \frac{q_R}{2}$$

(1.48)

However, in the real planets with the mass concentrated toward the center, we have $J_2 < q_R/2$ and (under hydrostatic equilibrium) the following relationship between the parameters defined above holds

$$f_{EP} = \frac{3}{2}J_2 + \frac{1}{2}q_R \tag{1.49}$$

According to this expression, the measurement of Ω, M, R_E, and f_{EP} allows for the calculation of J_2.

Another important parameter that characterizes the mass distribution within a planet comes from its moment of inertia about a given axis, defined as

$$I = \iiint \rho(\vec{r})r_c^2 \, d^3\vec{r} \tag{1.50}$$

where r_c is the distance from a mass element of the body to the rotation axis. For a spherical body with a radial dependence of the density, Equation 1.50 can be simplified as

$$I = \frac{8\pi}{3} \int_0^{R_p} \rho(r)r^4 \, dr = \alpha_I M R_p^2 \tag{1.51}$$

where $\alpha_I (= I/MR_p^2)$ is called the *inertial factor*. A spherical body with a uniform density $\rho(r) = \rho_0$ has, according to (1.51), a moment of inertia

$$I = \frac{2}{5}MR^2 \tag{1.52}$$

and therefore $\alpha_I = 0.4$. As an illustration, two extreme cases in mass distribution are a spherical shell ($\alpha_I = 2/3 = 0.667$) and a sphere with the mass concentrated in its center ($\alpha_I = 0$, equivalent to a point mass). In general, the process of mass differentiation within a body (occurring following planetary formation; see Section 5.2) makes the mass increase toward the center and then $\alpha_I < 0.4$. In a rapidly rotating body in hydrostatic equilibrium, the following relationship holds for the inertial factor

$$\alpha_I = \frac{(3/2)J_2}{J_2 + (q_R/2)} \tag{1.53}$$

Table 1.3 gives the physical properties of planets and satellites and in Table 1.5 we present the values of the *J*-coefficients and inertial factor for the planets and main satellites.

TABLE 1.5

Gravitational Moments and Inertial Moment Factor

Object	J_2 (×10⁻⁶)	J_3 (×10⁻⁶)	J_4 (×10⁻⁶)	J_5 (×10⁻⁶)	J_6 (×10⁻⁶)	α_I
Mercury	60					0.33
Venus	4.5	−1.9	−2.4			0.33
Earth	1,082.6	−2.5	−1.6	−0.2	0.65	0.33
Moon (S)	203.4					0.393
Mars	1,960.5	31.5	−15.5			0.366
Jupiter	14,736	0	−587	0	31	0.254
Io (S)	1,863					0.375
Europa (S)	438					0.348
Ganymede (S)	127					0.311
Callisto (S)	34					0.358
Saturn	16,298	0	−915	0	103	0.26
Uranus	3,343.4	0	−28.9	0		
Neptune	3,411	0	−35			

1.3.2 GRAVITATIONAL TIDAL FORCE AND TORQUE

The gravitational force acting on a body due to another close object varies from point to point due to the r^{-2} gravitational dependence and the resulting differential force between the points is called the tidal force. Tides and their torques can deform the bodies, determining their equilibrium shape, and can also act on their atmospheres. Time-variable tidal forces and associated torques on bodies describing eccentric orbits (as occurs with some moons and with the family of "hot Jupiter" extrasolar planets close to their stars) can result in their flexing producing internal heating (see Section 1.4.2.3). If they are strong enough, they can cause the body to break apart (see Section 3.1.3).

The tidal force exerted by a mass m on a body can be calculated using potential theory. Neglecting rotational effects, the gravitational potential produced by a mass at a distance r_0 from a planet at a point (r,φ) is

$$V_G(r) = -\frac{Gm}{r_0}\left(1+\frac{r}{r_0}\cos\varphi+\frac{r^2}{2r_0^2}(3\cos^2\varphi-1)+\cdots\right) \tag{1.54}$$

The third term in this series is called the tidal potential. The radial and orthoradial components of the gravitational field due to the tidal potential are given by

$$g_r = \frac{Gmr}{r_0^3}(3\cos^2\varphi-1) \tag{1.55a}$$

$$g_\varphi = -\frac{3Gmr}{r_0^3}\sin\varphi\cos\varphi \tag{1.55b}$$

The pull of the gravitational field on a mass element of the body produces a tidal bulge since at $\varphi=0°$ and $180°$ we have $g_r>0$ and $g_\varphi=0$ (corresponding to an elevation), but at $\varphi=90°$ we have $g_r<0$ and $g_\varphi=0$ (corresponding to a depression). Therefore, a deformable body will become elongated in the direction of the tidal force.

The tidal effect produced by a body of mass m placed at a distance r of another of mass M_p and radius R_p can be characterized by a parameter similar to that used for the rotational effect in Equation 1.46, called the tidal parameter

$$q_{Ti} = \frac{m}{M_p}\left(\frac{R_p}{r}\right)^3 \tag{1.56}$$

On the other hand, although the total angular momentum of a pair of orbiting bodies is conserved in the absence of an external torque (Equations 1.7 and 1.8), spin and orbital angular momentum can be interchanged through the tidal torques produced by the net forces on the opposite bulges raised by the tides (see Figure 1.11). These bulges are not aligned in the direction of the two bodies due to a temporal delay caused by the time required for the elastic deformation of the body. Depending on the relative direction of the spin of one body and the direction marked by the orbital motion of the other, the tidal torque modifies the orbit of the lower mass body (expanding or contracting it relative to the nontidal case), whereas it increases or decreases the rotation rate of the largest body. The rate of change of the orbital energy produced by this mechanism can be related to the applied torque ($T_T=dL_0/dt$) in the following way. Combining Equations 1.14 and 1.15 through m, we can express the orbital energy of the low mass body as

$$E = -\frac{GML_0}{2\sqrt{GMa^3(1-e^2)}} \tag{1.57}$$

and for M, L_0, and e constant, we get

$$\frac{dE}{dL_0} = -\frac{1}{2}\sqrt{\frac{GM/a^3}{1-e^2}} \tag{1.58}$$

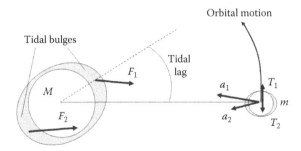

FIGURE 1.11 Tidal forces (F_1, F_2), accelerations (a_1, a_2), and torques ($T_1>T_2$) between a planet and a satellite.

The temporal change of the orbital energy is then given by

$$\frac{dE}{dt} = \frac{dE}{dL}\frac{dL}{dt} = -\frac{T_{Ti}}{2}\sqrt{\frac{GM/a^3}{1-e^2}} \tag{1.59}$$

There are several examples of the tidal effects in the solar system. The Moon is presently retreating from the Earth at a rate of 4 cm year^{-1} while the Earth's rotation rate is slowing down, as is evident from geological records. Tidal torques are also responsible for the observed synchronous spin–orbit state of most satellites (rotational period equal to orbital period) and solar gravitational tides are involved (in a more complex way) in the slow rotation rate of Mercury and Venus. However, the other planets of the solar system (Mars, Jupiter, Saturn, Uranus, and Neptune) have kept their primordial rotation rates due to their high mass when compared with that of their satellites.

1.3.3 ROCHE'S LIMIT

When tidal forces are strong enough to overcome the cohesion forces, they can disrupt the formation of a body as a unique entity or can break it apart. There is a limit distance between two bodies that can destroy the smaller one due to tides raised on it by the larger one, called the Roche limit (R_L) (Figure 1.10). This distance can be calculated in a simple way in the following approximation: (a) the secondary body describes a circular orbit around the main body, (b) the orbit is synchronized (the rotation and orbital periods are equal), (c) the secondary body is spherical and the main body can be treated as a point mass (homogeneous sphere), and (d) the secondary body is assumed to be held by its self-gravity. The Roche limit obtained under such approximation is given by

$$R_L = 1.44 R_p \left(\frac{\rho_p}{\rho_s}\right)^{1/3} \tag{1.60}$$

where
  R_p and ρ_p are the radius and density of the primary body (typically a planet)
  ρ_s is the density of the secondary body (typically a satellite)

If the secondary body is a liquid, fully deformable object, the factor 1.44 in Equation 1.60 must be replaced by 2.46 according to more detailed calculations. Examples of the relevance of the Roche limit are the planetary rings of the four giant planets, which lie within their planet's Roche limit distance.

1.3.4 PLANETARY ROTATION SYSTEM AND COORDINATES

There are two ways to define the rotation period of a body. The *sidereal period* (τ_{sid}) is the time employed for a fixed direction in the body to go through an angle of 360° relative to the stars (assumed fixed in the sky background for the periods involved).

The *synodic period* (τ_{syn}) is the time for this fixed direction to go through this angle relative to the central star. Both are related by

$$\frac{1}{\tau_{syn}} = \frac{1}{\tau_{sid}} \pm \frac{1}{\tau_{orb}}$$

(1.61)

where τ_{orb} is the orbital period (the planetary year). The ± sign applies to a planet rotating in the same or opposite direction as that of its orbit.

The longitude system of bodies with a rigid surface, such as the terrestrial planets and major satellites, is defined from a reference longitude (meridian 0°) selected arbitrarily (such as the Greenwich meridian on Earth) and their rotation period is that of their surfaces. In the case of the gaseous planets, the rotation period must be known in order to define the longitude system. The rotation rate of Jupiter, Saturn, Uranus, and Neptune is taken to be that of their magnetic fields, which are assumed to be tied to the deep planetary interiors (see Section 1.6) and represent the bulk rotation of the planet. The rotation of the magnetic field can be indirectly measured by means of the modulated radio emission produced by accelerated charged particles trapped in the field lines. This rotation system is known as System III in the four gaseous planets. For historical reasons, two other systems are employed in Jupiter, based on the averaged rotation rate of cloud markings located in the Equatorial region (System I) and outside the Equator (System II). It has been found recently that the radio rotation System III of Saturn varies with time (of the order of a few minutes) apparently due to a complex interaction of its magnetosphere with the satellite Enceladus. Until a definitive rotation rate is obtained, the recommended value for Saturn's rotation rate is that measured at the Voyager time (years 1980–1981). The rotation periods of the planets are given in Table 1.3. The zero longitude for System III on each planet is also arbitrary but is defined by the International Astronomical Union.

Spherical coordinates are employed to locate a point on the surface or in the atmosphere of the terrestrial planets and major satellites: vector radius (r), longitude (ϕ), and latitude (φ). The coordinate systems are defined relative to their mean axis of rotation and geometric center of the body. For the giant fluid planets, which have oblate spheroid shapes, no rigid observable surfaces exit, so oblate spherical coordinates are employed. There are two alternative definitions of latitude (see Figure 1.12): planetographic latitude (φ_g) is the angle from the perpendicular to the tangent to the

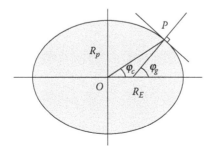

FIGURE 1.12   Planetographic (φ_g) and planetocentric (φ_c) on an ellipsoidal planet.

surface intersecting the semimajor axis of the ellipse and planetocentric latitude (φ_c), which is relative to the center of the ellipse. Both are related by the formula

$$\tan \varphi_g = \left(\frac{R_E}{R_P}\right)^2 \tan \varphi_c \qquad (1.62)$$

The vector radius that gives the position of a point at the surface spheroid can be expressed as a function of the planetocentric latitude and the equatorial and polar radius by

$$r(\varphi_c) = \frac{R_E R_P}{\sqrt{R_E^2 \sin^2\varphi_c + R_P^2 \cos^2\varphi_c}} \qquad (1.63a)$$

A useful approximation to this equation can be obtained employing the oblateness $f_{EP} = (R_E - R_P)/R_P$

$$r(\varphi_c) = R_E(1 - f_{EP} \sin^2\varphi_c) \qquad (1.63b)$$

The radii of curvature along the zonal direction (i.e., along parallels or in longitude) r_Z and the meridional direction r_M are given by

$$r_Z = \frac{R_E}{(1 + \varepsilon_p^{-2} \tan^2\varphi_g)^{1/2}} \qquad (1.64a)$$

$$r_M = \frac{R_E}{\varepsilon_p^2}\left(\frac{r_Z}{R_E \cos \varphi_g}\right)^3 \qquad (1.64b)$$

with $\varepsilon_p = R_E/R_P$.

A useful simplified expression for the acceleration of gravity that includes the planetary rotation and the deviation from the spherical shape can be obtained from the above formulas

$$g(\varphi_c) = g_0(\varphi_c) - \Omega^2 r(\varphi_c)\cos \varphi_c \qquad (1.65a)$$

where

$$g_0 = \frac{GM}{r(\varphi_c)^2} \qquad (1.65b)$$

1.4 PRIMARY ENERGY SOURCES ON PLANETS

Atmospheres are thermodynamic machines whose primary working mechanisms are the planetary rotation and the energy sources (internal and external heat). In planets with a surface (e.g., terrestrial planets and most satellites—the exceptions are Io and Enceladus), the global action of the internal heat energy source on the atmospheric properties can be neglected, except during its initial formation period. However, the internal energy sources produce substantial indirect effects; for example, generating magnetic fields that play an important role in the upper atmospheric layers in these planets or through volcanic and similar activity escaping through their surfaces, as occurs presently on the Earth and satellites Io, Enceladus, and Triton and probably on Venus and Mars in the past. On the contrary, in the giant planets internal energy sources become probably as important as sunlight in driving, for example, atmospheric motions (at least in the upper visible cloud layers).

In this section, we consider the nature of the primary energy sources that intervene in physical and chemical atmospheric processes. We do not consider secondary energy sources such as the internal friction heat (dissipation) within the atmosphere (e.g., through molecular and eddy viscosities) or the energy released from phase changes of condensing elements. These energy sources will be discussed in detail in other chapters of this book.

1.4.1 ROTATIONAL MECHANICAL ENERGY

The rotation kinetic energy of a planet is given by

$$E_{rot} = \frac{1}{2} I \Omega^2 \qquad (1.66)$$

where
  I is the moment of inertia of the planet about the rotation axis
  Ω is its angular velocity

1.4.2 INTERNAL ENERGY

We define the internal energy of a planet as that coming from its deep interior, that is, that directly related to processes occurring within the body. There are different possible production mechanisms.

1.4.2.1 Radioactive Decay

The presence of unstable radioactive isotopic elements in the interior of terrestrial type planets and satellites can generate heat by radioactive energy emission during their decay series. Each isotope has its own half-lifetime and those with very large temporal decay contribute to the planetary heating along a large part of the body lifetime. Examples are ^{235}U (0.71 Gy), ^{238}U (4.5 Gy), ^{232}Th (13.9 Gy), and ^{40}K (1.4 Gy). Short life isotopes, such as ^{26}Al (0.74 My), make an important contribution during the body formation stage. The radiogenic heating rate (in watts) from the unstable elements can be calculated as follows:

$$L_{rad} = m_{rock} \sum_{i=1}^{n} C_i X_i H_i \exp\left[-\tau_{in}(t-t_0)\right] \tag{1.67}$$

Here

m_{rock} is the total mass of the rocks containing the radioactive elements

C_i is the present day concentration of the radiogenic element (in parts per million, ppm)

X_i is the present fraction of the ith radioactive nuclei in kg of nuclei per kg of rock

H_i is the power released per kg of the ith nuclei upon decay (W kg^{-1})

τ_{in} is the half-lifetime of the nuclei

t_0 is the initial time (= 4.6 Ga for the solar system)

The energy flux (in W m^{-2}) generated by this radioactive decay at the planet surface is given by

$$F_{rad} = \frac{L_{rad}}{4\pi R_p^2} \tag{1.68}$$

where R_p is the planetary radius.

1.4.2.2 Gravitational Energy

Gravitational energy is accumulated as the planet grows during the accretion of the material that forms it. The gravitational potential energy gained by a growing sphere of mass $m(r)$ and radius r when an infinitesimal mass element dm is added from infinity is

$$dE_p = -G\frac{m(r)dm}{r} \tag{1.69a}$$

If we assume that the body is built maintaining a constant density ρ_0 until it gets a radius R_p, $dm = 4\pi r^2 \rho_0 dr$ and E_p can be calculated upon the integration of (1.69a)

$$E_p = -G \int_0^{R_p} \frac{m(r)dm}{r} = -3G\left(\frac{4\pi}{3}\right)^2 \rho_0^2 \frac{R_p^5}{5} = -\frac{3}{5}\frac{GM^2}{R_p} \tag{1.69b}$$

For any other arbitrary spherical density distribution $\rho(r)$, we have from dimensional analysis that $E_p \propto -GM^2/R$. In the case of the Earth, Equation 1.69b leads to $E_p = -2.25 \times 10^{32}$ J, only 10% over that obtained from a more detailed calculation.

If we assume that all this energy is converted into heat, then

$$E_p = -MC_p\Delta T \tag{1.70}$$

where C_p is the specific heat of the material, and combining with (1.69b) the increase in the internal temperature of the planet ΔT will be given by

$$\Delta T = \frac{3}{5} \frac{(GM/R_p)}{C_p} \tag{1.71}$$

Part of this energy is radiated away through the surface and a balance equation between the heat accumulated by accretion and the losses to space by radiation must be established. A large part of this energy is lost by radiation during the formation process since heat conduction inside the body and radiation to space are very efficient and fast processes as compared with the lifetime of a planet.

Further ongoing contraction (and compression) of the body and internal differentiation of the material with the sink of denser elements toward the center, also contribute to transform gravitational energy into heat. If the radius of a body decreases while maintaining its mass constant, its gravitational potential energy decreases since it becomes more negative according to (1.69). The rate of change of gravitational energy is given by

$$\frac{dE_p}{dt} \propto G \frac{M^2}{R_p^2} \frac{dR_p}{dt} \tag{1.72}$$

As has been mentioned, when contracting, $dR_p/dt<0$ making $dE_p/dt<0$. The decreasing gravitational energy increases the body's internal temperature.

The gaseous planets of the solar system currently have important internal heat sources as has been deduced from the radiance measurements in the infrared (Uranus is a exception). The energy accumulated as heat during the formation stage is lost to space by radiation during the planet's lifetime. If the energy per unit time released from the interior is L_{int} (in watts, called intrinsic internal luminosity), the internal heat flux F_{int} (W m^{-2}) emitted by the planet is given by

$$F_{int} = \frac{L_{int}(t)}{4\pi R_p^2} \tag{1.73}$$

In Table 1.4, we give the measured values of the intrinsic heat flow in the solar system bodies of interest. Jupiter's internal luminosity can be accounted for from the energy stored during the gravitational accretion phase with the additional effect of contraction, but Saturn (and probably Neptune) requires additional energy sources. The lower temperature in the interior of Saturn, as compared with that in Jupiter, suggests that the required additional heat could be produced by a differentiation process in the interior (see Section 1.5.5) with helium dropping toward Saturn's core ("raining out") within a metallic hydrogen layer. Uranus has practically lost its internal heat source whereas Neptune still emits important quantities of heat coming most probably from its initially very high accretion temperature.

1.4.2.3 Tidal Heating

As discussed previously in Section 1.3.2, tidal forces can lead to internal heating. Tides depend on the distance between the bodies under mutual interaction through the r^{-3} law, producing distortions (flexures) in the interior of the body and a relative movement of the mass elements and internal stresses generating an internal frictional heating source. Planets and satellites with elliptic orbits will suffer from such heating. The amount of tidal heating released is expressed in terms of a *quality tidal factor* Q_T (similar to that defined for a forced and damped oscillator):

$$Q_T = 2\pi \frac{\langle E \rangle}{\Delta E} \tag{1.74}$$

Here

    ΔE is the change in the tidal energy

    $\langle E \rangle$ is the average of the energy perturbation over one cycle of flexure; the lower the value of Q_T, the more energy is dissipated

For the Earth, the Moon, and Mars $Q_T \sim 30$–100 but for the giant planets Q_T can be as large as 10,000–60,000.

It has been proposed that tidal heating could also play an important role in giant extrasolar planets located close to their parent stars (the family of "hot Jupiters"). Tides raised on the planet in an eccentric orbit by the star with mass M_* would generate a power given by

$$L_{tid} = \frac{dE}{dt} = \frac{e^2 G M_p M_*}{a(1 - e^2) \tau_c} \tag{1.75a}$$

where apart from the orbital parameters a circularization time τ_c (i.e., the time necessary to damp out the orbital eccentricity) has been introduced

$$\tau_c = \frac{e}{de/dt} \approx \frac{2\tau_{orb} Q}{63\pi} \frac{M_p}{M_*} \left(\frac{a}{R_p} \right)^5 \tag{1.75b}$$

This energy source has been proposed to be the cause for the observed "inflated" radius of some extrasolar planets that possess a larger radius than predicted for the planet in a cold ambient.

Tidal heating is a very important mechanism for generating internal power in some satellites that possess thin atmospheres. The best example is the satellite Io of Jupiter whose huge volcanic activity (origin of its thin atmosphere) is due to internal tidal heating. Although the long-term effect of tides is to circularize the orbit, the eccentricity of Io's orbit is maintained by the perturbations introduced by the other Galilean satellites due to resonances (see Section 1.2.4). Tidal heating is also the primary energy source in Jupiter's satellites Europa and Ganymede—in both cases it is responsible for the presumed subsurface oceans. It is also probably responsible for the "plume" (geyser-like) activity found in satellites Enceladus and Triton.

1.4.3   EXTERNAL ENERGY

In the absence of internal energy, the main energy source for a planet or satellite atmosphere is the radiation coming from the Sun (or from the parent star for extrasolar planets) called *irradiation*. Radiation is absorbed by the body surface or atmosphere (mainly at visible wavelengths) and is reradiated to space (mainly at infrared as a black body). In the long-term, a balance is established between both channels (absorption and radiation) preventing the body from heating up or cooling off infinitely. The irradiation of a body depends primarily on the strength of the solar (stellar) luminosity, which can be intrinsically variable due to the stellar activity or, in the long-term scale, the star evolution. It also depends on the orbital properties (distance, eccentricity, and spin–orbit tilt) that lead to seasonal and long-term changes and on the planet reflectivity (more properly, its *albedo*) and rotation rate (spin) around its axis.

1.4.3.1   Global Insolation

Let L_* be the *star luminosity* (watts) that depends mainly on its spectral type and age (we denote the Sun's luminosity as $L_\odot = 3.85 \times 10^{26}$ W). The instantaneous stellar flux F_p (in W m^{-2}) reaching a planet that orbits its star at a distance r as it moves along its orbit is given by

$$F_p = \frac{L_*}{4\pi r^2} \tag{1.76a}$$

and for the solar system bodies

$$F_{\odot p} = \frac{L_\odot}{4\pi r^2} \tag{1.76b}$$

The solar (or stellar) flux reaching a planet can be expressed in terms of the solar (stellar) constant on the planet. It is defined as the flux of solar energy (covering the entire electromagnetic spectrum) across a surface of unit area normal to the solar beam at the mean distance between the Sun and the body

$$S_{\odot p} = \frac{L_\odot}{4\pi \langle r \rangle^2} \tag{1.77a}$$

where the mean distance is calculated using (1.13) as

$$\langle r \rangle^2 = \frac{1}{2\pi} \int_0^{2\pi} r^2 \, d\theta = a^2 \sqrt{1 - e^2} \tag{1.77b}$$

For a nearly circular orbit, as in most solar system planets, $\langle r \rangle \sim a$ and we write

$$S_{\odot p} = \frac{L_\odot}{4\pi a^2} \tag{1.77c}$$

For example, on Earth, $S_{\odot Earth}=1370\,\text{W m}^{-2}$ (mean distance is 1 AU) and on Mars, $S_{\odot Mars}=590\,\text{W m}^{-2}$ (mean distance 1.52 AU).

The variation of the solar flux incident on the surface of an airless body or at the top of a planetary atmosphere along the body's orbit can be written as

$$F_{\odot p}(r) = S_{\odot p}\left(\frac{a}{r}\right)^2 = S_{\odot p}\left(\frac{1+e\cos\theta_{an}}{1-e^2}\right)^2 \tag{1.78}$$

where r is the heliocentric distance (r and a in AU) and where we now use θ_{an} (instead of θ as in Equation 1.13) for the longitude angle that is called the *true anomaly of the Sun* from perihelion and is given by

$$\theta_{an} = \theta_{\odot} - \theta_p \tag{1.79}$$

where

$\theta_{\odot}$ is the longitude of the Sun (the solar longitude)
θ_p is the longitude of the planet perihelion, both measured from the intersection between the orbital and rotational planes

There is a greater flux (given by $F_{\odot p}(\max)=S_{\odot p}/(1-e)^2$) at perihelion than at aphelion (given by $F_{\odot p}(\min)=S_{\odot p}/(1+e)^2$) with important seasonal variations in objects with thin atmospheres (e.g., Mars and Pluto). Averaged over a year, the flux excess at perihelion compensates for the flux defect at aphelion and is a weak function of the eccentricity $\langle F_{\odot p}\rangle=S_{\odot p}/(1-e^2)^{1/2}$.

1.4.3.2 Insolation in a Planet or Satellite

The insolation distribution on a body varies with latitude and longitude, which in turn depend on the body rotation rate, orbital properties (in particular obliquity and eccentricity), and planet shape (spherical or oblate spheroid, see Section 1.3.4). This gives rise to the daily and seasonal changes and on a long-term basis to large-scale changes like the glaciation eras, a subject that is presented in Chapter 2. The resulting three-dimensional structure of the temperature patterns is the primary source in producing pressure gradients in the atmosphere to drive the motions.

The local and instantaneous solar irradiance onto the planet (surface if airless or top of the atmosphere) is

$$F_{\odot p}(\varphi,\phi) = F_{\odot p}\cos i_Z \tag{1.80}$$

where i_Z is the incidence angle (solar zenith angle) given by

$$\cos i_Z = \sin\varphi\sin\varphi_{\odot} + \cos\varphi\cos\varphi_{\odot}\cos h_{\odot} \tag{1.81}$$

where

$\varphi_{\odot}$ is the subsolar latitude
$h_{\odot}=(2\pi/\tau_d)t$ is the solar hour angle measured from midnight
τ_d is the solar (or stellar) day duration

The subsolar latitude is given by

$$\sin \varphi_{\odot}(t) = \sin i \sin \theta_{\odot} \tag{1.82}$$

where i is the obliquity angle (spin to orbit tilt).

The insolation at latitude φ during a given temporal interval Δt is given by

$$\langle F_{\odot p}(\varphi) \rangle_{\Delta t} = \int_{\Delta t} F_{\odot p}(t) \, dt = \int_{\Delta t} \frac{L_{\odot}}{4\pi r(t)^2} \left[\sin \varphi_{\odot}(t) \sin \varphi + \cos h_{\odot}(t) \cos \varphi_{\odot}(t) \cos \varphi \right] dt$$

$$\tag{1.83}$$

where $r(t)$ is given by (1.13).

For example, in a rapidly rotating planet ($\tau_d = \tau_{rot} \ll \tau_{orb}$) as in Earth, Mars, and the giant fluid planets, the daily insolation ($\Delta t = \tau_d$) is

$$\langle F_{\odot p} \rangle_{daily} = \frac{L_{\odot}}{4\pi r^2} \int_{day} \cos i_Z(t) \, dt = S_{\odot p} \left(\frac{\langle r \rangle}{r} \right)^2 \int_{sunrise}^{sunset} \cos i_Z(t) \, dt \tag{1.84}$$

since r is approximately constant on one rotation of the body. On the other hand, $\varphi_{\odot}$ is also approximately constant in this interval and by substituting time dependence by hour angle dependence ($dh/dt = \omega_d = 2\pi/\tau_d$), we obtain (adopting $\langle r \rangle = a$ for the solar system planets)

$$\langle F_{\odot p} \rangle_{daily} = S_{\odot p} \left(\frac{a}{r} \right)^2 \int_{-\eta_d}^{+\eta_d} \left[\sin \varphi_{\odot} \sin \varphi + \cos h_{\odot}(t) \cos \varphi_{\odot} \cos \varphi \right] \frac{dh_{\odot}}{\omega_d}$$

$$= S_{\odot p} \left(\frac{a}{r} \right)^2 \frac{1}{\pi} \left[\eta_d \sin \varphi_{\odot} \sin \varphi + \cos \varphi_{\odot} \cos \varphi \sin \eta_d \right] \tag{1.85}$$

where η_d is the rotational half-angle of daylight (from sunrise to noon or from noon to sunset, in radians) whose value is

(i) If $|\varphi| < 90° - |\varphi_{\odot}|$ (day and night)

$$\cos \eta_d = -\tan \varphi_{\odot} \tan \varphi$$

(ii) If $\varphi \geq 90° - \varphi_{\odot}$ or $\varphi \leq -90° - \varphi_{\odot}$ (polar day)

$$\cos \eta_d = -1 \quad \text{and} \quad \langle F_{\odot p} \rangle_{daily} = S_{\odot p} \left(\frac{\langle r \rangle}{r} \right)^2 \frac{1}{\pi} (\sin \varphi_{\odot} \sin \varphi)$$

(iii) If $\varphi \leq -90° + \varphi_\odot$ or $\varphi \geq 90° + \varphi_\odot$ (polar night)

$$\cos \eta_d = 1 \quad \text{and} \quad \langle F_{\odot p} \rangle_{daily} = 0$$

At the poles, $\varphi = \mp \pi/2$ so that $\cos i_z = \sin \varphi_\odot$ ($i_z = \pi/2 - \varphi_\odot$) and the solar elevation angle $\pi/2 - i_z$ is equal to the declination of the Sun. At the equator ($\varphi = 0°$) or at equinoxes ($\varphi_\odot = 0°$) $\cos \eta_d = 0$ and the length of a solar day is $\tau_d/2$. At solar noon at any latitude, the hour angle is $h_\odot = 0$ and $\cos i_z = \cos(\varphi - \varphi_\odot)$ or $i_z = \varphi - \varphi_\odot$.

Let us now consider the calculation of the average insolation over an orbital interval $\Delta t = t_2 - t_1$ (between epochs t_1 and t_2) at latitude φ. Using (1.85), Equation 1.83 can be written as

$$\langle F_{\odot p} \rangle_{\Delta t}(\varphi) = \frac{1}{\Delta t} \int_{t_1}^{t_2} \langle F_{\odot p} \rangle_{daily} \, dt$$

$$= \frac{1}{\Delta t} \int_{\Delta t} S_{\odot p} \left(\frac{a}{r(t)} \right)^2 \frac{1}{\pi} \left[\eta_d \sin \varphi_\odot \sin \varphi + \cos \varphi_\odot \cos \varphi \sin \eta_d \right] dt \quad (1.86)$$

The calculation is rather involved and will not be detailed here. It can be shown that for the annual interval, the yearly insolation at latitude φ is given by

$$\langle F_{\odot p} \rangle_{annual}(\varphi) = \frac{S_{\odot p}}{\pi} \frac{\tau_{orb} \sin \varphi \sin i}{(1-e^2)^{1/2}} \int_0^{2\pi} (\eta_d - \tan \eta_d) \sin \theta_\odot \, d\theta_\odot \quad (1.87)$$

where

$$\cos \eta_d = -\frac{\tan \varphi \sin i \sin \theta_\odot}{(1 - \sin^2 i \sin^2 \theta_\odot)^{1/2}} \quad (1.88)$$

The solution to Equations 1.87 and 1.88 requires a numerical integration.

Annual global insolation can be evaluated using the instantaneous insolation over the cross-section of the body $(L_\odot/4\pi r^2)(\pi R_p^2)$. If we assume that this energy is distributed over the whole surface, the mean insolation for 1 day becomes $\tau_{day}(L_\odot/4\pi r^2)(\pi R_p^2/4\pi R_p^2)$, which can be integrated over a planetary year using Kepler's second law integrating day to day, leading to

$$\langle F_{\odot p} \rangle_{\substack{global \\ annual}} \approx \frac{S_{\odot p}}{4} \tau_{orb}(1-e^2)^{-1/2} \quad (1.89)$$

Additional effects not discussed here are those introduced by the body shape (spheroid instead of sphere) and by the scattering and shadowing effects produced by the presence of dense equatorial rings (as is the case for Saturn). Figure 1.13 presents the insolation at the top of the atmosphere for various planets.

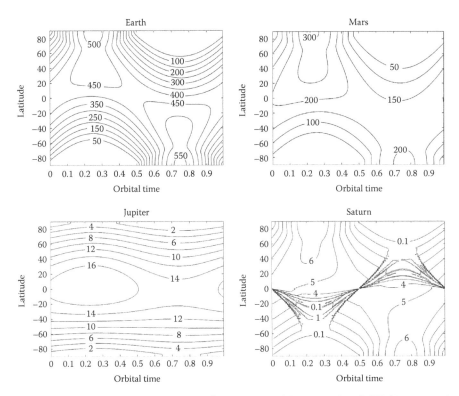

FIGURE 1.13 Daily insolation (in W m^{-2}) at the "top of the atmosphere" (TOA) on several planets: Earth, Mars, Jupiter, and Saturn, for this last case including the effects of ring shadowing and scattering. Note also the effect of the obliquity in insolation distribution. (Courtesy of S. Pérez-Hoyos, UPV-EHU, Spain.)

1.4.3.3 Albedo

The total power intercepted by a planet with radius R_p at a distance r from the Sun (or star) is given by

$$L_p = \frac{L_\odot}{4\pi r^2} \pi R_p^2 \tag{1.90}$$

The *Bond albedo* A_B (or spherical albedo) is defined as the ratio of the emergent (or reflected) power to the incident power:

$$L_{refl} = A_B L_p \tag{1.91}$$

Let us assume that the planet is observed at a distance Δ_{EP}. If radiation were reflected isotropically, the observed flux would be

$$F_{obs} = \frac{L_{refl}}{4\pi \Delta_{EP}^2} \tag{1.92}$$

However, radiation is reflected anisotropically and the flux must be corrected by a factor C_a $\Phi(\alpha)$ that depends on the phase angle α between the Sun (or star), the planet, and the observer. $\Phi(\alpha) = F_{obs}(\alpha)/F_{obs}(\alpha = 0°)$ is called the *phase function* and is defined as the ratio between the flux reflected by the body at angle α to that at $\alpha = 0°$. The normalization constant is given as

$$C_\alpha = \frac{2}{\int_0^\pi \Phi(\alpha)\sin\alpha \, d\alpha} \tag{1.93}$$

The observed flux density is therefore

$$F_{obs} = C_\alpha \Phi(\alpha) \frac{L_{refl}}{4\pi\Delta_{EP}^2} \tag{1.94}$$

The Bond albedo is then given as

$$A_B = p_0 q_\alpha \tag{1.95}$$

where q_α is the *phase integral* that contains the phase angle dependence of the sunlight scattered by the body, given by

$$q_\alpha = 2\int_0^\pi \frac{F(\alpha)}{F(\alpha = 0)} \sin\alpha \, d\alpha \tag{1.96}$$

and p_0 is the *geometric albedo* or "head on reflectance" given by

$$p_0 = \frac{r^2 F_{obs}(\alpha = 0°)}{S_{\odot Earth}} \tag{1.97}$$

with r given in AU The geometric albedo is the ratio between the radiation reflected from a body to that of a *Lambert surface*, which is a diffuse perfect reflector at all wavelengths (a "white surface" will have $A_B = 1$).

A parameter usually used in planetary photometry (surface, atmosphere, or a combination of the two) is the reflectivity I/F where I is the reflected intensity at a given wavelength at a given position (spatially resolved) and πF is the incident solar flux density at the planet at this wavelength. A flat *Lambertian* surface has $I/F = 1$ when observed at normal incidence and $I/F = p_0$ at a given wavelength when observed at phase angle 0° (Table 1.4).

1.4.4 EQUILIBRIUM AND EFFECTIVE TEMPERATURES

The illuminated hemisphere of a spherical body of radius R_p absorbs a power (in W) that depends on its cross-section and albedo given by

$$L_{abs} = (1 - A_B) \frac{L_\odot}{4\pi r^2} \pi R_p^2 \qquad (1.98)$$

with L_* instead of the solar luminosity in the case of an extrasolar planet. If we assume that the body is rotating rapidly, its temperature will be approximately constant and it will reradiate energy from its entire surface $(4\pi R_p^2)$ according to the Stephan–Boltzmann law with a power of

$$L_{out} = 4\pi R_p^2 \varepsilon \sigma_B T^4 \qquad (1.99)$$

where
ε is the emissivity of the body (integrated over the whole wavelength range)
σ_B is the Stephan–Boltzmann constant

The absorbed energy heats up the body and the equilibrium temperature (T_{eq}) can be deduced from the balance between the incoming and outward fluxes and the absorbed energy flux (F_{abs}), assuming it is spread uniformly over the full planet area:

$$F_{abs} = \frac{L_{abs}}{4\pi R_p^2} = \frac{(1 - A_B)}{4} \frac{L_\odot}{4\pi r^2} \qquad (1.100)$$

$$T_{eq} = \left(\frac{F_{abs}}{\varepsilon \sigma_B}\right)^{1/4} \qquad (1.101)$$

In the simplest case of a perfectly emitting body ($\varepsilon = 1$), the temperature dependence on the distance r to a star (see Equation 1.13 to account for eccentricity) is given by

$$T_{eq} = \left[\frac{(1 - A_B)L_\odot}{16\pi\sigma_B r^2}\right]^{1/4} \approx \frac{1}{\sqrt{r}} \qquad (1.102)$$

When internal energy sources are present, it becomes convenient to characterize the energy balance using the effective temperature, which for $\varepsilon = 1$ is given by

$$T_{eff} = \left(\frac{F_{abs} + F_{int}}{\sigma_B}\right)^{1/4} \qquad (1.103)$$

The power (in watts) due to the internal energy source can then be written in terms of these two temperatures:

$$L_{int} = 4\pi R_p^2 \sigma_B (T_{eff}^4 - T_{eq}^4) \qquad (1.104)$$

The ratio between the internal and external energy sources can then be quantified simply as $(T_{eff}/T_{eq})^4$, representing a measure of the importance of the internal energy source for atmospheric processes, as occurs in the giant planets Jupiter, Saturn, and Neptune (see Table 1.4).

1.5 INTERNAL STRUCTURE OF PLANETS AND SATELLITES

The internal structure of a body depends on its mass and composition and on the internal energy sources, if present. The density or the pressure distributions within a planet are the parameter employed for this purpose.

1.5.1 EQUATION OF STATE

The equation of state (EOS) gives the relationship between the pressure, density, and temperature for a given composition of a body in thermodynamic equilibrium. It is expressed by a function $P(\rho, T)$ that depends on the state of matter that, for a given mixture of compounds, is represented by a phase diagram (usually a pressure–temperature relationship). The phase diagram and EOS for a mixture of materials are derived from laboratory measurements and from theoretical models of the atomic, ionic, and molecular interactions. The most simple equation is that of gases at low pressures; the perfect gas law is given by

$$PV_0 = n_0 R_g T \tag{1.105a}$$

or as

$$P = \rho \left(\frac{R_g}{\mu} \right) T \tag{1.105b}$$

where
   V_0 is the volume occupied by the gas
   R_g is the gas constant
   n_0 is the mole number
   μ is the molecular weight

This equation is valid for densities below $\sim 0.02\,g\,cm^{-3}$, which on massive atmospheres, for example on giant planets occurs for $P < 1$ kbar. As the pressure increases, intermolecular forces are strong enough to modify this simple relationship and the van der Waals and other EOS must be used.

   For matter in solid and liquid states, the EOS is represented by a complex relationship between pressure and density. It involves the bulk modulus of the material defined as $K_B = \rho\,(dP/d\rho)$, which is in turn a function of the pressure. Experimental data and theoretical studies for solid materials typical of planetary interiors at $P < 2$ Mbar are available as series expansion of K_B.

In the case of hydrogen and helium mixtures, of interest in the study of the interior of the giant planets, a useful EOS is the polytropic equation

$$P = K_p \rho^{1+(1/\upsilon)} \tag{1.106}$$

where K_p and υ are the polytropic constant and index, respectively. At low pressures, $\upsilon \to \infty$ and $P \sim K_p \rho$ but at higher pressures $\upsilon \to 1$ and $P = K_p \rho^2$, a reasonable approximation for the giant planets deep atmospheres. This last EOS has the advantage of allowing an analytical solution for the density and pressure profiles for the planet's deep atmosphere, as will be shown below.

1.5.2 DENSITY PROFILES

Under hydrostatic equilibrium, the structure of a spherical body is determined by a balance between gravity and the gradient pressure force:

$$\frac{\nabla P}{\rho} = -\vec{g} \tag{1.107a}$$

Assuming radial symmetry within the planet, this equation becomes

$$\frac{\partial P(r)}{\partial r} = -g(r)\rho(r) \tag{1.107b}$$

If we know $\rho(r)$, we can integrate this equation to get the pressure distribution $P(r)$ inside the body.

For solid planets, the density gradient $\rho(r)$ can be obtained if the bulk modulus K_B of the material is known. Since $K_B = \rho(dP/d\rho)$, cross differentiation leads to

$$\frac{\partial \rho}{\partial r} = -\frac{\rho^2 g}{K_B} \tag{1.108}$$

A useful approach is given by the Murnaghan equation or some variant of it and then $K_B = A\rho^\upsilon$ with A constant. The above equation is then written as

$$\frac{\partial \rho}{\partial r} = -\frac{g}{A\rho^{\upsilon-2}} \tag{1.109}$$

with $\upsilon = 4$ for silicates. The distribution of mass with depth is given by Equation 1.29 and accordingly we have

$$M(r) = M_p - 4\pi \int_r^{R_p} \rho(r) r^2 \, dr \tag{1.110}$$

In the simpler case of a homogeneous planet with constant density through its interior ($\rho = \rho_0$), the mass and pressure distributions are given by

$$M(r) = \frac{4}{3} \pi r^3 \rho_0$$

$$(1.111)$$

$$P(r) \approx P_C - \frac{2}{3} \pi G r^2 \rho_0^2$$

where P_C is the pressure at the center (see Problem 1.13)

$$P_C = \frac{3 G M_p^2}{8 \pi R^4}$$

$$(1.112)$$

which gives only an order of magnitude estimate to the real central pressure since planets tend to be denser toward their center and therefore not homogeneous (Figures 1.14 and 1.15).

In order to develop realistic and accurate models of the interior of a body, it is necessary to know the constituent relationships for its composition. Specifically, we need their phase diagrams (temperature and pressure relations). In addition, we need to know the internal energy sources and heat transport processes. For the study of the atmospheres, our interest for the internal structure resides in the giant and icy planets made of hydrogen and helium, since they have deep atmospheres that extend to a substantial part of the body. We concentrate on them.

The equation of state for hydrogen (or more accurately of a mixture of hydrogen and helium) at the high pressures that prevail at the interiors of the giant and icy planets is not well known, but theoretical models and laboratory experiments lead

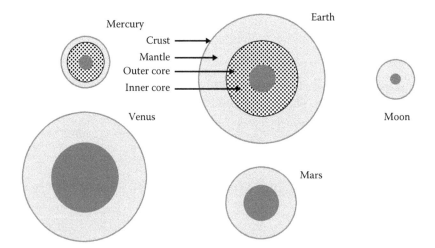

FIGURE 1.14 Differentiated internal structure and main layers in the terrestrial planets and the Moon.

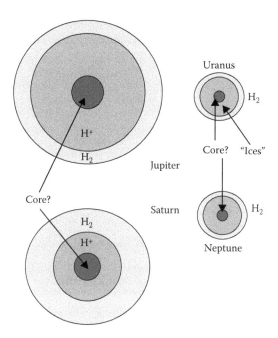

FIGURE 1.15 Differentiated internal structure and main layers in the gaseous giants and icy planets.

to constraints of the basic states for the hydrogen under such conditions. At low temperatures, the hydrogen is in solid state for a large range of pressures. For the temperature ranges typical of the upper cloud level of the giant planets of the solar system (50–150 K) and "hot Jupiters" (~1000 K), and up to the temperatures at the center of these planets (10,000–40,000 K), hydrogen is mostly in its molecular form (H_2) up to pressures of ~1 Mbar. At higher pressures (above ~1.4–3 Mbar), the separation between H_2 molecules is of the order of their size and hydrogen becomes an electrical conductor (a fluid metal). At $P \geq 3$ Mbar, dissociation of the molecule leads to an atomic metallic state (liquid). At much higher pressures, theoretical calculations predict that hydrogen becomes a degenerate plasma (this occurs at the plasma phase transition [PPT]). At the center of these planets, a rocky core is expected to exist according to some formation scenarios. Both theory and measurements of the gravitational field indicate the presence of a 5–10 Earth masses core for Jupiter (although a no core scenario is also possible) and 10–20 Earth masses core for Saturn. According to current models, the neutral atmosphere of Jupiter (a hydrogen and helium mixture) extends down to about ~15,000 km ($r/R_p \sim 0.8$) and in Saturn it extends down to ~30,000 km ($r/R_p \sim 0.5$). Uranus and Neptune are less massive than the Jovian planets and models for their interior indicate that below the molecular atmosphere of hydrogen and helium, at pressures around 0.1 Mbar ($T \sim 2000$ K), a transition occurs where a mixture of ices (water, ammonia, and methane) in ionic state is predicted, perhaps mixed with some heavier material. Accordingly, the neutral atmosphere of Uranus extends down to about ~5500 km ($r/R_p \sim 0.8$) and in Neptune it extends down to ~3500 km ($r/R_p \sim 0.85$) (Figure 1.15).

The approximate density profile in the hydrogen molecular atmospheres of the giant planets can be obtained using the hydrostatic equation (1.107a) and the polytropic equation

$$\frac{\nabla P}{\rho} = \frac{1}{\rho} \nabla \left[K_p \rho^{1+(1/\upsilon)} \right] \tag{1.113}$$

and from the gravitational field definition (1.30), we get

$$\nabla V_G = -(\upsilon+1)K_p \nabla(\rho^{-\upsilon}) \tag{1.114}$$

Taking the divergence in this equation and using Poisson's equation (1.31), we get

$$(\upsilon+1)K_p \nabla^2(\rho^{-\upsilon}) = -4\pi G\rho \tag{1.115}$$

If we now assume spherical symmetry, this equation reduces to

$$(\upsilon+1)K_p \frac{1}{r^2} \frac{d}{dr}\left[r^2 \frac{d}{dr}(\rho^{-\upsilon}) \right] = -4\pi G\rho \tag{1.116}$$

Using the polytropic index $\upsilon=1$ for the giant planets, the above differential equation has an analytical solution for the radial density profile

$$\rho(r) = \rho_0 \left[\frac{\sin \pi\eta_p}{\pi\eta_p} \right] \tag{1.117}$$

where $\rho_0 = (\pi M_p/4R_p^3)$ is the density at the center of the planet, $\eta_p = r/R_p$ and $R_p = (\pi K_p/2G)^{1/2}$. In this approximation, the radius of the planet is independent of the planetary mass although it depends on the polytropic constant (in the giant planets $K_p \sim 2.7 \times 10^{12}$ cm^5 g^{-1} s^{-2}).

From the polytropic equation, the radial distribution of pressure becomes

$$P(r) = \frac{2GR_p^2}{\pi} \rho(r)^2 \tag{1.118}$$

and the gravitational acceleration radial profile follows immediately upon differentiation

$$g(r) = -g_0 \frac{d}{d\eta_p}\left[\frac{\sin \pi\eta_p}{\pi\eta_p} \right] \tag{1.119}$$

Laboratory studies of the behavior of hydrogen at high pressures suggests that on the giant planets the transition between the molecular hydrogen state and the metallic phase occurs at $P \sim 2 \times 10^6$ bar. Solving for η_p in (1.117) and (1.119) for each particular planet (including extrasolar giants with known mass and radius), we can retrieve the vertical extent of the atmosphere (i.e., the radius of the molecular layer $r = \eta_p R_p$).

1.5.3   MASS–RADIUS RELATIONSHIP

Mass and composition are key parameters to determine the body radius (R_p). Metallic and rocky bodies have small radii whereas giant planets have 10 or more times the size of those planets. In the giant planets, the internal energy sources can "inflate" their size (relative to that expected from hydrostatic conditions of a cold planet), as observed in some extrasolar "hot Jupiter" planets. Since planets and satellites are differentiated bodies with internal energy sources, models of the mass–radius relationship can only be derived formally by resolving the full set of hydrostatic, thermodynamic, and mass and energy conservation equations, assuming that the equation of state for their constituents is known. Simple approaches can be used, however, to have a first-order estimate of the mass–radius relationship. The basic case is that of an incompressible body where the density is essentially constant (the best example are bodies made of rocks and metals). A body that has a spherical shape of radius R_p and mass M_p obeys simply

$$M_p = \frac{4}{3}\pi \langle \rho \rangle R_p^3 \propto R_p^3 \qquad (1.120)$$

The minimum size of a homogeneous solid body with a spherical shape can be derived from the assumption that the internal strength force S_F prevents substantial deformation

$$R_P(\text{min}) = \left(\frac{2S_F}{\pi G \rho^2} \right)^{1/2} \qquad (1.121)$$

For silicate and metallic materials ("rocky planets"), $S_F = 5 \times 10^6 - 2 \times 10^9$ N m^{-2}, which gives $R_p(\text{min}) \sim 350$ km (silicates) and $R_p(\text{min}) \sim 100$ km (metallic) bodies. Obviously, the real size of a metallic-rocky spherical body is in general expected to lie between this minimum value and an upper maximum value. Figure 1.16 shows the mass–radius relationship for the planets and major satellites of the solar system.

For planets with large masses ($M_p \gg 500$ Earth masses), the internal compression and pressure is high enough to ionize atoms. The degeneracy pressure of free electrons balances gravity in a hydrostatic equilibrium (full degeneration never occurs, this is reserved for stellar masses, as in highly compressed white dwarfs). Under such circumstances, an increase of mass produces a decrease in radius following a law of the type $M_p \sim R_p^{-3}$ (Figure 1.17).

A practical and useful expression for the relationship between the radius and mass that a particular spherical planetary body will reach can be derived from the balance between the degeneracy kinetic energy of the electrons and the combined electrostatic and gravitational energy of the atoms (Cole and Woolfson, 2002)

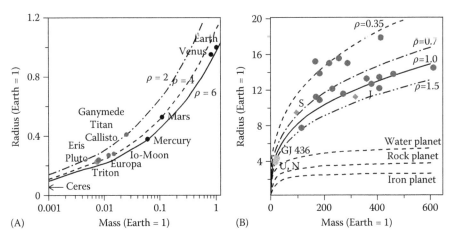

FIGURE 1.16 (A) Mass–radius relationship for terrestrial planets, dwarf planets, and major satellites. The lines show the density curves for the indicated values in g cm^{-3}; (B) Mass–radius relationship for giant gaseous planets and extrasolar planets up to 600 Earth masses. The lines show the density curves for the indicated values in g cm^{-3}.

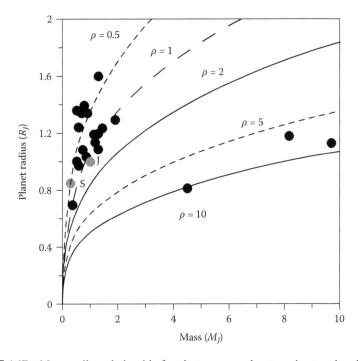

FIGURE 1.17 Mass–radius relationship for giant gaseous planets and extrasolar planets up to the brown dwarf mass limit (13 Jovian masses). The lines show the density curves for the indicated values in g cm^{-3}.

$$R_p = \left(\frac{C}{M_p^{1/3}} + C'M_p^{1/3} \right)^{-1} \tag{1.122}$$

with the parameters $C = 5.9A^{1/3}Z^{2/3}$ and $C' = 3 \times 10^{-18}(A/Z)^{5/3}$, where Z and A are the atomic and mass number and R_p and M_p are measured in m and kg, respectively. For small masses, the first term on the right-hand side dominates and we have

$$M_p \propto R_p^3 \tag{1.123a}$$

as above, whereas for very large masses, the second term dominates and a relationship of

$$M_p \propto R_p^{-3} \tag{1.123b}$$

holds. The maximum radius of a planet can be obtained making $dR_p/dM_p = 0$ in (1.122), leading to

$$R_p(\text{max}) \approx 1.19 \times 10^5 \frac{Z^{1/2}}{A} \ (\text{km}) \tag{1.123c}$$

For example, a giant planet made of hydrogen ($A = 1$, $Z = 1$) will have $R_p(\text{max}) \sim 10^5$ km. In the other extreme of planetary composition, a metallic sphere with the same mass as a giant planet (for instance 450 Earth masses, $1M_\oplus = 5.974 \times 10^{24}$ kg) but made of magnesium ($A = 12$, $Z = 24$) or iron ($A = 26$, $Z = 56$) would have a radius of $R_p(\text{max}) \sim 2 \times 10^4$ km (Figure 1.18).

1.5.4 ENERGY TRANSPORT IN A PLANETARY BODY

As stated in Section 1.4, a planetary body has energy sources located outside it and/or in its interior. The transport of this energy within the body can take place by three different mechanisms: conduction (that dominates transport in solid materials), radiation (dominating in gases and low opacity fluids), and convection (mass motion occurring in fluids). The intensity of the energy sources and the efficiency of the heat transport within the planetary body determine its temperature structure and its global thermal evolution. In the atmosphere, the dominant transport mechanisms are radiation and convection, with some contribution by conduction in their upper and less dense parts. They will be treated in other chapters in the book. Here we introduce the heat transport concepts relevant to the planet interior.

1.5.4.1 Conduction

Let F_Q be the heat flux (in W m^{-2}) within a region of a body. The Fourier law establishes the relationship between this heat flux and the temperature

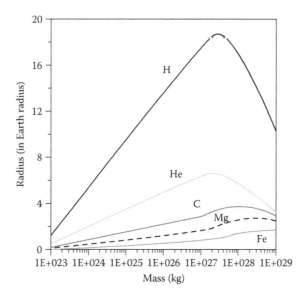

FIGURE 1.18 Theoretical mass–radius relationship for cold spheres of the indicated composition.

$$\vec{F}_Q = -K_T \nabla T \tag{1.124}$$

where the parameter K_T is the thermal conductivity (W m^{-1} K^{-1}), which characterizes the capability of a material to conduct this heat flux from one point to another within the body. As a simple useful example, let us assume a spherical homogeneous body with a radius R_p with thermal conductivity K_0 and an internal energy source at its center producing a heat flux F_0. Following (1.124), the temperature gradient within the body is given by

$$\frac{dT}{dr} = -\frac{F_0}{K_0} \tag{1.125a}$$

which upon integration gives the radial temperature profile

$$T(r) = T_0 + \frac{F_0}{K_0}(R_p - r) \tag{1.125b}$$

where T_0 is the surface temperature.

The heat transport mechanism is a time-dependent process and the rate at which the temperature changes in an isotropic and homogeneous body is controlled by the heat diffusion equation

$$\nabla^2 T(\vec{r},t) - \frac{1}{D_T}\frac{\partial T(\vec{r},t)}{\partial t} = -\frac{Q(\vec{r},t)}{K_T} \tag{1.126}$$

where

$D_T = K_T/(\rho C_p)$ is the thermal diffusivity (m² s⁻¹)
C_p the specific heat (J kg⁻¹ K⁻¹)
Q is the available heat power per unit volume (W m⁻³)

Consider, for example, the case of a body that suffers a periodic heating at the surface due to the absorption of sunlight (or stellar) radiation that we assume to be constant in time. The temporal dependence of the heat power deposited on any point of the body (characterized by position vector $\vec{r}$) relies on two characteristic periods (rotation and orbital revolution). Let us assume for simplicity that the body temperature is dominated by a single period (or frequency $\omega = 2\pi/\tau_{orb}$ or $2\pi/\tau_{rot}$), so the heat power at the surface can be expressed as

$$Q(\vec{r},t) = Q(\vec{r})[1 + e^{i\omega t}]$$ (1.127)

Introducing (1.127) in the diffusion equation (1.126) yields a *Helmholtz wave equation* for the temperature

$$\nabla^2 T(\vec{r}) - q_T^2 T(\vec{r}) = -\frac{Q(\vec{r})}{K_T}$$ (1.128)

where $q_T = (1 + i)/\mu_T$ is a thermal wave vector and $\mu_T = (2D_T/\omega)^{1/2}$ is the thermal diffusion length. If heating occurs at the surface of the body, for z the vertical coordinate (perpendicular to the surface), the one-dimensional solution to this equation is written as follows for the periodic component:

$$T(z,t) = T_0 \exp\left(-\frac{z}{\mu_T}\right) \exp\left[i\left(\omega t - z\mu_T^{-1} - \frac{\pi}{4}\right)\right]$$ (1.129)

This equation represents a strongly damped thermal wave propagating from the surface (where the temperature is T_0) to the interior ($z > 0$). Figure 1.19 shows the temperature variation at the surface of a planet as predicted by this equation. For this strongly damped thermal wave, the thermal wavelength represents the effective "skin depth penetration" below the surface.

1.5.4.2   Radiation

Heat transport by radiation becomes important in the interior of planets with fluid layers, as in the deep atmospheres of the giant planets. If the energy source in the interior of a planet has a power L_{int} radiated outward with an isotropic radiation field, the temperature gradient can be obtained from the Stephan–Boltzmann law as

$$\frac{dT}{dr} = -\frac{3\kappa\rho L_{int}}{64\pi\sigma T^3 r^2}$$ (1.130)

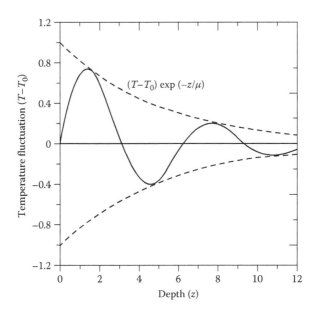

FIGURE 1.19   Structure of a thermal wave propagating beneath the surface of a body.

Here, κ (cm^2 g^{-1}) is the absorption coefficient that represents a measure of the opacity of the material to the radiative transport. The absorption coefficient κ depends on the wavelength according to the composition of the fluid layer, and is in general a function of the pressure and temperature. A first-order estimate of (1.130) can be obtained employing the *Rosseland mean opacity*, which is the wavelength-averaged value of opacity for a given composition. This heat transfer mechanism will be discussed in detail for the atmospheres in the following chapters, so we will not extend more here.

1.5.4.3   Convection

When the opacity to the radiation field is high, as expected in dense fluid media, the radiative gradient becomes large enough so heated parcels of the fluid can become lighter and therefore buoyant relative to their surroundings, moving upward and transporting heat as the warm masses ascend, a mechanism called convection. For this mechanism to be operative, the buoyant force must overcome other forces such as friction (e.g., due to viscosity). Many other factors intervene in this complex process and for the moment we only outline the basic ideas for convective heat transport in planetary interiors. A full treatment of the heat transport and the resulting patterns of motion must take into account the spherical geometry and the rotation of the body. This will be presented in the following chapters in the context of shallow and deep atmospheres.

To asses the capability for convective onset, the nondimensional *Rayleigh number* (*Ra*) is employed. This number compares the buoyancy force acting on a fluid parcel with the dissipative forces, and for convection to occur, it must be above a certain critical value (Ra_{crit}) determined in laboratory experiments. The Rayleigh number for a layer of thickness d and mean density ρ_0 is given by

$$Ra = \frac{\alpha_T \Delta T g d^3}{D_T v_v} > Ra_{crit} \approx 500 - 1000 \tag{1.131}$$

Here

α_T is the thermal expansion coefficient of the material (for perfect gases $\alpha_T = 1/T$)

ΔT (that must be positive) is the temperature difference across the layer

v and D_T are the viscosity and diffusivity coefficients characterizing the medium

Fluids have in general low values for these two parameters, so the denominator is low and the convective transport generally tends to dominate over conduction. The coefficients $D_T(T)$ and $v_v(T)$ are temperature functions and for layers with strong thermal gradients the Rayleigh number varies across it. In general, $D_T(T) \sim 1/T$ with a small contribution ($\sim T^{-3}$) occurring when the material is sufficiently transparent to transport heat by infrared radiation.

Convective motions produce the mixing of the material driving the medium toward an adiabatic state. Consider the case of a perfect gas. The temperature gradient established by convection can be obtained using the adiabatic relationship

$$T^\gamma P^{1-\gamma} = \text{Const} \tag{1.132}$$

where $\gamma = C_p/C_v$ is the adiabatic coefficient. Differentiating this expression with respect to the radial coordinate along the deep atmosphere, and using the hydrostatic approach in (1.107a) for the bulk interior, we get (see Chapter 4)

$$\frac{dT}{dr} = -\frac{g}{C_p} \tag{1.133}$$

More generally, for a fluid with a thermal expansion coefficient α_T, we have

$$\frac{dT}{dr} = -\frac{g \alpha_T T}{C_p} \tag{1.134}$$

An order of magnitude estimate of the convective upward velocity produced by the planet's internal heat source is provided by Prandtl *mixing length theory*. This theory proposes that convective motions are turbulent at a high Rayleigh number with a characteristic upward scale ℓ for a mass before it mixes with others (the "mixing length scale"). Taking $\ell = H$ (the density scale height; see Chapter 4) the vertical velocity is given by

$$w = \left(\frac{R_g^*}{\rho C_p} F_{int} \right)^{1/3} \tag{1.135}$$

where R_g^* is the specific gas constant (R_g/μ). Accordingly, the vertical velocities depend on the cubic root of the internal heat flux, and for the values of these parameters in the giant planets ($H \sim 30\,\text{km}$), we have $w \sim 1\,\text{m s}^{-1}$.

The relative importance of the mechanisms involved in the heat transport within a planet can be quantified by the nondimensional Nusselt number that compares the measured heat transport flux (F_{int}) with that expected by conduction from the Fourier law:

$$Nu = \frac{F_{int}}{F_Q} \tag{1.136}$$

For example, for a layer of thickness d, thermal conductivity K_T, and a temperature difference across it of ΔT, Equation 1.125 implies $F_Q = K_T(\Delta T/d)$. If $Nu = 1$ ($F_Q = F_{int}$), the heat transport is dominated by conduction and (1.131) states that in the layer $Ra < Ra_{crit}$, but if $Nu > 1$ convective transport dominates and $Ra > Ra_{crit}$.

1.5.5 PLANETARY STRUCTURE: INTERNAL DIFFERENTIATION OF SOLID BODIES

The interior of a solid planet (icy, rocky, and/or metallic) is evenly mixed in composition during its formation. However, gravity tends to separate the elements by weight and in a few million years, the heavy elements (metals) sink to the center and the lighter elements rise. As the planet solidifies, the internal structure becomes fractionated in shells, characterized by the density distribution of the materials with depth. As indicated before, the gravitational self-compression controls the EOS of matter at the high pressures and temperatures reached within each differentiated layer. The differentiation process depends on the strength of the internal heat source and on how the heat is transported in the interior of the planet. Solid bodies are therefore expected to be internally structured in layers and observations suggest that we can classify these layers in three parts (obviously each one can be divided into different sub-layers): (1) a central dense *core*, (2) an intermediate *mantle*, and (3) an outermost usually thin "boundary surface" (Figure 1.14 and Table 1.6). For example, the Earth has an inner (solid) and outer (liquid) core occupying 0.55 of the Earth's radius ($R_\oplus = 6371\,\text{km}$) where the density increases with depth from 10 to $13.5\,\text{g cm}^{-3}$, and a lower and upper mantle occupying $0.45 R_\oplus$ with density increasing with depth from 3 to $5\,\text{g cm}^{-3}$. Finally, it has a solid crust with a thickness ranging from 60 to $100\,\text{km}$ and density of $2.6\,\text{g cm}^{-3}$. Since this outer layer is thin, most models of planetary interiors can be reduced to two main zones: a *core* (size r_c, density ρ_c) and a *mantle* (density ρ_m). If the mean density of the body is $\langle \rho \rangle$, the mass–volume mixing relationship gives

$$\rho_m R_p^3 + (\rho_c - \rho_m) r_c^3 = \langle \rho \rangle R_p^3 \tag{1.137}$$

and the core radius is then

$$r_c = R_p \left(\frac{\langle \rho \rangle - \rho_m}{\rho_c - \rho_m} \right)^{1/3} \tag{1.138}$$

TABLE 1.6

Internal Composition of the Terrestrial Planets

Planet	Mantle (%)	Core (%)
Mercury	35 (MgO, SiO_2)	65 (Fe)
Venus	68 (SiO_2, MgO)	32 (Fe)
Earth	70 (SiO_2, MgO)	30 (Fe)
Mars	88 (SiO_2, MgO)	12 (Fe, S)

Note: Main composition in mass percentage. For a body with two basic components, e.g., rock-metal and ice, or rock and ice, the mass fraction m of component 1 follows from the equation

$$\frac{1}{\langle \rho \rangle} = \frac{m_1}{\rho_1} + \frac{1 - m_1}{\rho_2}$$

where

$\langle \rho \rangle$ is the measured mean density
ρ_1 is the density of the component 1
ρ_2 is the density of component 2

Typically, we have ice density $= 1200\,\mathrm{kg\,m^{-3}}$; hydrated silicates density $= 2500\,\mathrm{kg\,m^{-3}}$; anhydrous silicates and iron density $= 3500\,\mathrm{kg\,m^{-3}}$.

In general, the border of a planet with the outer space can be a thin solid *crust*, a liquid partial or global "ocean," or a gaseous atmosphere that can be deep (extending to a significant fraction of the radius) as in the giant planets, or thin as in the terrestrial planets. Geological structures form on the upper solid surface of the crust. *Tectonic* activity develops on it when the mantle is a fluid in motion due to the internal heat source. The presence of an active mantle and tectonic activity favors the recycling of atmospheric compounds. But, undoubtedly, the most conspicuous features of the planetary surfaces are the *impact craters* abundant in bodies without a significant atmosphere or internal heat source (Table 1.7). The crust is surrounded in most terrestrial planets and major satellites by a gaseous layer, the *atmosphere* that is the main subject of study in this book. The surface acts as a reservoir for volatiles (condensibles) and for dust that can be lifted to or removed from the atmosphere. Through the action of winds and precipitation, atmospheric processes erode the ground, a mechanism known as *weathering*, modifying the surface on a long-term scale. The surface topography also plays an important role in conditioning the atmospheric dynamics, for exampling forcing large-scale motions as waves.

Cores and mantles can be made of a variety of materials, depending on the nature of the planet. Recently, internal differentiation has even been proposed for the

TABLE 1.7
Surfaces of Terrestrial Planets and Major Satellites

Planet/Satellite	Surface Composition	Main Textures
Mercury	Silicates	Impact craters
Venus	Basalts, granites (?)	Basins and mountains
Earth	Basalts, granites, water	Oceans, Polar icy caps, continents and mountains (plate tectonics)
Mars	Basalts, clays, ice	Impact craters, basins, ridges, mountains, volcanoes (inactive)
Io	Sulfur, SO_2 deposits, silicates	Volcanic activity
Europa	Water ice	Cracks-ridges
Ganymede	Dirty water ice	Impact craters
Callisto	Dirty water ice	Impact craters
Titan	Dirty water ice, methane	Methane icy deposits, hills
Triton	Water ice, methane	Impact craters
Pluto	Water ice, methane and N_2 ice	—
Eris	Methane ice, mixture methane with water and N_2 ice	—

extrasolar planet around the star HD149026 that has a mass close to that of Saturn but a smaller radius, suggesting the presence of a large dense core occupying about half the mass of the planet.

1.6 MAGNETIC FIELD

Planetary magnetism is relevant to planetary atmospheres since it controls the interaction with the stellar (an interstellar) magnetic environment (or the planetary-satellite magnetic ambient), affecting the structure and evolution of the upper atmospheres, in particular to their non-neutral (charged) parts. The space around a body that is subjected to the influence of its magnetic field is known as the magnetosphere. The magnetic field is in the origin of an ample phenomenology that occurs in the upper atmospheres of the planets and will be studied in Chapter 4. Here, we outline the fundamentals of the magnetic field origin and structure in planetary bodies.

Magnetic fields have their origin in electrical charge motions. Moving free charges give raise to electrical currents that can exist in the vacuum space and within material systems. The Biot–Savart law gives a direct relationship between the current intensity I_E (in amperes, A) and its geometry and the magnetic field (sometimes called "magnetic flux density" and "magnetic induction") $\vec{B}(\vec{r},t)$ (in teslas, T). Charges are also present in atoms and their bounded motion as "spinning" charges produces microscopic currents that can adopt special permanent configurations, for example with aligned orientations within a material. This configuration depends on the temperature, and materials with such properties are called ferromagnetic—the magnets being the best known example. The temperature below which a magnetic material becomes ferromagnetic is called "Curie temperature" (T_C) (e.g., $T_C = 770°C$ for Fe, 365°C for Ni, and 1075°C for Co).

To understand the sources and nature of the planetary magnetic fields, we must start with the basic equations of the electromagnetism.

1.6.1 MAXWELL EQUATIONS

Electric $\vec{E}(\vec{r},t)$ and magnetic $\vec{B}(\vec{r},t)$ fields are related to their sources by the four Maxwell equations:

$$\nabla \cdot \vec{E} = \frac{\rho_E}{\varepsilon_0} \tag{1.139a}$$

$$\nabla \times \vec{E} = -\frac{\partial \vec{B}}{\partial t} \tag{1.139b}$$

$$\nabla \cdot \vec{B} = 0 \tag{1.139c}$$

$$\nabla \times \vec{B} = \mu_0 \vec{J} + \varepsilon_0 \mu_0 \frac{\partial \vec{E}}{\partial t} \tag{1.139d}$$

Here

   ρ_E is the charge density (free and bounded, C m^{-3})
   $\vec{J}$ is the current density (free and bounded, A m^{-2})
   ε_0 and μ_0 are the dielectric permittivity and magnetic permeability of the vacuum space

Planetary magnetism can have different sources. We can gain some insight on the planetary magnetism by performing some manipulation of the Maxwell's equations. In planets, $\partial \vec{E}/\partial t$ is observed to be small and can be neglected so the Ampere's law can be simplified to $\nabla \times \vec{B} = \mu_0 \vec{J}$. The current density in a moving medium is related to the electric and magnetic fields by the microscopic Ohm's law

$$\vec{J} = \sigma_c(\vec{E} + \vec{v} \times \vec{B}) \tag{1.140}$$

where σ_c is the conductivity (ampere per volt-meter, A V^{-1} m^{-1}). Now we can eliminate the electric field from (1.140) using the induction law (1.139b) and Ampere's law. Assuming that conductivity is constant, we have

$$\frac{\partial \vec{B}}{\partial t} = -\nabla \times \left(\frac{\vec{J}}{\sigma_c} - \vec{v} \times \vec{B} \right) = -\nabla \times \left(\frac{\nabla \times \vec{B}}{\mu_0 \sigma_c} - \vec{v} \times \vec{B} \right) = -\frac{1}{\mu_0 \sigma_c} \nabla \times (\nabla \times \vec{B}) + \nabla \times (\vec{v} \times \vec{B})$$

$$\tag{1.141a}$$

Using the vector identity $\nabla \times (\nabla \times \vec{B}) = \nabla (\nabla \cdot \vec{B}) - \nabla^2 \vec{B}$ and since $\nabla \cdot \vec{B} = 0$, we get the equation that gives the time-dependent variability of the magnetic field:

$$\frac{\partial \vec{B}}{\partial t} = \nabla \times (\vec{v} \times \vec{B}) + \lambda_m \nabla^2 \vec{B} \qquad (1.141b)$$

The temporal and spatial structure of this equation is typical of diffusion processes and is therefore called a magnetic diffusion equation, with $\lambda_m = (\mu_0 \sigma_c)^{-1}$ the magnetic diffusivity (in $m^2 \, s^{-1}$) of the medium. For a nonmoving medium ($\vec{v} = 0$), we have

$$\frac{\partial B}{\partial t} = \lambda_m \nabla^2 \vec{B} \sim \frac{\lambda_m B}{L_m^2} \qquad (1.142)$$

Here, L_m is a characteristic magnetic length in the medium, which allows us to introduce a magnetic diffusion time constant $\tau_m \sim L_m^2 / \lambda_m$ that gives the characteristic time for the magnetic field to spread or diffuse in the medium. Consider, for example, the case of a giant planet made of hydrogen. Laboratory measurements of the electrical conductivity of hydrogen at high pressures give $\lambda_m \sim 5 \, m^2 \, s^{-1}$, and taking as a characteristic length-scale the planetary radius ($L_m \sim R_p$), we get $\tau_m \sim 1 \, My$, a temporal scale well below the age of the solar system (4650 My). Similarly, in the interior of the Earth $\lambda_m \sim 2 \, m^2 \, s^{-1}$ and taking the size of the Earth's core as the magnetic length $L_m \sim R_c = 3500 \, km$, we get $\tau_m \sim 0.5 \, My$. These numbers imply that any primordial magnetic field trapped within a planet and without sources will be lost in a short time scale. Electrical currents decay rapidly in planetary interiors and any magnetic field inside a planetary body requires a regenerating mechanism to maintain it over a long time. The most accepted mechanism proposed to produce such a magnetic field is the *dynamo*, a complex and not well–understood process occurring in the fluid electrically conducting interiors of planetary mass bodies. In addition to it, remnant ferromagnetism could also contribute in some cases to a planetary magnetic field (as probably happens on Mars), although in general its decay time is also short compared with that of the lifetime of the planetary systems. Small planetary magnetic fields can also be produced by the induction mechanism (Equation 1.139d) when an electrically conducting body (or internal parts of it) move in the ambient magnetic field created by the Sun or a star. Such an induced magnetic field apparently occurs in Europa, whose fluid conducting interior generates a magnetic field as the satellite moves in the powerful field of Jupiter.

1.6.2 DYNAMO REGIME

Analytical and numerical magnetohydrodynamic models of the planetary interiors indicate that the magnetic field in the Earth and in the giant planets is generated by a dynamo process. The basic ingredients for this mechanism to operate require a rapid planetary rotation and an electrically conducting fluid in motion. Since the equations describing the magnetic field and fluid motions are coupled, we deal with a complex nonlinear problem. However, a simple scale-analysis can be used to get a first-order value for the magnetic field resulting from this mechanism. From our solar system experience, the planets with a self-excited dynamo have magnetic Reynolds numbers of the order

$$Re_m = \frac{V_m L_m}{\lambda_m} \approx 10 - 100 \tag{1.143}$$

where V_m is the characteristic velocity of the fluid. This number, which results from the ratio between the two terms to the right of Equation 1.142, measures the importance of the magnetic diffusion effect against the fluid motion. On the other hand, the relevance of the planetary rotation angular velocity Ω on the magnetic field of the planets is measured by the Elsasser number

$$\Lambda = \frac{B^2}{2\Omega\mu_0\rho\lambda_m} \tag{1.144}$$

which in the solar system has typically a value $\Lambda \geq 1$. Thus, if the body has high internal temperatures that produce fluid regions in motion with a scale-length L_m, the required velocities for dynamo action should be $V_m \sim 10$–100 (λ_m/L_m) according to (1.143). The magnetic field intensity is then of the order of

$$B \approx \sqrt{2\Omega\mu_0\rho_0\lambda} \tag{1.145}$$

In Table 1.8, we give the values for the magnetic fields measured in the solar system.

TABLE 1.8
Planetary Magnetic Properties

Planet	M_D[a]	B[b]	Dipole Tilt[c]	Dipole Offset[d]
Mercury	0.0004	0.0033	+14°	—
Venus	<0.0004	<0.00003	—	—
Earth	1	0.305	+10.80	0.08
Mars	<0.0002	<0.0005	—	—
Jupiter	20,000	4.28	−9.60	0.12
Saturn	600	0.22	<10	0.04
Uranus	50	0.23	−58.60	0.3
Neptune	25	0.14	−470	0.55

[a] Dipolar moment M_D relative to that of the Earth $(M_{Earth} = 7.91 \times 10^{25}\,G$ $cm^3 = 7.9 \times 10^{22}\,A\,m^2)$.
[b] Magnetic field at dipole equator (G).
[c] Angle between the magnetic field and rotation axis.
[d] Dipole offset in planetary radius (R_p).

1.6.3 Dipole Field

The structure of the magnetic field of the planets can be measured directly using magnetometers onboard spacecrafts or indirectly from the radio-emissions produced by trapped accelerated particles in the field lines or from observations of the aurora activity. In general, these observations show that planetary magnetic fields are nearly dipolar with slight deviations from this configuration.

If we assume that local free currents and time-variable electric fields have negligible effects, then from (1.139d) we have $\nabla \times \vec{B} = 0$, and since from Equation 1.139c $\nabla \cdot \vec{B} = 0$, the magnetic field can be expressed as the gradient of a magnetic scalar potential $\vec{B} = -\nabla V_M$. To get a representation of the magnetic field configuration, the magnetic potential V_M can be expanded in harmonic series in a theory formally similar to that of the gravitational field previously presented with the important difference that the dominating term in the spherical harmonics expansion is the dipolar term. The expansion has the form

$$V_M(r,\varphi,\phi) = R_p \sum_{n=1}^{\infty} \sum_{m=0}^{n} \left(\frac{R_p}{r}\right)^{n+1} P_n^m(\cos\varphi)\left\{g_n^m \cos m\phi + h_n^m \sin m\phi\right\} \qquad (1.146)$$

where
   $\varphi = 90° - \delta_M - \varphi_c$ is the magnetic colatitude (φ_c here is the latitude and δ_M is the magnetic tilt relative to the geographic pole)
   ϕ is the magnetic longitude
   $P_n^m(\cos\varphi)$ are the Schmidt-normalized associated Legendre polynomials
   g_n^m and h_n^m are the Gauss coefficients, which have the dimension of the magnetic field

The magnetic field configuration in the planets of the solar system is to a first order that of a dipole, tilted in general relative to the rotation axis and in some cases off-centered. It is therefore reasonable to assume a pure dipolar configuration as a first approach to the planetary field, independent of the magnetic longitude, and characterized by a dipolar magnetic moment M_D. This reduces Equation 1.146 to

$$V_M = \frac{M_D \cos\varphi}{r^2} \qquad (1.147)$$

where $M_D = B_E R_E^3$, where B_E is the value of the magnetic field at a distance R_E in the Equatorial plane. The magnetic dipolar moment is given in SI units in nT m^{-3} (nT = 10^{-9} T), although it is more common to give it in cgs units (G cm^{-3}, with 1 G (gauss) = 10^5 nT).

Spherical components of the magnetic field strength are obtained from the gradient of the scalar magnetic potential

$$B_r = -\frac{\partial V_M}{\partial r} = -\frac{2M_D}{r^3}\sin\varphi_c$$

$$B_\varphi = -\frac{1}{r}\frac{\partial V_M}{\partial \varphi_c} = \frac{M_D}{r^3}\cos\varphi_c \qquad (1.148)$$

$$B_\phi = -\frac{1}{r\sin\varphi_c}\frac{\partial V_M}{\partial \phi} = 0$$

expressed in terms of the latitudinal (to the north) and azimuthal (to the east) coordinates and the total magnetic field

$$|B| = \sqrt{B_r^2 + B_\varphi^2} = \frac{M_D}{r^3}(1+3\sin^2\varphi_c)^{1/2} \qquad (1.149)$$

This is a good representation for the magnetic field generated from internal electrical currents. However, electrical currents in the magnetic environment outside the planet (in the magnetosphere) can also contribute to the magnetic field with a term B_{ext}. Then the total magnetic field can be expressed as

$$\vec{B} = -\nabla V_M + \vec{B}_{ext} \qquad (1.150)$$

A useful empirical formula, deduced for solar system bodies with dynamo fields, that can be applied to other hypothetic planets relates the magnetic dipolar moment M_D to the angular momentum due to the planetary spin

$$M_D = 4\times10^{-9} L_0^{0.83} \qquad (1.151)$$

with $L_0 = I\Omega$ given in kg m² s⁻¹ and M_D given in G m³.

1.6.4 MAGNETIC FIELDS IN SOLAR SYSTEM BODIES

A number of planets and satellites have been observed to have magnetic fields of different origins. The most intense fields occur in the terrestrial planets Mercury and Earth and in the four giant and icy planets. All are internally generated by a dynamo mechanism; in the case of Mercury and Earth in their molten fluid cores, in the case of Jupiter and Saturn in their hydrogen metallic interior, and in the case of Uranus and Neptune in their ionic molecular layer. The complexity of the mechanisms intervening in the magnetic field generation by the dynamo action can be appreciated when comparing the variety of strengths, axial tilts, and offsets from the planet center observed in the dipole moments in each planet.

Venus probably possessed a magnetic field in its past resulting from an active dynamo after its formation, but now this mechanism has apparently ceased to be active and its weak magnetic field results from the induction due to the interaction between the solar wind and the charged particles moving in its ionosphere. Since the surface temperature is high (above the Curie temperature), no magnetic field is expected to exist in its crust. The Moon and Mars have localized crustal magnetic fields. They are residual, possibly resulting from a pass dynamo operating when their cores were hot and molten, although a local contribution coming from the heat generated following large impacts has also been proposed. Dynamo action in the cores of the Galilean satellites Io and Ganymede could also explain their weak magnetic field. As indicated before, the magnetic field of Europa could be activated when the satellite moves in the powerful magnetic field of Jupiter, interacting with its plasma environment, or it could be induced due to the presence of a salty subsurface liquid ocean.

1.7 PLANETODIVERSITY

The classification and evolution of planets using their physical properties in terms of a single diagram, as it is done with stars (luminosity versus effective temperature, the "Hertzsprung–Russell diagram"), is not evident. Each planet is probably a unique product of its initial formation conditions and its evolutionary history, physical and orbital. The variety of planets and satellites present in the solar system, as well as those discovered in "exotic" environments (around pulsars, white dwarfs, and brown dwarfs), or in "unexpected" orbits (giants with distances to the star 1/10th that of Mercury or in highly eccentric orbits), the existence of isolated "planets" (nick-named "free floating" objects), or those that are very different from what we predict to exist on theoretical grounds, tells us that we must keep our minds open to a large amount of possible types of planets and atmospheres.

As we have seen in the previous sections, the three basic parameters determining the physical structure of a planet are its mass (M_p), chemical composition (X), and available energy (E) (Figure 1.20). The mass is the fundamental parameter used to distinguish between a planet and a brown dwarf star (a sub-stellar object able to generate in its interior nonsustained deuterium thermonuclear fusion reactions). It is usually

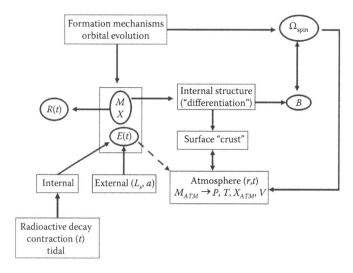

FIGURE 1.20 Flow diagram of the origin and relationships between the basic planetary parameters (M, mass; R, radius; X, composition; E, energy; Ω, angular rotation speed; and B, magnetic field), and the global structure of a planet.

accepted from theoretical models that the frontier separating both types of objects is 13 Jupiter masses (M_J) for brown dwarfs with a solar chemical composition. However, there is not an accepted criterion for the minimum mass of a planet since this will depend, among other considerations, on the orbital configuration discussed previously (e.g., cases B and C in Figure 1.1). The recent debate on the classification of the planets and dwarf planets of our solar system by the International Astronomical Union presented in Section 1.2 shows that it is not a simple issue to classify a body as a planet. The dwarf planet Pluto has a mass $2 \times 10^{-3} M_\oplus$ or equivalently $6 \times 10^{-6} M_J$, whereas the most massive satellite Ganymede has a mass of $2.4 \times 10^{-2} M_\oplus$ (about 12 times the mass of Pluto) and the dwarf planet Ceres has a mass $1.6 \times 10^{-4} M_\oplus$ (about 1/10th the mass of Pluto). The less massive extrasolar "planet" so far reported is around the pulsar PSR B1257+12b, with a mass $4 \times 10^{-4} M_\oplus$ (about 1/5th of the mass of Pluto).

The planet composition (X) depends on the initial inventory of volatile and refractory elements in the forming planetary system and of their gradients within the protoplanetary disk. Volatile and refractory elements are relative terms that describe those that condense and vaporize at low and high temperatures, respectively. Their abundance on the protoplanetary nebula will determine the formation of minerals and ices of different mixtures according to the condensation equilibrium curves of the compounds, which is a function of the distance to the star. Abundance measurements in the stars studied to search for planets suggest that there is a correlation between their metallicity (abundance of elements heavier than helium) and the presence of planets. The mass, composition, and available energy will determine the planetary radius (R_p), the internal structure of the planet, and the nature of the atmosphere.

All the physical and chemical processes so far described should produce a large variety of planetary types in the universe—a planetary zoo or "the physical planetodiversity." The classification of planets in simple types is not evident, but based on our solar system examples and on the above theoretical considerations, using the bulk composition as a guide, we can introduce the following five "fundamental taxonomic planetary bodies classes" for planets with masses $M_p \leq 13$ Jovian mass (the theoretical transition limit to brown dwarfs) characterized by their basic composition: (1) *metallic* (Fe, Ni, and others); (2) *silicate-rocky* (SiO_2, MgO, FeS, FeO); (3) *icy* (H_2O, CO_2, NH_3, CH_4); (4) *gaseous and fluid ionic* (H_2, He, H_2O, NH_3, CH_4); and (5) *gaseous* (H_2, He). However, the most common planets should be mixtures of the above basic types that we find in the solar system, that is, metallic-rocky, with different percentages of both materials (Mercury, Venus, Earth, and Mars); bulk silicate-rocky (Moon, Io, and Europa); ice-rock (Ganymede, Callisto, Titan, Triton, and Pluto); icygaseous (Uranus and Neptune); and bulk gaseous (Jupiter and Saturn). According to Equation 1.122, the mean density of these planets could vary strongly depending on their mass and radius. As an example, gaseous giants planets could have mean densities in the range $\rho \sim 0.5$–16 g cm^{-3} for masses ranging from 0.5 to $13 M_J$.

Which other possible kinds of planets could exist in the universe? The types that have been found so far include the nicknamed "Hot Jupiters" and "Very hot Jupiters" families of gaseous giants at distances <0.01 AU to their stars and low mass "Hot Uranus-Neptunes" with 14–$21 M_\oplus$. A new class of planets called "Super-Earths," with Earth-like mean compositions but masses above Earth but below that of Neptune, have been detected (see Table 1.9). A typical "Super-Earth" should

TABLE 1.9
Properties of Transiting Exoplanets—Planets Whose Mass and Radius Have Been Measured (January 2010)

Planet Name	Mass (M_J)	Radius (R_J)	a	e
WASP-19b	1.15	1.31	0.0164	0.02
CoRoT-7 b[a]	0.0151	0.15	0.0172	0
WASP-18 b	10.43	1.165	0.02047	0.0092
WASP-12 b	1.41	1.79	0.0229	0.049
OGLE-TR-56 b	1.29	1.3	0.0225	0
TrES-3	1.92	1.295	0.0226	0
WASP-4 b	1.1215	1.416	0.023	0
OGLE-TR-113 b	1.32	1.09	0.0229	0
CoRoT-1 b	1.03	1.49	0.0254	0
GJ 1214 b[a]	0.0179	0.2415	0.014	<0.27
WASP-5 b	1.637	1.171	0.02729	0
OGLE-TR-132 b	1.14	1.18	0.0306	0
CoRoT-2 b	3.31	1.465	0.0281	0
SWEEPS-11	9.7	1.13	0.03	—
WASP-3 b	1.76	1.31	0.0317	0
WASP-2 b	0.914	1.017	0.03138	0
HAT-P-7 b	1.8	1.421	0.0379	0
HD 189733 b	1.13	1.138	0.03099	0
WASP-14 b	7.725	1.259	0.037	0.0903
TrES-2	1.199	1.272	0.03556	0
OGLE2-TR-L9 b	4.5	1.61	—	—
WASP-1 b	0.89	1.358	0.0382	0
XO-2 b	0.57	0.973	0.0369	0
GJ 436 b	0.072	0.438	0.02872	0.15
HAT-P-5 b	1.06	1.26	0.04075	0
HD 149026 b	0.359	0.654	0.04313	0
HAT-P-3 b	0.599	0.89	0.03894	0
HAT-P-13 b	0.851	1.28	0.0426	0.021
TrES-1	0.61	1.081	0.0393	0
HAT-P-4 b	0.68	1.27	0.0446	0
HAT-P-8 b	1.52	1.5	0.0487	0
WASP-10 b	3.06	1.08	0.0371	0.057
OGLE-TR-10 b	0.63	1.26	0.04162	0
WASP-16 b	0.855	1.008	0.0421	0
XO-3 b	11.79	1.217	0.0454	0.26
HAT-P-12 b	0.211	0.959	0.0384	0
Kepler-4 b	0.077	0.357	0.0456	0
Kepler-6 b	0.669	1.323	0.04567	0
WASP-6 b	0.503	1.224	0.0421	0.054
Kepler-8 b	0.603	1.419	0.0483	0
HD 209458 b	0.685	1.32	0.04707	0.07
Kepler-5 b	2.114	1.431	0.05064	0

TABLE 1.9 (continued)
Properties of Transiting Exoplanets—Planets Whose Mass and Radius Have Been Measured (January 2010)

Planet Name	Mass (M_J)	Radius (R_J)	a	e
TrES-4	0.919	1.799	0.05091	0
OGLE-TR-211 b	1.03	1.36	0.051	0
WASP-11/HAT-P-10 b	0.46	1.045	0.0439	0
WASP-17 b	0.49	1.74	0.051	0.129
WASP-15 b	0.542	1.428	0.0499	0
HAT-P-6 b	1.057	1.33	0.05235	0
Lupus-TR-3 b	0.81	0.89	0.0464	0
HAT-P-9 b	0.78	1.4	0.053	0
XO-1 b	0.9	1.184	0.0488	0
OGLE-TR-182 b	1.01	1.13	0.051	0
OGLE-TR-111 b	0.53	1.067	0.047	0
CoRoT-5 b	0.467	1.388	0.04947	0.09
XO-4 b	1.72	1.34	0.0555	0
XO-5 b	1.077	1.089	0.0487	0
SWEEPS-04	<3.8	0.81	0.055	—
CoRoT-3 b	21.66	1.01	0.057	0
WASP-13 b	0.46	1.21	0.0527	0
HAT-P-1 b	0.524	1.225	0.0553	<0.067
Kepler-7 b	0.433	1.478	0.06224	0
HAT-P-11 b	0.081	0.452	0.053	0.198
WASP-7 b	0.96	0.915	0.0618	0
HAT-P-2 b	9.09	1.157	0.06878	0.5171
WASP-8 b	2.23	1.17	0.0793	—
CoRoT-6 b	2.96	1.166	0.0855	<0.1
CoRoT-4 b	0.72	1.19	0.09	0
HD 17156 b	3.212	1.023	0.1623	0.6753
HD 80606 b	3.94	1.029	0.449	0.93366

Source: Schneider, J., *The Extrasolar Planets Encyclopedia*, CNRS-LUTH, Paris Observatory, Paris, France, http://exoplanet.eu/catalog-transit.php.

Planets whose mass and radius have been measured (January 2010). Mass and radius in terms of those of Jupiter (Table 1.3).

[a] These planets are called "Super Earth" since they have masses above that of the Earth but below those of Uranus and Neptune. Specifically:

CoRoT-7 b: 4.8 ± 0.8 Earth mass, 1.68 ± 0.09 Earth radius, density $= 5.6 \pm 1.3 \, g \, cm^{-3}$.

GJ 1214 b: 6.55 ± 0.98 Earth mass, 2.68 ± 0.13 Earth radius, density $= 1.9 \pm 0.4 \, g \, cm^{-3}$.

have masses of ~ 5–$10 M_\oplus$ and radii of $\sim 2 R_\oplus$ according to our discussion in previous sections (Equation 1.122).

On theoretical grounds, other planetary types so far proposed include "Ocean planets," worlds with masses in the range of 1–$10 M_\oplus$ covered by liquid water (perhaps with global oceans $\sim 100 \, km$ in thickness), resulting from the inward migration and evolution of icy planets and "Carbon planets" with masses up to $60M_\oplus$, formed by carbon and carbides perhaps recalling in composition the carbonaceous chondrite meteorites. Carbon, CO planets, or even the densest "iron planets" (similar perhaps to the cores of the terrestrial planets), could be the type of planets found around pulsars or predicted to be orbiting around evolved stars (e.g., "white dwarf planets"). We can imagine other types of "liquid planets" fully covered with non-water oceans, for example of liquid methane, ammonia, or nitrogen and so on, to infinity. From the elemental bulk and radial compositional abundances in the galaxies and in nascent protoplanetary disks, we can imagine planets made of helium, nitrogen, oxygen, and other most abundant elements. We must be open to any possibility.

PROBLEMS

1.1 Calculate the energy for a body of mass m describing a circular orbit at a distance r_0 around a body of mass M.

Solution

$$E = -\frac{GMm}{2r_0}$$

1.2 Derive the generalized third Kepler law (Equation 1.24).

1.3 Derive an approximate expression for the radius of the Hill sphere (Equation 1.26) assuming that the orbital angular velocity of a body around a planet of mass M_2 (at distance R_H) and the orbital angular velocity of that planet around the star of mass M_1 (distance a) are equal.

1.4 (a) Calculate the acceleration of gravity due to the Earth (mass M) and the Moon (mass m) as a function of the distance r in the line joining their center of mass (assume they are both homogeneous spheres separated by a distance D). (b) At which point does the gravitational attraction they exert equal zero?

Solution

$$(a) \quad \vec{g} = \left[-\left(\frac{GM}{r^2}\right) + \left(\frac{Gm}{(D-r)^2}\right) \right] \hat{r}$$

(b) $r_0 = 0.90D$

1.5 Calculate the gravitational potential $V_G(r)$ and the force $F(r)$ produced by two masses M_1 and M_2 rotating around their common center of mass as a function of a (semimajor axis of the orbit) and a_2 (the distance between the center of mass of the system and the mass M_2). Then derive Equation 1.25.

Solution

$$V_G(r) = -\frac{GM_1}{(a-r)} - \frac{GM_2}{r} - (a_2 - r)^2 G \frac{(M_1 + M_2)}{2a^3}$$

$$F(r) = -\frac{\partial V_G}{\partial r} = \frac{GM_1}{(a-r)^2} - \frac{GM_2}{r^2} - (a_2 - r)G\frac{(M_1 + M_2)}{a^3}$$

Equation 1.25 is obtained making $F(r) = 0$.

1.6 Using the integral form for (1.30) and (1.31), the Gauss's theorem $\iint_S \vec{g} \cdot d\vec{S} = -4\pi GM(r)$, calculate the acceleration of gravity $\vec{g}$ as a function of the distance r to the center inside the following mass distributions: (a) a thin spherical shell and (b) a sphere of uniform density, both with radius R_p.

Solution

(a) $\vec{g} = 0$

(b) $\vec{g} = -\left(\frac{Gmr}{R_p^3}\right)\hat{r}$

1.7 Compare the relative intensity between the forces exerted on the Earth by the Sun and Jupiter for (a) the opposition and (b) the conjunction distances. Use the data from Table 1.1.

Solution

(a) 18,377.7

(b) 39,979.8

1.8 Calculate the moment of inertia and the moment of inertia factor α of a spherically symmetric planet whose density distribution as a function of distance r is given by $\rho(r) = \rho_0(1 - r/R_p)$.

Solution

$$I = \frac{4}{15} MR_p^2$$

1.9 Demonstrate that in an oblate spheroid body, an element of distance along the meridian (ds) changes with planetocentric latitude according to the equation

$$\frac{ds}{d\varphi_c} = \left[\left(\frac{r(2f_{EP} - f_{EP}^2)\sin\varphi_c \cos\varphi_c}{1 - (2f_{EP} - f_{EP}^2)\cos^2\varphi_c}\right)^2 + r^2\right]^{1/2}$$

1.10 Using the center of mass of a body with mass M as the coordinate origin, calculate the acceleration on a body of mass m due to the tidal force assuming that $r_0 \gg R$. Using the data from Table 1.1, demonstrate that the ratio between the accelerations on a point of the Earth's ocean surface due to the Moon and the Sun is 2.22.

Solution

$$a = \frac{F_T}{M} \approx \frac{2Gmr}{r_0^3}$$

1.11 Calculate the tidal strength (q_{Ti}) produced by the satellite on a planet for the following cases (see Table 1.2): Earth-Moon ($m = 7 \times 10^{22}$ kg, $r = 384.4 \times 10^3$ km), Pluto-Charon ($m = 0.16 \times 10^{22}$ kg, $r = 19.4 \times 10^3$ km), and Jupiter-Io ($m = 8.9 \times 10^{22}$ kg, $r = 421.8 \times 10^3$ km).

Solution

$q_{Ti} = 5.3 \times 10^{-7}$ (E-M), 2.2×10^{-5} (P-C), 2.2×10^{-7} (J-I)

1.12 Apply Equation 1.61 to find the synodic period of a planet relative to the Earth (i.e., the time elapsed between successive conjunctions with the Earth). Take the positive sign for inferior planets (Mercury and Venus) and the negative sign for superior planets (Mars, Jupiter, Saturn, Uranus, and Neptune).

Solution

$$\frac{1}{\tau_{sid(planet)}} = \frac{1}{\tau_{sid(Earth)}} \pm \frac{1}{\tau_{syn(planet)}}$$

$\tau_{syn(planet)}$ (days) = 116 (Mercury), 584 (Venus), 780 (Mars), 399 (Jupiter), 378 (Saturn), 370 (Uranus), 368 (Neptune).

1.13 Derive the potential energy of a planet with mass M_p and radius R_p that obeys the polytropic equation of the state of index $\upsilon = 1$.

Solution

$$E_p = -\frac{3}{4} G \frac{M_p^2}{R_p}$$

1.14 Calculate the pressure at a center of a body with uniform density (Equation 1.91b). Apply this result to the giant and icy planets to get a crude order of magnitude of the pressure at the center of the planet.

Solution

$P_c(J) = 11$ Mbar, $P_c(S) = 2$ Mbar, $P_c(U) = 1.4$ Mbar, $P_c(N) = 2.2$ Mbar.

1.15 It has been proposed theoretically that massive "Carbon planets" can exist, formed by carbon and carbides, perhaps remembering in composition the carbonaceous chondrite meteorites. What would be the maximum radius of a planet with a mass of $60M_\oplus$?

Solution

Since $A = 6$ and $Z = 12$ we get $R_p(\max) = 12{,}567\,\text{km}$.

1.16 Calculate the minimum radius of an incompressible iron body with a mean density of $8\,\text{g cm}^{-3}$ if the material has a strength of $S_F = 4 \times 10^8\,\text{N m}^{-2}$.

Solution

$R_p(\min) = 244\,\text{km}$

1.17 What is the skin depth in the surface of a planet with a rotation period of 12 h if it is made of: (a) iron ($K_T = 80\,\text{W m}^{-1}\,\text{K}^{-1}$, $\rho = 8000\,\text{kg m}^{-3}$, $C_p = 500\,\text{J kg}^{-1}\,\text{K}^{-1}$); (b) ices and rocks ($K_T = 20\,\text{W m}^{-1}\,\text{K}^{-1}$, $\rho = 3000\,\text{kg m}^{-3}$, $C_p = 800\,\text{J kg}^{-1}\,\text{K}^{-1}$)? If the temperature at the surface is 280 K, what is the temperature fluctuation at a depth of $z = 1\,\text{m}$ for each case?

Solution

(a) $\mu_T = 0.52\,\text{m}$, $T = 40.9\,\text{K}$
(b) $\mu_T = 0.338\,\text{m}$, $T = 14.6\,\text{K}$

1.18 The internal heat flux in Jupiter in $5.44\,\text{W m}^{-2}$. Assume that its metallic hydrogen interior has a thermal conductivity of $K_T = 10^3\,\text{W m}^{-1}\,\text{K}^{-1}$. (a) What would be the temperature gradient if heat conduction dominates? (b) What is the temperature difference between the upper boundary of this layer ($r = 0.9R_J$) and the center of Jupiter? (c) Since detailed models of the interior of the planet predict differences of 20,000 K, what can you say about heat transport within Jupiter?

Solution

(a) $5.44\,\text{K km}^{-1}$
(b) $\Delta T = 350{,}000\,\text{K}$
(c) Convection must dominate over conduction since Jupiter is not so hot.

1.19 A hypothetical type of extrasolar planet, a "Super-Earth," produces radiogenic heat in its interior at a rate of $5 \times 10^{-12}\,\text{W kg}^{-1}$. If it has a core with two times the radius of the Earth and with four Earth masses, (a) what is the heat flux at the surface of its core? If the planet has a solid mantle with thermal conductivity $K_T = 10\,\text{W m}^{-1}\,\text{K}^{-1}$, (b) what is the temperature gradient across the mantle? (c) What would be the temperature difference across the mantle if its size is 5000 km?

Solution

    (a) 0.058 W m^{-2}

    (b) 5.8 K km^{-1}

    (c) $\Delta T = 29,196$ K

1.20 Compare the radiative and convective temperature gradients in the fluid interior of Jupiter ($P = 7$ kbar, $T = 1700$ K, $g = 25$ m s^{-2}, $C_p = 12,360$ J K^{-1} kg^{-1}, $T_{eff} = 124$ K) assuming that the fluid follows the perfect gas law in hydrostatic equilibrium for the following two opacity situations for this gas: (1) $\kappa = 0.032$ cm^2 g^{-1} and (2) $\kappa = 1$ cm^2 g^{-1}. What kind of heat transport mechanism will dominate in each case?

Solution

    (a) $(dT/dr)_{conv} = 2$ K km^{-1}

    (b) (1) $(dT/dr)_{rad} = 0.316$ K km^{-1}, (2) $(dT/dr)_{rad} = 9.9$ K km^{-1}. For situation (1), the interior of Jupiter will transport heat by radiation.

1.21 Find an expression for the Rayleigh number in terms of the internal heat flux within a planet. Apply to the case of a planet with $g = 10$ m s^{-2}, mantle of depth $d = 200$ km, internal heat flux $F_{int} = 30$ mW m^{-2}, thermal diffusivity $D_T = 10^{-6}$ m^2 s^{-1}, viscosity $v_v = 10^{21}$ kg m^{-1} s^{-1}, thermal expansivity $\alpha_T = 2 \times 10^{-5}$ K, and specific heat $C_p = 1000$ J kg^{-1} K^{-1}.

Solution

$$Ra = \frac{\alpha_T F_{int} g d^4}{D_T^2 v_v C_p}, \quad Ra = 380$$

1.22 Find that the temperature gradient in the interior of an adiabatic gaseous planet obeying the perfect gas law is related to pressure gradient according to

$$\frac{dT}{dr} = \left(1 - \frac{1}{\gamma}\right) \frac{T}{P} \frac{dP}{dr}$$

1.23 Using the mixing length theory for convective motions, find an expression for the vertical velocity w in the interior of a giant planet in terms of the temperature and heat flux assuming that adiabatic conditions (adiabatic coefficient γ) and the perfect gas law applies.

Solution

$$w = \left(\frac{R_g}{c_p} \frac{F}{T_0^\gamma P_0^{1-\gamma}} T^{\frac{1}{1-\gamma}}\right)^{1/3}$$

1.24 Compare the value of the magnetic field predicted by Equations 1.134 and 1.140 for a giant extrasolar planet that is assumed to be a homogeneous sphere with mass $M_p = 1.5M_J$ and radius $R_p = 1.12R_J$, and has a rotation period of 10 h and magnetic diffusivity of 8 m^2 s^{-1}.

Solution

(1.134) $B = 20.5$ G; (1.140) $B = 2.2$ G.

2 Origin and Evolution of Planetary Atmospheres

We initially classify planetary atmospheres by a single bulk parameter, their mass, or equivalently by a measurable magnitude—the pressure they exert at their base (defined by a surface or a deep interior level in the fluid giants). The "surface pressure" is the weight of the mass column per unit surface ($P_s = Mg/S$; strictly in the Cartesian coordinate system). Accordingly, the atmospheres of the solar system planets and satellites can be divided into the following types: (1) thin atmospheres ($P_s < 10^{-5}$ bar): Mercury, Moon, Galilean satellites (Io, Europa, Ganymede, and Callisto), Enceladus, Triton, and Pluto; (2) intermediate atmospheres ($P_s = 7 \times 10^{-3}$ to 90 bar): Venus, Earth, Mars, and Titan; and (3) massive and deep atmospheres ($P_s >$ kbar–Mbar): Jupiter, Saturn, Uranus, and Neptune. The atmospheres originate with the planet or satellite and evolve, changing their mass and properties as they interact with their boundaries (the surface and the outer space). This chapter deals with the basics of the origin and evolutionary processes of the atmospheres of planets and satellites in the solar system. We do not intend to be exhaustive in details, but we will describe the fundamental, common processes affecting their lifetime on a long-term basis.

First, we start with a brief outline of a formation of a planetary system. Since the solar system provides the vast majority of data of any planetary system, models of planet formation are studied based on the solar system. However, astronomical observations of extrasolar planets, as presented in the previous chapter, and circumstellar disks around young stars have provided us with a large quantity of new data on star and planetary system formation. New ideas have emerged and tests on the models of planet formation can now be performed in different scenarios.

2.1 ORIGIN OF THE SOLAR SYSTEM

Any model of the solar system must explain its actual structure and properties. The main observational data sets are the following:

- Orbital properties of the planets (distances and mutual separations, eccentricities, motion close to a single plane)
- Orbital properties of satellites relative to their parent planet
- Orbital properties of many families of small bodies (asteroids, Kuiper Belt and trans-Neptunian objects [TNOs], dwarf planets, and comets)
- Angular momentum distribution (98% is contained in the orbital motions of the giant planets)

- Mass distribution within the solar system (as a whole) and within the planet–satellite systems
- Rotation state of the planets and satellites
- Bulk composition distribution among the planets structured in two basic classes: terrestrial and giants; a variety of satellite types and composition of small bodies
- Isotopic composition
- Age—the solar system age is fixed to be 4560×10^9 years based on radioisotope dating of primitive meteorites
- Surface structure, in particular cratering among most planets and satellites
- Magnetic field in planets and satellites

Stars form following the gravitational collapse of a mass of gas and dust within the nebular regions of the interstellar medium in galaxies (Figure 2.1). The solar system formed from the *Protosolar Nebula*, with the mass concentrated in the

Gaseous Pillars · M16   **HST · WFPC2**
PRC95-44a · ST ScI OPO · November 2, 1995

FIGURE 2.1   The Aquila Nebula (M16) is a gas and dust region of stellar (and probably planetary systems) formation. (Courtesy of J. Hester and P. Scowen, Arizona State University, Tempe, AZ.)

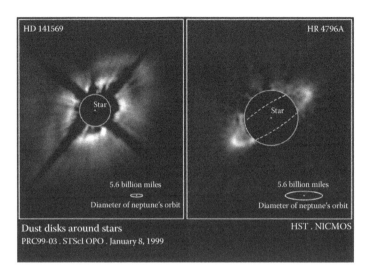

FIGURE 2.2 Examples of protoplanetary disks of dust and gas. (Courtesy of S. Smith, University of Hawaii, Honolulu, HI; G. Schnieder, University of Arizona, Tucson, AZ; E. Becklin and A. Weinberger, UCLA and NASA, Los Angeles, CA.)

proto-Sun growing by gravitational contraction at the center whereas conservation of angular momentum during this contraction gave rise to the formation of an extended flattened disk around it (Figure 2.2). The disk contained less than $0.01M_\odot$ and contained both gas and dust and where the planets and other bodies formed.

There are two main mechanisms that have been proposed by which planets form. The first is the fragmentation of the disk by gravitational instabilities, which leads to growing accretion of the gas nebula to form "giant and icy planet types" over timescales of few hundred thousand years. This process is favored in massive disks ($\sim 1M_\odot$). Once the proto-planet forms, the interaction with the disk could allow the planet to migrate inward from the external parts of the system to an inner orbit, or even to be engulfed by the star. Giant planets formed in this way are expected to be homogeneous in chemical composition (should not have an internal massive rocky core) and have regular compositional variation with distance to the star. In the second scenario, the planets form from the accretion of small solid bodies called *planetesimals* that grow from the dust (solid particles) in a less massive disk (~ 0.01–$0.02M_\odot$). The collision and gravitational attraction between the planetesimals is a slower process when compared with disk instability, but their accumulation leads to the formation of a proto-planet. There is a consensus that the terrestrial planets form in this way at distances close to the star. Their chemical composition is dominated by refractive materials (no volatiles) due to the high temperatures in this part of the disk. Far from the star, planetary mass embryos made of volatiles form by this mechanism (mass range of ~10 Earth masses or more). The accretion of massive quantities of gas from the nebula form an envelope over the embryo that becomes the core of a gaseous giant planet. Therefore, the distinctive sign against the "gravitational instability mechanism" is that giant planets formed in this way are expected to have a core in

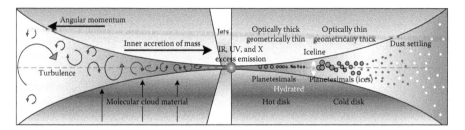

FIGURE 2.3   Scheme of some processes occurring in a disk and leading to planetesimal formation. (Courtesy of Ricardo Hueso, UPV-EHU, Spain.)

their center or at least to be enriched in heavy elements if the core is eroded. This is the so-called *standard model* for solar system formation.

The proposed processes occurring within the disk can be summarized in the following six stages (Figure 2.3):

1. The separation and settling of the dust into the mean plane of the disk.
2. The accretion and collisions of the small particles to form the planetesimals with sizes ranging from a hundred meters to a few kilometers. This is a process fairly less known because collisions can be constructive or destructive depending on the size of the particles and their relative velocities.
3. Growing through the gravitational attraction of the planetesimals to form the terrestrial planets and the possible cores of the giant planets. Typical timescales to form these bodies are ~10^6 years (1 gigayear [Gy]).
4. The large cores in the outer solar system acquire the gaseous material from the disk and form the envelopes of the giant planets. A core with 10 Earth masses will have an escape velocity for the gas of 18 km s^{-1}—10 times lower than the thermal escape for hydrogen atoms in a disk at 100 K (see Section 2.4). Therefore, attraction of the gas by the core will inevitably form an envelope (a deep atmosphere) in about ~10^5 years. This process is also dependent on the ability of the envelope to radiate the energy released in the gravitational capture of the gas and, depending on the models, can take up to a few million years.
5. The formation of the final inner planets by collisions of the planetary embryos and planetesimal left-overs over a timescale of 100 million years (My).
6. The small planetary migration of the outer and inner planets and the late heavy bombardment of the inner planets.

The formation of the regular satellites of the major planets could occur in a similar accretion process from the disk that surrounds the protoplanet during its formation stage. Smaller irregular satellites in tilted orbits, relative to the planet's equator, with retrograde motions are capture bodies (i.e., the satellite of Neptune, Triton). On the other hand, the orbital, physical, and compositional properties of the Moon are best explained from a collision of a large body (Mars-like mass) with the Earth in phase 5. More precisely, the Moon is formed from the accretion of the debris left by the impact.

One basic aspect in the formation of the different type of planets is the chemical composition of the disk. The initial material forming the star and disk (i.e., the abundance of weighted elements than hydrogen) comes from the gas and dust of the interstellar medium nebula, although the initial elemental composition undergoes chemical processing during the planetary formation. The chemical reactions within the disk depend on the elemental abundances, gas pressure, and temperature. From the thermochemical properties of the condensable components and the reactions between solid and gaseous phases, it is possible (under a chemical equilibrium assumption) to calculate the sequence of compound condensation, which, being a function of temperature, will result in a distance to the Sun sequence. Under chemical equilibrium, the partial pressure P_i of a molecular gas i is related to the partial pressures P_j of the monoatomic gases j by the law of mass action

$$P_i = K_i \prod_j P_j^{v_{ij}} \qquad (2.1)$$

where

K_i is the equilibrium constant that is a function of temperature

v_{ij} is the stoichiometric coefficient that is the number of atoms of element j in the phase i

Condensation into a solid phase occurs when

$$A_k < \frac{K_k \prod_j P_j^{v_{ij}}}{P_k} \qquad (2.2)$$

with A_k being the activity of the solid k.

Figure 2.4 schematically depicts the situation. Above 2000 K, most compounds are in atomic and ionic form. Below this temperature, the majority of atoms are neutral, not ionic, and the thermal energy $k_B T$ (k_B is the Boltzmann gas constant) is comparable with the strengths of the chemical bonds, and diatomic and polyatomic molecules may form. Therefore, as the nebula and the disk cool, the chemical reactions between the elements form molecules and mineral compounds when condensing below a characteristic temperature. The order of condensation of the refractory elements is in the order of decreasing temperatures (although, it depends on the oxidation state of the gas): Al, Ti, Ca, Mg, Si, Fe, and Na. For example, the condensation temperature for Al_2O_3 is 1743 K for a relative abundance of C/O = 0.55 but 1235 K for C/O = 1.2. At $T \sim 1400$ K, Fe–Si forms and at slightly lower temperatures magnesium minerals like fosferite (Mg_2SiO_4) and enstatite ($MgSiO_3$) can form. Sodium and calcium form minerals at $T < 1000$ K. Halides and sulfides (NaF, K_2S, etc.) form at lower temperatures. Below about 500 K, water plays an important role and the elements H, O, C, and N combine to form basic molecules in a process that

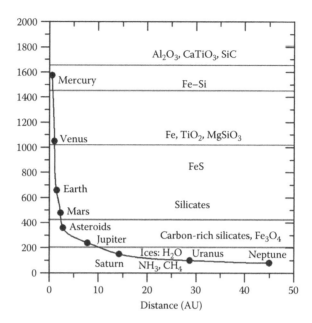

FIGURE 2.4 Condensation sequence of minerals and ice for a gas with the solar composition as a function of distance and temperature to the Sun. Refractory minerals form close to the Sun (distances closer than 3 AU) whereas ice forms at about 5 AU, the distance of Jupiter to the Sun.

depends on temperature: C to form CO at $T > 700\,K$ but methane (CH_4) at lower T, or N to form N_2 at $T > 300\,K$ but NH_3 at lower T (see Chapter 3). Below $200\,K$, pure water forms and when it mixes with ammonia and methane can condense to form hydrates and clathrates of the type NH_3-H_2O and CH_4-$6H_2O$. Below $100\,K$, these gases condense to form the simple ice of CH_4 below $40\,K$, and CO and N_2 form below $25\,K$. Hydrogen (H_2) and neon (Ne) condense at even lower temperatures but they can also be retained in clathrates at low temperatures of $40\,K$.

The approximate compositional sequence of compounds described above is expected in the bodies (planets and satellites) as a function of the separation from the Sun since the temperature decreases as $T \sim r^{1/2}$ (Equations 1.100 and 1.101). Terrestrial planets are found close to the Sun (Mercury, Venus, Earth, and Mars) with a bulk increasing ratio of metal/rocks from 65% to 35% in Mercury to 40%–60% in Mars. Below the "frost line" or "snow line" at about $150\,K$, the ice condenses and we find the giant fluid planets mainly composed of hydrogen and their satellites formed by mixtures of rock and ice. The rock/ice ratio decreases from 80% with increasing distance from the Sun.

2.2 THERMAL EVOLUTION OF THE PLANETS

The thermal evolution of a planet or satellite is basically the history of how its internal energy sources and temperature changes in time, from their formation period to the present. Obviously, it affects the history of its atmosphere. Essentially, a planetary body cools as its internal energy source decreases, radiating its heat to space from the surface, its atmosphere, or both. Theoretically, as the planet extinguishes its energy source, its temperature tends toward an equilibrium value in the stellar radiation ambient.

The thermal evolution of a planet depends on the strength of the energy source and of the planet structure (mass, size, composition, internal layering, and efficiency of heat transport). The energy equation for the whole planet gives its temperature evolution and, neglecting the external stellar flux, is written as (see (1.126))

$$M_p C_p \frac{\partial T}{\partial t} = M_p H_{int} - 4\pi R_p^2 F_s \tag{2.3}$$

where

$H_{int}(t)$ is the internal energy released per unit mass (W kg^{-1})
$F_s(t)$ is the surface heat flux

The temporal behavior of the internal heat source depends on its nature, and the surface heat flux depends on the internal transport mechanisms within the planet. A simple way to search for the temperature solution to (2.3) is to parameterize the right-hand term using *Newton's law of cooling* that states that the rate of heat loss of a body is proportional to the difference in temperatures between the body and its surroundings. The temperature evolution then follows the law of

$$\frac{dT}{dt} = -\frac{T - T_{eq}}{\tau} \tag{2.4}$$

whose solution is

$$T(t) = T_{eq} + [T(0) - T_{eq}]\exp\left(-\frac{t}{\tau}\right) \tag{2.5}$$

with τ being the characteristic cooling time of the body given to first order by

$$\tau \approx \frac{R_p^2}{D_T} \tag{2.6}$$

and D_T being the thermal diffusivity (Chapter 1). For example, for the Earth using $D_T = 10^{-5}\,m^2 s^{-1}$ and $R_p = 6.4 \times 10^6$ m, we get $\tau \sim 1.3 \times 10^{11}$ years. In Figure 2.5, we show representative cases for this simple cooling decaying law. In reality, the characteristic cooling time also depends on temperature, decreasing with increasing temperatures.

In a steady-state situation ($\partial T/\partial t = 0$), the internal energy production within the planetary body is balanced by the heat loss by radiation to space through the surface–atmosphere interface. According to (2.3), and neglecting again the absorption

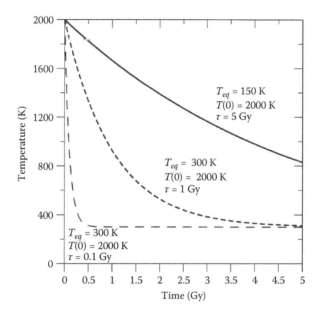

FIGURE 2.5 Temperature evolution of a hypothetical planet characterized by different thermal diffusivity and diffusion time constant properties and for two different equilibrium temperatures (Equation 2.5). The final asymptotic temperature evolution corresponds to the equilibrium temperature.

of energy from the star, the surface temperature of the body (T_s) can be obtained combining this equation and (1.101) to get

$$T_s = \sqrt[4]{\frac{4\pi R_p^2 \varepsilon \sigma_B}{M_p H_{int}}} \tag{2.7}$$

2.3   ORIGIN OF PLANETARY ATMOSPHERES

Planetary atmospheres can exist if the gravitational force of the body (its mass) is high enough to retain the gases against a variety of processes that favor their escape. The main process is related to temperature, i.e., to the kinetic energy of the atmospheric atoms and molecules. For a planet to retain an atom or molecule, it must be that

$$\frac{GM_p}{R_p} \geq k_B T \tag{2.8}$$

where
   M_p and R_p are the mass and radius of the body
   T is the absolute gas temperature

Obviously, the presence of sources for the atmospheric gases coming from the interior of the body or from its surface can maintain an atmosphere against the action-escape processes.

The terrestrial planets and major satellites had initial atmospheres formed with the planet. They were rapidly produced by the outgassing of volatiles at the high planet temperatures caused by the bombardment of planetesimals and large collisions. The primary composition of the primitive atmospheres was similar to their bulk composition, with volatiles outgassing proceeding, according to the body temperature, during the first ~800 My after their formation. The atmospheres transform in time and become of a secondary type evolving from the primordial state as they change their mass and chemical composition due to their interaction with the surface and external environment. The relative abundance of noble gas isotopes in Venus, Earth, and Mars, when compared with those found in meteorites, confirm that their actual atmospheres did not form from the primordial gas in the solar nebula. On the other hand, the giant planet atmospheres are so massive that they have suffered little evolution and they are considered as primary (they have retained their primordial composition of hydrogen in a 90% with an additional 10% of helium). We first discuss the sources for the atmospheric gases.

2.3.1 ELEMENTAL ABUNDANCES AND ISOTOPIC RATIOS

The measurement of the relative abundance of the elements and of the isotopic ratios of stable and radioactive atoms in solar system bodies provides a very important source for the knowledge of the initial formation and evolution of the planets and their atmospheres. In the first order, the differences in the chemical composition and relative abundances of the elements reflect primarily in the nature of their origin (e.g., comparing the volatile to refractory abundances). Important elements for this purpose are carbon (C), sulfur (S), nitrogen (N), and oxygen (O). Noble gases do not react with other elements (they are chemically inert) and should primarily reflect the original composition of the solar nebula. The original isotopic compositions of planetary systems depend on the stellar nucleosynthesis. Analysis of the isotopic ratios among these bodies and their comparison to the solar abundances provide information on the sources for the bodies and their atmospheres and on the processes and transformations they have undergone over time. For example, isotope fractionation occurs when isotopes become separated from each other due to chemical reactions or during physical processes as occurs when a gas moves in a porous medium. Unstable (radiogenic) isotopes, i.e., those that suffer a radioactive decay, serve to constrain the age of the bodies or are sensitive to physical parameters (e.g., temperature), serving as their measure. There are a number of "fingerprints" that are used for such tasks. The most used environment isotopes on Earth are deuterium, tritium, carbon-13, carbon-14, nitrogen-15, oxygen-18, silicon-29, and chlorine-36. Table 2.1 presents the isotopic ratios as measured in planetary atmospheres. Here, I summarize the most important cases:

1. The deuterium to hydrogen ratio (D/H). Deviations from the Big Bang original ratio (1.5×10^{-5}) in solar system bodies reflect the action of different

TABLE 2.1
Isotopic Ratios in Planetary Atmospheres

Ratio	Venus	Earth	Mars	Jupiter	Saturn	Titan	Uranus	Neptune
D/H	1.6×10^{-2}	1.5×10^{-4}	8×10^{-4}	2.2×10^{-5}	1.7×10^{-5}	7.5×10^{-5}	5.5×10^{-5}	6.5×10^{-5}
$^{12}C/^{13}C$	<84	89	90	92.6	90.9	84	—	78
$^{14}N/^{15}N$	—	272	170	435–450		56		
						168–211		
$^{16}O/^{18}O$	(500)	489	515			(346)		
$^{16}O/^{17}O$	—	2520	2655					
$^{36}Ar/^{38}Ar$	5.56	5.3	5.5	5.6				
$^{40}Ar/^{36}Ar$	1.1	296	3000					
$^{129}Xe/^{132}Xe$	—	0.97	2.5	0.98				
$^{22}Ne/^{20}Ne$	0.07	0.097	0.1	0.076				
Element	%	%	%					
^{78}Kr	—	0.61	0.64					
^{80}Kr	14.6	3.96	4.1					
^{82}Kr	47.9	20.2	20.5					
^{83}Kr	29.2	20.14	20.34					
^{84}Kr	100	100	100					
^{86}Kr	16.6	30.5	30.1					

Notes: These are averaged values with no r.m.s. from different sources presented. See references for further details.

D/H: (1) Protosolar ratio = $2–70 \times 10^{-5}$; (2) Comets = 3×10^{-4}.

$^{14}N/^{15}N$ (Solar) = 357; $^{12}C/^{13}C$ (Solar) = 84; $^{36}Ar/^{38}Ar$ (Solar) = 5.77.

References

For Earth and Mars: Owen, T., p. 824; Kieffer, H.H. et al., p. 30; Pepin, R.O. and Carr, M.H., in *Mars*, H.H. Kieffer, B.M. Jokosky, C.W. Snyder, and M.S. Matthews, eds., The Arizona University Press, Tucson, Arizona, 1992.

For Venus: Donahue, T. and Pollack, J., and von Zahn, U. et al., in *Venus*, D.M. Hunten, L. Colin, T.M. Donahue, and V.I. Moroz, eds., The Arizona University Press, Tucson, Arizona, 1983, pp. 422–429.

For the giant planets: Irwin, P.G.J., *Giant Planets of the Solar System: An Introduction*, Springer-Praxis, 2009, p. 35; Taylor, F.W. et al., in *Jupiter: The Planet, Satellites and Magnetosphere*, F. Bagenal, T. Dowling, and W. McKinnon, eds., Cambridge University Press, 2004, p. 68 and 71; Gautier, D. et al., in *Neptune*, D.P. Cruikshank, eds., The Arizona University Press, Tucson, Arizona, 1995, p. 588.

For Titan recent determinations: Vinatier, S. et al., *Icarus*, 191, 712, 2007; Jennings, D.E. et al., *Astrophys. J.*, 681, L109, 2008.

processes and serve to constrain the source for volatiles containing hydrogen in planetary atmospheres (e.g., water). Most of these processes tend to increase D/H (a good example is the preferential loss in atmospheres of lighter H with temperature, see below). As a reference, on Earth D/H = 1/6410 = 1.56×10^{-4}.

2. The concentrations and isotopic ratios of noble gases are good indications of the evolution of an atmosphere. The radiogenic helium (^4He) is of primordial origin, but it is also produced in stellar nucleosynthesis and by radioactive

decay of uranium and thorium. Argon has a stable isotope (argon-36) and radiogenic argon-40 is created only by the radioactive decay of potassium K-40 (with a half-life of 1.25 Gy). The ratio $^{40}Ar/^{36}Ar$ serves to constrain the volatile sources for planets and their atmospheres and is less affected by thermal escape due to its weight. The other useful nonradiogenic noble gas is neon (^{20}Ne) that is cosmically abundant and does not undergo thermal escape, and the more rare gases are krypton (^{84}Kr) and xenon (^{129}Xe and ^{132}Xe).

3. The measurement of the isotopic ratios of most common elements in solar system bodies is employed to study the solar nebula composition and origin of volatiles in different bodies. These are $^{15}N/^{14}N$, $^{13}C/^{12}C$, and $^{18}O/^{16}O$. The first two isotopic ratios have important astrophysical implications through the *CNO thermonuclear cycle* that occurs in the lifetime of main sequence stars. The oxygen isotopic ratio $^{18}O/^{16}O$ measured in the deep ice cores on Earth serves as a measure of the planet temperature. When the water ice forms, the evaporation rates of the oxygen isotopes are different depending on the temperature. O^{18} molecules are more difficult to evaporate (due to their weight) than O^{16}. Lower O^{18}/O^{16} ratios mean higher temperatures.

To characterize the isotopic ratio present in solid material, the *differential isotope composition* called the "δ-value" is usually employed

$$\delta_i = \left(\frac{R_s}{R_r} - 1 \right) \times 1000 \, (\text{in } ‰) \qquad (2.9)$$

where
  R_s is the isotopic ratio of the sample
  R_r is the ratio in a reference material

For example, for the oxygen case R_s and R_r, refer to $^{18}O/^{16}O$. For the elements S, C, N, and O, the average terrestrial abundance ratio of the heavy to the light isotope ranges from about 1/22 (S) to 1/500 (O).

Unstable radioactive decaying isotopes are used to date or determine the geological age. The isotope ^{14}C is used to determine the age of carbonaceous materials with an age limit that ranges between 58,000 and 62,000 years. Chlorine-36 is unstable with a half-life of 301,000 ± 4,000 years and is used to date soil and groundwater up to 1 My.

2.3.2 OUTGASSING PROCESSES

"Outgassing" (or degassing) is the primordial mechanism by which the atmospheres of the terrestrial-like planets form. A proto-atmosphere is formed during the accretion stage that originates these planets by releasing the gases as the body heats up. A second source of outgassing can result from the internal differentiation that forms the shell structure and core. The melting of the mantle injects through the surface vents and volcanic activity volatiles to the atmosphere. The temperature evolution of the outer layer of the body during this process can be determined to first order from its energy balance (1.126). The change of the total thermal energy (per unit mass) of the mantle layer must be equal to the energy deposited by the accreting mass (gravitational and thermal) minus the energy losses due to radiation:

$$\frac{d}{dt}\left(4\pi R_p^3 x_t C_p \rho T\right) = \frac{dM}{dt}\left[\frac{GM}{R_p} - C_p(T - T_0)\right] - 4\pi R_p^2 \sigma_B T^4 \qquad (2.10)$$

The term on the left-hand side of the equation is the total thermal energy of this layer (thickness $x_t R_p$ with $x_t < 1$) and dM/dt is the constant mass inflow with the term between brackets representing the gain in gravitational and thermal energy. T_0 is the temperature of the falling material and it is assumed that the planet mass and radius does not change significantly relative to their initial values during the process. The last term on the right represents the heat losses by radiation. For masses above ~$0.01 M_\oplus$, the impact of the accreted planetesimals is high enough to raise the temperatures to ~500–700 K, high enough to vaporize elements like H, C, N, S, Cl, and the rare gases. When the mass of the proto-planet is above ~$0.2 M_\oplus$, the proto-atmosphere can produce an increase of the surface temperatures up to ~1600 K by a blanketing effect, at which most materials vaporize.

These stages take up approximately the first hundred years; following these stages, the planet cools as discussed in Section 2.2 and the degassing ceases gradually. During cooling, only the low melting point compounds release volatiles. A primordial atmosphere forms from the remaining gases as some condense and settle on the surface at its temperature. The details depend on the complete thermal history of the planet that must be described by a more complex form of Equation 2.4, an issue beyond the purpose of this book.

2.3.3 CAPTURE PROCESSES

The capture of material (grains and gases) from the primordial nebula or from the debris of the planetary formation (solids that volatize when entering the planet) contributes significantly as a source for the atmosphere. As explained before, this is the main mechanism proposed to explain the giant planet atmospheres under the core accretion theory. Following this, on a long-term scale, the solar or stellar wind (the flow of fully ionized particles escaping from the star) sputters the surfaces of small mass planets and satellites with thin atmospheres, giving rise to a continuous source of neutrals for these bodies.

The mass captured by a planet embedded within the primordial nebula at temperature T_0 and pressure P_0 generates a proto-atmosphere whose pressure and temperature at the surface can be easily calculated with some simplified assumptions. For example, if during accretion the gas pressure follows the hydrostatic balance and the capture process is assumed to be adiabatic and the gas is ideal, the following equations hold

$$\frac{dP}{dr} = -\rho g = -\left(\frac{P}{R_g^* T}\right)\left(\frac{GM_p}{r^2}\right) \qquad (2.11)$$

$$C_p \, dT = V \, dP \qquad (2.12)$$

with V being the specific volume. Using the ideal gas equation $PV = R_g T$ and since $R_g^* = R_g/\mu$, substituting (2.6) into (2.5) gives

$$C_p \, dT = -\frac{\mu GM_p}{r^2} \, dr \qquad (2.13)$$

Integrating from T_0 (assumed at $r=\infty$), we get at the planet surface ($r=R_p$) a temperature T_s

$$T_s = T_0 + \frac{\mu GM_p}{C_p R_p} \qquad (2.14)$$

This equation further assumes that the accumulated atmosphere has a mass much lower than the planetary mass and does not change its molecular weight. It can therefore be applied to the case of the terrestrial planets and satellites, but not to the giant planets. The second term to the right of (2.14) is the adiabatic temperature increase produced by the compression of the gas. The surface pressure can be obtained using the adiabatic relationship

$$\left(\frac{P_s}{P_0}\right)^{\gamma-1} = \left(\frac{T_s}{T_0}\right)^{\gamma} \qquad (2.15)$$

with $\gamma = C_p/C_v$. Note also that this approach also neglects the opacity of the gas to the thermal radiation produced by increasing temperatures as accretion proceeds.

In the case of the giant planets, within the context of the core accretion theory, the captured mass is well above that of the proto-planet embryo (hundred masses compared with about $10M_\oplus$ for the core) and the hydrostatic equation (2.11) must be rewritten as

$$\frac{dP}{dr} = -\frac{\rho GM(r)}{r^2} \, dr \qquad (2.16)$$

For a body growing in thin shells (thickness dr), that preserves the spherical shape during accretion, the mass increment dM is given by

$$dM(r) = 4\pi r^2 \rho \, dr \qquad (2.17)$$

2.3.4 OVERVIEW OF THE ORIGIN OF PRIMORDIAL ATMOSPHERES FOR EACH BODY

In Table 2.1, we give the basic atmospheric composition for planets and satellites. The abundance of elements composing the atmospheres reflect that of the Sun and the solar nebula and their combination to form the simplest molecules according to the ambient temperature. But first, the planet must be able to retain the atmosphere. According to (2.8) the escape velocity v_{esc} for an atom or molecule (i.e., the velocity needed to abandon the body gravitational field) must be lower than its kinetic energy represented by the thermal velocity v_{th}

$$v_{esc} = \sqrt{\frac{2GM_p}{R_p}} < v_{th} = \sqrt{\frac{2k_B T}{m_a}} \tag{2.18}$$

with m_a being the mass of the particle. Since the thermal speed depends on the ratio T/m_a, the lighter and most abundant elements, hydrogen and helium, will have higher thermal velocities and only massive cold planets can retain them.

Figure 2.6 shows a scheme of the two main types of atmosphere in the solar system according to their origin. Venus, Earth, and Mars have secondary atmospheres derived from the evolution of primordial atmospheres formed by outgassing and/or by the accretion of asteroids and comets rich in volatiles. Subsequent evolution due to the interaction of the atmosphere with the surface and interior or to biochemical reactions (as in the case of the Earth) formed the secondary atmospheres. If these atmospheres were formed from the direct accretion of solar nebulae material, as those of the giant planets, they should contain significant quantities of noble gases

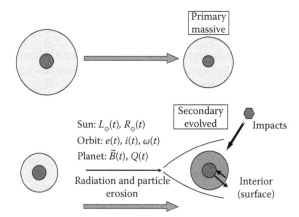

FIGURE 2.6 Scheme of the basic processes and parameters intervening in the evolution of planetary atmospheres. Primary atmospheres have suffered few modifications in their composition since the planet formation. This is the case of the giant and icy planets. Secondary atmospheres have suffered important changes in mass and chemical compositions due to a variety of processes: Sun luminosity and radius changes, astronomical cyclical changes (orbital eccentricity, planet axis tilt, and precession), and intrinsic planet changes (magnetic field, internal heat, and surface interactions, including life on Earth).

(Ne, Ar, Kr, and Xe) in solar proportions since these elements weigh enough to be retained by terrestrial planets. However, the observed proportions in these planets are a million times lower than in the Sun supporting the planetesimal-embryo scenario for the formation of terrestrial planets.

Carbon dioxide (CO_2) forms a massive atmosphere in Venus and dominates that of Mars although it is much less massive. On Earth, CO_2 is trapped beneath the oceans in the form of calcium carbonate. N_2 dominates the atmosphere of the Earth and is also abundant on Venus, both with origins in outgassing and volcanic processes. It is also the most abundant compound in the satellite Titan, in this last case due to the photochemical dissociation of ammonia. The large amount of O_2 on Earth is due to the existence of life. Mercury is small and too close to the Sun to retain a stable atmosphere. Its tenuous gaseous envelope is due to released compounds from the surface produced by the solar wind sputtering. The satellites Europa, Ganymedes, and Callisto have very tenuous oxygen atmospheres derived from the particle bombardment of their water ice surfaces by charged particles present in the Jovian environment. The satellite Io has a volcanic sulfur atmosphere. The satellite Triton and the dwarf planet Pluto form atmospheres from the sublimation of the surface ice (nitrogen and methane) as the temperature increases during the orbital cycle.

On the other hand, the solar system giant planets (Jupiter, Saturn, Uranus, and Neptune), as well as the extrasolar giants, have "primary" atmospheres composed basically of molecular hydrogen (about 85% in mass) with a smaller portion of helium (about 15%) and very small quantities of the other elements (C, N, O, etc.) present in reduced forms. However, C and N are enriched relative to the Sun (particularly carbon in Uranus and Neptune) and the noble gases (Ne, Ar, Kr, and Xe) are present with abundances within a factor of 10 to the solar composition, which complicates the detailed modeling of the formation of these atmospheres.

The dominance of one or the other type of molecule resulting from the combination of the main elements (C, H, N, O, S, etc.) depends on temperature and pressure. Under thermochemical equilibrium, the reactions governing the formation of these molecules are

$$CH_4 + H_2O \leftrightarrow CO + H_2$$

$$CH_4 \leftrightarrow C + 2H_2$$

$$2NH_3 \leftrightarrow N_2 + 3H_2$$

$$H_2S + 2H_2O \leftrightarrow SO_2 + 3H_2$$

with compounds on the left (ammonia, methane, water, and sulfur hydrogen) forming at low temperatures and high pressures, and CO and N_2, on the right, forming at high temperatures and low pressures. When CO and N_2 dominates, the reaction of CO with water produces carbon dioxide:

$$CO + H_2O \leftrightarrow CO_2 + H_2$$

From a broad point of view, this is what we actually find in the planets and satellites. The giant and icy cold planets have traces of C, O, N, and S in the form of CH_4, NH_3, H_2O, and H_2S. These atmospheres are *reduced* since they contain abundant quantities of hydrogen. On the contrary, the terrestrial planets and satellites are dominated by oxygen compounds CO_2, N_2, O_2, H_2O, and SO_2 and these atmospheres are *oxidizing*.

2.4 ATMOSPHERIC EVOLUTION PROCESSES

In Figure 2.6, we also summarize the main processes affecting the evolution of a planetary atmosphere. Atmospheres evolve on the one hand due to processes external to the planet (stellar wind and stellar variability, collisions with other bodies, and orbital changes). On the other hand, they evolve due to internal phenomena (heat flow escaping from the planet interior and surface interaction). In principle, all atmospheres are subjected to such processes, but the retention of gases in massive atmospheres that are also free of surface and interior interactions make them more stable on a long-term scale. However, extrasolar giant and ice-like planets that have migrated to orbits close to the star suffer a strong blowoff in their outer layers, giving rise to important losses. The timescales for these processes occur in a wide range from punctual phenomena like impact erosion, long-term continuous evolution scales of the order of Gy, or periodic phenomena as those related to the orbital cycles (of the order of kilo years [ky]). We divide the escape and evolution processes suffered by an atmosphere into five broad groups according to their external or internal origin.

2.4.1 EROSION AND ESCAPE PROCESSES

Atoms and molecules can escape from the top of planetary atmospheres through a variety of mechanisms. The erosion processes can be divided in three broad groups: (1) thermal or Jeans escape, (2) nonthermal or photochemical escape, and (3) hydrodynamics. The features that characterize them are, apart from the involved mechanism, the region of escape, the escape speed, and mass outflow. Some of these processes and related mechanisms are discussed in Chapter 6 in some detail.

The escape region of a planet or satellite is defined as the *exosphere* and its base is the *exobase*. It is defined as the altitude level at which the mean free path of a particle (the distance it can travel without colliding with another particle) equals the atmospheric-scale height:

$$\ell_{fp}(z_{ex}) = H_p \tag{2.19}$$

Here, $H_p = k_B T/(m_p g)$ is the scale height, the e-folding exponential density decay (m_p is the particle's mass). In bodies with thin atmospheres, the exobase is at the surface itself or very close to it and the atmosphere is in fact an exosphere since it is continuously subjected to the escape processes (as in Mercury and most satellites). In planets with massive atmospheres, the exobase is at an altitude z_{ex} from the surface or from an arbitrary level (usually 1 bar pressure) in the case of the giant planets (see Table 2.2). Below the exobase, the species are in thermodynamic equilibrium with collisions between particles driving them to a Maxwell distribution of velocities remaining trapped in the atmosphere. Above the exobase, atomic collisions are negligible and the atoms are moving on purely ballistic trajectories. If the particle has an upward motion in a direction away from the planet, it does not ideally collide with any other particle along this path and it has enough energy

TABLE 2.2
Escape Velocity and Exobase Data (Representative Values)

Planet	Basic Composition	Escape Velocity (km s^{-1})	Exobase z_{ex} (km)	Exobase Temperature (K)
Mercury[a]	O (0.52)	4.25	0	440
	Na (0.39)			
Venus	CO_2 (0.96)	10.36	200	275
	N_2 (0.035)			
Earth	N_2 (0.78)	11.18	500	1480
	O_2 (0.21)			
Mars	CO_2 (0.953)	5.02	250	300
	N_2 (0.027)			
Jupiter	H_2 (0.864)	59.54	2000	700
	He (0.136)			
Io (S)[a]	SO_2 (0.99)	2.6	200	1000
Europa (S)[a]	O_2 (0.99)	2.0	0	~100
Ganymede (S)[a]	O_2 (0.99)	2.86	0	~100
Callisto (S)[a]	CO_2 (0.99)	2.45	0	~100
Saturn	H_2 (0.85)	35.49	2000	420
	He (0.14)			
Titan (S)	N_2 (0.98)	2.6	1000	300
	CH_4 (0.016)			
Enceladus (S)	H_2O (0.91)	0.19	—	—
Uranus	H_2 (0.864)	21.26	2000	700
	He (0.136)			
Neptune	H_2 (0.85)	23.53	2000	700
	He (0.14)			
Triton (S)[a]	N_2 (0.99)	1.46	300	300
Pluto (dP)[a]	N_2 (0.99)	1.23	—	—

Sources: Chamberlain, J. and Hunten, D.M., *Theory of Planetary Atmospheres*, Academic Press, New York, 1987; Irwin, P., *Giant Planets of the Solar System: An Introduction*, Springer-Praxis, Chichester, U.K., 2009.

Notes: S is for satellite; dP is for dwarf planet.

[a] Thin atmospheres. For the giant and icy planets (J, S, U, and N), the exobase is taken relative to the radius of the planet at 1 bar pressure level. More complete data on composition can be found in Tables 5.1 and 6.9.

to abandon the body (Equations 2.3 and 2.12)—it will escape to the interplanetary medium. A net upward flow is then established and the atmosphere loses mass to the outer space.

2.4.1.1 Thermal or Jeans Escape

Thermal "evaporation" or Jeans escape occurs when $v_{es} > v_{th}$ (see Equation 2.18). The fraction of atoms that reach velocities in excess of the escape velocity can be calculated assuming they are in thermal equilibrium with a Maxwell velocity distribution $f(v)$ given by

$$f(v)\,dv = N_p \sqrt{\frac{2}{\pi}} \sqrt{\left(\frac{m_p}{k_B T}\right)^3} v^2 \left[\exp\left(-\frac{m_p v^2}{2k_B T}\right)\right] dv \qquad (2.20)$$

where
  v is the particle's velocity
  N_p is the number density (particles cm^{-3})

Although Gaussian in form, in practice, there are virtually no particles with speeds above v_{th}. In the framework of the Jeans approximation (that assumes hydrostatic equilibrium and no cooling of the atmosphere), the upward flux of particles can be obtained upon integration to this velocity distribution above the exobase, giving the *Jeans formula* for the particle's escape rate (atoms cm^{-2} s^{-1})

$$\Phi_{th} = N_{ex} v_{eff} = \frac{N_{ex} v_{th}}{2\sqrt{\pi}} (1 + \lambda_{esc}) \exp(-\lambda_{esc}) \qquad (2.21)$$

where
  N_{ex} is the exobase number density of particles
  v_{eff} is the "effusion" velocity
  λ_{esc} is the *escape parameter*

The ratio between the potential and kinetic energy of the particle is

$$\lambda_{esc} = \frac{GM_p m_p}{k_B T (R_p + z)} = \left(\frac{v_{esc}}{v_{th}}\right)^2 \qquad (2.22)$$

Since $\lambda_{esc} \sim m_p$ and $\Phi_{th} \sim \exp(-\lambda_{esc})$, only the lightest species will escape at appreciable rates through this process and will do so at a much faster rate than the heavier ones. The *mean loss time* is defined as

$$\tau_{esc} = \frac{H_p}{v_{eff}} = \frac{HN_p}{\Phi_{esc}} \qquad (2.23)$$

The net upward flux is limited by diffusion processes and cannot exceed the "limiting diffusion flux" as described later in Section 6.5.1.

2.4.1.2 Nonthermal or Photochemical Escape

The Jeans mechanism explains the escape of the lightest atoms. However, other nonthermal processes can dominate the escape flux of weighted atoms with speeds in excess of the escape velocity. In Chapter 6, we treat some of these processes with more detail. These are as follows.

2.4.1.2.1 Photochemical Escape

When a diatomic molecule splits (dissociates) due to UV radiation (*photodissociation*) or by impacting *photoelectrons*, it produces species with velocities above the escape velocity.

2.4.1.2.2 Charge Exchange

This mechanism involves a charge exchange between a very energetic "hot" proton or ion and a neutral atom. The new neutral atom that forms from the former ion could then gain sufficient energy to escape.

2.4.1.2.3 Solar Wind and Magnetospheric-Plasma-Driven "Sweeping"

The solar wind interacts directly with the atmosphere in bodies without a magnetic field, imparting to its atoms sufficient energy to escape. The pick up of atmospheric atoms by the solar wind is limited by the formation of a shock layer (in the ionosphere), and the escape flux due to this mechanism is proportional to the flux of solar wind protons. In nonmagnetized planets (such as Venus or Mars), the ion or proton is provided by the solar wind, and this loss mechanism becomes important for C and N on Mars and H on Venus. An upper limit to the total atmospheric mass loss rate due to solar wind sweeping can be obtained from the momentum transfer by the solar wind to an atmospheric shell with an area of $2\pi(R_p + z)H$

$$\frac{dM}{dt} \leq 2\pi(R_p + z)H\rho_{SW}v_{SW} \qquad (2.24)$$

where

H is the atmospheric-scale height at the sweeping level z above the surface
ρ_{SW} and v_{SW} are the solar wind density and speed on the planet, respectively

On magnetized planets (Earth and the giants), the magnetic field deflects the solar wind but charge-exchange reactions within the magnetic cavity can still provide energy to escape. Additionally, in nonmagnetized planets, the solar wind ions exchange their positive charges with neutral atmospheric atoms resulting in the capture of solar wind particles.

In the massive satellites of the giant planets that have tenuous atmospheres (Io, Europa, Ganymede, and Callisto in Jupiter), and in those with significant atmospheres (Titan in Saturn), the role of solar wind as a sputtering source is played by the high-energy magnetospheric particles (ions, electrons). Their low gravity puts the exobase at large distances from the surface, favoring the escape in particular on satellites close to the planet (such as Io and Europa).

2.4.1.2.4   Sputtering or Knock On

This process involves the collision of a fast atom or ion (for example, cosmic ray particles) with an atom in the atmosphere or surface that can gain sufficient energy to escape. The surface sputtering of crust atoms is in part of the origin of the tenuous and rarefied atmospheres of airless bodies, such as Mercury and the Moon. This mechanism also occurs in the satellites of the giant planets due to the collision of the high-energy magnetospheric particles with the surface, followed by a cascade of collisions that can give the atoms enough energy to escape (see the details in Section 6.7).

2.4.1.2.5   Ion-Neutral Reaction

In this process, a fast atom that can abandon the atmosphere and a molecular ion are created from a reaction between an atomic ion and a molecule.

2.4.1.2.6   Ion-Electric Field

Ions can be accelerated by the electric fields present in the outer atmosphere (for example, in the ionosphere). The ions themselves can reach the escape velocity and abandon the planet along magnetic field lines or collide with neutral particles imparting upon them the escape velocity.

2.4.1.3   Hydrodynamical or "Blowoff" Escape

This mechanism involves the mass outflow of the atmospheric gases at sonic speeds when lighter atoms escaping the atmosphere drag heavier atoms with them that cannot escape by a thermal process. The hydrodynamical loss occurs when $H_p > 0.5r_c$, with H_p being the scale height of the lightest gas and r_c being the distance from the center of the body to the critical level where the process begins. In general, $H_p < 0.5r_c$ so this mechanism requires the support of a large energy source. The upward flux of the heavier gas Φ_2 carried by the flux Φ_1 of the lighter gas is given by

$$\Phi_2 = \frac{N_2}{N_1}\left(\frac{m_c - m_2}{m_c - m_1}\right)\Phi_1 \tag{2.25}$$

where
  N_i is the number density
  m_i is the particle mass
  m_c is the crossover mass given by

$$m_c = m_1 + \frac{nk_BT}{b_cg}\Phi_1 \tag{2.26}$$

with $N = N_1 + N_2$ and b_c being the *binary collision parameter* that depends on the nature of the light and heavy gases involved and on temperature

$$b_c = A_D T^s \qquad (2.27)$$

See Section 6.5 for a definition and values of the coefficient A_D and exponent s. The blowoff mechanism has been proposed to have been an important process in the loss of the primitive atmospheres of Venus, Earth, and Mars during the first 100 My when planetesimal accretion supplied the energy source.

2.4.1.4 Geometrical "Blowoff" Escape

This mechanism is of particular interest in the case of "Hot Jupiter" extrasolar planets due to their proximity to the parent star. They suffer the effects of the intense radiation in particular at short wavelengths where a strong heating by X and UV radiation (XUV) produces a high hydrodynamic escape. The upper atmosphere will be continuously flowing to space being somewhat "inflated" by the thermal expansion, thus lacking a defined exobase. The escape rate (the total mass loss rate in g s^{-1}) is given as

$$\frac{dM}{dt} = 4\pi r^2 \rho v \qquad (2.28)$$

where

r is the distance to the center of the planet
ρ is the mean planet density
v is the vertical component of the atmospheric velocity

As a first approach, we can calculate the mass lost rate in (2.28) by considering only the XUV heating, without taking into account the tidal effect produced by the star-planet gravitational interaction. This gives

$$\frac{dM}{dt} = \varepsilon_m \frac{\pi R_{XUV}^2 F_{XUV}}{(GM_p/R_p)} \qquad (2.29)$$

where

F_{XUV} (in W m^{-2}) is the stellar flux averaged over the whole planet surface
ε_m is an efficiency factor that gives an account of the absorbed energy by the atmosphere
R_{XUV} is the altitude relative to the planet center where the stellar energy is deposited

However, with the distance between the planet and the star being small, the Lagrange point L_1 is close enough to the planet that its expanded atmosphere can fill the Roche lobe (see Sections 1.2.4 and 1.3.3 and Problem 1.5). If the exobase level (at temperatures ~10^4 K) reaches L_1, direct hydrodynamic escape will occur through this point. Then (2.29) is modified in the denominator (the potential energy) to take into account this effect and include the separation between the Roche lobe boundary and the planetary surface:

$$\frac{dM}{dt} - \frac{\pi R_{XUV}^2 F_{XUV}}{(GM_p/R_p)K_m} \tag{2.30}$$

Here, K_m is a geometric factor given as

$$K_m = 1 - \frac{3}{2r_{L_1}/R_p} \tag{2.31}$$

and r_{L_1} is the position of the Lagrange point L_1 given by

$$r_{L_1} \approx a\left[\left(\frac{M_p}{3M_*}\right)^{1/3} - \frac{1}{3}\left(\frac{M_p}{3M_*}\right)^{2/3}\right] \tag{2.32}$$

These equations are obtained after some manipulation of the expression for the gravitational potential of the star and the planet as they rotate about their common center of mass (see Problem 1.5), assuming that $a \gg r_{L_1} > R_p$ and $M_* \gg M_p$.

2.4.2 IMPACTS AND COLLISIONS

The collision of a large body with a planet or satellite with an atmosphere can impart enough energy to the gases so they can obtain the escape velocity. It is obviously assumed that the body is not disrupted (the size of the impacting body $R_c \ll R_p$). A measure of the effect of an impact is to compare its size with the atmospheric scale height H. If $R_c < H$, the kinetic energy of the impacting body that is converted into heat dissipates in a volume that is small relative to the atmospheric mass and the loss is small. However, if $R_c > H$, the heated volume of gas can be blown off. The atmospheric mass that is lost (M_c) depends on the impactor's kinetic energy (the impact velocity is $v_i = \sqrt{2GM_p/R_p}$) and can be simply estimated as the product of the impactor cross section πR_c^2 times the atmospheric mass per unit area (P/g) (the "plug" of the atmosphere), weighted by a factor that depends on the impactor's velocity. It is then given by

$$M_c = \pi R_c^2 \left(\frac{P}{g}\right)\left(\frac{v_i}{v_{esc}}\right)^2 \frac{1}{(1+v_i^2 L_{ev})} \tag{2.33}$$

where $L_{ev} \sim 6.2 \times 10^{-9}$ m^{-2} s^2 is a parameter that takes into account the additional gas produced by ablation and the heat of vaporization.

2.4.3 SURFACE PROCESSES

The surface of the terrestrial planets and satellites represents a lower boundary for the atmosphere, a place where mass exchange and chemical reactions occur. The surface through temperature, among others, is an important controlling agent for the atmospheric evolution along various timescales. The atmosphere and surface inter-action on short temporal scales (e.g., daily or yearly) compared to the age of the solar system will be presented in Chapter 5. Here we outline the basic surface processes affecting long-term evolution. As indicated previously, outgassing from the interior through surface vents and volcanoes is one of the major sources of atmospheric gases. Condensation and sublimation of volatile compounds from solid or liquid deposits at the surface to the atmosphere and adsorption and chemical weathering are the other important processes that change the atmospheric structure and compo-sition. The atmosphere tends to become stable in composition and mass as long as the atmosphere contains inert (nonreactive) gases against chemical and photochemi-cal actions and to condensation and sublimation.

2.4.3.1   Condensation

Condensation of atmospheric gases into the surface occurs when the partial pressure of the gas reaches its saturation vapor pressure. The condensation process will be discussed in Chapter 5 in relation to cloud formation, but, in advance, we can say that the key variable is the surface temperature since the saturation vapor pressure strongly depends on T. Examples of condensation and sublimation or evaporation from the surface are the cases of gases such as H_2O on Earth, CO_2 on Mars, CH_4 on Titan and Triton, and SO_2 in the case of Io (see Section 5.1.2). Changes in the surface albedo, increasing due to ice formation or decreasing due to sublimation, make the body more or less reflective to solar radiation, decreasing or increasing its temperatures.

2.4.3.2   Adsorption

Gases can be adsorbed onto a surface when fine-grained areas are present. Gases with a large polarizability are more easily adsorbed on sediments, a process that depends on the sediment's temperature and constitution and on the partial pressure of the gas. In the case of thin atmospheres, the UV solar radiation reaches the surface and induces photochemical transformation in the minerals that favors the adsorption of molecules.

2.4.3.3   Dissolution

Gases soluble in liquid surface deposits can be removed from the atmosphere even when their partial pressure is below the saturation vapor pressure over their pure condensed phases dissolving into the solvent. An important process in the Earth is the dissolution of CO_2 in water producing various salts.

2.4.3.4   Chemical Weathering

The chemical reactions between the atmosphere and the minerals brought to the surface from the interior by tectonic activity represent a sink and a source of modi-fication of the atmospheric composition.

2.4.3.5   Atmospheric Feedbacks

Changes in the temperature structure of an atmosphere can lead to increasing or decreasing radiation to space, tending to cool or warm the planet. This process is coupled to the planetary albedo with ice formation or sublimation, cloud-aerosol formation (spatial distribution and cloud properties), and increasing or decreasing degassing (e.g., water vapor).

2.4.3.6 Biological Processes

Life is known to exist in only one place in the solar system—the Earth. Biological processes at the surface drastically transformed the atmosphere of our planet during more than half of its lifetime. As will be discussed later, photosynthesis by green plants is the reason for the large amount of oxygen currently present in the Earth's atmosphere, which is essential for most living organisms, whereas other trace gases such as nitrous oxide and methane are produced by biological activity.

2.4.4 Solar (Stellar) Luminosity Variability

Once a planetary system is formed, the stellar evolution stages marked by long-term luminosity (radius and temperature) changes directly affect the evolution of planetary atmospheres. In general, most stars pass a large part of their lifetime in a stable stage called the *main sequence*, converting hydrogen to helium through thermonuclear fusion reactions. The details of the stellar evolution phases before, during, and after the main sequence depend basically on the star mass (M_*). During its initial formation stage, the star suffers a contraction period that comprises two phases: a first initial rapid collapse and a thermal period during which it radiates energy from the accumulated heat. The star also radiates large amounts of x-rays and ultraviolet radiation during these initial phases, simultaneously accompanied by an intense stellar wind of charged particles. In the case of the Sun, this period is estimated to have occurred during the first $\sim 20 \times 10^6$ years (20 My) and therefore corresponds to the phase when the planets were still in their formation stage, influencing the proto-atmospheres.

Once the star enters the main sequence, its lifetime becomes a function of its mass approximately as M_*^{-2} and its luminosity L_* depends on mass approximately as M_*^3 (for massive stars) and $M_*^{2.5}$ (for low mass stars). The Sun, currently in the main sequence stage, is expected to live for $\sim 10,000 \times 10^6$ years (10 Gy) (it has already passed 4.6 Gy on the main sequence), but a star with 15 times the solar mass will live only $\sim 10^6$ years and on the contrary one with 0.25 solar masses will spend $\sim 70,000 \times 10^6$ years (70 Gy), which is much longer than the current age of the Universe. The stellar luminosity shows variations in the main sequence stage due to a variety of processes as, for example, by the magnetic activity (e.g., "sunspots"). However, the luminosity changes are much more dramatic as the stellar evolution progresses after the main sequence stage. Again the changes are strongly dependent on mass. In a few words, low mass stars such as the Sun first pass through a *red giant star* stage (when their size and luminosity increase enormously) and then, ejecting the outer layers, end as a collapsed and dense *white dwarf* star that is in essence the stellar core at high temperatures and luminosity. On the other extreme, high mass stars end their life abruptly as a *supernovae* explosion during which the luminosity increases by a factor of 100×10^6 relative to the main sequence stage. Simultaneously, their core collapses and forms a high density *neutron star* (*pulsar* if magnetized and rapidly spinning) or for the most massive stars the collapse yields an even more compact object, a *black hole*. The planetary system as a whole can be disrupted during the development of such processes.

This synopsis shows that the long-term evolution of a planetary system and in particular that of a planet and its atmosphere are directly related to the compulsive phases of its parent star evolution. In the case of the solar system, the long-term luminosity variation of the Sun during its main sequence phase can be represented by the formula

$$L_\odot(t) \simeq L_\odot(t_0)\left[1+0.3\left(1-\frac{t}{t_0}\right)\right]^{-1} \qquad (2.34)$$

where $t_0 = 4.5\,\text{Gy}$ is the approximate current age of the Sun. More detailed models of the Sun's evolution predict that the luminosity was lower by about 30% of its present value during the first 2.5 Gy, affecting the atmospheric evolution of the terrestrial planets through their surface temperature. On Earth and Mars, the surface temperatures at this stage would have been below the water freezing point. However, geological evidence indicates that liquid water was present on both planets, so the temperatures were higher than predicted; this is called *the faint young Sun problem*. Most probably, higher concentrations of CO_2 and or clouds could have raised the temperatures by a greenhouse effect. However, there is a debate on the details and this represents an open problem and a challenge to explain the initial climate evolution of the atmospheres of these planets. The long-term evolution of the Sun's luminosity and of its radius is shown in Figure 2.7.

On much shorter timescales (from days to years), current changes in the solar luminosity related to its magnetic activity (e.g., sunspots and its 11 year cycle) affect the short wavelength radiation and energetic particle emissions (Figure 2.8). Variations in the UV irradiance can induce changes in the stratospheric chemistry. This variability can be coupled with intrinsic changes in the magnetic environment of the planet (its own magnetic field) and modulate the cosmic ray flux reaching a planetary atmosphere.

2.4.5 Orbital Cycles

As indicated in Chapter 1, the solar (or stellar) insolation reaching a planet or satellite depends primarily on the orbital eccentricity e and on the obliquity i (the tilt of the rotation axis relative to the orbital plane). This gives rise, on a short timescale (yearly), to the seasonal radiation and temperature variations as discussed in Section 1.4.3.

However, the mutual gravitational interaction between the planets leads to a long-term periodic or quasi-periodic variability of their orbital parameters. The geological evidence on Earth and Mars shows the long-term signature (timescales of 10^4–10^6 years) of cyclical temperature changes due to the variability in the incoming solar radiation produced by "orbital cycles." The basic ideas on the cyclical variability of the insolation due to orbital variability were introduced in the 1920s by

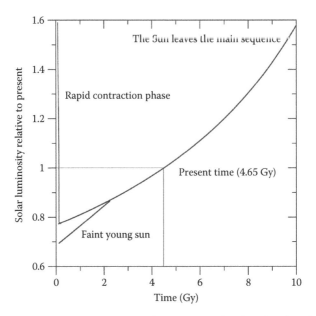

FIGURE 2.7   Calculation of the variation of the solar luminosity with time from the Sun's origin to the epoch when our star will abandon the main sequence. Two possible scenarios are represented during the first 2.3 Gy.

M. Milankovic. The quasi-periodic changes resulting from the gravitational perturbations on the Earth by the planets and the Moon affect the following orbital parameters on a long-term scale (Figure 2.9): eccentricity (e), obliquity of the spin axis (i), inclination ψ of the orbital plane relative to the ecliptic, and the longitude of the perihelion (ϑ) measured for a reference point (the vernal equinox), which depends on $e \sin \vartheta$. Mars suffers similar orbital cyclic long-term changes due to its proximity to the massive Jupiter (Figure 2.10) (see Section 2.5.3).

Including high-order terms, the quasi-periodic temporal variability of these parameters can be expressed as a series expansion:

$$e(t) = e_0 + \sum_j E_j \cos(f_{ej}t + \varphi_{ej}) \tag{2.35a}$$

$$i(t) = i_0 + \sum_j A_j \cos(f_{ij}t + \varphi_{ij}) \tag{2.35b}$$

$$e \sin \vartheta = \sum_j B_j \cos(f_{\vartheta j}t + \varphi_{\vartheta j}) \tag{2.35c}$$

where
  e_0 and i_0 are the mean values
  the terms E_j, A_j, and B_j are the amplitudes of the oscillation with related frequencies f_{ij} and phases φ_{ij}

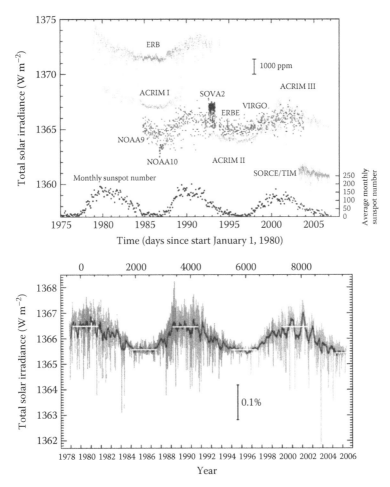

FIGURE 2.8 The solar constant in time. Sun's luminosity variability in recent times (from 1978 to 2006) as measured by satellites at the top of the Earth's atmosphere (TOA). (Upper graph from Mishckenko, M.I. et al., *Bull. Am. Meteol. Soc.*, 677, May 2007. With permission; lower graph combines these data from Foukal, P. et al., *Nature*, 443, 161, 2006. With permission.)

2.5 VARIABILITY OF THE EARTH'S ATMOSPHERE

The science that studies the evolution of the chemical characteristics of the atmosphere due to the chemical and biological processes as shown by geological records is called *biogeochemistry*. The study of the changes and evolution suffered by the Earth's atmosphere during its entire history is called *paleoclimatology*. Both areas of science have grown so rapidly during the last few years that by themselves they form ample disciplines. Here we will outline the basic processes that have affected the properties of our atmosphere, basically its chemical composition and surface temperature. The evidence for such variability resides in geological and fossil records, which give testimony to chemical and biological evolutions.

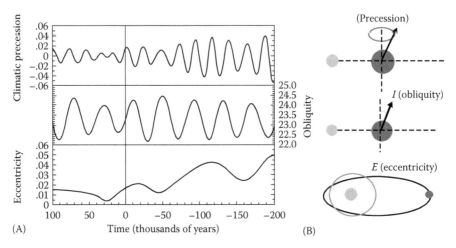

FIGURE 2.9 (A) Earth's orbital cycles. Orbital eccentricity (period 100 and 400 ky), Obliquity or rotation axis tilt (period 41 ky), Precession of the equinoxes (periods 19 and 23 ky); (B) Scheme of the periodic or quasi-periodic long-term scale variability of the main orbital parameters that affect the insolation of planetary atmospheres (in particular Earth and Mars).

The geological record goes back to about −3.8 Gy with the oldest rocks still present in our planet. The earlier signs of life go back to −3.5 Gy (methanogenic microbes, microbial mats, bacteria) with the oldest photosynthetic bacteria at −3 Gy (although a debate exists in the community on the confidence of these data). The earlier chemical evidence of complex cells dates to −2.7 Gy (cyanobacteria) and the presence of the first green algae (the eukaryotes) dates back to −2.3 Gy, which was followed at −2 Gy by the increasing presence of free oxygen in the atmosphere. By about −700 My, the Ediacara fauna developed (first skeletons) followed at −570 My by the great life outburst and at −6 to −2 My by the first hominids, and finally the last hundred millennia by the homo species (*Homo sapiens* appeared at about −150 ky).

Simultaneously, the plate tectonics of the crust and upper mantle produced a change in the volume and surface distribution (continents versus oceans) of the crust itself during the Earth's history, strongly influencing the atmosphere (see Section 2.5.3). A Precambrian supercontinent called Rodinia existed 750 My ago, which evolved toward Pangea. Another continent formed 550 My ago by different subcontinents. One of them, Gondwana, broke up 160 My ago in several sub-continents (Africa, South America, Antarctica, Australia, and India). At 65 My, Africa and Eurasia connected and the Atlantic Ocean opened. The plate tectonic motions continue and predictions indicate that in another 250 My from now the continents will once again join in another supercontinent.

There are a variety of techniques that allow for recovering the basic characteristics of the atmosphere back to about 800 ky and beyond. These are some of the proxy indicators of climate:

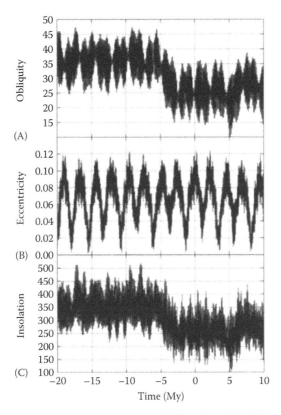

FIGURE 2.10   Calculations of the long-term variability of the (A) obliquity, (B) eccentricity, (C) insolation in Mars. (From Laskar, J. et al., *Icarus*, 170, 343, 2004. With permission.)

1. The isotopic ratios of oxygen O^{18}/O^{16} and deuterium to hydrogen (D/H) present in water molecules trapped, for example, in marine sediments. They represent a thermometer to retrieve the Earth's temperature as indicated in Section 2.1. The oxygen isotopic ratio is measured in $CaCO_3$ from marine organisms that incorporate ocean water. There are effects added to this analysis that introduce uncertainties, but it is a good long-term chronological technique.

2. Ice cores from Greenland and Antarctica contain pollen whose abundance is indicative of plant growth, and tiny air bubbles that serve to retrieve its composition. Other important ice comes from the Vostok core.

3. Tree-ring studies (*dendroclimatology*) allow for the retrieval of rainfall and temperatures going back to the last centuries and up to a few millennia.

4. The study of organic content and isotopic ratios in sediment layers in the bottom of lakes and oceans and the properties of sedimentary rocks.

5. *Paleomagnetism* in rocks (the study of the past variations in the orientations of Earth's magnetic field in minerals: magnetite, hematite, and in grains).

On the other hand, it must be noted that the rotation period of the Earth has changed in time, with important consequences for climate. During the Devonian period (400 My ago) there is geological evidence that the day was shorter by about 2 h and the year was longer by about 35 days. A linear extrapolation shows that the length of a day 1 Gy ago would be 19 h and that the Moon was at this time very close to the Earth (15,000–20,000 km separation, 1.2 Gy ago). Rotation would have influenced the intensity of the Earth's magnetic field (the magnetic moment is estimated to double its intensity for Earth's rotation reduced to half its current period). This would in turn have changed the environment protection of the Earth's atmosphere through the shield that the magnetic field provides.

2.5.1 Loss of CO_2

As indicated in Section 2.2.3, the primitive atmosphere of the Earth resembled in composition that of Mars and Venus, being mainly composed of CO_2 and N_2 with some H_2O, CH_4, and possibly NH_3 from the outgassing process (volcanic outgassing released ~80% of water vapor and 10% of CO_2). The H_2O, NH_3, and CH_4 became partially dissociated from UV radiation leaving C and N, whereas H was lost to space. C and O and on the other hand N combined with O to form an atmosphere of CO_2 and N_2. The presence of CO_2 can explain the "faint young Sun paradox," since increasing temperatures due to a carbon dioxide greenhouse effect could have counteracted the ~25%–30% reduction in solar luminosity corresponding to the early stages of the Sun as a star, but it is not evident that the atmosphere contained enough CO_2. Figure 2.11 shows the theoretical calculation and some empirical constraints on the atmospheric

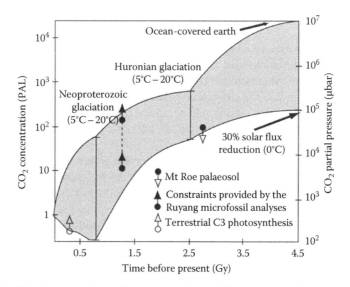

FIGURE 2.11   Atmospheric levels and evolution of CO_2 abundance in the Earth's atmosphere from 4.5 Gy ago (Archaean and Proterozoic) to present according to various indexes. (From Kaufman, A.J. and Xiao, S., *Nature*, 425, 279, 2003. With permission.)

CO_2 evolution. The initial mass of carbon dioxide (prior to the emergence of the continents) could have produced a surface pressure ~1–10 bar and surface temperatures of 100°C during several million years when water was not liquid but was a vapor forming a water-steam atmosphere mixed with the other gases. When the temperature dropped below 100°C, the water condensed and oceans formed.

CO_2 was removed from the atmosphere by *silicate weathering* (also called the *Urey reaction*) and partially dissolved on the oceans. This avoided an atmospheric "runaway effect" that would result from the CO_2 accumulation in the atmosphere (together with that of water vapor) that would elevate the temperatures and produce a further increase of these gases, leading to a Venus-like situation (see Section 2.5.2). CO_2 supplied by volcanic and metamorphic emanations is removed as sediment in the form of $CaCO_3$ (limestone) and organic matter (as CH_2O). This is a chemical reaction between the CO_2 dissolved in water with silicate minerals containing Ca and Mg. The atmospheric CO_2 forms carbonic acid rain (H_2CO_3), which dissolves in silicate rock by weathering (conversion to soil) since carbon dioxide, unlike other atmospheric gases, dissolves and reacts with water. For example, CO_2 reacting with calcium bearing silicates produces calcium carbonate and quartz (weathering on land)

$$CaSiO_3 + 2CO_2 + H_2O \rightarrow CaCO_3 + SiO_2 + H_2O$$

The carbonate sediments deposited in the ocean floor enter the mantle by tectonic activity where the high temperatures and pressures drive the above equation to the left (upper mantle metamorphosis) with the CO_2 being incorporated again in the atmosphere by volcanic activity. The process is referred to as "silicate weathering" and is regulated by the surface temperatures, which in turn depend on the greenhouse effect through the CO_2 abundance. The silicate weathering rate is faster in a hot and wet ambient and slower where cold and dry. Presently, most CO_2 is in the form of carbonate rocks and organic matter sediments. Figure 2.12 shows the measured evolution of CO_2 and surface temperatures in the last 400 ky, and in Figure 2.13 we show them with more detail for the last 20 ky, including the most recent industrial increasing period. Removing CO_2 favored the formation of an N_2 atmosphere.

2.5.2 THE RISE OF OXYGEN

However, about 2.4 Gy ago the transition started to the present oxidizing atmosphere, representing a profound change in the Earth's atmosphere and environment (oceans, crust, and life). Prior to the origin of life, the atmospheric molecular oxygen abundance was very low, about 10^{-12} present atmospheric level (PAL) at the surface (partial pressure of 0.21 bar). The *Great Oxidation Event* started 2.4 Gy ago when values at least 1%–10% of the present value were attained. This change was induced by the photosynthesis mechanism of blue-green algae and subsequent biological activity. This occurred ~300 My after cyanobacteria started oxygen production. Sugars are produced by the photosynthetic conversion of solar radiation ($h\nu$, h is the Planck constant and ν is the frequency of sunlight radiation), and oxygen escaped as a by-product into the atmosphere:

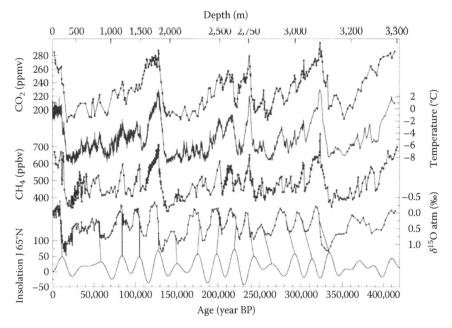

FIGURE 2.12 Evolution of CO_2 (a), surface temperate (b), CH_4 (c), [18]O index (d), and insolation (e) in the Earth during last 400 ky from measurements performed at Lake Vostok. (From Petit, J.R. et al., *Nature*, 399, 429, 1999. With permission.)

$$nCO_2 + nH_2O + hv \rightarrow (CH_2O)_n + nO_2$$

The injection of O_2 in the atmosphere depends on the burial of organic carbon in sediments. The released oxygen reacted immediately with iron and sulfide dissolved in the oceans followed by the deposition of sediments into "red beds." This is the best evidence of O_2 rise: the presence of "red beds" and "banded iron" formations at 1.7 Gy ago. Gradually, oxygen production became quicker than geological processes giving rise to its atmospheric accumulation. At the same time, ozone (O_3) formed by oxygen photolysis by ultraviolet radiation and catalysis started to accumulate, and a protecting ozone layer formed:

$$O_2 + hv(\lambda < 242 \, nm) \rightarrow O + O$$

$$O_2 + O + M \rightarrow O_3 + M$$

The O_2 rise was accompanied by a gradual decrease of CO_2 by biological processes, trapped as carbonates by the formation and accumulation of seashells in marine sediments. At about one-tenth the actual oxygen content, a biological screening was reached and it is suggested that this was the reason for the Cambrian life animal explosion 500 My ago and the emergence of land life 350–390 My ago. Figure 2.14 shows the events accompanying the growing rate of atmospheric oxygen.

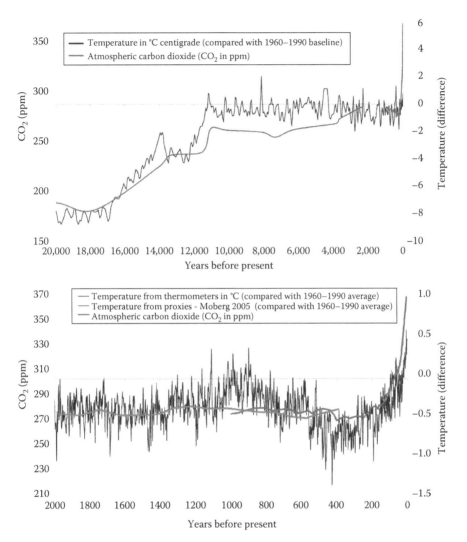

FIGURE 2.13   Evolution of CO_2 and surface temperate in the Earth during last 20 ky. (Upper figure: Data from http://www.ngdc.noaa.gov/paleo/icecore/antarctica/vostok/vostok.html, http://www.ngdc.noaa.gov/paleo/icecore/antarctica/law/law.html, http://cdiac.esd.ornl.gov/trends/co2/sio-mlo.htm; Lower figure: Adapted from Moberg, A. et al., *Nature*, 433, 613, 2005.)

2.5.3 SNOWBALL EARTH

"Snowball Earth" refers to the extreme cold climate periods that are supposed to have occurred in the Proterozoic eon (2.2 Gy ago) and Precambrian epoch (850–630 My ago). In these periods, the coldest global environment occurred on Earth with glacial ice covering most of the planet that reached a global mean temperature of −50°C and equatorial averaged temperatures of −20°C. The evidence points to a reduction in the content of "greenhouse" gases in the atmosphere, principally CO_2 and CH_4 in

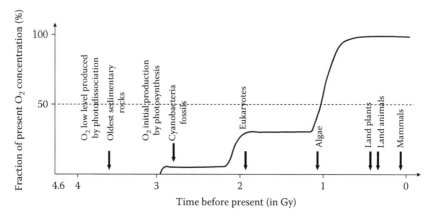

FIGURE 2.14   Evolution of the oxygen levels during the Earth's history and related biological events. (Adapted from Xiong, J. and Bauer, C.E., *Annu. Rev. Plant Biol.*, 53, 503, 2002.)

its origin. The reduction of CO_2 abundance was due to the alteration of the silicate weathering cycle previously described that was produced by the combined effects of a unique distribution of continents in the tropics, the break-up of Rodinia (the only supercontinent that existed at this time) that began ~830 My and continued for nearly 200 My, and the massive eruption of basalt lava at 723 My in Arctic Canada (which was close to the equator at this time). In addition, methane gas, which is ~30 times more powerful as a greenhouse gas than CO_2, was removed by oxidation when the O_2 levels elevated in the primitive Earth 2.2 Gy ago. The high albedo of fresh snow (about 0.9) deposited at the surface as frozen water and reflected solar radiation efficiently back to space, further decreasing temperatures. In addition, the decrease in the ocean water content and its heat transport and redistribution across the Earth produced enhanced temperature fluctuations of the day and yearly (seasonal) cycles.

The fingerprints of these glacial periods are (a) the sediments deposited by glacial flows in tills, moraines, eskers, etc.; (b) the paleomagnetic records that denote variability in the Earth's magnetic field; and (c) and the anomalous $^{12}C/^{13}C$ isotopic ratios. The Precambrian snowball period ended very rapidly as water vapor and CO_2 was incorporated into the atmosphere increasing the greenhouse effect, and perhaps producing the conditions that favored the emergence of the Cambrian explosion and Ediacara fauna at about 550 My ago.

2.5.4   CATASTROPHIC IMPACTS

During the first 1 Gy of Earth's history, impacts from the debris of the solar system formation were common, as indicated by the crater record in the Moon and other planets and satellites. This bombardment participated in the formation and erosion mechanism of the Earth's primitive atmosphere. Here we concentrate on the impact effects on the atmosphere during the last 600 My. The extinctions produced major environmental and climatic changes. There is evidence from the fossil records that massive biological extinctions have occurred during the Earth's history since the Cambrian period (570 My ago) when fossils are easily retrieved. These extinctions have affected a large variety of fauna (dinosaurs, marine and flying reptiles, etc.) and flora (plants and most species of plankton, etc.). Two major extinctions assigned to impact processes are well documented: the Permian–Triassic impact (−251.4 My, known as PT) and the Cretaceous and Tertiary boundary period impact (−65 My, known as KT). The PT extinction killed about 96% of the marine species and 70% of the vertebrates. There is some debate about the impact origin for the PT extinction. However, there is a general consensus that this was the reason for the KT extinction that abruptly killed the dinosaurs. The main reason is the presence of an iridium excess (a material abundant in meteorites and presumably in their parent asteroids) in the layer that corresponds to the geological chronology with this period.

Calculations indicate that the −65.5 My impact was produced by a 10 km asteroid that struck the Earth at 25 km s^{-1}, liberating an energy of ~4 × 10^{23} J equivalent to 10^8 Mt of TNT (see Section 2.3.2). The large Chicxulub crater found in the submarine area in the Yucatan peninsula in Mexico is probably the footprint of the impact. The wide spread of iridium around the world suggests that the most significant environmental perturbations were the direct and indirect result of ejected debris that rained through the atmosphere. Material that evaporated from the land and oceans following the impact was carried in a vapor-rich plume that expanded on a ballistic trajectory around the whole Earth as it fell back into the stratosphere, being dispersed around the world by the atmospheric circulation in a year or so. There are various proposals for the subsequent extinction mechanism. The most favored is the Sun's shadowing due to the increased optical depth by the injected atmospheric submicron aerosols, including aerosol-producing SO_2 and SO_3 (the enhanced stratospheric S could have been of the order of 10^5–10^6 times relative to present abundances). This in turn produced the cessation of photosynthesis during the darkness period (plankton and plants), breaking the feeding channel for animals. Another consequence of the impact was a strong temperature change produced by an increasing greenhouse effect from water vapor injection by evaporation from oceans and of CO_2 by combustion and fires. Additionally, acid rain due to chemical products resulting from the impact fireball, in particular those derived from nitrogen oxides (N_2O_5 forming nitric acid $2HNO_3$) and sulfur dioxide (SO_2 forming H_2SO_4), has been proposed as another possible killer mechanism.

2.5.5 Ice Ages (Glaciations)

The ice ages were the largest and most intense cyclic changes in the Earth's climate. During these periods, the annual average temperature of the Earth descended by a few degrees as revealed by a variety of geological records. There have been about 10 such ice ages during the last 730 ky (Figure 2.12). The last one occurred 18 ky ago producing a descent in the temperature of 5°C. According to the *Milankovitch theory*, the glaciations occurred as a response of the climate system to Earth's orbital cycles (see Figure 2.9). The main cycles involved are as follows:

1. The eccentricity change $e(t)$ affects to a first order the temperature by an amount ~eT_{eq} (see 1.102) from aphelion to perihelion (2.35a). The present eccentricity is 0.017 and over the past 1 My varied within the range of

0.001–0.054. The dominant term in the series expansion in (2.35a) has a period of 413 ky but the other terms produce on average an oscillation in e with a period of 100 ky. The first cycle was missed in the climatic data and the second cycle produces small variations unable to explain the climatic changes detected within the 100 ky period; therefore, its origin remains an unsolved problem.

2. More important is the variation in obliquity $i(t)$ (the rotation spin-axis tilt relative to the orbital plane), which determines the way solar radiation is meridionally distributed during the year (2.35b). This is the main mechanism originating the Earth's glaciation periods. For the Earth, the change in obliquity is on the order of $1°$ around a mean value of $23°26'$. The gravitational action of the Moon is complex and plays an important role in keeping this low variability. The spectrum of obliquity is dominated by a period of 41 ky.

3. The precession of equinoxes produced by the spin-orbit coupling between the Moon and Sun on Earth's oblate spheroid has a period of 26 ky and makes the hemispheres periodically tilt away and toward the Sun at aphelion with a small modulation in the obliquity of ~$0.9°$ (2.35c). The variation of $e \sin \omega$ has two peaks with periods of 19 and 23 ky (on the average this gives a single peak with a period of 22 ky).

Combining all these cycles, the averaged summer insolation in the northern hemisphere (top of the atmosphere) varied during the last 500 ky from ~450 to 550 W m^{-2}. Climate models are constructed on the basis of this energetic variability.

2.5.6 WARM PERIODS

During the Cretaceous period (65–100 My ago) the average global temperatures reached high values most probably due to a favorable configuration of the continents that allowed for a large oceanic circulation and temperature mixing that impeded the formation of a large scale ice sheet. Among the registered notable atmospheric warm period events are the *Paleocene–Eocene thermal maximum* (PETM at −55 My), which could have been related to the rapid increases in the CO_2 concentration, the middle Pliocene (−3 My), the Late Paleocene (58 My), the Cretaceous (130–65 My), and the Early Jurassic (180 My).

2.5.7 LAST MILLENNIA CLIMATE AND MODERN TIMES: "CLIMATE CHANGE"

The last millennia started at year 1000 with the *Medieval Warm Period* probably related to a change in the North Atlantic oscillation (Figure 2.13). The *Maunder Minimum* in the sunspot number with the disappearance of the 11 year cycle is apparently connected to the coldest period of the *Little Ice Age*, which began in the early seventeenth century with the rapid expansion of the mountain glaciers in Europe (temperatures were on average lower than the 1961–1990 mean by about 0.6°C).

Around 1780, the increase in the atmospheric concentration of greenhouse gases due to industrial human activities took place. Detailed information on temperature data exists since 1860 when the thermometer-based records began. Vertical temperature measurements by means of balloon-borne radiosounders began with global coverage in the 1950s and radiometric temperature measurements of the troposphere by means of satellites have been obtained since 1979. According to these measurements, the global mean surface temperature of the Earth rose by about 0.6°C between 1900 and 2000 (Figure 2.13).

2.5.7.1 Global Warming

The analysis of the satellite data indicates a current temperature trend increase in the troposphere in the range between +0.16°C per decade and +0.18°C per decade (according to the International Panel on Climate Change [IPCC] although an ampler limit between +0.14°C per decade and +0.26°C per decade is also defended by others). An increasing greenhouse effect due to the increase of CO_2 in the atmosphere at a rate of about 3.8% per decade is the main reason. Different types of models predict than on the average doubling the CO_2 levels will produce a global temperature increase in the range of 2°C–4.5°C, although a considerable debate exists on the exact value. Other greenhouse gases contribute significantly to the enhancement of the natural greenhouse effect: N_2O, CH_4, and O_3 and the halomethanes CCl_4, CF_2Cl_2, and $CFCl_3$. The presence of aerosols in suspension affects the atmosphere in the opposite direction, i.e., cooling it due to the backscattering of the solar radiation. All these effects will be quantified in Chapters 4 and 5.

2.5.7.2 Pollution

Atmospheric chemical pollution occurs globally but in particular in largely populated urban areas. In addition to CO_2, the principal primary pollutants are hydrocarbons, aldehydes, nitrogen oxides, and carbon monoxide. The major oxidant agent called *smog* (from smoke and fog) is ozone (O_3) that forms from the photochemical dissociation of NO_2 followed by a reaction with O_2. Smog in urban areas results from emissions due to a mixture of sources (transportation, residential, and industrial). Besides pollutants involved in oxidant atmospheres like organics, nitrogen oxides, and organic compounds, other common pollutants found in polluted areas are particulate matter and sulfur dioxide.

2.5.7.3 Acid Rain

Acid rain affects industrial regions and involves the precipitation of water containing sulfuric and nitric acids (H_2SO_4 and HNO_3) with the dry deposition of sulfates also playing an important role.

2.5.7.4 Ozone Hole

The name ozone hole refers to the seasonal decrease in the total column of stratospheric ozone (O_3) over the Earth's polar region (altitudes ~25 km). It has been observed since around 1980. The process starts with the photodissociation by UV radiation of chlorofluorocarbons (CFC), also known as freons, and of bromofluorocarbon compounds formed by human activities at the surface and transported to the polar stratosphere. As a result, chlorine and bromide atoms are produced that destroy the ozone layer through a catalytic reaction. The series of reactions can be summarized as follows for the chlorine case:

$$CFCl_3 + h\nu \rightarrow CFCl_2 + Cl$$

$$Cl + O_3 \rightarrow ClO + O_2$$

$$ClO + O \rightarrow Cl + O_2$$

The lifetime of CFC molecules is long and their destroying effect is also large. On average, a CFC molecule takes around 15 years to be transported from the surface to the upper atmosphere where it can reside for about a century, destroying up to 10^5 molecules of O_3 during this time. The ozone depletion is measured by the reduction in the column ozone abundance using the "Dobson unit" (DU). A total of 1 DU corresponds to a thickness of 0.01 mm if all the ozone above a place was compressed to a pressure of 1 bar at a temperature of 0°C. Typical values of the ozone column range from 250 to 600 DU, depending on the place and the season. Reductions up to 70% in the ozone column in the southern pole over Antarctica during spring were first reported in 1985 and through the 1990s the total column ozone in September and October decreased by 40%–50%, with values of the total ozone column well below 200 DU over Antarctica. The chemical destruction is favored by the presence of polar stratospheric clouds (PSC) formed by droplets of water-ice with nitric acid dissolved in them during the cold winter time (see Chapter 5). Ozone destruction is favored by the development of westerly winds around the polar continent in the middle to lower stratosphere that form a "polar vortex" (Chapter 9). The vortex confines the stratospheric air masses containing the contaminants that destroy the ozone.

2.6   OBSERVED ATMOSPHERIC CHANGES IN OTHER PLANETS

2.6.1   Venus

The high temperatures in the lower atmosphere and surface of Venus are due to the intense *runaway greenhouse* effect produced by CO_2. The lack of CO_2 weathering in Venus is the reason for this effect. Although water is currently by a factor 10^5 less abundant on Venus than in the Earth, the D/H ratio is ~150 times larger on Venus, indicating that the planet was wetter in the past but that water has been lost since then. The standard explanation for the water loss on Venus is the runaway greenhouse effect, while vapor evaporation in the atmosphere increased due to Venus' proximity to the Sun. The surface temperature increased due to an enhanced greenhouse effect implying a positive feedback between temperature and opacity.

On the other hand, the crater record in the surface of Venus is low and dating of the current crust indicates it has an age of ~750 My. A global resurface event should have occurred in the past between ~350 and 1000 My producing a major change in the atmosphere's structure.

2.6.2 MARS

Mars' rich geological features (craters, topography, volcanoes, basins, valleys, etc.) represent the fingerprints of its evolution as a planet and have been used to divide the timescales in Mars into three major periods (called *eons*): (1) the Noachian eon (from 4.6 to 3.5 Gy ago), (2) the Hesperian eon (from 3.5 to 2.5 Gy ago), and (3) the Amazonian eon (from 2.5 Gy ago to the present Gy).

The geological records obtained from orbiting spacecraft and landers on Mars show that the planet atmosphere was in the past warm and dense enough to have allowed water to be liquid and flow on the surface. The numerous channels observed in old cratered terrains are suggestive of the outflows produced by running water. The mineralogical analysis performed from orbit and in situ by the landers Spirit and Opportunity point in this way to the existence of a dense atmosphere during at least the Noachian eon. In situ surface studies by the rover Opportunity have revealed the presence of spherules (called "blueberries") of hematite (Fe_2O_3), a mineral that requires water to form. The analysis of a group of meteorites coming from Mars called SNC from the initials of Shergotty, Nakhla, and Chassigny, the places where the first Mars' stones were found, reinforces this point. The evidence points toward an early warm Mars, perhaps like the Earth with oceans or lakes (in the Northern Hemisphere), although there is a debate on the spread and quantification of the magnitude of this effect. The geological record remains confusing and contradictory in some places that apparently had not been exposed to these warm, wet conditions.

During this period, the atmosphere could have contained 1–2 bar (or more) of CO_2 and a lesser quantity of H_2O (released from the outgassing and volcanic activity) as indicated by the measurements of a high atmospheric D/H ratio. The estimated amount of water released by volcanism during the first 3.8 Gy was equivalent to a layer of 50 m depth covering the surface of the planet (a layer of water 1 m thick is equivalent to a water volume of 0.144×10^6 km^3). The resulting greenhouse effect from the release of both gases could have elevated the temperatures to 300 K. However, the condensation of CO_2 would produce a reduction in the vertical temperature gradient, there by lowering the greenhouse effect. In addition, CO_2 is a good Rayleigh scatterer (2.5 times better than air), producing an increase in the albedo that outweighs the greenhouse effect. An alternative hypothesis that has been proposed for the warming of Mars in the past is that the planet was fully covered by clouds of CO_2 ice crystals 10–50 μm in size, comparable to thermal-IR wavelengths, so outgoing thermal-IR radiation from the surface was backscattered more effectively than the incoming (visible/near-IR) solar radiation, producing the surface warming. However, this hypothesis has its own problems (the need for full cloud coverage; low or thick clouds can also cool as observed on Earth, the localized heating they create can destroy them). The recent detection of methane CH_4 suggests that it may also have helped to warm the planet through the greenhouse effect.

The isotopic ratios of $^{13}C/^{12}C$, $^{18}O/^{16}O$, $^{38}Ar/^{36}Ar$, and $^{15}N/^{14}N$ have higher values than on Earth, which has been interpreted to indicate that 50%–90% of the initial CO_2, N_2, and Argon were lost over the last 3.5 Gy. The atmosphere lost was probably produced by impact erosion from asteroid fragments, by hydrodynamic and nonthermal escape processes, and by the surface adsorption and weathering of the CO_2 to produce carbonaceous minerals. But there is a problem with these minerals still not found in the required abundance. The lack of tectonic activity does not allow the recycling of CO_2 as on Earth, and only a small fraction of this gas was retained in the atmosphere, part of it condensing in winter time, and with primitive H_2O mostly lost or perhaps accumulated in buried reservoirs. There have also been claims of recent ground ice or ground water activity on Mars (<10 My) from the observations of gullies in some crater walls (including alcoves and aprons), polygonal ground structures, rapid changes in bright surface features, and neutron spectrometric data (from the Mars Odyssey spacecraft).

Mars' lack of large satellites and its closeness to the massive Jupiter, makes it to support large chaotic changes in its obliquity (from ~15° to 45°) and orbital eccentricity giving rise to important climatic response variations with ice ages occurring with a periodicity ~10^4–10^6 years according to numerical calculations (Figure 2.10). The large volcanoes that are concentrated in the Tharsis area, such as Olympus Mons with a base of ~550 km in width and a height of ~27 km, are today extinct but when active ~3.5–2.5 Gy ago, should have ejected large quantities of aerosols and gases that are enough to produce important climatic variations.

2.6.3 Solar System Giants and Extrasolar Planets

The giant planets gradually cooled from their hot formation stage as presented in Section 2.2. The cooling of the outer atmospheres kept their interior at sufficiently low temperatures so that chemical constituents became immiscible and started to separate. One interesting process that could occur is the immiscibility of helium into metallic hydrogen (abundant in the interiors of Jupiter and Saturn) when the temperature drops to ~10^4 K. This process could in fact have occurred in Saturn's metallic hydrogen interior (~2 Mbar, T ~ 8000 K; see Section 1.5), colder than Jupiter, with two effects: the depletion of helium in the atmosphere and the gravitational energy released by He separation (a source for Saturn's intrinsic luminosity).

Transient changes can occur in the atmospheres of the outer planets of the solar system due to the impacts of large bodies attracted from the high gravity intensity of these bodies (see Section 2.4.3). In recent times (July 1994), we assisted for the first time in the direct impact of the fragmented comet "Shoemaker-Levy 9" (SL9) with Jupiter. The effects of the impact were impressive producing planetary-scale changes in the Jovian atmosphere that are described in detail in Section 9.8. A similar situation, although on a lower scale, occurred on July 2009 with the impact of a new comet on the planet suggesting these are relatively frequent events on Jupiter.

Due to a bias in the detection methods, most of the extrasolar planets so far discovered pertain to the so-called Hot Jupiter, planets highly irradiated due to their proximity to the star (see Table 1.8). Model calculations indicate that they cool at a slower rate than the cold giants of the solar system due to the presence of an extended radiative zone extending up to pressures ~1 kbar. Some of these planets are "inflated" showing a larger radius than expected, and different mechanisms have been invoked to explain this anomaly (energy excess transported by dynamics, tidal effects, etc.). Being so close to the star, these planets should suffer important mass loss due to thermal and nonthermal processes (blowoff mechanism) including the tidal deformation of the planet due to the star proximity making possible the escape from the Roche lobe through the Lagrange point L_1 as discussed in Section 2.4.1.

2.6.4 TITAN

Titan formed with Saturn in the outer parts of the solar system where the temperatures are low and icy materials such as NH_3 and CH_4 are abundant as *clathrate hydrates*. Measurements by the Huygens probe of the atmospheric composition and isotopic ratio of primordial noble gases indicate that nitrogen (the main component) was captured as NH_3 and in other non-N_2-bearing compounds. Subsequent photolysis in a hot proto-atmosphere generated by the accreting Titan or possibly the impact-driven chemistry of NH_3 led to the actual nitrogen atmosphere. Note that NH_3 is easily photodissociated into N_2 and H_2 by sunlight. The N_2 is retained by Titan's gravity but H_2 escapes. Thus, over time, Titan has built up a N_2 atmosphere like the Earth's from an original secondary atmosphere that was rich in NH_3. Calculations using the measured HCN abundance and $^{14}N/^{15}N$ ratio predict a primitive atmosphere between 2 and 10 times today's value in mass. Therefore, it is estimated that perhaps several times the present mass of the atmosphere was lost over geologic time.

On the other hand, chemical reactions induced by sunlight on CH_4 build hydrocarbons such as ethane, acetylene, and propane, all of which have been detected in Titan's atmosphere. Hydrocarbons can form long molecular chains (polymers) that remain suspended in the atmosphere forming an aerosol haze layer whereas others will sink to coat Titan's surface. Thus, photochemistry destroys methane irreversibly on Titan, so that its lifetime in the atmosphere is only 10–100 My and methane must be continually or periodically replenished on Titan. A geological source with a possible clathrate reservoir as storage in the interior of Titan is most probably the source.

PROBLEMS

2.1 Find an expression for the heat balance at the surface of a planet equivalent to Equation 2.3 but including the absorption by the body of the stellar radiation heat flux.

Solution

$$M_p Q_{int} = 4\pi R_p^2 \varepsilon \sigma T_s^4 + (1 - A_B)\frac{L*}{4\pi a^2}\pi R^2$$

2.2 An iron planet orbiting around a pulsar formed at the melting point of iron ($T = 1810\,K$) and has a radius of 200 km. Its mean density is $7.8\,g\ cm^{-3}$, its thermal conductivity $K_T = 80\,W\ m^{-1}\ K^{-1}$, and its specific heat is $500\,J\ kg^{-1}\ K^{-1}$. Calculate the time necessary to cool to a temperature of 300 K in the ambient of the star with luminosity $L* = 0.1 L_0$ if placed at a distance of 1 AU with a constant albedo of 0.5 and emissivity of 0.9.

Solution

$t = 152\,My$

2.3 Calculate the cooling time rate of a planet with mass M, constant internal luminosity L_{int}, and specific heat c_v for a change in its internal temperature ΔT_i. Applying to the case of Jupiter, how long will it take for its current luminosity $(L_{int}=33.5 \times 10^{16}$ W) to change its internal temperature in 100 K $(C_v=500$ J kg^{-1} K^{-1})?

Solution

$$\Delta t \approx \frac{Mc_v\Delta T_i}{L_i}, \quad \Delta t = 9 \text{ My}$$

2.4 Calculate the pressure and temperature surface produced by the capture of hydrogen from the primordial nebula on the Earth and Titan assuming that at their positions the nebula gas temperature and pressure are respectively: $T_0=600$ and 100 K and $P_0=7 \times 10^{-5}$ and 2×10^{-8} bar. What is the mass that has been accreted on each body?

  Data: $C_p/\mu = 20,800$ J kg^{-1} K^{-1}.

Solution

T_s (K)$=3605$ K (Earth), 274 K (Titan); P_s (bar)$=6.2 \times 10^{-3}$ bar (Earth), 2.5×10^{-7} bar (Titan); $M_{atm}=3.2 \times 10^{16}$ kg (Earth), 1.5×10^7 kg (Titan).

2.5 Calculate the luminosity of the Sun when it had 1 Gy as compared to its actual value. Then evaluate the equilibrium temperature for Earth and Mars at these two times assuming that the emissivity of both planets was 0.9 and their albedo was 0.37 for Earth and 0.15 for Mars. Compare with the expected current equilibrium temperature for the same parameters.

Solution

$0.81L_{\odot}$, Earth: $T_{eq}=236$ K; $(1L_{\odot}, T_{eq}=236$ K). Mars: $T_{eq}=206$ K; $(1L_{\odot}, T_{eq}=217$ K).

2.6 Mass spectrometry measurements of an Earth basaltic rock has $^{18}O/^{16}O=1/496$. (1) Calculate $\delta^{18}O$ if the reference isotopic standard is 1/499. (2) What information gives the positive or negative sign? For the ocean, $\delta^{18}O=0.0$ but pure water has $^{18}O/^{16}O=0$. (3) What is $\delta^{18}O$ for pure water?

Solution

  (1) 6‰
  (2) A positive value means that the sample contains more heavy isotopes than the standard (the contrary the negative)
  (3) -1000‰

2.7 Calculate the escape parameter λ_{esc} for H_2 molecules in the atmosphere of Jupiter and in exoplanet HD 209458b assuming that their exospheric temperatures are 700 and 1800 K, respectively both at 500 km altitude above the reference radius. What can you conclude from this result?

  Data: Tables in Chapter 1. Hydrogen atomic mass: 1.0079 amu.

Solution

$\lambda_{esc} = 203$ (Jupiter), 46 (HD 209458b).

2.8 Calculate the escape mass (in g s^{-1}) and its characteristic time for hydrogen atoms in Earth's atmosphere as predicted by Jeans' theory if the exospheric temperature is 900 K and its level is at 200 km above the surface where the measured exobase number density of particles is 10^5 cm^{-3}.

Solution

544 g s^{-1}, 20,000 s

2.9 In some planets, the thermal escape may be limited by the diffusion of the escaping constituent across a well mixed atmosphere (see details in Section 6.5). This occurs at the *homopause* where molecular and eddy diffusion processes become equally important. At this level, the upward flux of the escaping species is approximately given by $\Phi \sim N_i(D_i/H)$, where N_i is the particle density, D_i is the diffusion coefficient, and H is the atmospheric scale height. Calculate the diffusion velocity of hydrogen atoms in Mars homopause located at $z = 120$ km ($T = 120$ K, $D_i = 10^8$ cm^2 s^{-1}) and compare it with the effusion velocity predicted by Jeans theory at exobase. What can you conclude from these numbers?

Solution

$v_{dif} = 166$ cm s^{-1}, $v_{eff} = 100$ cm s^{-1}

2.10 Calculate the hydrodynamic (blowoff) escape rate (in cm^{-2} s^{-1}) of CO_2 molecules due to a flux of 10^8 cm^{-2} s^{-1} of hydrogen atoms in primitive Earth for the following conditions: temperature $= 500$ K, binary collision parameter $C_b = 84 \times 10^{-6}$ (cgs units), and temperature exponent $\alpha = 1.6$.

Solution

10^6 cm^{-2} s^{-1}

2.11 Measurements of the minimum mass escape rate in the extrasolar planet HD 209458b give 10^{10} g s^{-1}. Calculate (1) the approximate lifetime of this planet assuming this constant escape rate, (2) where the Lagrangian point L_1 is located for this planet, and (3) the escape rate due to the "geometrical blowoff" assuming that $R_{XUV} = 1.3 R_p$ in the absorption of stellar radiation and $F_{*XUV} = 8.5 \times 10^{-4}$ W m^{-2} at 1 A.U. and scales as the square of the distance.

  Data: HD 209458b table in Chapter 1.

Solution

   (1) 4.15 Gy
   (2) 4.3 R_p
   (3) 2×10^{11} g s^{-1}

2.12 Assume that the impact of an asteroid with a radius of 10 km moving with a velocity of 20 km s^{-1} occurs on planets Venus, Earth, and Mars. Calculate the mass of the atmosphere blown into space and compare it with the present atmospheric mass in each of these planets.

Solution

Venus: $M_{esc} = 2.9 \times 10^{14}$ kg, $M_{esc}/M_{atm} = 6.3 \times 10^{-7}$; Earth: $M_{esc} = 2.5 \times 10^{12}$ kg, $M_{esc}/M_{atm} = 4.9 \times 10^{-7}$; Mars: $M_{esc} = 2.3 \times 10^{11}$ kg, $M_{esc}/M_{atm} = 8.5 \times 10^{-6}$

2.13 (1) Calculate the mass escape rate of hydrogen atoms on Mars produced by the solar wind sweeping at an altitude of 200 km if the temperature is 300 K. (2) Compare it with the Jeans escape rate if the particle density at the exobase is 10^5 particles cm^{-3}. Assume that the solar wind on Mars has a particle density of 2 protons cm^{-3} moving with a velocity of 400 km s^{-1}.

Solution

(1) 1 g s^{-1}

(2) 2.6×10^5 g s^{-1}

2.14 Calculate the insolation change (in %) in the polar regions of Earth and Mars due to the long-term cyclic variations of their eccentricity (e) and obliquity (ε) if for the Earth ε changes from 20° to 25° and e changes from 0 to 0.045, and for Mars ε changes from 15° to 45° and e between 0 and 0.15.

Hint: From Section 1.4.3 and Equation 1.89, use

$$\langle F_{\odot p} \rangle_{pole \atop annual} \approx S_{\odot p} \tau_{orb} \sin i / 4(1 - e^2)^{1/2}$$

Solution

(1) Earth : $\Delta \langle F_{\odot p} \rangle_{pole \atop annual} (e) = 0.1\%$, $\Delta \langle F_{\odot p} \rangle_{pole \atop annual} (i) = 8\%$

(2) Mars : $\Delta \langle F_{\odot p} \rangle_{pole \atop annual} (e) = 1.1\%$, $\Delta \langle F_{\odot p} \rangle_{pole \atop annual} (i) = 45.8\%$

3 Spectroscopy and Composition

3.1 ATMOSPHERIC COMPOSITION: FUNDAMENTALS

In the previous chapter, we have seen how the atmospheres originated in planets and satellites. Since the atmospheric mass in the less massive bodies changes, to a variable degree, in response to the external forcing or to the interaction with the surface and interior of the body volatile reservoirs, we discuss the mean atmospheric composition of the actual (present?) atmospheres here. The atmosphere is composed of a variety of gases that can be in atomic, molecular, or ionic form: some are active to sunlight and thermal radiation (absorb and emit radiation), others are inert compounds (as the noble gases), others chemically active (reacting with others species), and some are condensable gases that suffer phase changes (condensation, evaporation, or sublimation) depending on temperature and pressure.

The atmospheric composition varies spatially and temporarily in response to a variety of factors. This determines the distribution of the energy sources within the atmosphere (that affects the composition), for example, by vertical heating due to solar radiation deposition, or cooling by infrared thermal radiation, or in the horizontal distribution by land–sea or topographic albedo contrasts. Gases that suffer phase changes intervene in the atmospheric energy budget through the latent heat associated to the transformation, forming, for example, a variety of condensable clouds and hazes that in turn modify the radiation budget of the atmosphere. This chapter deals with the fundamentals of the spectroscopy, which, as we will see, plays a key role in determining the composition of atmospheres. Its influence on the energy sources (thermal structure in Chapter 4) and in the cloud and haze formation (Chapter 5) is discussed in other chapters.

The composition of a planetary atmosphere can be determined by *remote sensing* techniques using *spectroscopy* and *occultation methods* (Sun, stars, or using radio-occultation experiments from spacecrafts) from ground-based telescopes and from orbiting or fly-by spacecrafts, and from in situ measurements using *mass spectrometers* onboard probe descent modules and surface landers that determine the atomic weight and number density of the atmospheric gases. The methods and techniques are described in Appendix. Combining in situ and remote sensing methods (when possible) allows the composition to be determined precisely (even at the isotopic level). When an adequate sampling of measurements is available (as for example with orbital vehicles), it is possible to determine the spatial and temporal variability of the active compounds.

3.2 PLANETARY SPECTRUM: THE CONTINUUM

The spectrum of a planet or satellite body (the x-rays to radio wavelength distribution of the radiative flux) originates, in general, from a combination of its surface and atmosphere. On the one hand, it contains the reflected spectrum of the sunlight radiation (or from star for extrasolar planets) by the gases and particles in the atmosphere (aerosols) and by the surface (in the case of terrestrial planets and satellites). This contribution broadly falls from UV to near infrared wavelengths (~1–10 μm) where the Sun and stars have their emission peak. On the other hand, the continuum also contains the thermal radiation spectrum due to the radiative flux emitted by the planet whose origin is the Sun's or stellar radiation (i.e., arising from the conversion to heat of the absorbed solar radiation) and/or the own internal energy of the body (as in the giant planets). This second contribution occurs in the infrared and in microwave regions (wavelengths > 1–10 μm), and depends on the body temperature. These two parts comprise the continuum spectrum that, to a first approach, can be obtained by the addition of a diffuse blackbody mirror of the star spectrum plus the blackbody of the thermal spectrum. Superimposed on this continuum we have a discrete line and band absorption and emission spectrum. Figure 3.1 shows schematically the structure of a planetary spectrum. Figure 3.2 shows the processes contributing to the radiative flux of the reflected and thermal spectrum for the Earth. Figure 3.3 shows the real case of the reflected sunlight and thermal emission spectrum for Jupiter.

At short wavelengths (mainly UV and visible), an emission spectrum can occur in the atmospheres due to the excitations of atoms, ions, or molecules by the absorption of solar photons or through the precipitation of particles, and

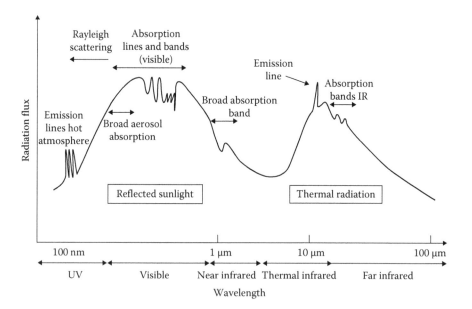

FIGURE 3.1   Sketch of the typical spectrum of a hypothetical planet with atmosphere.

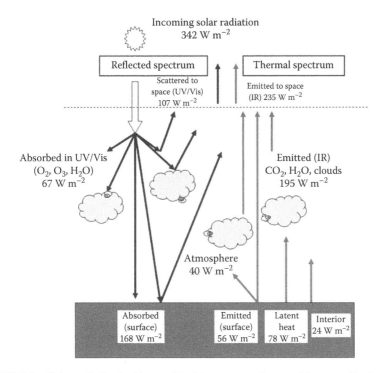

FIGURE 3.2 Solar radiation incident on Earth's top atmosphere and flux contribution to the reflected (left) and emission spectra (right) from the atmosphere and surface.

the subsequent de-excitation and emission of a photon at the same wavelength or nearby wavelengths from a cascade series (e.g., from fluorescent emission and aurora phenomena). These processes will be presented in detail in Chapter 6. At the other extreme, at wavelengths longer than microwaves (radiofrequencies at about 1–10 cm), the planetary emission is dominated by nonthermal processes occurring in their magnetic field (Section 1.6), mainly by electrons gyrating in spirals along magnetic field lines (cyclotron emission and also synchrotron emission in the case of Jupiter).

The discrete spectrum contains (a) emission lines produced by ions, atoms, and molecules excited by a variety of processes (in an atmosphere they occur typically at high temperatures), (b) absorption lines and bands produced by atoms and molecules when they absorb the solar reflected or the thermally emitted continuum. The reflected spectrum also contains the *Fraunhofer lines* originated in the Sun's photosphere. The absorption bands are usually broad in wavelength and formed by the superposition of individual rovibrational lines, which are typical of atmospheric molecules. The study at high spectral resolution of the planetary spectrum contains a lot of information of its atmosphere (and surface), allowing to retrieve its composition (identification of major and minor atmospheric constituents), the physical and chemical processes occurring in it, and the temperature distribution.

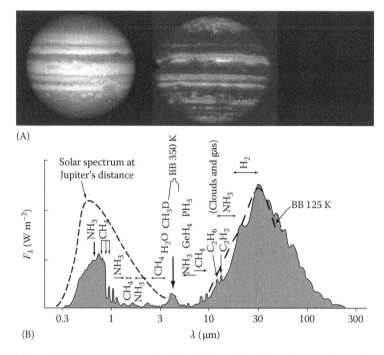

(A)

(B)

FIGURE 3.3 (A) Images of Jupiter in the visible (left, reflected sunlight by clouds from 350 nm to 1 μm) and in the thermal infrared (right) produced by the cloud opacity to the planet emission at 5 μm. (B) The corresponding reflected and thermal spectrum of Jupiter. (Adapted from Hanel et al., *J. Geophys. Res.*, 86, 8705, 1981.)

3.2.1 REFLECTED SPECTRUM

Stars can be considered, to a first approach, as blackbody radiators at the temperature of their photospheres, which is typically in the range of 3,000 K (red "M-spectral type" stars) to 30,000 K (white-blue "O-type" stars). The Sun is a G2 spectral type yellow star with a blackbody temperature of 5800 K and peak wavelength in the visible at ~560 nm. As shown in Section 1.4.3, the reflected spectrum depends on a global scale, on the albedo (reflectivity by the clouds, by the surface, or both). For example, on Earth and Mars, the reflected spectrum comes from a combination of the radiation reflected by the surface and clouds and suspended aerosols, whereas on Venus, Titan, and the giant planets, it comes essentially from the clouds and hazes.

The reflected spectrum is usually described in terms of the geometric albedo $p_0(\lambda)$ when considering the reflectivity from the integrated disk, or in terms of the reflectivity I/F when referring to the spatially resolved reflectivity at a particular position in the disk. The reflectivity I/F depends on wavelength because of the scattering and absorption processes that suffer the incoming solar radiation in the atmosphere and/or surface, and on the geometry of the reflection process (incident and reflected—viewing angle—directions). Scattering processes are produced by the gases (Rayleigh scattering dominating at short wavelengths), aerosols, and cloud particles (that have a more flat scattered spectrum), and by the surface when accessible. All this produces a diffuse reflection spectrum with the approximate shape of the Sun's blackbody photosphere emission but smaller than that would be produced by a perfect reflective mirror. This is mainly due to the absorption produced at selected wavelengths by gases (in lines and bands) and by particles. In addition, a gas line emission spectrum by the gases is also observed at particular wavelengths superimposed on the spectrum. Here we present the theory for gas (atoms and molecules). The scattering and absorption processes by gases and particles will be presented in detail in Chapter 5. Figures 3.4 through 3.8 show representative reflection spectra for Venus, Earth, Jupiter (see also Figure 3.3), Saturn, Titan, Uranus, and Neptune.

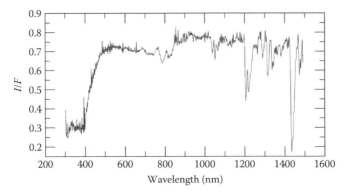

FIGURE 3.4 Venus-reflected spectrum (daytime) as captured by the Messenger spacecraft fly-by in June 2007 during its route to Mercury. (Courtesy of G. Holsclaw, W. McClintock, prepared by S. Pérez-Hoyos, UPV-EHU, Spain.)

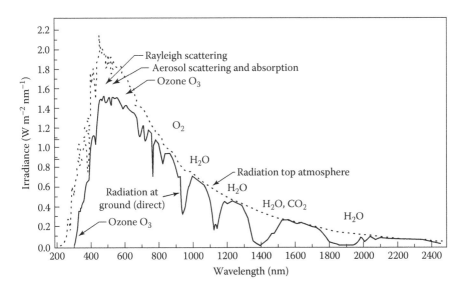

FIGURE 3.5 Earth-reflected spectrum. Radiation incident at the top of the atmosphere is indicated by the dashed line (normal incidence). The continuous line shows the radiation at the ground after suffering scattering and absorption processes (gas absorbers, as indicated).

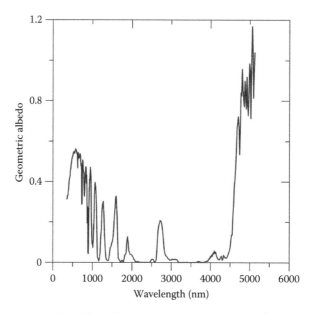

FIGURE 3.6 A detailed visible and near infrared spectrum of Jupiter. The thermal spectrum begins to dominate the reflected one longward of about 3 μm. Data for the visible spectrum taken from Karkoschka (1998) and by Baines and Orton for the near infrared. (Courtesy of K. Baines.)

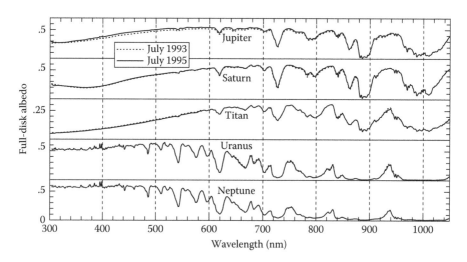

FIGURE 3.7 The reflected spectra of Jupiter, Saturn, Titan, Uranus, and Neptune. Absorption bands due to methane are centered around the wavelengths 619 nm, 725 nm, 890 nm, and 1.1 μm. (From Karkoschka, E., *Icarus*, 133, 134, 1998. With permission.)

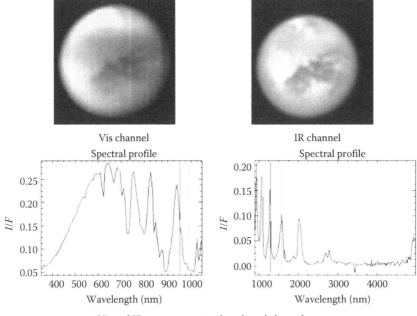

Vis and IR average spectra: the selected channel are
identified by vertical RGB lines

FIGURE 3.8   Titan-reflected spectrum from 350 nm to 5 μm with corresponding images of the satellite in the visible and near infrared channels. Images and spectra obtained with the VIMS instrument onboard the Cassini spacecraft (NASA-ESA mission). (From Coradini, A., Piccioni, G., and Filacchione, G., Cassini's VIMS instrument, at http://vims.artov.rm.cnr.it/data/res-tit.html, NASA and ESA, 2004.)

3.2.2   THERMAL SPECTRUM

The thermal part of a planet or satellite spectrum can be assumed, to a first approach, to be that of a blackbody that emits according to the *Plank radiation law*:

$$B_\lambda(T) = \frac{2hc^2}{\lambda^5 \left(e^{hc/\lambda k_B T} - 1\right)} \tag{3.1a}$$

where
  $B_\lambda(T)$ is the specific intensity or brightness (units of W m^{-2} sr^{-1} μm^{-1})
  h and k_B are, respectively, the Planck and Boltzmann constants
  c is the speed of light in vacuum
  T is the temperature that can be taken to be the equilibrium or the effective temperature of the atmosphere (see Section 1.4.4)

Since the frequency ν and wavelength are related by the expression $\nu = c/\lambda$, the Planck function in terms of the frequency can be written as

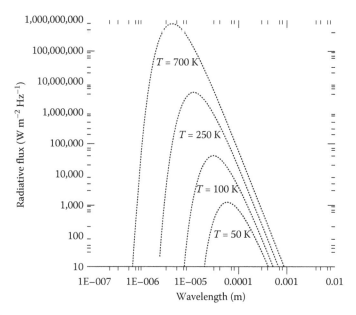

FIGURE 3.9 The blackbody Planck curve emission corresponding to temperatures typical in planetary atmospheres.

$$B_\nu(T) = \frac{2h\nu^3}{c^2} \frac{1}{\left(e^{h\nu/k_B T} - 1\right)} \tag{3.1b}$$

Figure 3.9 shows examples of the Planck function for different temperatures.

The wavelength at which the brightness function $B_\lambda(T)$ has a maximum is obtained by setting $\partial B_\lambda / \partial \lambda = 0$ and is known as the *Wien displacement law*:

$$\lambda_{max} = \frac{2897.8}{T} \tag{3.2a}$$

$$\nu_{max} = \frac{5.88 \times 10^{10}}{T} \tag{3.2b}$$

with the wavelength given in micrometer in Equation 3.2a, the frequency in hertz (in Equation 3.2b), and temperature in kelvin.

The radiating flux density (W m^{-2} µm^{-1}) emitted from a blackbody is given by

$$F_\lambda = \Omega_s B_\lambda(T) \tag{3.3}$$

where Ω_s is the solid angle into which the radiation is emitted. The solid angle (unit steradian, sr) is defined such that, integrated over a sphere (see Figure 3.10), it gives

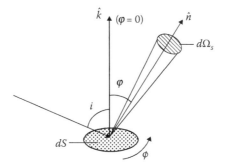

FIGURE 3.10   Geometry for the incident and emitted radiation from a surface element (dS). The incident (i), emission angles (φ, ϕ), and the solid angle ($d\Omega_s$) (see Equation 3.4) are drawn.

$$\oiint d\Omega_s = \int_0^{2\pi} d\phi \int_0^\pi \operatorname{sen} \varphi \, d\varphi = 4\pi \text{ (sr)} \tag{3.4}$$

If $B_\lambda(T)$ is integrated over all wavelengths and solid angle, we obtain the total flux F (in W m^{-2}) from a blackbody called the *Stephan–Boltzmann law*:

$$F = \int_0^\infty B_\lambda(T) d\lambda = \frac{\sigma_B}{\pi} T^4 \tag{3.5}$$

Measuring the thermal spectral radiance of a planet and fitting it to a blackbody curve with temperature as a free parameter allows us to retrieve the temperature at the surface or at the corresponding emission level in the atmosphere. For longer wavelengths (short frequencies), Equation 3.1 permits to retrieve the brightness temperature T_B using the simple formula (see Problem 3.1)

$$T_B = \frac{c^2 B_v}{2 k_B v^2} \tag{3.6}$$

Obviously the blackbody temperature is an idealization of the real temperature of a body since it emits less radiation than that of a pure blackbody (see Equations 1.99 through 1.101). In reality, the atmospheres absorb, reflect, and transmit radiation. To quantify these processes, we define absorptivity as the fraction of the absorbed intensity to the incident monochromatic radiation I_λ ($a_\lambda = I_{\lambda a}/I_\lambda$). Similarly, reflectivity is defined as the fraction of the reflected intensity ($r_\lambda = I_{\lambda r}/I_\lambda$), and transmissivity ($t_\lambda = I_{\lambda t}/I_\lambda$) as the fraction of the transmitted intensity. Energy conservation requires that

$$a_\lambda + r_\lambda + t_\lambda = 1 \tag{3.7}$$

The emissivity ε_λ of a body is defined as the ratio of the emitted intensity to that of a blackbody $B_\lambda(T)$ and obviously $\varepsilon_\lambda \lesssim 1$. Kirchhoff's law states that under thermodynamic equilibrium (at a given temperature) and at a given wavelength, the emissivity must equal the absorptivity:

$$\varepsilon_\lambda = a_\lambda \tag{3.8}$$

In Figures 3.11 through 3.15, we show a family of representative thermal emission spectra for planets with atmospheres: Venus, Earth, and the giant planets Jupiter (see also Figure 3.3), Saturn, and Uranus (Neptune is virtually the same as Uranus).

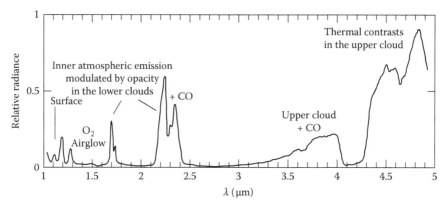

FIGURE 3.11 Night-side emission spectra from Venus in the near and thermal infrared obtained by VIRTIS instrument onboard Venus Express (ESA) spacecraft. (Courtesy of Ricardo Hueso, UPV-EHU, Spain.)

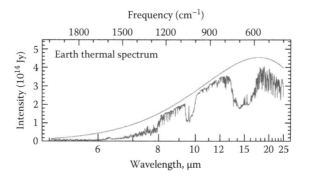

FIGURE 3.12 Earth thermal emission spectra and the corresponding blackbody emission curve for a temperature of 280 K. Note the strong absorption band due to CO_2 at 15 μm. (Courtesy of NASA, Washington, DC.)

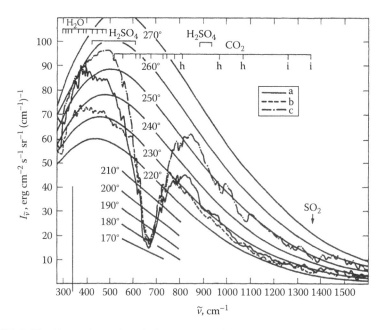

FIGURE 3.13 Venus thermal emission spectra with the corresponding blackbody emission curves for different temperatures and with the indication of the absorption bands due to H_2O, CO_2, and SO_4H_2. The spectra correspond to three different regions: (a) latitude 48°, (b) latitude 67°, and (c) latitude 82° ("dipole area"). Data obtained with the Venera 15 spacecraft. (From Moroz, V.I., *Adv. Space Res.*, 5(11), 197, 1985. With permission.)

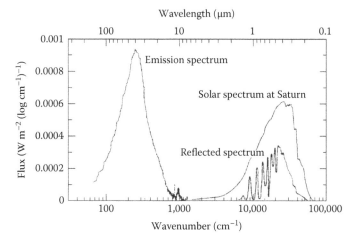

FIGURE 3.14 Saturn's reflected and thermal spectrum obtained with the IRIS instrument onboard the Voyager spacecraft. Note the abscissa axis is in increasing wavenumber (cm^{-1}). (From Hanel, V.G. et al., *Icarus*, 53, 262, 1983. With permission.)

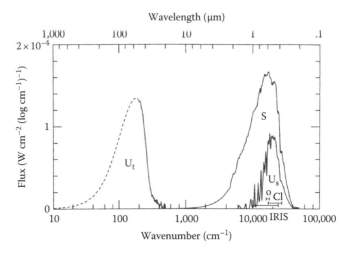

FIGURE 3.15 Uranus's visible reflected and thermal infrared spectrum obtained with the IRIS instrument onboard the Voyager spacecraft. (From Pearl, J.C. et al., *Icarus*, 84, 12, 1990. With permission.)

3.3 ATOMIC SPECTRUM

According to quantum mechanics, atoms and molecules absorb and emit radiation at discrete wavelengths due to transitions between the energy levels characterizing their state, producing emission and absorption lines in the spectrum. A line is characterized by a wavelength-dependent *intensity* $I(\lambda)$ (unit W m^{-2} sr^{-1} μm^{-1}) and it is normally observed only in a high-resolution spectra. For absorption lines, $I < I_0$, where I is the intensity at the center of the line and I_0 the intensity of the continuum or the reference level outside the line; in emission lines, $I > I_0$. It is customary to represent the planetary spectra normalized to the continuum background as defined by the reflected or thermal spectrum presented in the previous section.

3.3.1 ATOMIC STRUCTURE: BASICS

Emission and absorption of photons by atoms and molecules occurs when there is a change in the energy state of the particle. In atoms, the nucleus is formed by protons and neutrons that are surrounded by a cloud of electrons. According to Bohr semi-classical atomic theory (sufficient for our purposes at the moment, without entering in full quantum mechanics), the electrons orbit the nucleus under the action of the electrostatic Coulomb force:

$$\vec{F}_E = -\frac{Zq_e^2}{4\pi\varepsilon_0 r^2}\hat{r}$$

(3.9)

where
  Z is the atomic number
  q_e is the electron charge
  ε_0 is the vacuum permittivity
  r is the distance (orbit radius)

The total energy of the electron (mass m_e) in a circular orbit can be obtained using the same arguments as that in Section 1.2 for planetary orbits (changing the gravitational Newton's law by Coulomb law),

$$E = \frac{1}{2}m_e v^2 - \frac{Zq_e^2}{4\pi\varepsilon_0 r} = -\frac{Zq_e^2}{4\pi\varepsilon_0(2r)} \tag{3.10}$$

since from Newton's second law (1.2), we have $m_e v^2/r = F_E$. Energy and angular momentum relationship for the electron about the nucleus follows simply from the definition $L_0 = m_e vr$:

$$E = -\frac{m_e q_e^4 Z^2}{2L_0^2(4\pi\varepsilon_0)^2} \tag{3.11}$$

which is valid only for circular orbits. However, comparison with the atomic spectra and experimentally determined energy levels in the hydrogen atom shows that the angular momentum can only acquire discrete values, that is, it is quantized. According to Bohr's theory, the possible values of the angular momentum for circular orbits are given by

$$L_0 = n\hbar \quad (n = 1, 2, 3, \ldots) \tag{3.12}$$

where
  n is the principal quantum number and $\hbar = (h/2\pi) = 1.0545 \times 10^{-34}$ J s
  h is the Planck constant

It also follows that the radius for the circular orbit is given by

$$r = \frac{n^2 h^2 \varepsilon_0}{\pi m_e Z q_e^2} = \frac{n^2}{Z}a_0 \tag{3.13}$$

where $a_0 = (h^2\varepsilon_0/\pi m_e q_e^2) = 5.2917 \times 10^{-11}$ m, the Bohr radius, which corresponds to the radius of the hydrogen atom ($Z=1$) in its ground state ($n=1$). Introducing (3.13) into (3.11), we get the energy of the orbit characterized by the quantum number n (energy levels E_n):

$$E_n = -\frac{m_e q_e^4 Z^2}{8\varepsilon_0^2 h^2 n^2} = -\frac{R_R hc Z^2}{n^2} = -\frac{2.18\times10^{-18}Z^2}{n^2}(\text{J}) = -\frac{13.6Z^2}{n^2}(\text{eV}) \tag{3.14}$$

where R_R is the Rydberg constant for the hydrogen atom and the energy levels are in electron-volt (eV) unit. When considering the mass of the nucleus, and therefore the motion of the electron and nucleus around their common center of mass, we must substitute the Rydberg constant by its modified value $R_R^* = R_R/(1 + m_e/m_N)$, where m_N is the mass of the nucleus. In general, since $m_e \ll m_N$, it is a good approximation to take R_R.

According to Planck's theory, the energy of a photon is $h\nu$, where $\nu = c/\lambda$, its frequency and c the speed of light. If an atom absorbs the energy of the photon, the electron "will change to the orbit" (energy level) that has an additional energy

$$\Delta E = h\nu \tag{3.15}$$

known as the Bohr condition. If the atom emits a photon, it decreases its energy by the amount of the photon energy. The simplest atomic system is the hydrogen atom formed by an electron and a nucleus with one proton. The difference in energy between two levels n_1 and n_2 $(n_2 > n_1)$ is given by

$$E_2 - E_1 = R_R h c Z^2 \left(\frac{1}{n_1^2} - \frac{1}{n_2^2} \right) \tag{3.16}$$

and using the Bohr condition and the numerical values for R_R and c, we get the frequency corresponding to the transition between these two energy levels

$$\nu(\text{Hz}) = \frac{E_2 - E_1}{h} = 3.2899 \times 10^{15} Z^2 \left(\frac{1}{n_1^2} - \frac{1}{n_2^2} \right) \tag{3.17}$$

which is called Balmer's formula. If instead of the frequency we use the wavenumber $\tilde{\nu} = (\nu/c) = (1/\lambda)$ (unit in cm^{-1}), it suffices to change in Equation 3.17 the numerical constant by 1.0974×10^5. Figure 3.16 shows the transition series for the hydrogen atom. When applying the quantum mechanics theory, it is found that the orbital angular momentum of an electron $\vec{L}$ is quantized such that the allowed values of its modulus are $L^2 = \ell(\ell + 1)\hbar^2$ with the quantum number $\ell = 0, 1, \ldots, n - 1$ (customarily referred to as s, p, d, f, ... states). The angular momentum is also directional quantized so its z-direction (perpendicular to the orbital plane) takes the value $L_z = m_\ell \hbar$ with the magnetic quantum number taking the values $m_\ell = 0, \pm 1, \pm 2, \ldots, \pm \ell$. In addition, the electrons possess an intrinsic angular momentum, the quantized *spin* $S^2 = s(s + 1)\hbar^2$ with quantum number $s = 1/2$. For the z-component, we have $S_z = m_s \hbar$ with $m_s = \pm 1/2$. The total angular momentum of the electron is given by $\vec{J} = \vec{L} + \vec{S}$ and is quantized as in the previous cases so $J^2 = j(j + 1)\hbar^2$ and $J_z = m_j \hbar$, with $j = \ell \pm (1/2)$ and $m_j = \pm j, \pm(j - 1), \ldots$. In summary, the electronic energies are determined by the orbital distances of the electrons, their orbital angular momentum, and their spin.

Pauli's exclusion principle states that two electrons in an atom cannot have the same set of quantum numbers (n, ℓ, j, m). This implies that the total number of energy sublevels of the energy level n of an electron in an atom (i.e., those sublevels corresponding to the different quantum numbers (ℓ, j, m for a given n)) is given by

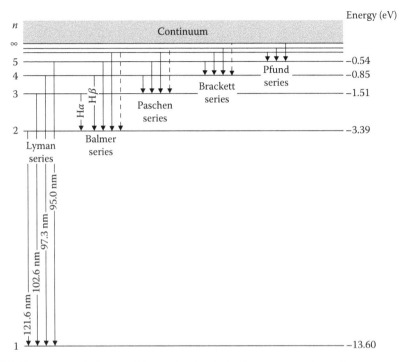

FIGURE 3.16   Radiative transition series for the hydrogen atom. The series correspond to the following energy levels transitions (Equation 3.16): Lyman series ($n_1 = 1$, $n_2 = 2$, 3, 4, ...), Balmer series ($n_1 = 2$, $n_2 = 3$, 4, ...), Paschen series ($n_1 = 3$, $n_2 = 4$, 5, ...), Brackett series ($n_1 = 4$, $n_2 = 5$, 6, ...), Pfund series ($n_1 = 5$, $n_2 = 6$, 7, ...).

$$2n^2 = g_n \qquad (3.18)$$

where g_n is known as the statistical weight or degeneracy of level n. These sublevels can be further subdivided in presence of magnetic and electric fields since their coupling to the angular momentum add energy to the electron (the Zeeman and Stark effects, respectively). The coupling between the orbital angular momentum and spin of the electron produces additional splitting of the energy levels (the spin–orbit coupling). The spin of the nucleus is also quantized introducing further multiplet levels when coupling with the spin of the electron depending of their relative orientation (parallel or antiparallel states), giving rise to the hyperfine structure.

The electronic transitions between these energy sublevels produce the line spectrum and obey the selection rules since not all the transitions are possible. During the transition, the atom behaves as an oscillating electric or magnetic multipole, and transitions are then classified as dipolar, quadrupolar, electric, or magnetic, each with their own selection rules and with a certain probability to occur. Normally, the most probable transition is the dipolar electric, followed by the dipolar magnetic and the quadrupolar electric. Since each atom or molecule has its own unique set of energy transitions, measurements of the wavelength position of the absorption or emission spectral lines can be used to identify the species present in the system, that is, in a planetary atmosphere. This is the first immediate application of the spectral analysis.

Consider now a system formed by a large quantity of atoms (as, for example, the atmosphere). According to statistical mechanics, the energy levels of the atoms in equilibrium at temperature T will be populated according to the Boltzmann distribution law:

$$\frac{N_1}{N_2} = \frac{g_1}{g_2} \exp\left[-\frac{(E_2 - E_1)}{k_B T}\right] \tag{3.19}$$

where

N_1 and N_2 are the number of atoms (or molecules) with energy states E_1 and E_2, respectively

g_1 and g_2 are their degeneracies (statistical weights)

At low temperatures, most atoms are in the ground state, but occupy higher energy levels at higher temperatures. For a gas, this equilibrium ratio is maintained by collisions between the particles (atoms or molecules) and the system is said to be under thermodynamic equilibrium conditions. The interaction between the radiation and the gas is then weak, and the equilibrium is maintained. This occurs, for example, when the mean time between collisions in the gas (which is inversely proportional to the gas pressure) is much shorter than the lifetime for the radiative transition between two energy levels. However, the atmosphere is a gas that does not have a uniform temperature, so strictly it cannot be in thermodynamic equilibrium. However, for high-enough pressures, the molecular collisions are sufficiently rapid to keep the atmosphere in what is called local thermodynamic equilibrium (LTE). The necessary gas pressure for applying LTE conditions depends on the radiative process. Under low pressures (i.e., in the case of tenuous atmospheres or at high altitudes in a dense atmosphere), departures from this condition must be considered and the gas is said to be under non-LTE conditions.

3.3.2 LINE PROFILES

Quantum mechanics shows that the strength of an absorption line of a given element depends on both the transition probability and on the population of the lower energy level. It is characterized by the line strength s_ℓ (sometimes also called line intensity), which describes the coupling of two states through the electromagnetic perturbation. The finite time over which a photon is absorbed or emitted by an atom or molecule is related to the energy of the level by the Heisenberg uncertainty principle:

$$\Delta E \, \Delta t \approx h \tag{3.20}$$

where ΔE and Δt are the uncertainties in the values of energy and time determination for the system. The principle applies to all the energy levels except for the ground state ($\Delta E=0$). Typically, the electronic transition time between two levels is ~10^{-8} s. The energy levels of an atomic system are naturally broadened according to this relationship. According to the Bohr condition (3.15), the photon emitted during a transition between two energy levels does not have a precise wavelength, but instead falls within a possible range of values that depend on ΔE. For example, if we assume an electronic transition from the ground state (energy E_1) to an excited state (energy E_2) that has an "energy-broadening" ΔE due to (3.20), the emitted or absorbed photon will have, according to (3.15), a frequency within the range

$$\Delta v = \frac{(E_2 - E_1 \pm (1/2)\Delta E)}{h} \tag{3.21}$$

This translates to a natural broadening of the absorption or emission line where the emitted or absorbed energy is distributed over a narrow frequency interval $\Delta v \approx \Delta t^{-1}$. This is usually small when compared with other broadening mechanisms, as will be shown below. For example, in terms of wavenumber, the natural broadening is $\Delta \tilde{v} \approx (c\Delta t)^{-1} \approx 10^{-7}$ cm^{-1}. Usually, the spectral information is given in terms of any of the three magnitudes (wavelength, frequency, wavenumber), so it is convenient to take into account the following relationships between them:

$$v = \frac{c}{\lambda}$$

$$\tilde{v} = \frac{1}{\lambda} = \frac{v}{c}$$

$$dv = -\frac{c\,d\lambda}{\lambda^2}$$

$$d\tilde{v} = \frac{dv}{c}$$

The shape of the spectral line, called the line profile, is represented by the function $I(\lambda)$ or $I(v)$ and, apart from the natural broadening, it contains information on a variety of other processes occurring in the atmosphere. The line profile is basically determined in an atmosphere by the abundance of the element producing the line and by the ambient pressure and temperature. High-resolution spectroscopy can be used to retrieve not only composition but other atmospheric properties. The main mechanisms contributing to the line shape profile are given in the following sections.

3.3.2.1   Collision or Pressure Broadening

Collisions between the gas elements (atoms or molecules) can occur during the absorption or emission of the photons, perturbing the energy levels of the electrons and spreading the range of wavelengths in the line profile. The effect will become more important as the gas pressure increases since collisions will be more abundant. Simple theory for the line profile leads to represent the process in terms of an absorption coefficient k_ν that depends on the frequency ν relative to the frequency at the center of the line ν_0, according to a Lorentz line shape profile:

$$k_\nu = \frac{s_\ell \gamma_L}{\pi\left[\left(\nu - \nu_0\right)^2 + \gamma_L^2\right]} \tag{3.22}$$

where

the line strength is $s_\ell = \int_0^\infty k_\nu\, d\nu$

$\gamma_L = 1/(2\pi t)$ is half width of the line

t is the mean time between collisions

The absorption coefficient k_ν can be defined in terms of a mass absorption coefficient (unit $cm^2\ g^{-1}$) or in terms of a volume absorption coefficient, which is the mass absorption coefficient multiplied by the density (unit cm^{-1}). For k_ν expressed in cm^{-1} (volume absorption coefficient), there are two possibilities in the usage of ν in Equation 3.22. If ν represents the frequency (unit s^{-1}), then γ_L must be expressed in s^{-1} and s_ℓ in $cm^{-1}\ s^{-1}$. If it represents the wavenumber, that is, ν is replaced by $\tilde{\nu}$ (unit cm^{-1}) in Equation 3.22, then γ_L must be expressed in cm^{-1} (in the above definition $\gamma_L = 1/(2\pi ct)$) and s_ℓ in cm^{-2}. Note that the absorption coefficient has the same shape in terms of frequency or wavenumber since c (the speed of light in vacuum) cancels out in Equation 3.22. For k_ν expressed in $cm^2\ g^{-1}$ (mass absorption coefficient), if ν and γ_L are expressed as frequencies (s^{-1}), then s_ℓ unit is $g^{-1}\ m^2\ s^{-1}$. If ν and γ_L are expressed as wavenumbers (unit cm^{-1}) in Equation 3.22, then s_ℓ unit is $g^{-1}\ cm$. It is also customary to use as unit for s_ℓ the wavenumbers per column density. The half-width depends on the temperature and pressure according to

$$\gamma_L = \gamma_{L0}\frac{P}{P_0}\left(\frac{T_0}{T}\right)^{1/2} \tag{3.23}$$

from a reference value γ_{L0} at P_0 and T_0, the standard pressure and temperature (STP). A typical collision broadened line width value is $\gamma_{L0} = 0.1\ cm^{-1}$.

3.3.2.2 Doppler Broadening

During the emission and absorption processes, the gas atoms or molecules are in motion. The characteristic speed is given by the Maxwell–Boltzmann distribution (Equation 2.20) that depends on the temperature and molecular weight or mass of the atoms or molecules. This velocity distribution produces a Doppler shift in the frequencies of the absorbed or emitted photons, that is, a frequency spread within the line that can be characterized, as above, by the absorption coefficient

$$k_\nu = \frac{s_l}{\sqrt{\pi}\gamma_D}\exp\left[-\left(\frac{\nu - \nu_0}{\gamma_D}\right)^2\right] \tag{3.24}$$

where

$$\gamma_D = \frac{\nu_0}{c}\left(\frac{2R_g T}{\mu}\right)^{1/2} = \gamma_{D0}\left(\frac{T}{T_0}\right)^{1/2} \qquad (3.25)$$

is the Doppler line half-width and μ is the molecular weight. The half-width at half maximum is $\gamma_D\,(\ln 2)^{1/2}$ and is proportional to $T^{1/2}$ (but independent of pressure when compared to (3.23)). To have an order of magnitude, for methane ($\mu = 16\,\text{g}$) at $T = 293\,\text{K}$ at the absorption or emission line at $7.7\,\mu\text{m}$ (wavenumber $1300\,\text{cm}^{-1}$), the Doppler width is $\gamma_{D0} = 0.002\,\text{cm}^{-1}$.

3.3.2.3 Voigt Broadening

The above relationships show that at high pressures the line broadening is dominated by the collision mechanism, but at low pressures, Doppler broadening dominates. For intermediate situations of temperature and pressure, both mechanisms intervene and the line shape results from their combination producing the so-called Voigt line shape:

$$k_V = \frac{s_\ell \beta}{\gamma_D \pi^{3/2}} \int_{-\infty}^{\infty} \frac{\exp(-y^2)}{(x-y)^2 + \beta^2}\,dy \qquad (3.26)$$

where $\beta = \gamma_L/\gamma_D$ and $x = (\nu - \nu_0)/(c\gamma_D)$. This equation must be solved numerically.

3.3.2.4 Equivalent Width, Transmittance, and the Curve of Growth

When the line shape cannot be resolved, a useful spectral parameter is the equivalent width (EW) defined by the integral

$$\text{EW} = \int_{-\infty}^{\infty}\left(1 - \frac{I(\nu)}{I_0}\right)d\nu \qquad (3.27)$$

where the line intensity $I(\nu)$ is measured relative to the background continuum I_0 and EW is given in hertz if we use the frequency for the spectral representation (or in cm^{-1} if we use the wavenumber or in nanometer, for example, if we use wavelength) (Figure 3.17A). The EW is a measure of the intensity removed from the continuum by the line and geometrically it represents the width of a rectangular absorbing spectral line with the same area as the real line. In order to estimate what information can be derived from this magnitude, assume, for simplicity, that a beam of radiation with intensity, I, that traverses an elementary slab of the atmosphere of density, ρ_a, and thickness, dz, is being absorbed without suffering any scattering process (this will be treated later). According to Lambert's or Bouguet's absorption law (see Chapter 4 for details), the intensity varies across the layer:

$$\frac{dI_\nu}{dz} = -I_\nu k_\nu \rho_a \qquad (3.28)$$

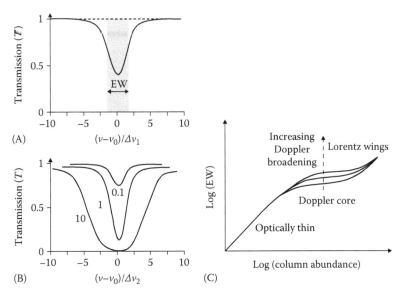

FIGURE 3.17 (A) Line profile and EW as a function of relative frequency difference. The profile is assumed to be of a Lorentz type; (B) transmission profile as a function of the optical depth (gas abundance); (C) curve of growth for a generic spectral line that is linear at low column abundance.

which upon integration, gives

$$I_v = I_v(0)\exp(-\tau_v) \tag{3.29}$$

where

$\tau_v = \int k_v \rho_a \, dz$ is the optical depth of the layer (a nondimensional magnitude)

k_v is the mass absorption coefficient in $m^2 \, kg^{-1}$

The function $\exp(-\tau_v)$ is called the transmittance T_v of the layer and, for a beam of radiation traversing a layer located at altitude levels z_1 and z_2, written as

$$T_v(z_1, z_2) = \exp\left(-\left|\int_{z_1}^{z_2} k_v(z)\rho_a(z)\,dz\right|\right) = \exp\left[-\left|\tau_v(z_2) - \tau_v(z_1)\right|\right] \tag{3.30}$$

with the absorptance defined as $A_v = 1 - T_v$ (Figure 3.17B). Both A_v and T_v are nondimensional quantities. Neglecting scattering processes (see Chapter 5), the absorption coefficient depends on the altitude z of the atmosphere through the pressure and temperature altitude variations. Assuming it is constant along the optical path (e.g., as measured in the laboratory at a constant P and T), the above equation becomes

$$T_v(z_1, z_2) = \exp\left[-k_v(P,T)u_M(z_1, z_2)\right] \tag{3.31}$$

where

$$u_M(z_1, z_2) = \left|\int_{z_1}^{z_2} \rho_a(z)\,dz\right| \tag{3.32}$$

is the mass of the absorber gas along the path of the radiation beam per unit of the traversed cross area. From Equation 3.32, if the path has a length ℓ and the absorbing gas has a constant density, then $u_M = \rho_a \ell$ (kg m^{-2}) and the optical depth τ_v is given by

$$\tau_v = \int_0^{\ell} k_v \rho_a \, dz = k_v \rho_a \ell \tag{3.33}$$

and the EW (3.27) becomes

$$EW = \int_{-\infty}^{\infty} (1 - e^{-\tau_v})\,dv = \int_{-\infty}^{\infty} \left[1 - e^{-k_v \rho_a \ell}\right]dv \tag{3.34}$$

Two approximations for EW are useful. In the weak line approximation $\tau_v \ll 1$, we have $e^{-\tau_v} \approx 1 - \tau_v$ and assuming a Lorentz lineshape for k_v, for example, Equation 3.22, we get

$$EW = s_\ell \rho_a \ell \tag{3.35a}$$

(unit for s_ℓ cm g^{-1} then EW in cm^{-1}). In such a situation, the EW increases linearly with the column density. That expression is also valid for the Doppler lines and hence, in the weak limit, the EW is independent of the line shape.

On the other hand, as the optical depth increases, the line profile becomes saturated. When $\tau_v \gg 1$, we are under the strong-line approximation, and then it can be shown that for a Lorentz line profile

$$EW = 2(s_\ell \gamma_{LP} \rho_a \ell)^{1/2} \tag{3.35b}$$

whereas for a Doppler line profile we have

$$EW = 2\gamma_D \left[\ln\left[\frac{s_\ell \rho_a \ell}{(\gamma_D \sqrt{\pi})}\right]\right]^{1/2} \tag{3.35c}$$

A graph of the EW as a function of the amount of absorber (column density $\rho_a \ell$) is called the curve of growth and is different for different line profiles and parameters (Figure 3.17C). At small optical depth, the absorption occurs in the core of the line profile, and the EW varies linearly with τ. At large optical depths, the core of the line becomes saturated (all the energy is removed) and absorption in the weaker wings becomes important. It can be used to determine the abundance of an element in a planetary atmosphere from the observed line width.

3.4 MOLECULAR SPECTRUM

The molecular spectrum is more complex due obviously to the existence of many atoms with a large variety of geometrical configurations. There are two basic types of molecular signatures in planetary spectra. On the one hand, there are very broad absorptions that correspond to molecules with no permanent dipole moment. In dense atmospheres, collision-induced dipole transitions are observed to occur for CO_2–CO_2 in Venus, N_2–N_2 in Titan, and H_2–H_2 and H_2–He in the giant planets. On the other hand, there are discrete absorptions due to rotational, vibrational-rotational, and electronic transitions. In general, transitions involve changes in the electronic state, accompanied by corresponding changes in the vibrational and rotational energies. The lines may be grouped close together or intensely enough to form large absorption bands or, in some cases, as for methane in Uranus and Neptune, a pseudocontinuum because of their very high density.

The electronic transitions in a molecule occur between energy levels that have similar energy to those of their constituent atoms, although the energy level structure is different. The potential energy of the molecule must include the terms due to the attraction between electrons and protons and those due to the repulsion between the electrons (themselves) and the protons (themselves). For diatomic molecules, there are different empirical or semiclassical approaches for the potential energy of the system. A useful empirical expression is that represented by the Morse potential

$$E_p(r) = E_D \left\{ 1 - \exp\left[-C_0(r - r_0) \right] \right\}^2 \tag{3.36}$$

where
    r is the interatomic distance
    the constants E_D, C_0, and r_0 are adjustable parameters for each molecule
    E_D is the dissociation energy for the molecule when it is in a minimum state of
        potential energy
    r_0 is the atomic equilibrium distance

For diatomic molecules with ionic binding (e.g., NaCl, HCl,...) another useful expression is

$$E_p(r) = -\frac{q_e^2}{4\pi\varepsilon_0 r} + \frac{C_1}{r^9} \tag{3.37}$$

TABLE 3.1

Atomic Parameters for Some Diatomic Molecules

Molecule	Type	E_D(eV)	r_0 (Å)	$h^2/8\pi^2 I$(eV)	$h\nu_0$(eV)
H_2	(1, 3)	4.48	0.74	8×10^{-3}	0.543
O_2	(1, 3)	5.08	1.21	1.78×10^{-4}	0.194
N_2	(1, 3)	7.37	1.09	2.48×10^{-4}	0.292
CO	(2, 3)	11.11	1.13	2.38×10^{-4}	0.268
HCl	(2, 4)	4.43	1.27	1.31×10^{-3}	0.369

Source: Adapted from Alonso, M. and Finn, E.J., *Fundamental University Physics*, Vol. III (*Quantum and Statistical Physics*), Addison-Wesley, Reading, MA, 1968.

1, Homonuclear; 2, permanent dipole; 3, covalent binding; 4, ionic binding.

E_D = dissociation energy; r_0 = atomic equilibrium distance.

The first term represents the Coulomb attraction between the ions and the second accounts for the repulsion between the nucleus and electronic charges. C_1 is an experimentally adjustable constant. The minimum in the potential energy corresponds to the dissociation energy for the ground state. Table 3.1 lists the dissociation energies for some molecules of interest in planetary atmospheres.

The electronic transitions occur in the ultraviolet-visible part of the spectrum, but in addition to it, the atoms forming a molecule can rotate and vibrate with different modes, and the energy associated with these motions is also quantized forming a system of discrete levels. In what follows, we use a simple description to interpret the molecular spectra.

3.4.1 VIBRATIONAL ENERGY LEVELS

Since the nuclei in atoms have masses (m_N) much larger than those of the electrons (m_e), the vibrational motions of the atoms involve energies that are smaller than the electronic ones by a factor $(m_e/m_N)^{1/2}$ and the transitions between energy levels giving rise to the vibrational spectrum fall in the near- and mid-infrared.

Consider the simple case of a diatomic molecule composed of two atoms with masses m_1 and m_2 such that they vibrate like a spring characterized by a constant K. The force acting on the atom masses is proportional to their separation x from an equilibrium value x_0, and therefore the potential energy of the system is that of an harmonic oscillator,

$$E_p(x) = \frac{1}{2}K(x - x_0)^2 \tag{3.38}$$

and the oscillation frequency of the system is given by

$$v_0 = \frac{1}{2\pi} \left(\frac{K}{\mu_M} \right)^{1/2} \qquad (3.39)$$

where $\mu_M = m_1 m_2 / (m_1 + m_2)$ is the reduced mass. The quantum mechanics theory shows that the allowed energy levels corresponding to this oscillator are quantized according to

$$E_v = h v_0 \left(v + \frac{1}{2} \right) \qquad (3.40)$$

where v is the vibrational quantum number $v = 0, 1, 2,....$ The vibration energy levels are nondegenerate with only one molecular state corresponding to each energy level E_v. The selection rule corresponding to dipolar transitions is $\Delta v = \pm 1$ and thus corresponds to an energy change $\Delta E_v = \pm h v_0$. The transition from the vibration ground state (lowest energy) to the next highest vibrational state is called the fundamental transition. From the typical values involved in the molecular parameters in Equation 3.37, the transitions give rise to vibration lines that occur in the infrared wavelength range (1–20 μm) of the spectrum. See Table 3.1 for values of these parameters for diatomic molecules.

For molecules with more than two atoms, the treatment of the vibrations is more complicated, but the spectrum reflects the possible normal modes of vibration. For example, a linear molecule composed of N atoms has $3N - 5$ vibrational degrees of freedom, but a nonlinear molecule has $3N - 6$ vibrational modes (see Figure 3.18).

3.4.2 ROTATIONAL ENERGY LEVELS

Atomic rotations within a molecule involve energies that are a factor of (m_e/m_N) times smaller than electronic transitions, and therefore the spectrum of rotational transitions fall in the far-infrared and millimeter-wave region.

For the simplest case of a diatomic molecule, the rotation of the atoms around the center of mass of the system is described in terms of their angular momentum and the moment of inertia of the atoms $I = m_N x_0$ (x_0 is the distance to the center of mass). Since the atomic angular momentum is quantized, we can write $L^2 = J(J+1)\hbar^2$, where $J = 0,1,2,...$ and the corresponding allowed rotational energy levels are given by

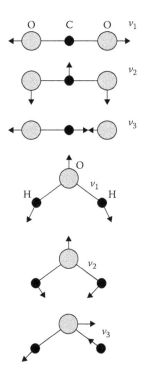

FIGURE 3.18 Scheme of the vibrational modes of two types of triatomic molecules. Upper series is for CO_2: ν_1 and ν_2 are the symmetric modes and ν_3 the asymmetric mode. The lower series is for H_2O, a nonlinear triatomic molecule.

$$E_J = \frac{L^2}{2I} = \frac{1}{2I}J(J+1)\hbar^2 \qquad (3.41a)$$

The energy levels are degenerate with $2J+1$ states (called statistical weight) cor-responding to the same value of the energy E_J. In Table 3.2, we show the classifica-tion of molecules according to their symmetry and values of the moments of inertia along the three principal axes of rotation (x, y, z), which, by convention, are ordered as $I_1 \le I_2 \le I_3$. Equation 3.41a is valid for linear rotors and spherical top molecules, but it can be modified conveniently in the case of molecules with three or more atoms such that they have different moments of inertia. The rotational energy levels of sym-metric rotors can be written as

$$E_J = \frac{L^2}{2I_1} + \frac{1}{2}\left(\frac{1}{I_2} - \frac{1}{I_1}\right)L_z^2 \qquad (3.41b)$$

where I_1 corresponds to the x- and y-axis, I_2 to the z-axis, $L^2 = J(J+1)\hbar^2$, and $L_z = m_J\hbar$. The rotational energy levels of asymmetric rotors are much more complex, and they have apparently a random distribution of energy levels (e.g., water molecule). Additional effects that complicate the rotational level structure originate from cen-trifugal forces that stretch the bonds between atoms. Empirically, for a diatomic molecule (linear rotor) Equation 3.41a is modified according to

$$E_J = \frac{\hbar^2}{2I}\left\{J(J+1) - \delta_s\left[J(J+1)\right]^2\right\} \qquad (3.41c)$$

where δ_s is the stretching constant.

The molecules must have a permanent dipole moment to allow for pure rota-tional transitions, which means that neglecting the electron contribution, the nuclear charges must be displaced from the center of mass. Homonuclear (symmetric)

TABLE 3.2

Molecule Types according to Rotational Structure

Type	Moments of Inertia	Examples
Linear rotor[a]	$I_1=0, I_2=I_3$	CO_2, C_2H_2
Symmetric rotor or top	Prolate: $I_1=I_A, I_2=I_3=I_B$	C_2H_6, C_2H_4
	Oblate: $I_3=I_C, I_1=I_2=I_B$	NH_3, AsH_3, PH_3
Asymmetric rotor or top	$I_1<I_2<I_3$	H_2O, O_3, H_2S, C_3H_8
Spherical top	$I_1=I_2=I_3$	CH_4, GeH_4

Source: Adapted from Irwin, P.G.J., *Giant Planets of Our Solar System: An Introduction*, Springer, Berlin Heidelberg, Germany; Praxis, Chichester, U.K., 2009.

[a] Diatomic molecules are by definition *linear rotors*.

---

diatomic molecules such as H_2, O_2, etc., and symmetric linear polyatomic molecules do not have permanent dipolar electric momentum and therefore they do not have pure rotational transitions (see Table 3.1). The selection rule for a dipolar transition is $\Delta J = \pm 1$. The energy difference for a rotational transition between two consecutive energy levels is given by

$$\Delta E_J = \frac{\hbar^2}{I}(J+1) \tag{3.42}$$

The line strength of the rotational lines depends on the population of the lower rotational state (dominant factor for rotational bands) and the transition probability derived from quantum mechanics. Under thermodynamic equilibrium, the population of states varies according to the Boltzmann distribution (see Equation 3.19):

$$N_J = N_0 g_J \exp\left(-\frac{E_J}{k_B T}\right) \tag{3.43a}$$

where
  g_J is the degeneracy of the Jth level
  E_J is the energy of the Jth level

Using Equation 3.41a for linear rotors and setting the degeneracy to $2J+1$, the number of molecules in the Jth rotational energy state is given by

$$N_J = (2J+1)N_0 \exp\left(-\frac{J(J+1)\hbar^2}{2Ik_B T}\right) \tag{3.43b}$$

From the rotational parameters of the molecules, the spacing between rotational levels is smaller than that between the vibrational levels. The rotational lines usually occur in the far infrared and microwave wavelength range (100–10^4 μm). See Table 3.1 for the parameter values.

3.4.3 VIBRATION–ROTATION BANDS

For solar photons with a short wavelength (<1 μm), the electronic transitions are excited. In addition, since the energy of the rotational transitions are much smaller than the vibrational ones, the rotational transitions can also be excited when the vibrational ones are forming a combined rotation–vibration spectra.

As a first approach, the molecular internal energy can be expressed as the sum of the electronic, vibrational, and rotational terms: $E = E_e + E_v + E_J$. The frequency of the radiation emitted or absorbed during the electronic transition is given by

$$v = \frac{\Delta E}{h} = \frac{\Delta E_e + \Delta E_v + \Delta E_J}{h} = v_e + v_v(v'',v') + v_\ell(J'',J') \tag{3.44}$$

where v_e, v_v, and v_ℓ are the frequencies corresponding to the transitions between the electronic, vibrational, and rotational states, respectively.

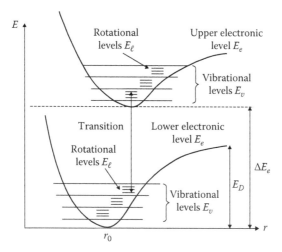

FIGURE 3.19 Rotational and vibrational energy levels of a molecule associated to two electronic states. A transition between two rotovibrational levels is indicated.

For a given electronic transition, the spectra consist of a series of bands, with each band corresponding to given values of v'' and v', and to all possible values of J'' and J'. The selection rules determined by quantum mechanics indicate the numbers permitted for the transitions. The whole molecular spectra is formed by the vibration–rotation bands denoted as *branches* (O, P, Q, R, S) with peaks corresponding to transitions $\Delta J = -2, -1, 0, +1, +2$ and a band structure that depends on the rotational symmetry of the molecule. In Figure 3.19, we show schematically the energy level structure for a molecule and in Figures 3.20 and 3.21 the simulations of pure rotational and rotational–vibrational spectra, respectively. In Figures 3.22 through 3.28, we present a selection of representative spectra at high spectral resolution to show the band structure (in absorption cold temperatures and in emission hot temperatures) in planetary atmospheres.

3.4.4 SPECIAL CASES

3.4.4.1 Overtones Bands

Anharmonic oscillations of the atoms in a molecule (large displacements that give rise to deviations from the harmonic oscillator) relax the selection rules for dipolar electric transitions to $\Delta v = \pm 2, \pm 3, \ldots$, and give rise to the so-called overtone bands for which $|\Delta v| > 1$.

3.4.4.2 Inversion Bands

The inversion band of ammonia (a trigonal pyramidal molecule) observed in the microwave spectra is the most interesting case of inversion bands for planetary atmospheres and occurs when the N-atom transits above and below the plane formed by the three hydrogen atoms (quantum mechanic tunnel effect). The vibration energy

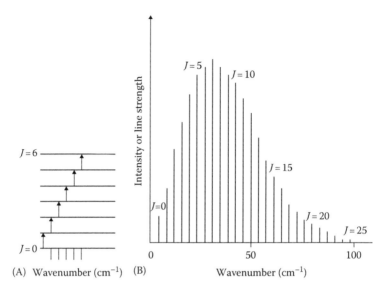

FIGURE 3.20 (A) Some energy levels and transitions are schematically indicated. (B) Simulated purely rotational spectrum of CO for line intensities corresponding approximately to a temperature of 300 K.

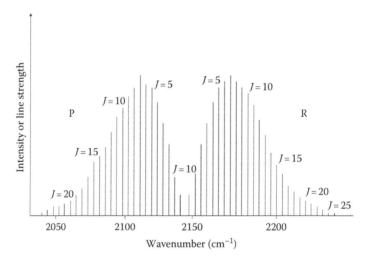

FIGURE 3.21 Simulated rotational-vibrational band ($v = 1 - 0$) of CO for line intensities corresponding approximately to a temperature of 300 K.

levels relative to the ground state for the axial motion of the N-atom correspond to wavelengths from 3.5 μm to 12.66 mm. The fundamental transition from the ground state to the first level originates absorption lines around 12.658 mm. This is a well-known feature in the microwave spectrum of the atmospheres of Jupiter and Saturn.

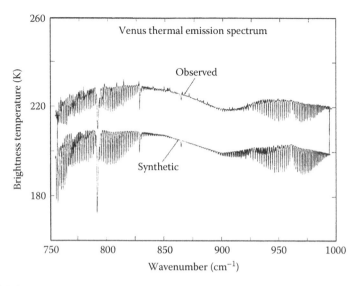

FIGURE 3.22 Measured and synthetic thermal emission spectra of Venus at high resolution obtained with a Fourier spectrometer. The bands correspond to the CO_2 absorption, but the spectrum also shows broad absorption features of liquid sulfuric acid clouds (H_2SO_4) near $900\,cm^{-1}$. (From Kunde, V.G. et al., *Icarus*, 32, 210, 1977. With permission.)

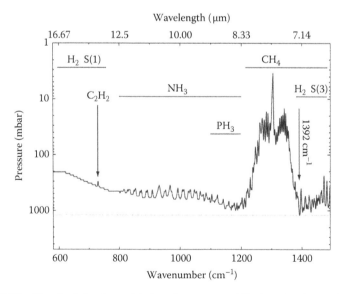

FIGURE 3.23 Detail of the thermal infrared spectrum of Jupiter showing emission lines from hydrocarbons (at high stratospheric temperatures) and absorption bands due to other compounds, including collision-induced absorption (CIA) from H_2. Measurements obtained using the CIRS instrument onboard Cassini spacecraft. (From Matcheva, K.I. et al., *Icarus*, 179, 432, 2005. With permission.)

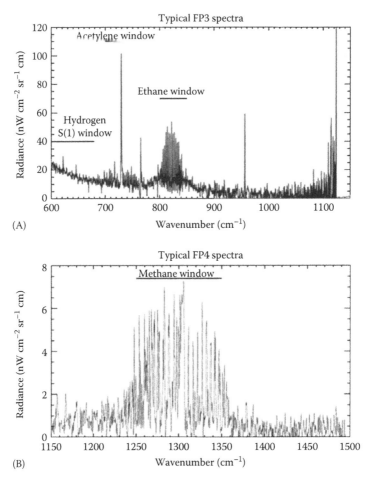

FIGURE 3.24 Details of the thermal infrared spectrum of Saturn showing emission lines from hydrocarbons (at high stratospheric temperatures) and absorption bands due to other compounds, including CIA from H_2. Measurements obtained using the CIRS instrument onboard Cassini spacecraft. (From Howett, C.J.A. et al., *Icarus*, 190, 556, 2007. With permission.)

3.4.4.3 Collision-Induced Absorptions and Quadrupolar Electric Transitions

Homonuclear molecules are formed by one type of element and can contain a number of atoms (two or more, see Table 3.1). The symmetric charge distribution in these molecules impedes the existence of a permanent electric dipole and therefore no dipolar electric transitions can exist. However, a collision-induced dipole can be formed transitorily in dense atmospheres, as occurs with H_2 on the giant planets. The absorption it leads is weak, but the large abundances of H_2 in these planets form broad absorption bands (CIAs) due to H_2–H_2 and H_2–He collisions that dominate the spectra (Figures 3.13, 3.23, and 3.24). Other examples of CIA are the N_2–N_2 and N_2–CH_4 CIAs observed in Titan.

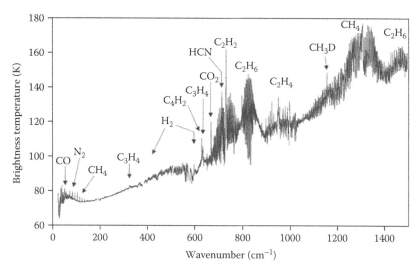

FIGURE 3.25 Details of the thermal infrared spectrum of Titan showing emission lines from hydrocarbons (at high stratospheric temperatures) and absorption bands due to other compounds. Measurements obtained using the CIRS instrument onboard Cassini spacecraft. (From Coustenis, A. et al., *Icarus*, 189, 35, 2007. With permission.)

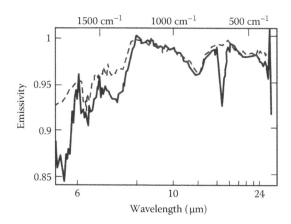

FIGURE 3.26 The spectrum of the Martian dust, soil, and atmosphere taken by the Mars Exploration Rover Spirit's mini-thermal emission spectrometer (continuous line) compared to that obtained by the Mars Global Surveyor's orbital vehicle thermal emission spectrometer (dashed line). Note the prominent atmospheric CO_2 absorption band at 15 μm. From image PIA05030. (Courtesy of NASA-JPL, Los Angeles and ASU, Tempe, AZ.)

The CIA by hydrogen molecules is particularly relevant since it dominates the far infrared spectra of the Jovian planets (10 μm < λ < 1000 μm). The nuclear spin of the two hydrogen atoms in the molecule can be aligned (hydrogen ortho-state) or antialigned (hydrogen para-state) known as spin isomers. Transitions between ortho-hydrogen and parahydrogen are highly forbidden but occur in the giant planets over

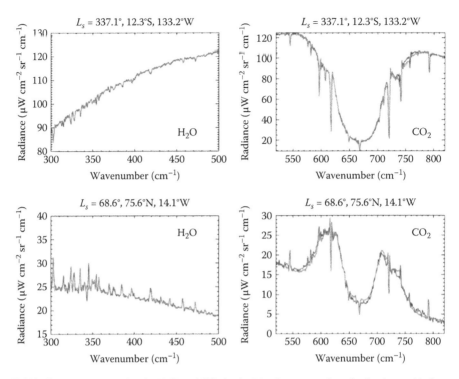

FIGURE 3.27 Absorption by H_2O and CO_2 in the Martian atmosphere in the thermal infrared for two different latitudes and seasons, measured using the PFS/LW instrument onboard Mars Express. (From Fouchet, T. et al., *Icarus*, 190, 32, 2007. With permission.)

long periods of time due to interactions with atomic hydrogen and other species. At high temperatures, the ratio of orthohydrogen to parahydrogen hydrogen is 3:1 (75% ortho and 25% para) as determined by the nuclear statistical weights and is called "a normal distribution." At lower temperatures (as in the giant and icy planets), the ratio decreases to 1:1 and is called "equilibrium distribution" since a tendency to similar values of both forms of hydrogen occurs on long temporal scales due to collisions with other species. The *ortho* and *para* hydrogens act as two different gases, and the specific heat and thermal behavior of the atmosphere depends on the distribution. Their importance for the giant planets will be discussed in more detail in the following chapters.

In addition, although a permanent electric dipole is not present in these molecules, a quadrupole charge distribution can occur allowing for weak quadrupole electric transitions with selection rules $\Delta J = \pm 2$ and $\Delta \upsilon = \pm 1, \pm 2, \pm 3, \dots$. They are called O-branch ($\Delta J = -2$) and S-branch ($\Delta J = 2$) as observed, for example, with H_2 in the red and near infrared spectra of the giant planets. Table 3.3 summarizes the wavelength ranges for the molecular bands of the main compounds in planetary atmospheres.

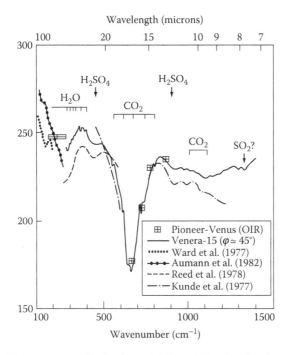

FIGURE 3.28 Venus spectrum in the thermal infrared showing the absorption due to different compounds but dominated by the prominent CO_2 absorption band at $15\,\mu m$. (From Moroz, V.I. *Adv. Space Res.*, 5(11), 197, 1985. With permission.)

3.4.5 Band Transmittance, Absorptance, and Equivalent Width

Measurements of the atomic and molecular absorption coefficients are carried out in laboratory experiments trying to simulate the atmospheric conditions. In the laboratory, however, the gas paths are homogeneous and do not allow the inclusion of large temperature and pressure variations of the atmosphere. Hence, significant errors can be introduced when calculating the transmittances if those variations are not included. The individual line strengths and line widths forming a molecular band are functions of temperature and pressure, and their spectral separation can be narrow. In addition, in atmospheres with multiple composition lines and bands, the lines can blend and superimpose.

When the individual lines of the molecular bands are blended, a *line-by-line* integration over frequency (or wavelength) can be performed by calculating the absorption coefficient for each gas under the specific atmospheric conditions. The monochromatic absorption of a single line (Equation 3.30) must be summed for all absorption lines of all gases (weighted by their abundance) that contribute to the optical path at the given wavelength. The monochromatic transmittance traversing a plane-parallel layer located at altitude levels z_1 and z_2 must include the contribution of all these compounds, and the fact that the temperature and pressure vary across this layer,

TABLE 3.3

Molecular Bands of the Main Compounds in Planetary Atmospheres

Species	Bands (Wavelength, μm)	Planets/Satellites
H_2	VR (0.62, 0.82, 1.2, 2.4), R-CIA (17, 28)	J, S, U, N, T
CO_2	VR (0.78–4.3, 15)	V, E, M
O_2	0.175–0.2, 0.63, 0.69, 0.76, 1.06, 1.27, 1,58	E, M
O_3	C (0.2–0.3, 0.5–0.7), E (0.3–0.34), 3.3, 4.74, VR (9.6)	E, M
CH_4	VR (0.4, 0.619, 0.725, 0.89, 1.1–2.3, 3.3, 7.6)	E, J, S, U, N, T, P, Tr
NH_3	0.65, 0.79, VR (8–12), R (40–503), I (13 mm)	E, J, S, U, U, N
H_2O	VR (0.66–2.66, 2.73, 6.3), C (8–12), R(10–1000)	V, E, M, J, S, U, N
SO_2	E (0.2–0.3), VR (7.35, 8.7, 19.3)	V, I
N_2O	VR (2.11–4.5, 7.8, 17)	E
C_nH_n	VR (10–45)	E, J, S, T, U, N

Notes: C, continuum; E, electronic; VR, vibration–rotation; R, rotation; I, inversion; CIA collision, induced absorptions. For hydrocarbons, see detected species in Tables 3.5 and 3.6. Planets and satellites by their initials: V, Venus; E, Earth; M, Mars; J, Jupiter; I, Io; S, Saturn; T, Titan; U, Uranus; N, Neptune; Tr, Triton; P, Pluto.

$$T_v(z_1, z_2) = \exp\left(-\int_{z_1}^{z_2} N(z)\left(\sum_j x_{Vj}(z) \sum_i k_{ij}(v, P(z), T(z))\right) dz\right) \qquad (3.45)$$

where

$N(z)$ is the number density of the atmosphere at altitude level z (molecules per cm^3)

$P(z)$ and $T(z)$ are the pressure and temperature in the layer, respectively

x_{Vj} is the mole fraction of gas j at altitude z (see below)

The summations are over all the absorption lines i of all gas j contributing to the transmission at this wavelength. For this purpose, one can use, for example, available spectroscopic databases such as HITRAN (extensively used for terrestrial atmospheric calculations) and GEISA (that includes more exotic gases than those found on Earth). This is computationally costly, depending obviously on the band structure.

Alternatively, there are different theoretical band models that are idealized representations of the blended lines (usually corresponding to the Lorentz shape profile)

along a given spectral interval. The choice of a band model depends on the type of absorber and on its spectral characteristics. A full discussion of these models is beyond the scope of this introductory book and they are presented in detail in a number of spectroscopy and radiative transfer books given in the references. Below, we outline the fundamentals of the most commonly employed methods. To illustrate the large-scale band structure in planetary atmospheres, we show in Figures 3.27 and 3.28 the varying shape of the CO_2 15 µm absorption band in Venus and Mars.

It is often useful to calculate the transmittance function (Equation 3.45) over a spectral band or interval Δv that contains many spectral lines, for example, the *average band transmittance* and *band absorptance*:

$$\langle T \rangle = \frac{1}{\Delta v} \int_{\Delta v} T_v \, dv \tag{3.46a}$$

$$\langle A \rangle = 1 - \langle T \rangle \tag{3.46b}$$

According to Equation 3.45, the integration over frequency requires the knowledge of the frequency dependence of the absorption coefficients $k_{ij}(v)$ along the band and along the optical path. The band EW is then given by

$$\langle EW \rangle = \int_{\Delta v} (1 - T_v) \, dv = \Delta v (1 - \langle T \rangle) = \Delta v \langle A \rangle \tag{3.47}$$

If lines in a band or interval Δv do not overlap, then the average transmission over the spectral interval Δv will be

$$\langle T \rangle = 1 - \frac{\sum_i EW_i}{\Delta v} \tag{3.48}$$

where EW_i is the equivalent width of the ith line.

3.4.5.1 Band Models

Consider first the case of a single absorbing compound. The averaged transmission in a wavenumber interval Δv (Equation 3.46a) computed line by line (sometimes labeled LBL) between two pressure levels P_1 and P_2 in a plane-parallel atmosphere (Equation 3.45) under hydrostatic equilibrium (taking for zero zenith angle) is given by

$$\langle T(P_1, P_2) \rangle = \frac{1}{\Delta v} \int_{\Delta v} dv \exp \left[-\frac{1}{g} \int_{P_1}^{P_2} x_M \, dP \left(\sum_{i=1}^{N} k_i(v, P, T) \right) \right] \tag{3.49}$$

where
  x_M is the mass mixing ratio (see below)
  k_i is the monochromatic absorption coefficient due to the ith line
  N is the total number of lines contributing to absorption at this wavenumber

The principal advantage of this method is that all the available information about the line strengths, shapes, and positions, as well as their dependence with pressure and temperature can be incorporated. However, this integration must be evaluated numerically, which is costly, as indicated above. Here are some alternative band models:

1. Elsasser model represents a good approximation to linear molecules (such as CO_2) whose lines are quasiregularly spaced and with nearly identical shapes. The model substitutes the absorption band by an array of infinite number of evenly spaced lines with the same shape (Lorentz line profile) and strength. The absorption coefficient of the band is given by the superposition of all these lines, and then

$$k_{vB} = \sum_{n=-\infty}^{\infty} \frac{s_\ell}{\pi} \frac{\gamma_L}{\left[(v-n\delta)^2 + \gamma_L^2\right]} \tag{3.50}$$

for n spectral lines separated by the distance δ. It can be demonstrated that such infinite sum can be expressed as the sum of periodic and hyperbolic functions

$$k(p) = \frac{s_\ell}{\delta} \frac{\sinh \beta}{\cosh \beta - \cos p} \tag{3.51}$$

where $p = 2\pi v/\delta$ and $\beta = 2\pi \gamma_L/\delta$. Taking into account the periodicity, the average transmittance can be calculated as

$$\langle T(u_M) \rangle = \frac{1}{\delta} \int_{-\delta/2}^{\delta/2} \exp\left(-k_v u_M\right) dv = \frac{1}{2\pi} \int_{-\pi}^{\pi} \exp\left(-k_v(p)u_M\right) dp \tag{3.52}$$

This expression does not have a representation in closed form, although asymptotic limits can be derived. For large values of β (small distances between spectral lines), it can be shown that

$$\langle T(u_M) \rangle = \exp\left(-\frac{s_\ell u_M}{\delta}\right) \tag{3.53}$$

and the term s_ℓ/δ may be considered as an absorption coefficient of a continuous spectrum. For small values of $\beta (\delta \gg \gamma_L)$, the distance between lines is greater than the half-width, and the overlap effect is negligible for weak lines. However, considerable overlap will occur in the wings of strong lines. In this case, $\cosh \beta \sim 1$ and $\sinh \beta \sim \beta$ and (3.52) can be written as

$$\langle T(u_M) \rangle = \frac{1}{\pi} \int_1^\infty \frac{\exp(-my)}{y\sqrt{y-1}} \, dy \tag{3.54}$$

where $m = s_\ell u_M \beta / 2\delta$ and $y = \sin^{-2}(p/2)$.

2. *Goody random band model* applies to the cases where the lines appears to be spaced "randomly" (as, for example, in triatomic molecules such as H_2O). In this case, their line strength is assumed to follow some statistical law for a given line profile. The model assumes that the lines are separated by a mean spacing δ, have a Lorentz profile, and their intensities follow a Poisson distribution $P(s_\ell) = (1/\sigma_l)\exp(-s_\ell/\sigma_l)$, where σ_l is the average line intensity in the given spectral interval. For this model, the transmittance is given by

$$\langle T(u_M) \rangle = \exp\left[-\frac{2\pi \langle u_M \rangle y}{\sqrt{1 + 2\langle u_M \rangle}} \right] \tag{3.55}$$

where $\langle u_M \rangle = \sigma_l u_M / (2\pi \gamma_L)$, $y = \gamma_L / \delta$. This equation can be rewritten in terms of physically meaningful parameters such as the average absorption coefficient $k_\nu = \sigma/\delta$, given in unit (cm-amagat)$^{-1}$ instead of cm^{-1} (see definition in Section 3.5) and the pressure coefficient $y_\nu = \gamma_{L0}/\delta = \gamma_L/(P\delta)$ (in bar^{-1}), as

$$\langle T(u_M) \rangle = \exp\left[-\frac{k_\nu u_M}{\sqrt{\left[k_\nu u_M / \pi y_\nu P \right] + 1}} \right] \tag{3.56}$$

In terms of the equivalent width EW_i of each line, the average transmission of a spectral interval of width $\Delta\nu$ containing a large number of independent (no overlapping) lines is given according to this model as

$$\langle T \rangle = \exp\left(-\frac{\sum_i EW_i}{\Delta\nu} \right) \tag{3.57}$$

3. *Curtis–Godson model* is a useful approximation for the transmittance in inhomogeneous atmospheric paths where the pressure, temperature, and absorber density vary. It may be shown that the mean transmission may be approximated by the mean transmission of an equivalent homogeneous path whose mean pressure $\langle P \rangle$ and temperature $\langle T \rangle$ weighted by the absorber density ρ_a are given by

$$\langle P \rangle = \frac{1}{u_{Ma}} \int_{z_1}^{z_2} P(z)\rho_a(z) \, dz, \quad \langle T \rangle = \frac{1}{u_{Ma}} \int_{z_1}^{z_2} T(z)\rho_a(z) \, dz \tag{3.58}$$

where $u_{Ma} = \displaystyle\int_{z_1}^{z_2} \rho_a(z)\,dz$ is the integrated mass per unit cross-sectional area along the path. In practice, the inhomogeneous atmosphere is represented by a series of equivalent homogeneous paths, and assuming that no correlation exists between the absorption lines of the different intervening molecules, the total transmission is simply found by multiplying the individual gas transmissions or by summing their optical thicknesses.

3.4.5.2 The Correlated-k Approximation

This method is useful for calculations of the thermal emission and spectra in scattering atmospheres (see Chapters 4 and 5). According to (3.31), (3.45), and (3.46), the average transmission along a uniform path (in pressure and temperature) varies with the absorber amount as

$$\langle T(u_M) \rangle = \frac{1}{\Delta v} \int_{v}^{v+\Delta v} \exp\left[-k_v u_M\right] dv \tag{3.59}$$

where the absorption coefficient at a particular frequency is the summation of all the individual line contributions. This coefficient is a rapidly varying function of wavenumber, and the calculation of (3.59) requires a very small integration step. The correlated k-approximation involves dividing the spectral interval into a large number of very narrow subintervals and rearranging them according to their absorption coefficient. A k-distribution probability density function $f(k)$ is then introduced as

$$f(k) = \frac{1}{v_2 - v_1} \sum_i \Delta v_i \tag{3.60}$$

where
  v_2 and v_1 are the limits of the spectral interval considered
  Δv_i is the width of a subinterval within which the absorption coefficient is between k and $k+dk$

Then, $f(k)dk$ represents the total fraction of the spectral interval where the absorption coefficient is between k and $k+dk$, within a designated spectral bin. This probability treatment means that the transmission computation is transformed from the wavenumber or frequency space (v-space) into the probability space of k-values (k-space). A cumulative k-distribution function is defined as

$$g(k) = \int_0^k f(k)\,dk \tag{3.61}$$

where its inverse k-distribution $k(g)$ is a smoothly varying function, which can be integrated much faster and more accurately in the g-space by using a few quadrature points. The mean transmission is approximately calculated as

$$\langle T(u_M) \rangle = \int_0^1 \exp\left[-k(g)u_M\right] dg \approx \sum_{i=1}^{N} \exp\left(-k_i u_M\right) \Delta g_i \qquad (3.62)$$

where
   k_i is the k-distribution at each of the N quadrature points
   Δg_i are the quadrature weights

The k-distributions for a number of gases at the temperatures and pressures found in planetary atmospheres have been calculated and are available for transmission calculations.

The transmission along a vertically inhomogeneous atmospheric path is obtained by dividing it into a stack of layers having varying temperature, and pressures, and the averaged transmission down to the Mth atmospheric layer is computed as an exponential sum (the correlated-k approximation):

$$\langle T \rangle \approx \sum_{i=1}^{N} \Delta g_i \exp\left(\sum_{j=1}^{M} k_{ij} u_{Mj}\right) \qquad (3.63)$$

where k_{ij} (T,P) and u_{Mj} are the values for the jth layer. The averaged absorption optical depth is obtained as

$$\langle \tau \rangle = -\ln\langle T \rangle \qquad (3.64)$$

Details of the practical use of this method can be found in the references.

3.5 ATMOSPHERIC COMPOSITION DEFINITIONS

The first fundamental application of a planetary spectrum is the identification of the gases in the planetary atmosphere that have originated it, and to determine their abundances. In order to compare the compounds' abundances in atmospheres with those in the Sun, we first present in Table 3.4 the solar abundances of the major elements. Then in Tables 3.5 through 3.7 we summarize the measured atmospheric composition of planets and satellites with atmosphere. In general, the values in these tables refer to the bulk composition (except when indicated). However, the different physical and chemical processes produce temporal and spatial variabilities from the mean composition. The main deviations occur, in general, in the vertical abundances due to condensation processes (and haze and cloud formation), photochemistry, and transport mechanisms (e.g., diffusion) (Figure 3.29). All these mechanisms will be presented in detail in Chapters 5 and 6. Here, we focus on the bulk composition, introducing the basic definitions for the composition of an atmosphere that will extensively be used throughout the book.

Let us denote by N_i the number of molecules of species i in a given parcel of gas and by n_i the number of moles of such species defined as $n_i = N_i/N_A$, where N_A is the Avogadro number (6.022×10^{23} molecules per mol). The molecular weight for this species is defined as μ_i, which is the sum of the weights of the atoms of which it is made (μ_i is given in g mol^{-1} or in kg kmol^{-1}). The total number of molecules and moles of the parcel of gas are $N_T = \sum_i N_i$, $n = \sum_i n_i$. We can rewrite $n_i = m_i/\mu_i$, where m_i is the mass (in kg) of gas species i, so the total mass for the gas parcel is $m = \sum_i m_i$.

The abundance of one compound in a mixture can be specified by its mass mixing ratio:

$$x_{Mi} = \frac{m_i}{m} = \frac{\rho_i}{\rho} \tag{3.65}$$

which is a nondimensional quantity.

As discussed in Chapter 1, the atmospheres of the planets can be assumed to behave like an ideal or perfect gas except at the high pressures found in the deep atmospheres of giant and icy planets. Accordingly, the gas mixture obeys

$$PV = nR_gT \tag{3.66a}$$

where
R_g is the universal gas constant
P is the total gas pressure

Defining the gas constant per unit mass, $R_g^* = R_g/\bar{\mu} = N_A k_B/\bar{\mu}$, where $\bar{\mu}$ is the mean molecular weight of the mixture $\bar{\mu} = \sum_i n_i\mu_i/n$, and k_B is Boltzmann's constant, the ideal gas law can be written as

$$P = \rho R_g^* T \tag{3.66b}$$

and

$$PV = N_T k_B T \tag{3.66c}$$

TABLE 3.4
Solar Abundances of the Major Elements Relative to Hydrogen

Species X	X/H
C	2.91×10^{-4}
N	8.02×10^{-5}
O	5.81×10^{-4}
S	1.83×10^{-5}
Ar	4.22×10^{-6}
Kr	2.27×10^{-9}
Xe	2.22×10^{-10}
As	2.50×10^{-10}
P	3.44×10^{-7}

Source: Lodders, K., *Astrophys. J.*, 591, 1220, 2003.

The hydrogen (H) volume mixing ratio in the Sun is 0.835 and that of helium (He) is 0.195.

TABLE 3.5
Composition of the Atmospheres of Earth, Venus, Mars, and Titan

Species	Earth	Venus	Mars	Titan
N_2	0.7808	0.035	0.027	0.98
O_2	0.2095	0–20 ppm	0.13 ppm	
CO_2	385 ppm (var)	0.965	0.953 (cond)	10 ppb
CH_4	3 ppm (var)		33 ppb (var)	0.016 (cond)
H_2O	<0.03 (var) (cond)	50 ppm	0–300 ppm (cond)	0.4 ppb
Ar	0.009	70 ppm	0.016	30 ppm (Ar^{40})
CO	0.2 ppm	50 ppm	700 ppm	10 ppm
O_3	10 ppm		0.01 ppm	
HCN				0.1 ppm
HC_3N				10–100 ppb
C_2H_2	8.7 ppb			2 ppm
C_2H_6	13.6 ppb			10 ppm
C_3H_8	18.7 ppb			0.5 ppm
C_2H_4	11.2 ppb			0.1 ppm
C_4H_2				1 ppb
CH_3C_2H				30 ppb
C_2N_2				10–100 ppb
NO	<0.01 ppm		3 ppm	
N_2O	0.35 ppm			
NO_2	15 ppb			
SO_2	<2 ppb (var)	60 ppm		
H_2	0.5 ppm		10 ppm	0.002
HCl		0.5 ppm		
HF		5 ppb		
COS		250 ppb		
He	5 ppm	12 ppm		
Ne	18 ppm	7 ppm	2.5 ppm	<0.01
Kr	1 ppm	0.2 ppm	0.3 ppm	
Xe	0.09 ppm	<0.1 ppm	0.08 ppm	

Sources: de Pater, I. and Lissauer, J., *Planetary Sciences*, Cambridge University Press, Cambridge, U.K., 2001 and references therein. Completed with data from Niemann, H.B. et al., *Nature*, 438, 779, 2005; Mumma, M.M. et al., *Science*, 323, 1041, 2009.

Notes: All numbers are given in volume mixing ratios, as a fraction or ppm (part per million, 10^{-6}) or ppb (part per billion, 10^{-9}). For the Earth, there are other species at the level <1 ppb. Largely variable species (in space and time) are marked as (*var*) and major condensable as (*cond*). CO_2 last data from: http://www.mlo.noaa.gov/programs/esrl/co2/co2.html.

TABLE 3.6
Composition of the Atmospheres of Jupiter, Saturn, Uranus, and Neptune

Species	Jupiter	Saturn	Uranus	Neptune
H_2	0.864	0.881	0.85	0.85
He	0.136	0.11–0.16[b]	0.18	0.18
H_2O	2–20×10^{-9} ($P < 50$ mbar)	1.7×10^{-7} (strat)	Detected (strat)	Detected (strat)
	6×10^{-4} (19 bar, cond)[a]	(cond)	(cond)	(cond)
CH_4	2.1×10^{-3a}	4.5×10^{-3}	24×10^{-3} (cond)	35×10^{-3} (cond)
NH_3	2.6×10^{-4} (P) (cond)	5×10^{-4} (cond)	$<2.2 \times 10^{-4}$	$<2.2 \times 10^{-4}$
	8×10^{-4} (8 bar)[a]			
H_2S	7.7×10^{-5} (16 bar) (reac)[a]	(4×10^{-4}) (reac)	(3.7×10^{-4}) (reac)	(10^{-4}) (reac)
^{20}Ne	(2.3×10^{-5})			
^{36}Ar	(1.5×10^{-5})			
^{84}Kr	(5×10^{-9})			
^{132}Xe	(2.3×10^{-10})			
PH_3	6×10^{-7} (dis)	7×10^{-6} (dis)		
GeH_4	7×10^{-10} (dis)	4×10^{-10} (dis)		
AsH_3	2.2×10^{-10} (dis)	3×10^{-9} (dis)		
CO	2×10^{-9} (dis)	10^{-9} (dis)	$<10^{-8}$ (dis)	10^{-6} (dis)
CO_2	Detected	3×10^{-10}		Detected
HCN			$<10^{-10}$ (dis)	$<10^{-9}$ (dis)
C_2H_2	3–20×10^{-8} (phot)	2.1×10^{-8} (phot)	10^{-8} (phot)	6×10^{-8} (phot)
C_2H_4	7×10^{-9} (phot)			Detected
C_2H_6	1–5×10^{-6} (phot)	3×10^{-6} (phot)	$<10^{-8}$ (phot)	2×10^{-6} (phot)
C_3H_4	2.5×10^{-9} (phot)			
C_3H_8	Detected			
C_4H_2	9×10^{-11} (phot)	9×10^{-11} (phot)		
C_6H_6	2×10^{-11} (phot)	2.5×10^{-10} (phot)		
CH_3C_2H		6×10^{-10} (phot)		

Sources: From de Pater, I. and Lissauer, J., *Planetary Sciences*, Cambridge University Press, Cambridge, U.K., 2001 and references therein; Atreya, S.K. et al., *Planet. Space Sci.*, 51, 105, 2003; Fletcher, L.N. et al., *Icarus*, 199, 351, 2009.

Notes: All numbers are given in volume mixing ratios (mole fraction). Some species are given as a function of altitude (pressure level, P). Detected (strat) means low value measurement in the stratosphere. The following marked compounds show variability in space (in particular, vertically) and time: condensable species are marked as (cond); disequilibrium species are marked as (dis); chemical reactive species are marked as (reac), in particular H_2S (see Chapter 6); photochemical species are marked as (phot).

[a] Reanalysis of Galileo data indicates the following deep Jovian values: ratio to solar (element relative to hydrogen): $CH_4 = 3.27 \pm 0.8$; $NH_3 = 2.96 \pm 1.3$; $H_2S = 2.75 \pm 0.66$; $H_2O = 0.29 \pm 0.1$.

[b] There is an uncertainty in the helium value for Saturn, affecting the hydrogen mixing ratio.

TABLE 3.7

Composition of the Atmospheres of Planets and Satellites with Tenuous Atmospheres

Planet	Species	Abundance
Mercury	O	4×10^4 (cm^{-3})
	Na	3×10^4 (cm^{-3})
	He	6×10^3 (cm^{-3})
	K	500 (cm^{-3})
	H	23 (*suprathermal*)–230 (*thermal*) (cm^{-3})
	Ca	30 (cm^{-3})
Moon	He	2×10^3 (day)–4×10^4 (night) (cm^{-3})
	Ar	1.6×10^3 (day)–4×10^4 (night) (cm^{-3})
	Na	70 (cm^{-3})
	K	16 (cm^{-3})
Pluto	N_2	99% (10^{15} cm^{-3}; *estimated*) (*cond*)
	CH_4	0.5%
	CO	(10^{12} cm^{-3}; *estimated*)
Triton	N_2	99% (10^{15} cm^{-3}; *estimated*) (*cond*)
	CH_4	(10^{10} cm^{-3}; *estimated*)
Io[a]	SO_2	10^{11}–10^{12} (cm^{-3})
	SO	*Trace*
	Na	
	K	
	O	
Europa	O_2	99%
	Na	
	K	
Ganymedes	O_2	99%
	H	
Callisto	CO_2	
	O_2	
Enceladus	H_2O	91%
	N_2 or CO	4%
	CO_2	3%
	CH_4	

Source: Data taken from de Pater, I. and Lissauer, J., *Planetary Sciences*, Cambridge University Press, Cambridge, U.K., 2001 and references therein. See also Table 6.9.

Note: Units used are the number density (atoms or molecules cm^{-3}) and relative percentage (%).

[a] Variable spatial and temporarily due to volcano activity. Ionized species in the plasma torus are not listed.

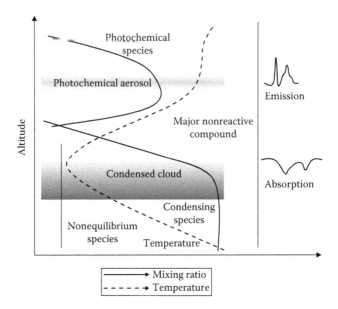

FIGURE 3.29 Scheme of a typical vertical mixing-ratio distribution of species in a dense planetary atmosphere and the basic mechanisms governing it. The vertical temperature structure (see Chapter 4) is also indicated as a reference with molecular emission and absorption occurring at high and low temperatures, respectively. The scheme adjusts approximately to that of the giant and icy fluid planets.

The partial pressure of the gas i is defined as that it would exert if alone, but occupying the volume V at temperature T. Thus, according to (3.66c), it is given by

$$P_i = N_i \frac{k_B T}{V} \tag{3.67}$$

These definitions lead naturally to Dalton's law of partial pressures and volumes:

$$P = \sum_i P_i \tag{3.68}$$

The mass mixing ratio and the partial pressure are then related as

$$x_{Mi} = \frac{\mu_i}{\bar{\mu}} \frac{P_i}{P} \tag{3.69}$$

The volume mixing ratio or mole fraction is given by

$$x_{Vi} = \frac{N_i}{N_T} = \frac{P_i}{P} \tag{3.70}$$

so the mixing ratios are simply related as

$$x_{Mi} = \frac{\mu_i}{\mu} x_{Vi}$$ (3.71)

From the above relationships, the mean molecular weight of the mixture can also be obtained as

$$\bar{\mu} = \sum_i \mu_i x_{Vi}$$ (3.72)

The use of one or another parameter to define the concentration of a given gas in atmospheres depends on the case/problem we are dealing with. Sometimes the abundance of a compound is given relative to the major atmospheric component (e.g., relative to hydrogen in the case of the giant planets). We leave the full definition of the gas concentration when a condensable is present to Chapter 5.

In planetary atmospheres, it is usual to employ as a measure of the abundance the *amagat*. An amagat represents the ratio of the number density of particles at a given temperature and pressure to the number density at the reference standard temperature and pressure (STP). We introduce then the Loschmidt number N_0

$$N_0 = \frac{P_{STP}}{k_B T_{STP}} = 2.687 \times 10^{19} \, cm^{-3}$$ (3.73)

so the number density (molecules per cm³) is given by

$$N = N_0 \frac{\rho}{\rho_{STP}}$$ (3.74)

Since in planetary atmospheres the number density is a function of height, it is customary to calculate the thickness of an equivalent atmospheric column at STP conditions, called the equivalent normalized altitude or the column abundance (unit cm-amagat or km-amagat) given by

$$Z_A = \frac{1}{N_0} \int_z^\infty N(z) \, dz$$ (3.75a)

If the atmosphere is in hydrostatic equilibrium and assuming that the volume mixing ratio and gravity acceleration do not change with altitude, the equivalent altitude is given by

$$Z_A = \frac{N_A}{N_0} \frac{x_{VV}}{\bar{\mu} g} P = V_0 \frac{x_{VV}}{\bar{\mu} g} P = H_0 P$$ (3.75b)

where $V_0 = 22.4 \times 10^3$ cm^3 is the volume occupied by one mole of gas at STP, $H_0 = 2.24 \times 10^5 (x_V / g \bar{\mu})$ is given in km-amagat bar^{-1} when g given in cm s^{-1} and P in bar. We denote the generic volume mixing ratio x_{Vi} as x_{VV}. Thus, if we use the mass absorption coefficient k_ν in kg^{-1} m^2, the absorber abundance u_M is given in kg m^{-2}, whereas when using k_ν in (km-amagat)$^{-1}$, we use for u_M the column abundance Z_A in km-amagat.

PROBLEMS

3.1 From the Planck blackbody radiation law, obtain the expressions for the two limiting cases corresponding to (a) $h\nu \ll kT$ (Rayleigh–Jeans law) and (b) $h\nu \gg kT$ (Wien law).

Solution

(a) $B_\nu(T) \approx (2\nu^2/c^2)kT$
(b) $B_\nu(T) \approx (2h\nu^3/c^2)e^{-h\nu/kT}$

3.2 Consider the hydrogen atom (mass number $A=1$). (a) Calculate the percent of energy correction ($\Delta E/E$) due to the use of the Rydberg constant R_R or R. (b) How much is it for the cases of the hydrogen isotopes deuterium (D, $A=2$) and tritium (T, $A=3$)? (c) What is the expected experimental significance of this energy difference? (d) Calculate the frequency and wavelength corresponding to the first excitation between hydrogen levels ($n=1$ and 2). (e) Calculate the photon energy and corresponding wavelength necessary to ionize the hydrogen atom.

Data: $m_e/m_N = 5.45 \times 10^{-4}/A$
Hint: Since $m_e/m_N \ll 1$, use the series expansion $(1+x)^{-1} \sim 1 - x$.

Solution

(a) $(\Delta E/E) = -5.45 \times 10^{-2}/A$ (%)
(b) $(\Delta E/E)$ (H)=0.0545, $(\Delta E/E)$ (D)=0.0275, $(\Delta E/E)$ (T)=0.0182
(c) It introduces differences in the position of the spectral lines between these isotopes
(d) $\nu = 2.47 \times 10^{15}$ Hz or $\lambda = 121.6$ nm
(e) $E = 13.6$ eV and $\lambda = 91.2$ nm

3.3 Measurements of the pressure-broadened line width of nitrogen in the laboratory gives a value of 0.09 cm^{-1} at STP conditions ($P_0 = 1$ bar, $T_0 = 273$ K). Calculate the mean time between nitrogen molecules collisions at the surfaces of Venus, Earth, Mars, and Triton. What can you say about the application of the LTE conditions if a typical electronic transition takes a time of 10^{-8} s? Use data from Chapter 4.

Solution
Venus (10^{-12} s), Earth (6 $\times$ 10^{-11} s), Mars (7.5 $\times$ 10^{-9} s), Triton (2.2 $\times$ 10^{-6} s). LTE conditions are applicable in this case in Venus, Earth, and Mars (at the limit) but not in Triton.

3.4 Find an expression for the pressure at which the half-width due to a collision broadening is equal to that due to Doppler broadening. Apply it to calculate the pressure for the following cases: (1) CO_2 line at $15\,\mu m$ for $T=500\,K$ in the atmosphere of Venus, (2) H_2O line at $6.25\,\mu m$ at $T=300\,K$ on Earth, (3) CH_4 line at $890\,nm$ at $T=125\,K$ in Jupiter. Assume that all lines have collision-broadened half-widths of $0.1\,cm^{-1}$ at STP.

Data: $R_g/\mu = 190\,J\ kg^{-1}\ K^{-1}$ (Venus), $287.1\,J\ kg^{-1}\ K^{-1}$ (Earth), $3745\,J\ kg^{-1}\ K^{-1}$ (Jupiter).

Solution

$$P = P_0\left[\frac{\tilde{\nu}_0}{c\gamma_{L0}}\left(\frac{2R_gT}{\mu}\right)^{1/2}\left(\frac{T}{T_0}\right)^{1/2}\right]; \quad 0.13\ \text{mbar}, 0.23\ \text{mbar}, 2.5\ \text{mbar}.$$

3.5 Obtain the expression for the equivalent width for a single line under the weak approximation (Equation 3.35a).

3.6 Obtain the expression for the equivalent width for a single Lorentz line under the strong approximation (Equation 3.35b).

Hint: Use $q=s_\ell\rho_a\ell/(\pi\gamma_L) \gg 1$; $x=(\nu - \nu_0)/\gamma_L > q$, and make the necessary approximations and integrate for EW by parts.

3.7 Obtain the expression for the equivalent width for a single Doppler line under the strong approximation (Equation 3.33c).

Hint: Use the definitions $q=s_\ell\rho_a\ell/(\pi\gamma_D)^{1/2} \gg 1$, $x=(\nu - \nu_0)/\gamma_L$, $|x| < \ln q$; make the necessary approximations and integrate for EW.

3.8 From the Morse potential equation (3.36), find the position for the minimum of the electronic potential energy. What is its value at that point? Compare with the vibrational energy of the ground state and derive the dissociation energy for the molecule (E_{De}) considering the vibrational energy.

Solution

$$r = r_0; \quad E_p(r = r_0) = 0; \quad E_{De} = E_D - (1/2)h\nu_0$$

3.9 Derive the constant C_1 in the potential energy of a diatomic ionic molecule (Equation 3.37) if the minimum energy occurs at position r_0. Find the dissociation energy for the molecule $E_{Di}=E_p(r=r_0)$.

Solution

$$C_1 = \frac{q_e^2r_0^8}{36\pi\varepsilon_0}; \quad E_{Di} = -\frac{(8/9)q_e^2}{4\pi\varepsilon_0r_0}.$$

3.10 Calculate the wavelengths of the transitions of a molecule with (1) a dipolar electric transition such that $\Delta E_e=4\,eV$; (2) a vibrational transition between states with $\nu''=4$ and $\nu'=0$ assuming that the molecular vibrational constant

$h\nu_0 = 0.3$ eV is the same for the two states; (3) a rotational transition between states with $J''-1$ and $J'=0$ assuming that the molecular rotational constant $h^2/(8\pi^2\nu_0) = 5 \times 10^{-4}$ eV is the same for the two states.

Solution

(1) 310 nm
(2) 1.03 μm
(3) 1.24 mm

3.11 Find an expression for the band transmittance using the Curtis–Godson model under the strong line broadening approximation for an atmosphere with uniform composition between pressure levels P_1 and P_2 assuming hydrostatic conditions and a fractional mass abundance for the absorber f_C. Assume that the individual lines have parameters $s_{\ell i}$ and γ_{L0i}.

Solution

$$\langle T \rangle = 1 - \frac{1}{\Delta\nu}\left[\frac{2f_C}{gP_0}\left(P_1^2 - P_2^2 \right) \right]^{1/2} \sum_i \left(s_{\ell i}\gamma_{L0i} \right)^{1/2}$$

3.12 Apply the above solution to the case of the atmosphere on Earth. Calculate the transmission for CO_2 (take volume mixing ratio 3.3×10^{-4}) and H_2O (take average volume mixing ratio 10^{-3}) from the top of the atmosphere to the surface (300 K) in the spectral range 775–800 cm^{-1} where

$$\sum_i (s_{\ell i}\gamma_{L0i})^{1/2}(CO_2) = 9.86(g^{-1/2}), \quad \sum_i (s_{\ell i}\gamma_{L0i})^{1/2}(H_2O) = 3.62\,(g^{-1/2}).$$

Solution

$$\langle T \rangle = 0.67(CO_2), \quad \langle T \rangle = 0.79(H_2O).$$

3.13 A model for the near infrared methane absorption bands (1–4 μm) by Owen and Cess (*Astrophys. J.*, 197, L37–L40, 1975) gives the following formula for the equivalent width:

$$EW = 2A_0\ln\left\{ 1 + \frac{x}{2 + \left[x\left(1+\beta^{-1}\right)^{1/2}\right]} \right\}$$

where A_0 is the bandwidth parameter, $x = (\eta S u_M/A_0)$, and $\beta = (4\gamma_0 P/\delta)$, where η is the air mass path factor, S is the band strength, γ_0 is the band half width at STP, δ is the mean line spacing, P is the effective pressure, and u_M is the column abundance in cm-amagat. Spectroscopic observations of the 2.36 μm methane band of Jupiter for observing conditions such that $\eta = 2$ give EW=403 cm^{-1}. From the laboratory measurements, the following values for the parameters have been obtained: $S = 20$ cm^{-2} amagat^{-1}, A_0 (cm^{-1}) = $124(T/300)^{1/2}$, γ_0 (cm^{-1}) = $0.075(300/T)^{1/2}$, and $\delta = 10.5$ cm^{-1}. Assuming that atmospheric conditions in Jupiter are $P = 1$ atm, $T = 165$ K, calculate the methane column abundance from these measurements. Also, find an approximate expression for the error in deriving the column abundance as a function of the uncertainty in measuring EW.

Solution

$$u_M(CH_4) = 40 \text{ m-amagat}, \quad \sigma(u_M) = \left(1 + \beta^{-1}\right)^{1/2} (2\eta S)^{-1} \exp\left(EW/2A_0\right)\sigma(EW)$$

3.14 Find an expression for the optical depth due to a gas absorption band across a vertical path in an atmosphere from its top to the pressure level P, assuming hydrostatic equilibrium, in terms of V_0 and of the averaged absorption coefficient for the band (see also Chapter 5). Apply it to methane in the Jovian atmosphere for the absorption bands at 619 nm, $k_{619} = 6 \times 10^{-6}$ (cm-amagat)$^{-1}$; 725 nm, $k_{725} = 4 \times 10^{-5}$ (cm-amagat)$^{-1}$; 890 nm, $k_{890} = 8 \times 10^{-4}$ (cm-amagat)$^{-1}$, at the pressure level of 200 mbar. Take x_V $(CH_4) = 3 \times 10^{-3}$ and other required data from the tables in the book.

Solution

$$\tau = k_\nu \frac{x_{VV} V_0}{\bar\mu} \frac{P}{g}; \quad \tau(619 \text{ nm}) = 0.0166, \tau(725 \text{ nm}) = 0.11, \tau(890 \text{ nm}) = 2.21.$$

3.15 Laboratory measurements of methane absorption bands show that at 890 nm, the transmittance is 0.68 for a given abundance at a pressure of 0.5 bar. The absorption coefficient $k_\nu = 20.9$ (km-amagat)$^{-1}$ and the pressure coefficient $y_\nu = 14$ (bar)$^{-1}$. Using the Goody random band model (Equation 3.56), obtain the methane abundance. Assume that this abundance corresponds to that found in Jupiter and Saturn. What would be the methane volume mixing ratio in both planets?

Solution
$Z_A = 0.069$ km-amagat; $x_{VV} = 3 \times 10^{-3}$ (Jupiter), $x_{VV} = 1.16 \times 10^{-3}$ (Saturn).

3.16 Find Equation 3.75b and calculate Z_A at 1 bar of hydrogen in the atmospheres of the giant and icy planets using the data from the tables in the book.

Solution

Z_A (km-amagat) for Jupiter = 37.45, Saturn = 100.5, Uranus = 94.08, Neptune = 74.56.

4 Vertical Temperature Structure

This chapter is devoted to exploring the thermal structure of planetary atmospheres. We only deal here with the case of bodies with massive or substantial atmospheres where the heat transport is dominated primarily by radiation, and, secondly, by convection and heat transport by mass motion. In Chapter 6, we will present the thermal structure of bodies with tenuous atmospheres, including the upper part of massive atmospheres due to the similarity of the processes occurring on both, for example, the heat transport by conduction. We consider here the case of dry atmospheres, that is, we first assume that there are no phase transformations in the atmospheric composition (that is, condensation, evaporation, or sublimation) contributing as heat sources and sinks. The existence of condensable compounds and their effects on the vertical thermal structure of the atmospheres are discussed in Chapter 5, where they will be treated in conjunction with the cloud and aerosol formation phenomena.

4.1 VERTICAL STRUCTURE OF THE ATMOSPHERES

The vertical structure of a planetary atmosphere is primarily determined by its mass (or density) distribution with altitude. The gravitational force causes the atmospheric density to decrease with altitude, with density and pressure related via the hydrostatic equation since this balance globally dominates the atmosphere. The vertical temperature structure depends on the heating and cooling sources from the exterior (at the surface and/or at cloud layers) and the interior of the planet, and/or the atmosphere itself. Dynamical transport of heat modulates the thermal structure. Temperature, pressure, and density are related by the equation of state of the ideal gas.

The vertical temperature structure varies both horizontally and in time, although the dominance of stratification and thermal inertia allows defining a mean structure (averaged in time and location). Detailed local and temporal measurements are available not only from the Earth, but also for some bodies where descending probes and landers have been used (Venus, Mars, Jupiter, and Titan). In these and in other bodies, remote sensing observations are used to retrieve the vertical profile, by two methods: (1a) occultation of a star by the planet and measurement of the refractivity profile of the atmosphere across the limbs (UV and visible wavelengths); (1b) radio-occultation by the planet when a spacecraft (making a fly-by or orbiting the body) emitting radio signals is occulted by the planet, sounding the atmosphere across the limb; (2) inversion of the thermal infrared spectra.

4.1.1 HYDROSTATIC EQUILIBRIUM: PRESSURE AND DENSITY SCALE HEIGHTS

Globally, the vertical distribution of pressure and density are determined by the hydrostatic equilibrium with the gravitational force balancing the pressure gradient force (Section 1.5.2). If instead of the radial coordinate r we use the vertical distance z (altitude), valid for vertical distances $\ll$ planetary radius, then (1.107) becomes

$$\frac{dP}{dz} = -\rho g \tag{4.1}$$

Using the equation of state for an ideal gas $(P = \rho R_g T/\mu)$, integration of Equation 4.1 gives

$$P(z) = P(0)\exp\left(-\frac{1}{R_g}\int_0^z \frac{g(z)\mu(z)\,dz}{T(z)}\right) = P(0)\exp\left(-\int_0^z \frac{dz}{H(z)}\right) \tag{4.2}$$

where the *pressure scale height* (the e-folding scale for pressure) is given by

$$H(z) = \frac{R_g}{\mu(z)}\frac{T(z)}{g(z)} \tag{4.3}$$

Similarly, we have for the density,

$$\rho(z) = \rho(0)\exp\left(-\int_0^z \frac{dz}{H^*(z)}\right) \tag{4.4}$$

where

$$\frac{1}{H^*(z)} = \frac{1}{T(z)}\frac{dT(z)}{dz} + \frac{g(z)\mu(z)}{R_g T(z)} \tag{4.5}$$

In the case of an *isothermal* region of the atmosphere where temperature is constant with altitude ($dT/dz = 0$, and then $T = T_0$), both definitions (4.3) and (4.5) are equivalent. The scale height is then $H = H^* = R_g T_0/\mu g$.

From the hydrostatic condition and the gas ideal law, we can derive that $g\,dz = -R_g^* T d(\ln P)$, where, as usual, $R_g^* = R_g/\mu$. Integrating it, we can obtain the thickness of a layer between two surfaces of constant pressure. If the surface with pressure $P = P_1$ locates at the altitude $z = z_1$ (and similarly for pressure $P = P_2$ at altitude $z = z_2$), the thickness is simply given by

$$z_2 - z_1 = -\frac{R_g^*}{g}\int_{P_1}^{P_2} T\,d(\ln P) \tag{4.6}$$

It should be noted that pressure decreases with altitude, whereas height increases upward. For an isothermal layer ($T = T_0$), its thickness is simply

$$z_2 - z_1 = \frac{R_g^* T_0}{g} \ln\left(\frac{P_1}{P_2}\right) \tag{4.7}$$

If the temperature varies across the layer, we first define the averaged temperature

$$\langle T \rangle = \frac{\displaystyle\int_{P_2}^{P_1} T d(\ln P)}{\displaystyle\int_{P_2}^{P_1} d(\ln P)} = \frac{\displaystyle\int_{P_2}^{P_1} T d(\ln P)}{\ln(P_1/P_2)} \tag{4.8}$$

and the thickness can then be calculated as

$$z_2 - z_1 = \frac{R_g^* \langle T \rangle}{g} \ln\left(\frac{P_1}{P_2}\right) \tag{4.9}$$

Note also that in both cases the thickness of a layer between two pressure surfaces is proportional to the mean temperature of that layer.

4.1.2 OBSERVED TEMPERATURE STRUCTURE

The vertical temperature structure $T(z)$ or $T(P)$ is mainly determined by its chemical composition and molecular weight, and by the thermal energy sources and the efficiency of the different mechanisms that intervene in heat transport. An important parameter for the characterization of the energy sources is the rate of change of the temperature with altitude, known as the lapse rate $\Gamma(z) = dT/dz$ (usually in K km^{-1}) since it represents a diagnostic of the dominating energy transport mechanism in the vertical (see Section 1.5.4). It is also customary to use the lapse rate in terms of the temperature change with pressure (dT/dP) or the temperature gradient given by $\nabla = d \ln T/d \ln P$. The use of both terms is found to be mixed or undistinguished in the literature. Here we will keep both definitions.

The vertical temperature structure in the atmospheres of terrestrial planets and satellites with substantial atmospheres (but still thinner when compared to those of fluid giants) is primarily determined by solar radiation (Section 1.4.3). In Figure 3.2, we schematically showed the sources and sinks contributing to heating and cooling in the radiation budget of the Earth (averaged over longitude and over a complete year). However, the flux imbalances in the equatorial–tropical and in the temperate–polar regions between the absorbed and emitted energy are at the end responsible for the heat transport through atmospheric mass motions (dynamics). On the giant planets, the heat emanating from the interior (Section 1.4.2) is added to the solar radiation field (see the measured thermal spectrums in Figures 3.3, 3.14, and 3.15). In Jupiter, Saturn, and Neptune, the absorbed solar flux follows a latitudinal trend similar to

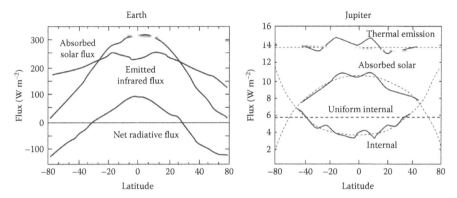

FIGURE 4.1 Radiation budget in the Earth and in Jupiter: comparison of the absorbed solar energy and emitted infrared radiation, averaged with respect to longitude, season, and time of day. (Adapted from Liou, K.N., *An Introduction to Atmospheric Radiation*, Elsevier, Orlando, FL, 2002 and references therein (Earth case); from Pirraglia, J.A., *Icarus*, 59, 169, 1984, for Jupiter.)

that on Earth. But the heat flux emitted from the interior, with a blackbody temperature corresponding to infrared wavelengths, overcomes the absorbed solar radiation (Figure 4.1). Most importantly, this infrared flux does not depend on latitude, which suggests that a meridional heat transport (i.e., across latitude circles) must take place at some unknown level in these atmospheres. In the case of Uranus, things are different since, on one hand, its rotation axis presents an extreme tilt (about 90° relative to the orbital plane, see Table 1.1) resulting in strong seasonal variations, and, on the other hand, the planet lacks an internal heat source (see Table 1.4).

Although the chemical composition, mass, and mean temperature differ strongly among the atmospheres, as we have seen in previous chapters, a basic similar vertical structure prevails in all of them. Atmospheres can be divided in layers parallel to the surface or reference levels that are more or less present in all the planets. Let us briefly describe this layered structure, where each part will be later treated in detail (Figure 4.2). The outer part of the neutral atmosphere is referred to as the *exosphere* (described in detail in Chapter 6) and it represents the external "border" of the atmosphere with the outer space. Collisions between atoms and molecules are rare, so the long path they can describe allows them to escape into space. Its lower limit defines the exobase. Below it, the UV shortwave solar radiation heats up the thermosphere, a region where the temperature increases with altitude and where diffusion and conduction mechanisms intervene significantly, showing high temporal variability in its temperature structure. Below it, we find the mesosphere (its upper frontier is the mesopause) where the temperature is nearly constant (the layer is isothermal) or decreases slightly with the altitude. Further below, we find the stratosphere (with frontier at the stratopause), a region where the temperature increases with altitude due to heat absorption of the solar radiation by molecules as ozone on Earth and a family of hydrocarbons in the giant planets, reinforced there by heating on thin aerosol particle layers. In this region, the radiative heat transport dominates, and due to the strong stratification, wavy motion develops

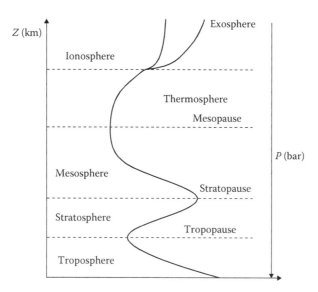

FIGURE 4.2 Scheme of the vertical temperature profile of an archetype atmosphere and its layered structure (named).

easily. These motions also transport heat. Photochemical reactions induced by solar radiation are important in transforming species in all these layers. For most planets, a frontier is found at a pressure level of around 100 mbar (the tropopause), where the temperature reaches a minimum value (the lapse rate is zero). It separates the stratosphere from the troposphere, which is the most massive part of the atmosphere and the lower layer in the planets with a surface. In the troposphere, the temperature decreases (on the average) with altitude due to the dominance in the absorbed sunlight by the lower massive atmosphere and by the planet surface. In the giant planets, the internal heat power source increases with depth and plays a similar role. Convection and advective mass motion become important in redistributing heat in the troposphere. Cloud formation due to condensation (liquid or solid particles) also contributes to modify the vertical thermal structure by means of changes in opacity to solar radiation or to latent heat release. This is the weather layer where most meteorological phenomena occur. On the giant planets, since they lack a surface, the troposphere does not have a lower limit and what is usually defined is an upper weather layer, which is the part where cloud condensation occurs, its upper part being accessible to remote sensing observations (in Jupiter and Saturn, it extends from ~0.1 to 5–10 bar).

We now describe the observed (real) vertical temperature distribution in planets and the satellite Titan that possesses a substantial atmosphere and follow the above layered scheme. The following sections will describe the fundamental theories explaining this structure. The relevant data on the thermal structure of planetary atmospheres are summarized in Tables 4.1 and 4.2. Tables 4.A.1 through 4.A.8 give the tabulated data for the mean (representative) vertical temperature structure of planetary atmospheres.

TABLE 4.1

Mean Specific Heat at Constant Pressure and Specific Gas Constant in Atmospheres

Planet/Satellite	Main Components (Abundance)	$\langle\mu\rangle$ (g mol⁻¹)	R_g^* (J g⁻¹ K⁻¹)	C_p (J g⁻¹ K⁻¹)
Venus	CO_2 (0.96)	44.01	0.19	0.85
	N_2 (0.035)			
Earth	N_2 (0.78)	28.97	0.29	1.00
	O_2 (0.21)			
Mars	CO_2 (0.953)	44.01	0.19	0.83
	N_2 (0.027)			
Jupiter[a]	H_2 (0.864)	2.22	3.75	12.36
	He (0.136)			
Saturn[a]	H_2 (0.85)	2.14	3.89	14.01
	He (0.14)			
Titan	N_2 (0.99)	28.67	0.29	1.04
	CH_4 (0.01)			
Uranus[a]	H_2 (0.85)	2.30	3.61	13.01
	He (0.15)			
Neptune[a]	H_2 (0.79)	2.30	3.61	13.01
	He (0.21)			
HD 209458[b]	H_2 (1.0)	2.00	4.16	14.00

Source: From Sánchez-Lavega, A. et al., Am. J. Phys., 72, 767, 2004.

C_p is a function of temperature. A third-order polynomial dependence of the type. $C_p(T)=a_0+a_1T+a_2T^2$ is usually sufficient for atmospheric calculations within a given temperature range. See Figure 4.21.

[a] For these planets c_p/R_g^* depends on the ortho–para hydrogen ratio. See Figure 4.22.

[b] Representative extrasolar planet of the type "Hot Jupiter."

TABLE 4.2

Thermal Properties and Tropopause Location in Atmospheres

Planet/Satellite	P_0 (bar)	T_0 (K)	$\langle dT/dz\rangle$ (K km⁻¹)	Γ_d (Adiabatic) (K km⁻¹)	H (km)	P_{Tr} (bar)	T_{Tr} (K)
Venus	92	731	−7.7	10.5	16	0.1	250
Earth	1.013	288	−6	9.8	8.5	0.1	217
Mars	0.07	214	−2.5	4.5	18	10⁻⁵	140
Jupiter	1.00	165	−1.9	2.1–2.45[a]	20	0.14	110
Saturn	1.00	134	−0.85	0.7–1.1[a,b]	45	0.08	85
Titan	1.50	94	−1.4 to −1.0	1.3	20	0.1	71
Uranus	1.00	76	−0.75	0.7–1.1[a,b]	30	0.11	53
Neptune	1.00	76	−0.95	0.85–1.34[a,b]	25	0.2	54

Data are for the reference level (P_0, T_0), which is the surface (terrestrial planets and Titan) and the 1 bar level for the giant fluid planets where the upper clouds reside. The approximate tropopause location is at (P_{Tr}, T_{Tr}) (last two columns).

[a] Range from equator to pole due to the dependence of g with latitude.

[b] Depends on the ortho–para hydrogen ratio assumed. See Figure 4.22.

4.1.2.1   Venus

The solar constant at Venus is 2621 W m^{-2}. The incoming solar flux for energy balance is 1/4 due to the ratio between the cross-section area for the intercept of solar radiation and the area of the spherical planet (see Section 1.4.3 for definitions). More than 75% is reflected back to space by the high-albedo SO_4H_2 clouds. About 50% of the sunlight received by the planet is absorbed by the CO_2 dense atmosphere at altitudes $z \sim 65$ km and by an unknown UV-absorber that is mixed with the clouds at altitudes $z \sim 35$–70 km. Another 47% is absorbed by CO_2 between the surface and $z \sim 65$ km. Only 2.6% of the solar flux incident at the top of the atmosphere reaches the surface (about 17 W m^{-2} are absorbed at the surface; see Figure 4.3A).

The high abundance of CO_2 produces a surface pressure of 92 bar and a temperature of 733 K due to the gigantic greenhouse effect (note that the equilibrium temperature due to high albedo clouds is ~ 240 K, Table 1.4). In the troposphere (from the surface to the cloud layers) the lapse rate is ~ 7.7 K km^{-1}. Lacking a stratosphere, a mesosphere extends up to $z \sim 100$ km above the troposphere. A thermosphere, at nighttime, or cryosphere, in daytime, extends upward (Figure 4.4).

4.1.2.2   Earth

In Figure 3.2, we summarized the heat flux balance on Earth. The solar constant is 1370 W m^{-2} so the incoming solar flux for the energy balance is 342 W m^{-2}. The mean surface pressure is 1.013 bar and the averaged temperature is 288 K, which

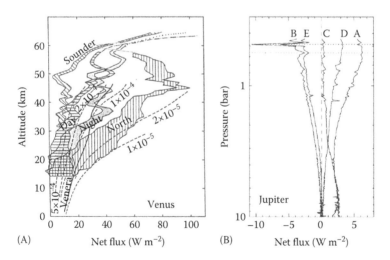

FIGURE 4.3   (A) Net thermal fluxes in Venus' atmosphere as a function of altitude derived from measurements of the Pioneer-Venus probes and Venera spacecrafts. The dashed curves show the modelled thermal fluxes for different water mixing ratios (values given in the curves). (From Revercomb, H.E. et al., *Icarus*, 61, 521, 1985. With permission.) (B) Net thermal flux in Jupiter's atmosphere as a function of altitude as measured in different spectral channels A–E by the Galileo probe in December 1995 ($A = 3$–200 μm; $B = 0.3$–5 μm; $C = 3.5$–5.8 μm; $D = 14$–200 μm; $E = 0.6$–5 μm). (From Sromovsky, L.A. and Fry, P.M., *Icarus*, 157, 373, 2002. With permission.)

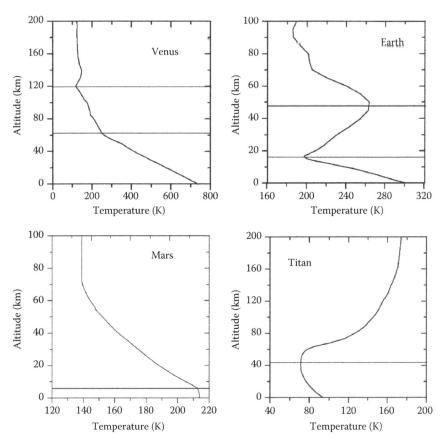

FIGURE 4.4 Vertical temperature structure of the terrestrial planets (Venus, Earth, Mars) and satellite Titan. Data are given in Tables 4.A.1 through 4.A.4.

is $\sim 35\,K$ above the equilibrium temperature due to a greenhouse effect produced by water vapor, CO_2, and CH_4 (Figure 4.4). The troposphere extends from the surface to the level $z \sim 10$–$20\,km$ where the lapse rate is $\sim 6\,K\ km^{-1}$. Above the tropopause in the stratosphere ($z \sim 20$ and $50\,km$), the temperature increases with altitude due to the absorption of UV and IR radiation by the gas ozone (O_3). In the mesosphere, up to $z \sim 90\,km$, the temperature decreases with altitude due to cooling rate by CO_2. Above it in the thermosphere, the temperature increases again with altitude due to the absorption of UV radiation and to the cooling inefficiency by gases at this layer. At $z \sim 150\,km$, the temperature reaches $\sim 400\,K$ where the air density is very low ($< 10^{-5}$ bar). Upward, an ionosphere can be found and the temperature strongly varies in time and with the solar cycle.

4.1.2.3 Mars

Most solar radiation usually reaches the surface of Mars due to the thinness of the atmosphere (surface pressure 6–7 mbar). The exception is the periods when dust storms develop on the surface, blocking the solar radiation. Since the atmosphere is so thin, it is mainly under radiative control and large temporal changes occur in response to the day–night cycle (with corresponding temperature changes of ~ 200 and 300 K, respectively) and to the yearly insolation cycle. The lower ~ 45 km define the troposphere where the lapse rate is ~ 2.5 K km^{-1}. As in Venus, Mars also lacks a stratosphere and a large mesospheric layer extends up to ~ 110 km surrounded by a highly variable thermosphere and an ionosphere (Figure 4.4).

4.1.2.4 Titan

The dense hydrocarbon haze layers present in the atmosphere of Titan between $z \sim 30$ and 300 km have a profound influence on the vertical temperature profile. The solar constant at Titan is ~ 15 W m^{-2}, and of the ~ 4 W m^{-2} available for the energy balance at the top of Titan's atmosphere only 0.1–0.5 W m^{-2} reaches the surface. There, the pressure is 1.46 bar and the temperature 94 K. Close to the surface, the lapse rate is 1.4 K km^{-1} and then decreases to ~ 1 K km^{-1} at $z \sim 4$ km and decreases further to 40 km where the tropopause is located (Figure 4.4). A large inversion then occurs defining an extended stratosphere up to $z \sim 200$ km produced by sunlight absorption by the hazes and hydrocarbons. The temperature then becomes nearly constant with altitude, defining the mesosphere; then a ionosphere follows with the exobase located at $z \sim 1600$ km altitude.

4.1.2.5 Jupiter, Saturn, Uranus, and Neptune

As indicated above, all the giant planets, with the exception of Uranus, differ from the terrestrial planets in that they have an internal energy source of heat that adds to the absorbed sunlight radiation, and in the lack of a solid surface. The heat flux as a function of depth was measured in Jupiter by the Galileo probe and compares reasonably well with what was expected from the models (Figure 4.3B). The atmospheric layer where condensate clouds form defines the troposphere that extends from tens of bars up to ~ 100 mbar where the tropopause is located. Its temperature ranges from 105 K in Jupiter to 50 K in Uranus and Neptune. The lapse rate in the troposphere ranges from ~ 2 K km^{-1} in Jupiter to ~ 0.9 K km^{-1} in the other planets. Above the tropopause, a stratosphere forms due to the hydrocarbon and haze heating extending up to ~ 1 mbar. Upward, a nearly isothermal mesosphere with temperature ~ 150 K extends up to 1 μbar (Figure 4.5).

4.2 RADIATIVE TRANSFER

We have seen in the previous chapters how the radiation fluxes coming from the Sun (or a star in the case of exoplanets) or from the interior of a planet determine its global temperature (equilibrium and effective temperature). We have also described how the insolation above the top of the atmospheres (TOA), or at the surface in airless bodies, can be obtained for any planet from their basic physical properties and orbital characteristics. The heat transport mechanisms were outlined in Section 1.5.4. In this section, we focus on the fundamentals of electromagnetic waves (radiation)

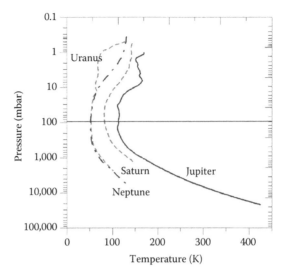

FIGURE 4.5 Vertical temperature structure of the fluid planets (Jupiter, Saturn, Uranus, and Neptune). Data are given in Tables 4.A.5 through 4.A.8.

propagation in planetary atmospheres. We follow a simplified treatment without entering into details of the radiative transfer problem, which is complex enough to be treated comprehensively in a large group of specialized books (see references). In this chapter, we will describe the radiative transfer in a purely gaseous atmosphere, without considering the presence of particulates (aerosols, hazes, and clouds), which will be treated in Chapter 5. This represents a first-order treatment to the interpretation of the temperature structure but is accurate enough since it depends primarily on radiative cooling and heating.

The mixture of neutral gases that forms the massive part of the atmosphere is mostly in molecular form. Some species are strongly active in their interaction with radiation (absorbing and/or emitting it), others are less active (in particular those without a permanent electric dipole), all of them scatter radiation, others suffer chemical reactions (photolysis in the less dense, lighter parts with solar radiation, thermochemistry, or through surface processes), and, finally, others suffer phase changes. These processes, of second-order type for the interpretation of the temperature profile, will be discussed in Chapters 5 and 6.

4.2.1 RADIATIVE TRANSPORT: DEFINITIONS

The propagation of the electromagnetic waves (or photons) along an atmospheric path depends on the geometry between the incident radiation beam, the interacting surface element, and the emergent radiation beam. In Chapters 1 and 3, we introduced the radiative concepts of energy flux F, flux density F_ν or F_λ, intensity of radiation I_ν, and optical depth τ_ν, and the geometry configuration (angles and solid angle) relevant for the incident and transmitted radiation are given in Figure 3.10.

The amount of energy within the frequency interval dv crossing per unit time a differential element of area $\hat{n} \cdot d\vec{S} = dS \cos \varphi$ propagating into a solid angle $d\Omega_s$ along the direction φ is given by

$$dE = I_v \cos \varphi \, dt \, dS \, d\Omega_s \, dv \qquad (4.10)$$

where the intensity is called usually "specific intensity" or "brightness" (in W m^{-2} sr^{-1} Hz^{-1}), and, as can be seen in Figure 3.10, is direction dependent. For example, the specific intensity emitted by a blackbody (Section 3.2.2) is

$$I_v = B_v(T) \qquad (4.11)$$

The *mean intensity* J_v, which is the average upon all the angles, is obtained following the angular integration ($\mu_\varphi = \cos \varphi$)

$$J_v = \frac{\oiint I_v \, d\Omega_s}{\oiint d\Omega_s} = \frac{1}{2} \int_{-1}^{1} I_v \, d\mu_\varphi \qquad (4.12)$$

The flux density is obtained integrating by all solid angles

$$F_v = \oiint I_v \cos \varphi \, d\Omega_s = 2\pi \int_{-1}^{1} I_v \mu_\varphi \, d\mu_\varphi \qquad (4.13)$$

4.2.2 EQUATION OF RADIATIVE TRANSFER

Conservation of energy within an atmospheric parcel governed primarily by photon absorption, emission, and scattering processes conduces to the equation of radiative transfer. The change of intensity dI_v along the path $d\ell$ must result from the balance between the emission and the extinction processes, which can be written as

$$\frac{dI_v}{d\ell} = j_v \rho - I_v \alpha_v \rho \qquad (4.14)$$

In this equation, the propagation of radiation along a path involves the processes of absorption, emission, and scattering by the medium being characterized by the following mass coefficients: (a) absorption k_v (defined in Section 3.3.2, unit cm^2 kg^{-1}); (b) emission j_v (W kg^{-1} sr^{-1} Hz^{-1}); (c) scattering σ_v (cm^2 kg^{-1}), with the first and third giving the mass extinction coefficient $\alpha_v = k_v + \sigma_v$. The optical depth can now be rewritten as

$$\tau_v = \int_0^\infty \alpha_v(\ell)\rho(\ell)d\ell \qquad (4.15)$$

From Figure 3.10, taking $d\ell$ in the direction of $\hat{n}$ we see that $dz = d\ell \cos \varphi = \mu_\varphi \, d\ell$ and then (4.14) can be rewritten as

$$\mu_\varphi \frac{dI_v}{d\tau_v} = S_v - I_v \tag{4.16}$$

where now the optical depth is specifically given in terms of the direction z normal to the surface $\tau_v = \int_{z1}^{z2} \alpha_v(z)\rho(z)dz$ and the source function is defined as

$$S_v = \frac{j_v}{\alpha_v} \tag{4.17}$$

which has the same units as the intensity. Note that it is also customary to use the extinction coefficients (absorption and scattering) defined by their products with the density so the unit is cm^{-1} (Section 3.3.2). The medium is defined as optically thick when $\tau_v \geq 1$ and optically thin when $\tau_v < 1$.

For simplicity, we replace $\tau/\cos \varphi$ by τ (or assuming that ray propagation takes place in the vertical direction, that is, that $\mu_\varphi = \cos \varphi = 1$), the formal solution to (4.16) can be written as

$$I_v(\tau_v) = I_v(0)e^{-\tau_v} + \int_0^{\tau_v} S_v(\tau_v')e^{-(\tau_v - \tau_v')}d\tau_v' \tag{4.18a}$$

If the source function S_v is known, the intensity of the radiation field can be obtained upon integration of Equation 4.18a. In planetary atmospheres, the situation is usually complicated since the scattering processes makes S_v to be a function of I_v or of temperature, which in turn depends also on I_v. If the compounds intervening in the optical depth have a constant density and extinction coefficients (as occurs for thin layers), then Equation 4.18a becomes

$$I_v(z) = I_v(0)e^{-\alpha_v \rho z} + \alpha_v \rho \int_0^z S_v(z')e^{-\alpha_v \rho(z-z')}dz' \tag{4.18b}$$

The radiation field is composed of two terms: the first representing the intensity being attenuated by the exponential factor due to extinction along the distance z, while the integral term represents the sums of the contributions of the emitted radiation from elements along the infinitesimal distance dz' for different distances z' along the path that are themselves attenuated by an exponential factor due to the extinction over the distance $z - z'$. In the following, we examine different simplified situations for Equation 4.18b in order to gain an insight into this equation.

4.2.2.1   Source Function Independent of the Optical Depth

In this case, Equation 4.18 is rewritten as

$$I_v(\tau_v) = (I_v(0) - S_v)e^{-\tau_v} + S_v \qquad (4.19)$$

Two extreme situations are as follows: (1) a very thick optically medium $\tau_v \gg 1$ and the radiation field is entirely determined by the source function since $I_v = S_v$; (2) for a very thin optically medium $\tau_v \ll 1$ and $I_v \rightarrow I_v(0)$, the intensity is determined by the incident radiation field.

4.2.2.2   Purely Absorbing Medium

A medium where only absorption occurs has $S_v = 0$ ($j_v = 0$) and $\alpha_v = k_v$, and then we obtain the simplified Lambert's law (see Equation 3.28) whose solution is

$$I_v(\tau_v) = I_v(0)e^{-\tau_v} \qquad (4.20)$$

For an optically thick medium $\tau_v \gg 1$ and $I_v \rightarrow 0$, no radiation is transmitted through the atmospheric layer. If the medium is optically thin ($\tau_v \ll 1$), then I_v decreases linearly with optical depth since $I_v(\tau_v) \approx I_v(0)(1 - \tau_v)$.

4.2.2.3   Local Thermodynamic Equilibrium

If the atmosphere is in equilibrium with the radiation field, we know that according to Kirchkoff's law (Section 3.2.2) an absorbing layer in the atmosphere emits radiation proportionally to the absorption at the same frequency. In this case, no scattering is present ($\sigma_v = 0$) and $\alpha_v = k_v$. Equation 3.8 can be rewritten in terms of the current definitions as

$$j_v = k_v B_v(T) \qquad (4.21)$$

The Planck function describes the radiation field in thermodynamic equilibrium and is the source function ($S_v = B_v(T)$, what is called "thermal radiation"). The radiative transfer in this case is called the Schwarzschild equation. Note that blackbody radiation corresponds to the situation $I_v = B_v(T)$.

4.2.2.4   Absorption and Emission Lines under LTE

When the source function is independent of the optical depth, the following simplification can be used for optically thin situations $\tau_v < 1 \rightarrow e^{-\tau_v} \approx 1 - \tau_v$; then (4.19) can be written as

$$I_v(\tau_v) = (I_v(0) - S_v)(1 - \tau_v) + S_v = I_v(0)(1 - \tau_v) + \tau_v S_v = I_v(0) + \tau_v(S_v - I_v(0)) \quad (4.22)$$

The sign of the second term in the last equality depends on the values of the source function and incident radiation field. If $I_v(0) > S_v$, the sign is negative, and for sufficient optical depth, the emergent intensity decreases and we expect the formation of absorption spectral lines. If $I_\nu(0)<S_\nu$, the sign is positive, and for sufficient optical depth, the emergent intensity increases and we expect the formation of emission spectral lines. For local thermodynamic equilibrium (LTE) conditions we have $S_\nu=B_\nu(T)$, which is a function that increases with increasing temperature. An atmospheric layer where $dT/dz<0$ has $I_\nu(0)>S_\nu$ and active radiative species will form absorption lines. On the contrary, when $dT/dz>0$ we have $I_\nu(0)<S_\nu$ and they will form emission lines.

Chapter 5 discusses the radiative transfer equation when scattering processes by the gas and/or by suspended particles (aerosols and clouds) intervene in the source function. This is specially relevant for the photometric measurements and the interpretation of the diffuse radiation of clouds and aerosols from the planets.

4.3   RADIATIVE EQUILIBRIUM IN ATMOSPHERES

Bodies with a tenuous atmosphere (or airless bodies) will have a surface temperature equal to their equilibrium temperature, as discussed in Section 1.4.4. For denser atmospheres, the atmosphere can increase the surface temperature by a greenhouse effect. If an internal energy sources is present, as in giant planets, then the effective temperature should be used instead of the equilibrium temperature. As already discussed, in planets with thick atmospheres, the temperature structure is governed primarily by the radiation sources, external (solar or stellar radiative flux) and/or internal heat sources. The energy balance of an atmospheric slab is obtained from the thermodynamic equation (1.126) (which is derived from the first law of thermodynamics, see Section 4.5 for more details) that gives the change of temperature of the slab from its volumetric heating rate Q (or heat power per unit volume, units W m^{-3}):

$$\frac{dT}{dt} = \frac{Q}{\rho C_p} \tag{4.23}$$

where dT/dt is for the total derivative. If only radiation contributes to the heating rate of a plane parallel atmospheric slab that is perpendicular to the vertical direction z, then $Q=dF/dz$ and Equation 4.23 can be written as

$$\frac{\partial T}{\partial t} = \frac{1}{\rho C_p}\frac{dF}{dz} \tag{4.24}$$

where the net radiative flux in the slab is defined as

$$F = \int_0^\infty \left(F_\nu(\uparrow) - F_\nu(\downarrow)\right) d\nu \tag{4.25}$$

where $F_\nu(\uparrow)$ and $F_\nu(\downarrow)$ are the upward and downward spectral density fluxes in the layer, respectively. Equation 4.25 states that the net rate of temperature change in the slab is due to the divergence of the upward and downward radiation fluxes. The equilibrium situation requires $\partial T/\partial t = 0$, which means that the net flux F is constant with depth. In other words, an atmosphere that is in radiative equilibrium obeys the equation

$$\frac{dF}{dz} = 0 \tag{4.26}$$

4.3.1 VERTICAL TEMPERATURE PROFILE IN RADIATIVE EQUILIBRIUM

A simple way to obtain the vertical temperature structure of an atmosphere is called the diffusion approximation, which is what we employ here. Consider the atmosphere under LTE conditions, that is, $S_\nu = B_\nu(T)$; then from the previous section the radiative transfer equation (4.16) becomes

$$\mu_\varphi \frac{dI_\nu}{d\tau_\nu} = B_\nu - I_\nu \tag{4.27}$$

A thick atmosphere (a deep and dense layer) in thermal equilibrium with the radiation field being isotropic has $I_\nu \approx B_\nu(T)$, so we rewrite this equation as

$$I_\nu = B_\nu - \mu_\varphi \frac{dB_\nu}{d\tau_\nu} \tag{4.28}$$

Substituting (4.28) in (4.13) and taking into account that the first term is zero (it does not carry flux), and using the definition for the optical depth (4.15), we get

$$F_\nu(z) = -2\pi \int_{-1}^{1} \frac{dB_\nu}{d\tau_\nu} \mu_\varphi^2 \, d\mu_\varphi = -\frac{4\pi}{3\rho\alpha_\nu} \frac{dB_\nu}{dT} \frac{dT}{dz} \tag{4.29}$$

The total flux is obtained upon integration over frequency:

$$F(z) = \int_0^\infty F_\nu(z) \, d\nu = -\frac{4\pi}{3\rho} \frac{dT}{dz} \int_0^\infty \frac{1}{\alpha_\nu} \frac{dB_\nu}{dT} \, d\nu \tag{4.30}$$

Defining an absorption coefficient α_R called the Rosseland mean opacity as

$$\frac{1}{\alpha_R} = \frac{\int_0^\infty (1/\alpha_\nu)(dB_\nu/dT) \, d\nu}{\int_0^\infty (dB_\nu/dT) \, d\nu} \tag{4.31}$$

and using Equation 3.5 for the frequency integrated flux, Equations 4.30 and 4.31 combine to give the radiative diffusion equation

$$F(z) = -\frac{16}{3} \frac{\sigma_B T^3}{\rho \alpha_R} \frac{dT}{dz} = -K_R \frac{dT}{dz} \tag{4.32}$$

where K_R is the radiative thermal conductivity coefficient (equivalent to the classical conductivity, see Section 1.5.4) that varies as T^{-3}. A net upward flux occurs when $dT/dz < 0$, that is, when temperature decreases with altitude. Introducing the equilibrium temperature definition (Equation 1.101) with emissivity $= 1$ (or equivalently the effective temperature to include the flux from the internal heat source, Equation 1.103), we get from Equation 4.32 the radiative lapse rate

$$\frac{dT}{dz} = -\frac{3}{16} \frac{\rho \alpha_R}{T^3} T_{eq}^4 \tag{4.33}$$

It is also usual to give the lapse rate in terms of pressure as vertical coordinate. Assuming hydrostatic equilibrium and the ideal gas law, Equation 4.33 can be rewritten as a nondimensional temperature gradient

$$\nabla_{rad} = \frac{d \ln T}{d \ln P} = \frac{3 \alpha_R}{16} \frac{P}{g} \left(\frac{T_{eq}}{T} \right)^4 \tag{4.34}$$

A more accurate calculation of the lapse rate or the temperature gradient requires resolving the radiative transfer equation (4.16) at all the frequencies with the condition (4.26), taking into account that $B_\nu(T)$ and the abundances of some compounds vary with altitude.

4.3.2 GREENHOUSE EFFECT

One important observation in Venus, Earth, Mars, and the satellite Titan is that their surface temperature is higher than expected from equilibrium according to their albedo and distance from the Sun. This is due to the so-called greenhouse effect. The atmosphere in these planets is (in the absence of clouds) quasitransparent to a large part of the UV-visible-near infrared wavelengths. Thus, sunlight heats the surface and lower, massive atmosphere that reradiate as a blackbody at the heated temperature (infrared wavelengths, i.e., in the thermal part of the spectrum). The greenhouse effect occurs when some species in the atmosphere (including aerosol particles) become semiopaque to this infrared radiation. The most important radiatively active gases in the terrestrial planets for the greenhouse effect; are CO_2, H_2O, and CH_4 that have intense absorption bands in the infrared (see Table 3.3). Venus, which has a massive CO_2 atmosphere, has a very strong greenhouse effect; Earth shows a moderate effect, and Mars and Titan have a reduced effect.

To explain the magnitude of this effect, an approximate solution can be found if we assume that the atmosphere is heated from the surface and that it is in radiative balance. In this simple estimation, any other effect (aerosols, dynamics) on temperature are ignored. In such a situation, the radiative transfer equation reduces to (Chamberlain and Hunten, 1987)

$$\frac{dB_v}{d\tau_v} = \frac{3}{4} F_v = \text{Const.} \tag{4.35}$$

A solution to this equation can be found by the method called the two-stream approximation in which the radiant intensity across an atmospheric layer is decomposed in an upward component $I_v^{\uparrow}(\mu_{\varphi}, \tau)$ and a downward $I_v^{\downarrow}(\mu_{\varphi}, \tau)$ component, so the net flux density is given by

$$F_v = \pi (I_v^{\uparrow} - I_v^{\downarrow}) \tag{4.36}$$

As a boundary condition we can use that at the top of the atmosphere is heated from below and then $I_v^{\downarrow}(\tau = 0) = 0$. We also assume that the ground emits as a blackbody at temperature T_g so $I_{vg}^{\uparrow} = B_v(T_g)$. Then the solution to (4.35) is

$$B_v(\tau_v) = B_v(T_0)\left(1 + \frac{3}{2}\tau_v\right) \tag{4.37}$$

where T_0 is the air temperature at $\tau_v = 0$, usually referred to as the skin temperature. To further simplify the situation, assume that the absorption coefficient is independent of wavelength (gray atmosphere) so with $\alpha_v = k_v = k = \text{constant}$. Integration over frequency (4.37) and using the Stephan–Boltzmann law equation (3.5) gives the temperature

$$T^4(\tau) = T_0^4\left(1 + \frac{3}{2}\tau\right) \tag{4.38}$$

The two stream approximation also gives the relationship between the equilibrium temperature (or the effective temperature if internal heat is present) and the skin temperature: $T_e^4 = 2T_0^4$. This finally allows writing the ground temperature as

$$T_g^4 = T_e^4\left(1 + \frac{3}{4}\tau_g\right) \tag{4.39}$$

where τ_g is the optical depth at the ground. For Venus, $T_g = 733\,\text{K}$ whereas $T_e = 240\,\text{K}$ implying, according to this approximation, that $\tau_g = 114$. The main contributors to the huge greenhouse effect are CO_2 (~65%), sulfuric acid clouds (~21%), and water vapor (~11%). On Earth, $T_g = 288\,\text{K}$ and $T_e = 255\,\text{K}$, giving $\tau_g = 0.8$ with the main opacity sources H_2O and CO_2. These are also the main contributors to Mars greenhouse that amounts to a few degrees kelvin. However, a greater warming occurs in Mars when airborne aerosols are injected in the atmosphere during global dust storms, increasing the temperature by 30 K at the surface and by 100 K at the tropopause.

4.3.3 HEATING RATES

The volumetric heating rate Q of an atmospheric slab due only to radiation is given by Equations 4.23 and 4.24, and depends on the atmospheric gas composition (vertical distribution of active radiative species) and on the absorption and scattering properties of suspended particles (aerosols and clouds). This second contribution is discussed in more detail in Chapter 5. The heating rate term in Equations 4.23 and 4.24 (unit K s^{-1}) at a given frequency can be written in different ways for hydrostatic conditions:

$$\frac{Q}{\rho C_p} = -\frac{1}{\rho C_p}\frac{dF_v}{dz} = \frac{g}{C_p}\frac{dF_v}{dP} = \frac{g}{C_p}\frac{d\tau_v}{dP}\frac{dF_v}{d\tau_v} \qquad (4.40)$$

The main contribution to the heating rate is due to the absorption of the incident solar radiation by radiatively active species and this will be denoted as Q_{solar}, whereas the principal cooling mechanism is the thermal emission in the infrared Q_{IR}. The solar heating rate is a function of time (daily and seasonal solar flux density variability) whereas the cooling and heating terms in the infrared are more dependent on the actual temperature. Thus, it is convenient to separate Equation 4.40 in two contributions

$$\frac{Q}{\rho C_p} = \frac{Q_{solar} + Q_{IR}}{\rho C_p} \quad \text{(unit K s}^{-1}) \qquad (4.41a)$$

$$\frac{Q_{solar}(t)}{\rho C_p} = \frac{g}{C_p}\left(\frac{d\tau_v}{dP}\frac{dF_v}{d\tau_v}\right)_{v>v_0} \qquad (4.41b)$$

$$\frac{Q_{IR}(T)}{\rho C_p} = \frac{g}{C_p}\left(\frac{d\tau_v}{dP}\frac{dF_v}{d\tau_v}\right)_{v<v_0} \qquad (4.41c)$$

where the cutting frequency v_0 is approximately given at the intersection between the reflected solar spectrum and the planetary thermal emission as presented in Section 3.2.2. Both heating rate terms depend on the particular location on the planet, that is, they are implicit functions of latitude, longitude, and height (that is, pressure). Usually they are presented as a function of pressure, but averaged in time over a characteristic period (day or year). Their calculation involves the flux divergence ($dF_v/d\tau_v$), which requires the resolution of the radiative transfer equation solved including the vertical abundance distribution of all the gaseous active species.

The solar heating term can be approximated in the case of a plane parallel atmosphere as (Problem 4.2)

$$\frac{Q_{solar}(t)}{\rho C_p} = \frac{g}{C_p}\left(\frac{R_\odot}{r_\odot(t)}\right)^2 [\pi B_v (5800 \text{ K})]\frac{d\tau_v}{dP}\exp\left(-\frac{\tau_v}{\mu_\varphi(h)}\right) \qquad (4.42)$$

where

$R_\odot$ is the radius of the Sun

$r_\odot(t)$ is the planet distance from the Sun (expression that can be generalized to extrasolar planets using their parent star radius and distance instead of those of the Sun)

The cosine of the solar elevation angle (μ_φ) is a function of the hour angle $h(t)$ (see Section 1.4.3); so to obtain the total heating rate, this expression must be integrated over frequency (for a given spectral range) and hour angle (for a daily or yearly mean).

The cooling term needs to fully solve the radiative transfer equation. To gain insight into its form, we give here the solution obtained by the method called "exchange formulation." The intensity (Equation 4.18) and the flux divergence ($dF_v/d\tau_v$) in (4.41) are calculated from the exchange of energy between a layer and the other atmospheric slabs assuming that the source function is represented by a blackbody Planck function. In the exchange integral formulation, the rate of energy transfer between layers is determined by the vertical temperature gradient through the Planck function. The "cooling to space approximation" applies to an isothermal atmosphere and gives good results for strong molecular absorption bands. The cooling term in the thermal infrared is obtained from

$$\frac{Q_{IR}(T)}{\rho C_p} = -2\pi\frac{d\tau_v}{dP}[B_v(T(\tau_v))]\frac{g}{C_p}\int_0^1 \exp\left(-\frac{\tau_v}{\mu_\varphi}\right)d\mu_\varphi \qquad (4.43)$$

To obtain the total cooling rate, this expression must be integrated over frequency and time.

The next step is to calculate the dependence of the optical depth as a function of atmospheric pressure (height). The formulation of band models presented in Section 3.4.6 is now relevant. Under hydrostatic conditions and for a uniformly vertical distributed compound with mixing ratio x_v, the optical depth of a thin slab is given by

$$d\tau_v = -x_v k_v \rho\, dz = x_v \frac{k_v}{g}dP \rightarrow \frac{d\tau_v}{dP} = \frac{x_v k_v}{g} \qquad (4.44\text{a})$$

To illustrate in a simple way the form of this term let us assume that the absorption coefficient of the gas is dominated by collisions. Then it is reasonable to assume that its value is proportional to the gas density $k_v = x_v a_v \rho$, where a_v (units cm^5 g^{-2}) is an opacity factor considered to be constant (in reality it is usually dependent on temperature and pressure). Using the ideal gas law, $P = \rho R_g^* T$ yields

$$\frac{d\tau_v}{dP} = x_V^2 a_v \frac{\rho}{g} = \frac{x_V^2 a_v}{g R_g^* T} P \tag{4.44b}$$

and

$$\tau_v = \frac{x_V^2 a_v}{2 g R_g^* T} P^2 = \frac{P^2}{P_{eff}^2} \tag{4.45}$$

where we have introduced an effective pressure P_{eff} defined as the pressure where the gas optical depth is 1, and gives the level where the temperature has a minimum value

$$P_{eff}^2 = \frac{2 g R_g^* T_{eff}}{x_V^2 a_v} = \frac{2 g R_g^* T_{eff}}{x_V^2 (a_{Vv}/\rho_{STP}^2)} \tag{4.46}$$

Here a_{Vv} is called the volumetric opacity and is given in units of cm^{-1} amagat^{-2} (see Problem 4.4). It is important not to mistake this pressure with that defined by the effective temperature (see Section 1.4.4), that is, $P(T_{eff})$, which is a different concept.

Figure 4.6 shows the heating/cooling rates as a function of altitude in Venus. Above the clouds ($z > 70\,km$), the CO_2 bands are responsible for the energy exchange. The heating rate increases rapidly from $1.5\,K\,day^{-1}$ at $65\,km$ up to $50\,K\,day^{-1}$ at

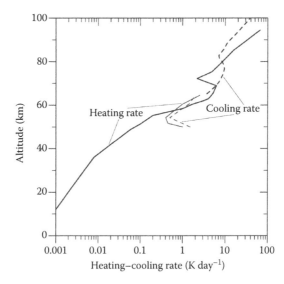

FIGURE 4.6 Vertical solar heating and thermal cooling rates on Venus' atmosphere. (Data taken from Moroz, V.I., *Adv. Space Res.*, 5(11), 224, 1985. With permission.)

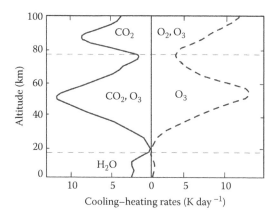

FIGURE 4.7 Heating rate profile at short wavelength (left) and cooling rate at long wavelength (right) on Earth's atmosphere. The main contributing compounds are shown at each altitude level. (Adapted from Andrews, D.G., *An Introduction to Atmospheric Physics*, Cambridge University Press, New York, 2000.)

the top of the mesosphere (90 km). The unknown UV absorber present in Venus' clouds (above 57 km) contributes with a heating rate of 8 K day^{-1}. Figure 4.7 shows the heating/cooling rates as a function of altitude on Earth. The short wavelength heating rate is dominated by the absorption by H_2O bands in the troposphere and by 200–300 nm ultraviolet radiation by ozone O_3 reaching a maximum of 12 K day^{-1} at 50 km. The cooling rate is dominated at long wavelengths by the 15 μm band of CO_2 and by the 9.6 μm band of O_3. Clouds and aerosols have a strong influence in the heating/cooling of the atmosphere with additional rates up to 4 K day^{-1}. This last aspect will be discussed in Chapter 5. The heating/cooling rates on Mars are dominated by CO_2 bands as in Venus but with the notable addition of the heating produced by suspended particles during the development of global-scale dust storms.

On the giant planets, the heating is provided by the absorption in the CH_4 bands in the wavelength range 0.45–3.3 μm and in the aerosol and haze layers, whereas the cooling occurs due to infrared emission (8–50 μm) by H_2–H_2 enhanced by H_2–He collisions, and in the far infrared bands ($\lambda > 7$ μm) of CH_4, C_2H_2, and C_2H_6. For example, in Saturn's stratosphere, typical heating rates (that include the aerosol absorption) are ~ 0.01 K day^{-1} and in Uranus' stratosphere, this value is 2×10^{-4} K day^{-1} (see Figure 4.8).

4.3.4 RADIATIVE TIME CONSTANT

When studying the radiative processes in the atmospheres it is important to know the rate at which the temperature changes are expected to occur. Atmospheres have a thermal inertia that depend on the mass, heat capacity, and temperature. To characterize the temporal thermal behavior of an atmosphere, we introduce here a characteristic response time—the radiative time constant. We start with the thermodynamic equation (4.23) separating the two main contributions,

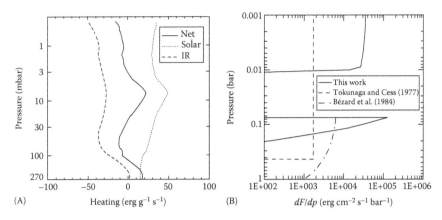

FIGURE 4.8 (A) Heating rates in Jupiter and (B) Saturn, including the aerosol contribution. Jupiter (From West, R. et al., *Icarus*, 100, 245, 1992. With permission.), Saturn (From Pérez-Hoyos, S. and Sánchez-Lavega, A., *Icarus*, 180, 368, 2006. With permission.)

$$\rho C_p \frac{dT}{dt} = Q_{solar} + Q_{IR} \tag{4.47}$$

obtained by combining (4.23) and (4.41a). Consider an atmospheric layer that is iso-thermal and whose emissivity $\varepsilon(z)$ varies with altitude. The Stephan–Boltzmann law allows writing the infrared term

$$Q_{IR}(T) = \sigma T^4 \frac{d\varepsilon}{dz} \tag{4.48}$$

Assume that the slab is heated up to a temperature T above a certain relaxing tem-perature T_0, the temperature difference being small ($\Delta T = T - T_0 \ll T_0$), and assume at the same time that the emissivity is insensitive to this temperature perturbation. The heating rate can then be calculated as the sum of a heating term (at temperature T_0) plus a departure temperature perturbation:

$$Q_{IR}(T) \approx Q_{IR}(T_0) + \frac{dQ_{IR}(T)}{dT} \Delta T = Q_{IR}(T_0) + 4\sigma T_0^3 \Delta T \frac{d\varepsilon}{dz} \tag{4.49}$$

Substituting (4.49) into (4.47) and changing the sign in the temperature difference, we get

$$\frac{dT}{dt} = \frac{Q_{solar} + Q_{IR}(T_0)}{\rho C_p} - \frac{4\sigma T_0^3}{\rho C_p} \frac{d\varepsilon}{dz} (T_0 - T) \tag{4.50}$$

The radiative time constant is defined according to the second term on the right:

$$\frac{1}{\tau_{rad}} = -\frac{4\sigma T_0^3}{\rho C_p}\frac{d\varepsilon}{dz} \tag{4.51}$$

This is an important parameter in atmospheres since it allows to estimate how long it will take for a temperature perturbation to relax to its unperturbed state radiating the heat excess in the infrared. A useful operative expression for the radiative time constant is obtained from (4.51) upon integration on an altitude H (e.g., the scale height for the isothermal layer, $(d\varepsilon/dz) \approx (\varepsilon/H)$) and taking $T_0 = T_{eff}$ as the effective radiating temperature,

$$\tau_{rad} = \frac{\rho C_p H}{4\varepsilon\sigma T_{eff}^3} \approx \frac{C_p PT}{\sigma g T_{eff}^4} = \frac{C_p (P/g)T}{(S_{\odot p}/4)(1 - A_B)} \tag{4.52}$$

where we have used the ideal gas law, with $S_{\odot p}$ being the solar (stellar) constant at the planet (see Section 1.4.3). Note that this expression can be read as the ratio between the heat stored per unit area of atmosphere above the level P and the outgoing thermal radiation flux. A more real calculation for τ_{rad} shows that it has a stronger dependence with altitude due to the vertical variability in the atmospheric emissivity produced by the different radiatively active species, and needs the radiative transfer equation to be solved. An order of magnitude estimation of this parameter for planets with a substantial atmosphere is given in Table 4.3.

TABLE 4.3
Radiative Time Constant in the Solar
System Atmospheres

Planet/Satellite	P (bar)	T (K)	$\langle\tau_{rad}\rangle$
Venus	0.1	250	9.5 days
	1	360	19.5 days
	92 (surface)	731	0.95 year
Earth	0.1	217	15.6 days
	1 (surface)	288	67.1 days
Mars	0.007 (surface)	214	4.8 days
Jupiter	0.1	120	3.3 years
	1	165	4.8 years
Saturn	0.1	85	15.2 years
	1	134	24.7 years
Titan	0.1	70	9.5 years
	1.5 (surface)	94	58.9 years
Uranus	0.1	56	84.4 years
	1	76	142.9 years
Neptune	0.1	61	62.5 years
	1	76	113.2 years

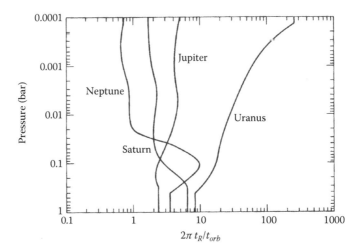

FIGURE 4.9 Radiative time constant as a function of altitude in the giant fluid planets. (From Conrath, B.J. et al., *Icarus*, 83, 255, 1990. With permission.)

In Figure 4.9, we show how the radiative time constant varies with altitude in giant fluid planets.

4.3.5 RADIATIVE TEMPERATURE: TEMPORAL VARIABILITY

Equation 4.50 can be rewritten using (4.51) as

$$\tau_{rad}\frac{dT}{dt} + T = T_0 + \tau_{rad}\frac{Q^*_{solar}}{\rho C_p} \tag{4.53}$$

In this equation, the temperature T_0 results from the mean annual insolation $\langle Q_{solar}\rangle$, such that $\langle Q_{solar}\rangle + Q_{IR}(T_{eq}) = 0$, and Q^*_{solar} (that contains the orbital and obliquity terms) represents the departure from the annual mean value. Since the insolation changes in the atmospheres are periodic, as presented in Section 1.4.3, let us assume that they produce a periodic temperature response $T(t)$ on it (that is, in steady state). Let us combine the terms on the right-hand side of Equation 4.53 into a theoretical forcing function called the equilibrium temperature $T_e(t)$:

$$T_e(t) = T_0 + \frac{Q^*_{solar}}{\rho C_p}\tau_{rad} \tag{4.54}$$

This function is time and spatially dependent such that its annual average value is the equilibrium temperature T_{eq} (Equation 1.102). Accordingly, Equation 4.53 is rewritten as

$$\frac{dT}{dt} = \frac{T_e - T}{\tau_{rad}} \tag{4.55}$$

The term $T_e(t)$ contains the rotational or the orbital and obliquity terms discussed in Section 1.4.3 and can be represented as periodic functions of the temporal frequency of interest (diurnal $\omega_{day} = 2\pi/\tau_{day}$ or orbital $\omega_{orb} = 2\pi/\tau_{orb}$). In addition, note that $T_e(t)$ is also a spatially variable function, dependent on altitude (pressure P), latitude (φ), and longitude (ϕ). Then the term $T_e(t)$ is represented by a Fourier expansion with amplitude coefficients T_e^n:

$$T_e(\varphi, \phi, P, t) = \sum_{n=0}^{\infty} T_e^n(\varphi, \phi, P) \exp(in\omega t) \tag{4.56}$$

Substituting this in (4.55) yields

$$T(\varphi, \phi, P, t) = \sum_{n=0}^{\infty} \frac{T_e^n(\varphi, \phi, P)}{[1 + (n\omega\tau_{rad})^2]^{1/2}} e^{-i(n\omega t - \psi_n)} \tag{4.57}$$

where

$$\tan(\psi_n) = n\omega\tau_{rad} \tag{4.58a}$$

The quantity $\omega\tau_{rad}$ is a measure of the thermal inertia of the atmosphere in response to the daily or seasonal radiative forcing. The steady component term ($n=0$) in Equation 4.57 is independent of $\omega\tau_{rad}$ and is equal to the instantaneous radiative equilibrium solution (T_{eq}) averaged over a period $2\pi/\omega$. From Equation 4.58 we can see that the resulting "temperature wave" lags the insolation forcing by a phase shift in the first harmonic ($n=1$), $\psi_1 = \tan^{-1}(\omega\tau_{rad})$ and that the amplitude varies as $[1 + (\omega\tau_{rad})^2]^{-1/2}$.

A useful parameter that can be derived from Equation 4.58 is the radiative time lag

$$\delta t = \frac{\tau_i}{2\pi} \tan^{-1}\left(\frac{2\pi\tau_{rad}}{\tau_i}\right) \tag{4.58b}$$

where τ_i represents the daily or yearly periods. For the daily frequency (day–night period), we have that in the upper atmospheres of Earth and Venus and at Martian surface, $\omega_{day}\tau_{rad} < 1$, which means that daily temperature fluctuation should occur with a small phase lag or $\delta t < \tau_{day}$. On the contrary, τ_{rad} is of the order of years in the giant planets tropospheres and at the surface of Titan, which means that $\omega_{day}\tau_{rad}$ and $\omega_{orb}\tau_{rad} \gg 1$. No daily temperature variations are expected on these parts of their atmospheres, and a long period of time must pass for a seasonal temperature response to occur (phase lag $\sim 90°$).

4.4 THERMAL TIDES: TEMPERATURE OSCILLATIONS

The most conspicuous radiatively controlled thermal phenomena in atmospheres
are the temperature oscillations resulting from the apparent motion of the Sun
across the sky. Thermal tides are global oscillations of the atmosphere forced by the
day to night variations of the heating/cooling produced by the deposition of solar
radiation. Seasonal thermal waves refer to those forced by the yearly insolation
cycle. Since different frequencies are involved, as shown in the previous section,
we speak of a diurnal tide when referring to that with a period of 1 day (mode $n = 1$
or wavenumber one with 1 cycle day^{-1}) and semidiurnal tide when referring to a
wave with a period of half a day ($n = 2$ or wavenumber two with 2 cycles day^{-1}). In
what follows, we present examples of the observational data set of the temperature
oscillations due to the radiative heating. The dynamical behavior of thermal tides is
presented in Section 8.6.

Let us discuss the expected radiative response atmospheres in terms of the first
harmonic term in Equation 4.57 using the data in Table 4.3. A crude estimation of the
day–night temperature difference in a rapidly rotating planet is $T_{night} \approx T_{day} [1 - (\tau_{rot}/\tau_{rad})]$. This can be understood by assuming that all the incident solar radiation above
a pressure level P_0 is deposited as heat. According to what we have previously dis-
cussed in Section 4.3, the relative amplitude of the temperature oscillations can be
calculated from the balance

$$\frac{P_0}{g} C_p \Delta T 4\pi R_p^2 = \pi R_p^2 (1 - A_B) \frac{S_{\odot p}}{a^2 (AU)} \tau_{syn} \qquad (4.59a)$$

where we have introduced τ_{syn} for the length of the day in a rapidly rotating planet
(sidereal period, $\tau_{syn} = \tau_{rot}$), except for Venus (see Equation 1.55). Then the fractional
amplitude of the temperature oscillation is

$$\frac{\Delta T}{T} = \frac{S_{\odot p}(1 - A_B)g\tau_{syn}}{4a^2 (AU)P_0 T C_p} \qquad (4.59b)$$

From the data in this table, we get as examples for T_{day} (Earth) $= 300\,K$ gives T_{night}
(Earth) $= 280\,K$, T_{day} (Mars) $= 214\,K$ gives T_{night} (Mars) $= 169\,K$, T_{day} (Jupiter) $= 165\,K$
gives T_{night} (Jupiter) $= 164.9\,K$. The amplitude of the thermal tide is small in the densest parts of atmospheres due to the increasing mass per unit area (P_0/g). For example, at Venus' surface, $\Delta T/T \sim 0.4\%$, but on Mars $\Delta T/T \sim 38\%$, and on Jupiter's clouds
$\Delta T/T \sim 0.002\%$ (Problem 8.2). However, $\Delta T/T$ becomes important in the upper atmosphere of planets, as on the thermospheres of Venus and Earth.

Observations show that diurnal thermal tides form down to the surface in the
atmospheres of Earth and Mars, and in the upper atmosphere of Venus. In Venus,
wavenumber two dominates in the thermal tide with amplitude ~ 5–$10\,K$ that
arises above the upper cloud level between altitudes ~ 65–$90\,km$ (Figure 4.10). At
higher exospheric altitudes, the temperature amplitude reaches $200\,K$. Note that in
Venus the solar day is about 117 days, resulting from a combination of the spin

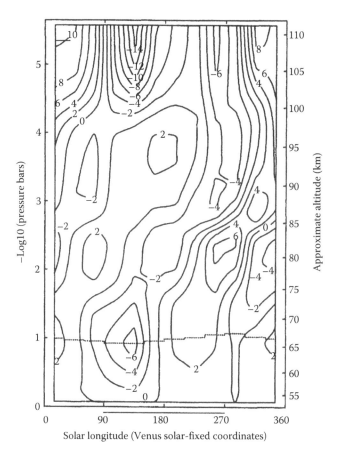

FIGURE 4.10 Deviations from zonal mean temperatures in Venus' atmosphere as a function of the solar longitude (Venus' thermal tide), for latitudes < 30°. The subsolar longitude is 0° and the morning terminator is at 90°. (From Seiff, A. et al., *Adv. Space Res.*, 5(11), 3, 1985. With permission.)

and translation period around the Sun (see Equation 1.61). On Earth, the afternoon temperature maximum and predawn minimum in the mean air temperature near the surface is a well-known example of a thermal tide (Figure 4.11). The amplitude variations are ~ 10 K. At higher levels in the atmosphere (~ 70–100 km), the tide shows both wavenumbers one and two, and the amplitude can reach maximum values ~ 100 K. The low atmospheric mass of Mars makes the pressure at the surface and the atmospheric temperatures to follow the daily insolation cycle with a high regularity (Figures 4.12 and 4.13). The measured daily surface temperature fluctuations are of ~ 20 K. Due to the high rotation rate of the giant planets (10–17 h) and to the large radiative time constant (~ years), diurnal or semidiurnal thermal tides have not been observed in these planets.

The seasonal temperature oscillation is present on Earth and Mars, but is not obvious on Venus due to the similarity of the rotation and orbital period and to the

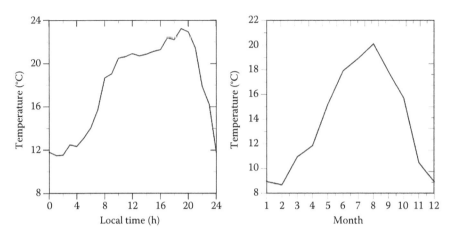

FIGURE 4.11 Ground temperatures at the city of Bilbao (Spain, latitude 43.3° North) as representative of temperature variations due to the day–night and annual insolation cycles on Earth. Daily (hourly) for June and monthly (averaged). Data from Euskalmet. (Courtesy of J. Saenz, Universidad del País Vasco, Spain.)

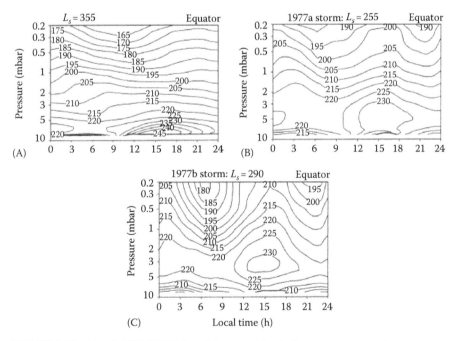

FIGURE 4.12 Mars's NOAA/GCM model maps of the daily temperatures at the equator as a function of altitude and local time. They show the Martian thermal tide for different orbital longitudes (L_s). The 1997a and 1997b panels include the effects of a dust storm development. (From Forbes, J.M., *Adv. Space Res.*, 33, 125, 2004. With permission.)

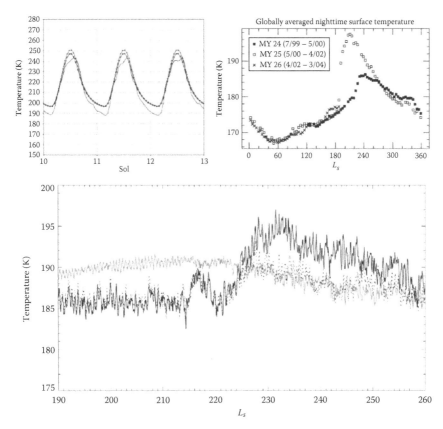

FIGURE 4.13 Temperature variations at different time scales: daily ("sol" is the Martian day), yearly averaged (as a function of solar longitude, L_s), and during a part of a Martian year showing the effect on the temperature field of the development of a dust storm. (From Moudden, Y. et al., *Adv. Space Res.*, 36, 2169, 2005; Smith, M.D., *Icarus*, 167, 148, 2004; Lewis, S.R. et al., *Icarus*, 192, 327, 2007. With permission.)

small tilt of its spin axis. On Earth, the yearly surface temperatures at a given latitude oscillate with an amplitude of ~ 20–$40\,K$ (Figure 4.11). In Mars, the annual temperature oscillation is large ($\sim 60\,K$, depending on latitude) (Figure 4.13) inducing large dynamical changes in the atmosphere. The most dramatic effect in this planet is the seasonal sublimation–condensation cycle of the North and South polar ice deposits of CO_2 and water that produce a global atmospheric pressure wave change in the whole planet. The other radiatively dependent cycle is the seasonal global episodes of dust storms that severely reduce the temperature oscillations to $\sim 40\,K$ (Figures 4.12 and 4.13). These phenomena will be discussed in the chapters regarding atmospheric dynamics.

In giant planets, the seasonal effects produced by the tilt of the rotation axis relative to the orbital plane (in the cases of Saturn, Uranus, and Neptune) and by the orbital eccentricity (Jupiter) generate temperature waves at pressure levels <0.3 bar where the radiative time constant is half that of the planet orbital period. The

annually predicted temperature oscillations have maximum amplitudes ~3 K at different atmospheric levels in Jupiter. North–South temperature profiles at a pressure
level of 20 mbar (stratosphere) retrieved from remote sensing observations show
the seasonal fluctuations coupled to a more complex wave dynamical phenomenon
(Figure 4.14, see also Figure 8.18 and Section 8.7). Similar temperature oscillations

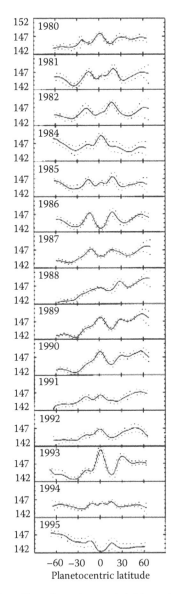

FIGURE 4.14   Meridional profiles of zonal mean brightness temperatures (in kelvin) from
emission due to methane at 7.8 μm in Jupiter's stratosphere for the years 1980–1995. See also
Figure 8.18. (From Friedson, A.J., *Icarus*, 137, 34, 1999. With permission.)

in latitude with the annual cycle have been detected in Saturn where the seasonal cycle is prominent due to the rotational axis tilt. At the 150 mbar pressure level, the yearly oscillations have an amplitude of ~6 K, in good agreement with simple radiative model predictions as those of Equation 4.57. The models predict large seasonal changes in the stratosphere, strongly dependent on latitude because other effects are added, as, for example, the periodic shadowing produced by the rings on the equatorial band. Typically over the poles, the yearly oscillation at 5 mbar level has an amplitude of ~25 K but reduces to a few degrees at the equator. In Uranus, at low temperatures a seasonal effect must occur due to the large axis tilt of ~90° and the models predict maximum temperature fluctuations at the poles of ~4 K. In Neptune, the North–South temperature contrast is ~3 K near the tropopause ($P = 100$ mbar) but reaches ~12 K near the 1 mbar level.

In Figures 4.15 through 4.18, we show measured vertical-meridional cross sections of the temperature distribution in the atmospheres of the terrestrial planets and Titan. Similar cross sections are shown in Figure 4.19 for Jupiter and Saturn. The temperature fluctuations across an isobar at selected altitude levels in the atmospheres of these two planets are shown in Figure 4.20. A possible correlation between the ammonia meridional distribution, temperatures, and zonal winds in Jupiter is also shown in Figure 4.20C. The altitude–latitude maps of the temperature field can be employed to extract the wind field, as will be discussed in Chapter 7.

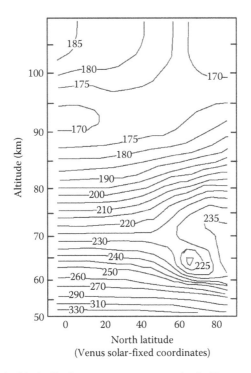

FIGURE 4.15 Vertical-latitudinal temperature cross section in Venus, according to Pioneer-Venus studies. (From Seiff, A. et al., *Adv. Space Res.*, 5(11), 3, 1985. With permission.)

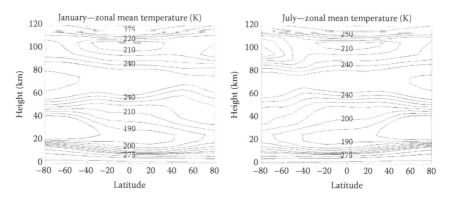

FIGURE 4.16 Vertical-latitudinal temperature cross section in Earth. (From British Atmospheric Data Center, COSPAR International Reference Atmosphere (CIRA-86), 2006, http://badc.nerc.ac.uk/data/cira)

4.5   ADIABATIC LAPSE RATE

We already presented in Section 1.5.4 the basic aspects of the heat transfer by convective motions. In the densest parts of the atmospheres, close to the heated surface of Venus, Earth, and Mars or at cloud level in giant planets, the opacity becomes large enough to make the heat transport by radiation inefficient, and convective transport becomes dominating. The vertical temperature profile in the atmosphere can be obtained from the thermodynamic principles. Two basic situations can be expected: a dry atmosphere, where no condensation of species occurs, and the wet or moist atmosphere, where phase transformations occur with latent heat released. This last case will be treated in detail in Chapter 5 in the context of cloud formation.

4.5.1   THERMODYNAMICS: ADIABATIC LAPSE RATE

The first law of thermodynamics is the expression of the energy conservation in terms of thermal processes. The heat (thermal energy) δQ supplied to an atmospheric parcel (assumed to be a closed system) at pressure P, temperature T, and occupying a volume V is equal to the increase in its internal energy dU plus the work done on the parcel $\delta W = P\, dV$:

$$\delta Q = dU + \delta W = dU + P\, dV \qquad (4.60)$$

The internal energy of a gas is represented by a macroscopic magnitude, the heat capacity at constant volume (unit J K^{-1}):

$$C_V = \left(\frac{\delta Q}{dT} \right)_V \qquad (4.61)$$

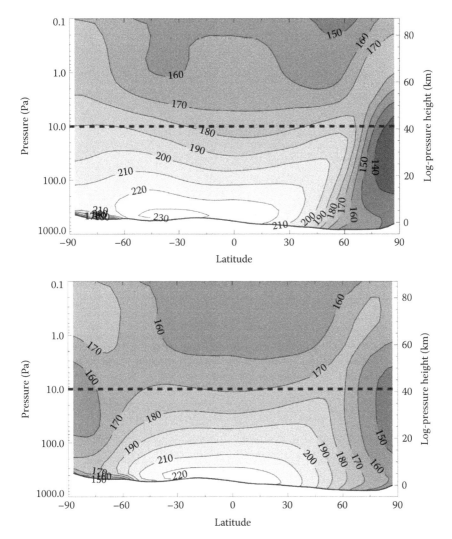

FIGURE 4.17 Vertical-latitudinal temperature cross section in Mars, according to data obtained with the Mars Global Surveyor. (From Lewis, S.R. et al., *Icarus*, 192, 237, 2007.)

Usually this magnitude is given per unit mass introducing the specific heat at constant volume $c_V = C_V/m$ (unit J K^{-1} kg^{-1}) or per mole C_V/n (unit J K^{-1} mol^{-1}). Since it can be shown that $dU = c_V\, dT$, Equation 4.60 can be expressed as

$$\delta Q = c_V\, dT + P\, dV \tag{4.62}$$

Similarly, we define the heat capacity at constant pressure C_P, which can also be normalized per mass or per mole to get their specific values. Both are related through the Mayer relationship $C_P - C_V = nR_g$ for a perfect gas. Differentiating the ideal gas equation of state $PV = nR_g T$ and using the Mayer relation we get

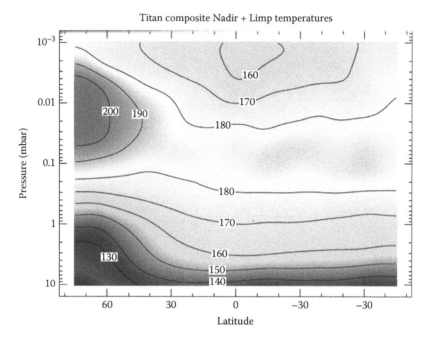

FIGURE 4.18 Vertical-latitudinal temperature cross section in Titan according to data obtained by the Cassini CIRS instrument for years 2004–2006. (From Achterberg, R.K. et al., *Icarus*, 194, 263, 2008. With permission.)

$$P \, dV + V \, dP = nR_g \, dT = (C_P - C_V)dT \tag{4.63}$$

Substituting from Equation 4.60 $P \, dV$ into (4.62) and using $dU = c_V \, dT$, we get

$$C_P \, dT - V \, dP = \delta Q \tag{4.64}$$

If no heat enters or is lost from the parcel, the process is said to be *adiabatic* and then $\delta Q = 0$, and from Equation 4.64 we have $C_P \, dT = V \, dP$. Assuming hydrostatic conditions $dP/dz = -\rho g$, this relationship yields the *adiabatic lapse rate* of a dry atmosphere:

$$\frac{dT}{dz} = -\frac{g}{c_P} = -\Gamma_d \tag{4.65}$$

where $c_P = C_P/m$ (unit J K^{-1} kg^{-1}). In Table 4.1, we give the values for c_P and R_g^* for average tropospheric temperatures and in Table 4.2, the derived dry adiabatic lapse rate for massive atmospheres. Figure 4.21 shows for comparison the lapse rate in the tropospheric layer of Jupiter, Saturn, and Uranus, and in the whole atmosphere and troposphere of Titan.

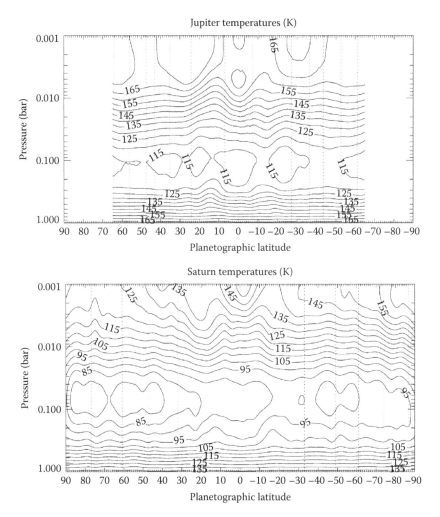

FIGURE 4.19 Vertical-latitudinal temperature cross sections for Jupiter and Saturn according to data obtained with the Cassini CIRS instrument. (From Fletcher, L.N. et al., *Icarus*, 202, 543, 2009. With permission.)

4.5.2 POTENTIAL TEMPERATURE

Consider again the case of an adiabatic process in the atmosphere ($\delta Q=0$). From (4.62) using the perfect gas law, we have

$$C_V\, dT = -P\, dV \rightarrow \frac{C_V}{nR_g} \frac{dT}{T} + \frac{dV}{V} = 0$$

and using the Mayer relationship we get

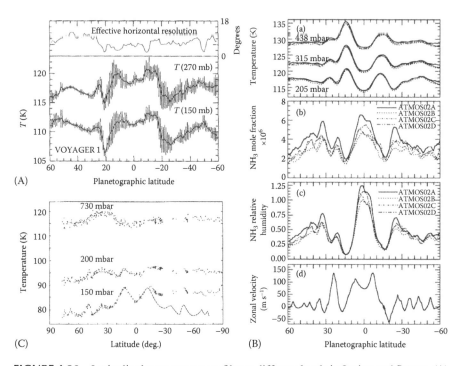

FIGURE 4.20 Latitudinal temperature profiles at different levels in Jupiter and Saturn. (A) Jupiter profiles according to Voyager IRIS instrument in the year 1979. (From Gierasch, P.J. et al., *Icarus*, 67, 456, 1986. With permission.) (B) Saturn profiles according from Voyager IRIS instrument in the year 1980. (From Conrath, B.J. and Pirraglia, J.A., *Icarus*, 53, 286, 1983. With permission.) (C) Jupiter meridional profiles of temperature, ammonia, and zonal winds at cloud level from measurements with the Cassini CIRS instrument. (From Achterberg, R.K. et al., *Icarus*, 182, 169, 2006. With permission.)

$$\frac{1}{\gamma-1}\frac{dT}{T}+\frac{dV}{V}=0$$

where $\gamma=C_P/C_V$ is the adiabatic coefficient already introduced in Chapter 1. Integration of this equation gives the following relationship (known as Poisson's adiabatic equations):

$$TV^{\gamma-1}=\text{Const} \tag{4.66a}$$

and employing the equation of state

$$PV^{\gamma}=\text{Const} \tag{4.66b}$$

$$T^{\gamma}P^{1-\gamma}=\text{Const} \tag{4.66c}$$

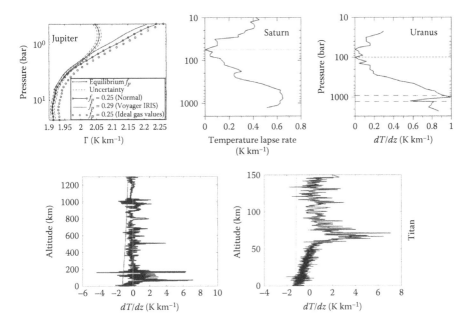

FIGURE 4.21 Temperature lapse rate in the atmospheres of the giant fluid planets and Titan. Jupiter at the equatorial latitude: different cases of the hydrogen ortho to para distributions are considered. Saturn: Lapse rate computed from the radio occultation temperature profile measured by the Voyager 2 spacecraft (tropopause indicated by the continuous line). Data taken from Table 4.A.5. Uranus: Lapse rate computed from the radio occultation profile measured by the Voyager 2 spacecraft (tropopause indicated by the short dashed line, clouds in the layer between the long dashed lines). Data taken from Table 4.A.7. Titan: Lapse rate for the whole atmosphere (left) and for the troposphere (right) as measured during the descent of the Huygens probe (adiabatic gradient indicated by the vertical dotted line). (From Magalhaes, J.A. et al., *Icarus*, 158, 410, 2002; Fulchignoni, M. et al., *Nature*, 438, 785, 2005. With permission.)

This last equation is particularly interesting in planetary atmospheres since it relates temperature with pressure, which is usually employed as vertical coordinate. Accordingly, a parcel moving adiabatically from a point with pressure and temperature (P_1, T_1) to a point with (P_2, T_2) will follow the relationship

$$T_1 = T_2 \left(\frac{P_1}{P_2} \right)^{\kappa} \tag{4.67}$$

where we have the adiabatic index $\kappa = (\gamma - 1)/\gamma$. This equation defines the potential temperature θ, which is the temperature a parcel would have if it is moved adiabatically from a level with (T, P) to a reference level with pressure P_0,

$$\theta = T \left(\frac{P_0}{P} \right)^{\kappa} \tag{4.68}$$

The potential temperature of a mass of air remains constant when it suffers an adiabatic process ($\theta(z)$=constant), or, in other words, the potential temperature changes if subjected to nonadiabatic processes (usually denoted diabatic).

It is also useful to derive this relationship from the second law of thermodynamics in order to use a function of state (the entropy) that applies to both reversible and irreversible changes. For a reversible process, the change in entropy of the parcel is given by $dS=\delta Q/T$ (unit J K^{-1}) and can be used to replace δQ by $T\,dS$ in Equation 4.64. Using the equation of state for the ideal gas $V=nR_g T/P$, we get

$$dS = C_P \frac{dT}{T} - nR_g \frac{dP}{P} \qquad (4.69a)$$

which upon integration gives the entropy

$$S = C_P \ln T - nR_g \ln P + \text{Const} \qquad (4.69b)$$

For an adiabatic process, $\delta Q=0 \rightarrow dS=0$ and S=constant. Integration of Equation 4.69a allows retrieving again the definition (Equation 4.68), and at the same time Equation 4.69b gives $S=C_P \ln \theta$+constant, which is a useful expression that relates the entropy of the parcel to its potential temperature.

4.5.3 Potential Temperature and the Brunt–Väisälä Frequency

In this section, we introduce the concept of potential temperature that is relevant for the study of vertical stability in the atmosphere. In Chapter 9, we will present in a general framework the problem of dynamical instability. Here, we will introduce the use of the concept of potential temperature to the study of vertical stability in the atmosphere.

The response of a parcel to a vertical displacement in the atmosphere with a given temperature profile can be understood when it is compared with the adiabatic lapse rate given by Equation 4.65. The adiabatic profile is represented in a temperature–altitude $T(z)$ graph by a straight line with a constant slope $-g/C_P$ (temperature decreasing with altitude, see Figure 4.22). Assume that the actual atmospheric profile is represented by the parabolic curve. The parcel (adiabatic profile) is warmer than its surroundings from altitude level A to level B. In A, the slope of the actual profile $T(z)$ is $dT/dz<-g/C_P$ and is said to be superadiabatic, and the decrease of temperature with altitude is greater than for the adiabatic profile. The atmosphere becomes unstable to vertical displacements and the parcel being buoyant rises. In B, the profile inverts its slope and $dT/dz>0$ so the temperature increases with altitude. An inversion occurs in the temperature profile and the atmosphere becomes stable against the vertical displacements.

To further investigate the vertical stability of a parcel, we calculate the buoyancy force due to the density difference between the parcel and surroundings (Archimedes'

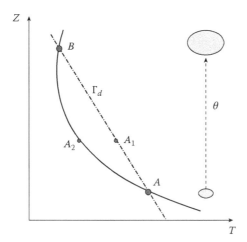

FIGURE 4.22 Temperature profiles and vertical stability of atmospheres. The linear profile is the adiabatic lapse rate and a parcel initially at A will rise adiabatically. Its potential temperature θ remains constant. The point A in the parabolic profile is unstable. In A_1 the parcel will be warmer than surrounding (A_2) and will continue to rise. Point B is stable.

principle). From the ideal gas law, the density of an atmospheric parcel ρ' at temperature T', and that of its environment ρ at the temperature T, initially at the same pressure level P_0 (the pressure of the parcel is assumed to adjust instantaneously to the environment), are given by

$$\rho' = \frac{P_0}{R_g^* T'}, \quad \rho = \frac{P_0}{R_g^* T'} \tag{4.70}$$

If $T' > T$, which occurs when the environment temperature falls more rapidly with height than the adiabatic lapse rate, a parcel displaced adiabatically upward is lighter than its environment ($\rho' < \rho$) and becomes buoyant. The buoyant force acting on a parcel with a volume V is $gV(\rho - \rho')$, and if the parcel is displaced vertically a small distance z' without perturbing its environment, from Newton's second law we have

$$\rho' V \frac{d^2 z'}{dt^2} = gV(\rho - \rho') \tag{4.71}$$

For the small displacement starting at a reference temperature T_0 of the undisturbed atmosphere, and using the atmospheric lapse rate Γ for the environment and a dry adiabatic for the parcel Γ_d, we can write $T' = T_0 - \Gamma_d z'$ and $T = T_0 - \Gamma z'$. Here T and $\Gamma = dT/dz$ are the temperature and lapse rate of the environment, respectively. Upon substituting the densities by temperatures using (4.70) and (4.65) we get

$$\frac{d^2z'}{dt^2} = g\left(\frac{T'}{T} - 1\right) - g\left(\frac{T_0 - \Gamma_d z'}{T_0 - \Gamma z'} - 1\right) - g\left(\frac{\Gamma - \Gamma_d}{T - \Gamma z'}\right)z' \sim -\frac{g}{T}(\Gamma_d - \Gamma)z' \quad (4.72)$$

where we have approximated this expression to the leading order of the small displacement z' (that is, $\Gamma z' \sim 0$). This equation can be rewritten as

$$\frac{d^2z'}{dt^2} + N_B^2 z' = 0 \tag{4.73}$$

where

$$N_B^2 = \frac{g}{T}(\Gamma_d - \Gamma) = \frac{g}{T}\left(\frac{dT}{dz} + \frac{g}{C_P}\right) \tag{4.74}$$

According to what we discussed previously, the atmosphere is statically stable as long as $\Gamma_d - \Gamma > 0$, which means that $N_B^2 > 0$. Equation 4.74 is the motion equation of a simple harmonic oscillator where N_B is called the buoyancy frequency or Brunt–Väisälä frequency (unit s^{-1}) of a dry atmosphere. Equation 4.73 has solutions of the form $z'(t) = A \exp(iN_B t) + B \exp(-iN_B t)$ indicating that the parcel oscillates about its undisturbed level with a period $\tau = 2\pi/N_B$ exchanging kinetic and potential energy. For $N_B = 0$, no force acts on the parcel, no energy exchange occurs, and the parcel is said to be in neutral equilibrium. Finally, the atmosphere becomes unstable when $\Gamma_d - \Gamma < 0$ and $N_B^2 < 0$, which means that N_B is imaginary. This leads to exponential solutions for Equation 4.73 of the type $z'(t) = A \exp(N_B' t) + B \exp(-N_B' t)$, where $N_B^2 = -N_B'^2$, and the parcel's displacement increases exponentially with time. Using Equation 4.68, the Brunt–Väisälä frequency and the potential temperature become related through (Problem 4.8)

$$N_B^2 = \frac{g}{\theta}\frac{d\theta}{dz} \tag{4.75}$$

Accordingly, the vertical stability conditions for dry air can be written as

$$\frac{d\theta}{dz} > 0 \rightarrow \text{statically stable}$$

$$\frac{d\theta}{dz} = 0 \rightarrow \text{statically neutral}$$

$$\frac{d\theta}{dz} < 0 \rightarrow \text{statically unstable}$$

Another useful magnitude to characterize the stability of the atmosphere to vertical displacements is the thermal static stability, defined simply as

$$S_T = \frac{dT}{dz} + \frac{g}{C_P} \tag{4.76}$$

S_T is generally given in K km^{-1} so the atmosphere is stable as long as $S_T > 0$. In Figure 4.23, we present the vertical structure of the static stability in the atmospheres of Venus, Earth, Mars, Jupiter, Saturn (theoretical), Titan, and Uranus. These are representative values, since temporal and spatial (latitude and longitude) dependences are obviously present. In any case, we can see how the dT/dz (Figure 4.21) and $S_T(z)$ (Figure 4.23) vertical profiles inform us on the stability of the atmosphere to vertical displacements.

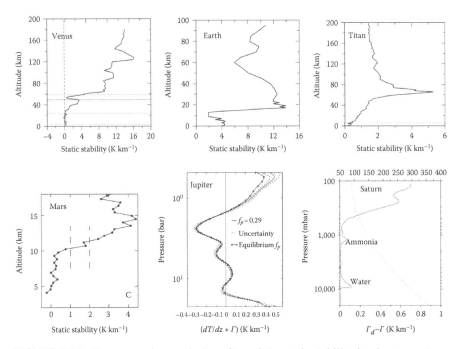

FIGURE 4.23 Representative vertical profiles of the static stability in planetary atmospheres: Venus, Earth, and Titan drawn using data taken from Tables 4.A.1, 4.A.2, and 4.A.4. Mars: Retrieved profiles from the radio occultation experiment onboard Mars Express. (From Hinson, D.P. et al., *Icarus*, 198, 57, 2008. With permission.) Jupiter: Retrieved from direct temperature measurements by the Galileo probe in December 1995. (From Magalhaes, J.A. et al., *Icarus*, 158, 410, 2002. With permission.) Saturn: Theoretical calculation of the static stability from the mean temperature profile (dashed line). It includes the effects of cloud condensation due to ammonia and water (see Chapter 5). (Courtesy of R. Hueso, Universidad del Pais Vasco, Spain.)

Equations 4.74 and 4.76 assume that the mean molecular weight $\bar{\mu}$ of the environment do not vary with altitude. Including this vertical dependence, Equation 4.76 can be rewritten as (Problem 4.16)

$$S_T = \frac{dT}{dz} + \frac{g}{C_P} - \frac{T}{\bar{\mu}} \frac{d\bar{\mu}}{dz} \tag{4.77}$$

It is important first to remember that the atmospheric value for C_p must be obtained as a weighted mean from the contribution to the specific heat of the different atmospheric species (C_{pi}). Using the mass mixing ratio x_{Mi} for each species, we have

$$C_p = \sum_i x_{Mi} C_{pi} \tag{4.78}$$

In addition, note that the specific heat at constant pressure $C_p(T)$ is a function of the temperature T, a dependence that must be taken into account in atmospheres where the temperature varies strongly with altitude (Figures 4.24 and 4.25). The sensibility to temperature is particularly important in the upper tropospheres of the coldest planets (Saturn, Uranus, and Neptune, less in Jupiter) since at their low temperatures (<200 K) the specific heat of hydrogen H_2 depends on the *ortho* to *para* hydrogen distribution, which can deviate from its "normal" 3:1 distribution value due to dynamics and conversion processes (Figure 4.25 and Problem 4.17).

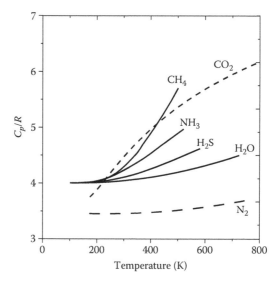

FIGURE 4.24   Specific heat normalized to specific gas constant of gaseous species of interest in planetary atmospheres as a function of temperature.

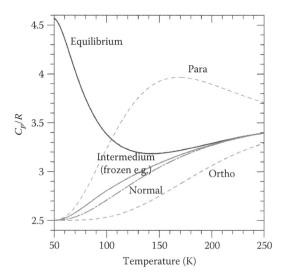

FIGURE 4.25 Specific heat normalized to specific gas constant for hydrogen gas as a function of temperature for different ortho to para distributions (equilibrium, intermediate, frozen, ortho, and para).

PROBLEMS

4.1 Obtain the radiative lapse rate equation (4.34). Calculate the radiative temperature gradient in Jupiter at $P=1$ bar ($T=165$ K) for a Rosseland mean opacity $\alpha_R=0.01$ cm^2 g^{-1}, and taking the other necessary data from tables.

Solution

$$\frac{dT}{dz} = 1.6 \text{ K km}^{-1}, \quad \nabla_{rad} = 0.23$$

4.2 Close to the Earth's surface the net flux divergence by long (LW) and short (SW) wave radiation are −0.035 and +0.007 W m^{-3}. Calculate the corresponding cooling and heating rates in °C day^{-1}. What can you say about the maintenance of a steady-state energy balance?

Solution
LW=+2.5°C day^{-1} and SW=−0.5°C day^{-1}. Since the radiative fluxes are not compensated, the steady-state maintenance requires other forms of heat transfer (e.g., sensible heat by advection or latent heat release by condensation–evaporation).

4.3 Derive Equation 4.42 for the volumetric solar heat deposition assuming plane parallel geometry and a pure absorbing medium (Equation 4.20).

Hint: At a given time and seen from a planet, the Sun subtends a narrow solid angle about its average position given by $\mu_\varphi=-\mu_\varphi$.

4.4 From the data in different tables in the book, calculate the effective pressure of the hydrogen emission as defined by Equation 4.46 in the giant and icy planets if the average volumetric opacity is $\langle a_{v_L}\rangle$ (cm^{-1} amagat^{-2}) $=2.1\times 10^{-6}$ (J), 1.6×10^{-6} (S), 9×10^{-7} (U and N).

Solution

$P_{eff}=340$ mbar (J), 200 mbar (S), 230 mbar (U), 260 mbar (N)

4.5 The use of the direct cooling to space approximation allows to use a simple expression for the averaged infrared cooling rate by H$_2$ in the giant planets (Equation 4.43) that takes into account the internal heat source: $(dF/dP) = A\sigma_B(T^4-T_0^4)$, where A is the cooling coefficient for hydrogen and T_0 is the radiative equilibrium temperature that would be attained in the absence of solar heating at a given latitude. For a gray atmosphere $T_0 = T_{int}^4 f(\tau)$, where T_{int}^4 is the temperature corresponding to the internal heat flux. Assuming that $f(\tau) \approx 1+(3/4)\tau$ (see Equation 4.39), calculate dF/dP in the giant and icy planets at a pressure level 0.15 bar if the optical depth is given by Equation 4.45 and $A=2.1$ bar^{-1}. Take at this level $T=11.7, 86.5, 53.5$, and 55 K for Jupiter, Saturn, Uranus, and Neptune, and use the other necessary data from the tables in the book. Give also the corresponding heating rate in K day^{-1}.

Solution

$$\frac{dF}{dP}(\mathrm{W\,m^{-2}\,bar^{-1}}) = 1.36(\mathrm{J}),0.16(\mathrm{S}),0.20(\mathrm{U}),0.06(\mathrm{N})$$

$$\frac{Q}{rC_p}(\mathrm{K\,day^{-1}}) = 2.2\times10^{-3}(\mathrm{J}),10^{-4}(\mathrm{S}),1.15\times10^{-4}(\mathrm{U}),0.4\times10^{-4}(\mathrm{N})$$

4.6 The radiative time constant in the upper stratosphere of Titan at $P=1$ mbar is 3×10^7 s. (1) Compare the seasonal phase and temporal lag in the thermal response with that expected at the surface (see Table 4.1). (2) What would be the ratio between the first Fourier harmonic terms at the two levels? (3) Which is the Fourier term in the upper stratosphere giving the same response as that at the surface? (4) Since Titan's obliquity is~27°, assume that at any given time the solar constant at the pole is about half that at the equator. What is the first-order temperature difference between pole and equator? The measured difference is less than 2 K, what conclusion do you reach?

Solution

(1) ψ(strat)$=11.3°$, ψ(surface)$=85.4°$, δt(strat)$=0.9$ years, δt(surface)$=7$ years

(2) $\dfrac{T_{surface}^{(1)}}{T_{strat}^{(1)}} \sim 12.2$

(3) $n=62$

(4) T(Pole)$=T$(Equator)-15 K

Since the temperature difference is larger than predicted, dynamical heat transport must be involved.

4.7 Calculate the radiative time constant in the atmosphere of the extrasolar giant "hot Jupiter" type planet HD209458b whose properties are given in Table 1.8 at the level $P=0.5$ bar where $T=1350$ K. Assume that its composition is mostly molecular hydrogen similar to Jupiter and its albedo is 0.3. Assuming that its spin and orbital time are synchronized, obtain the radiative phase and temporal lags. Compare these numbers with those of Jupiter, a similar much colder planet.

Solution

$\tau_{rad}=65$ days, $\psi=116°$, $\delta t=1.13$ days. These values are much lower than those of Jupiter.

4.8 Calculate the fractional change in temperature $\Delta T/T$ between local noon and midnight on (a) Venus' surface ($T_0=734$, $P_0=92$ bar); (b) Venus' clouds ($T_0=250$, $P_0=0.22$ bar); (c) Earth's surface ($T_0=285$, $P_0=1$ bar); (d) Earth's tropopause ($T_0=217$, $P_0=0.1$ bar); (e) Mars' surface ($T_0=240$, $P_0=0.007$ bar); Mars' upper atmosphere ($T_0=140$, $P_0=0.0001$ bar); Jupiter's cloud ($T_0=127$, $P_0=0.4$ bar).

Solution: $\Delta T/T$ in

   (a) Venus' surface $=0.015$
   (b) Venus' clouds $=18.7$
   (c) Earth's surface $=0.01$
   (d) Earth's tropopause $=0.13$
   (e) Mars' surface $=0.33$, Mars' upper atmosphere $=40.2$, Jupiter's cloud $=0.00017$

4.9 Derive the adiabatic lapse rate (Equation 4.65) for an atmosphere under hydrostatic conditions from the definition of the potential temperature (Equation 4.68).

4.10 Find an expression for the dependence of the pressure with altitude and temperature in an atmosphere with a constant adiabatic temperature gradient Γ_d. If at $z=z_0$ the temperature and pressure are T_0 and P_0, what is the vertical profile of the potential temperature?

Solution

$$P(z) = P_0\left[1-\frac{\Gamma_d(z-z_0)}{T_0}\right]^{g/\Gamma_d R_g^*}, \quad P(T) = P_0\left(\frac{T}{T_0}\right)^{g/\Gamma_d R_g^*}$$

$$\theta(z) = \left[T_0 - \Gamma_d(z-z_0)\right]\left[1-\frac{\Gamma_d(z-z_0)}{T_0}\right]^{-(g/\Gamma_d R_g^*)\kappa}$$

4.11 An air parcel in the deep troposphere of Jupiter has initially a temperature $T = 300$ K at the 7 bar pressure level. If the parcel lifts dry adiabatically, what is its change in density when it reaches the 1 bar level? What is its potential temperature?

Solution

$$\Delta\rho = 0.45 \text{ kg m}^{-3}, \quad \theta = 172 \text{ K}$$

4.12 Using the data in Table 4.3 (1) calculate the oscillation period of an air parcel in Venus ($g = 8.9$ m s^{-2}, $T = 270$ K), Earth ($g = 9.8$ m s^{-2}, $T = 270$ K), and Mars ($g = 3.7$ m s^{-2}, $T = 200$ K). (2) What can you say about the static stability in the tropospheres of the giant planets? (3) What is the oscillation period for a parcel in the tropopause of Jupiter? Compare it with the oscillation period of a parcel in the equatorial stratosphere of Saturn if at the 5 mbar level $T = 123.5$ K and $dT/dz = 0.6$ K km^{-1} for the range of possible values of Γ_d in Table 4.2?

Solution

(1) $\tau(\text{Venus}) = 104$ s, $\tau(\text{Earth}) = 85$ s, $\tau(\text{Mars}) = 164$ s
(2) On the giant planets, the tropospheres are close to neutral stability ($N_B = 0$)
(3) $dT/dz = 0$ and $\tau(\text{Jupiter tropopause}) = 65.5$ s. For Saturn, $\tau(\text{Saturn}) = 2.9$–$3.3$ s

4.13 Find the relationship between altitude z and pressure P (Equation 4.6) for an atmosphere that has an adiabatic lapse rate (that is, $\theta = $ constant) if for $z = 0$, $P = P_0$. What is the altitude difference between the 10 and 20 bar levels ($T = 338$ and 416 K, respectively) sounded by Galileo in 1995? What would be the altitude difference in the height separation between these levels if we use the isothermal definition and the mean layer approximation for the temperature given in this pressure range by the equation $T(P) = a_0 + a_1 T + a_2 T^2$?
    Data: $g_{EQ} = 23.1$ m s^{-2}, $R_g^* = 3890$ J kg^{-1} K^{-1}, $a_0 = 216.1$ K, $a_1 = 14.23$ K bar^{-1}, and $a_2 = -0.21$ K bar^{-1}.

Solution

$$z = \frac{R_g^* \theta}{g\kappa}\left[1 - \left(\frac{P}{P_0}\right)^{\kappa}\right], \quad 44 \text{ km (adiabatic)}, \quad 43.9 \text{ (mean isothermal)}$$

4.14 During the morning, the layer adjacent to the surface in the Martian atmosphere has a potential temperature profile given by $(\theta/\theta_0) = \exp((z/h_1) - (z/h_2))$ with $h_1 > h_2$. Characterize the stability of a layer and derive its Brunt–Väisälä frequency profile N_B^2. What is the oscillation period for a parcel at the altitude level $z = 1.5$ km for $h_1 = 2$ km and $h_2 = 1$ km?

Solution: Stable if

$$z > z_C = h_1 \ln\left(\frac{h_1}{h_2}\right), \quad N_B^2(z) = \frac{g\left[\exp(z/h_1) - 1/h_2/h_1\right]}{\exp(z/h_1) - z/h_2}, \quad \tau = 0.54 \text{ s}$$

4.15 The temperature close to the surface of the Earth is observed to decrease according to the equation $T(z)=T_0h/(h+z)$ with $T_0>h\Gamma_d$. (1) What is the critical altitude for neutral condition? (2) What is the Brunt–Väisälä frequency profile N_B^2? Calculate this critical level and N_B at 5 km above the surface in the Earth's lower atmosphere if $h=1$ km and $T_0=300$ K.

Solution

(1) $\quad z_C = \sqrt{\dfrac{hT_0}{\Gamma_d}} - h$

(2) $\quad N_B^2(z) = g\left[\dfrac{\Gamma_d}{T_0}\left(1+\dfrac{z}{h}\right)-\dfrac{1}{z+h}\right], \quad z_C = 4.5\ \text{km}; \quad N = 0.17\ \text{s}^{-1}$

4.16 A sounding of the atmosphere gives a variation of pressure with altitude according to $P(z)=P_0/[1+(z/h)^2]$. (1) Determine the temperature variation with altitude. (2) What can you say about the surface temperature? (3) Where is located the tropopause? (4) Determine $\theta(z)$ and $N_B^2(z)$.

Solution

(1) $\quad T(z) = \dfrac{gh^2}{2zR_g^*}\left[1+\left(\dfrac{z}{h}\right)^2\right]$

(2) At $z=0$, there is a singularity and we take from the ideal gas law
$T_0 = P_0/\rho_0 R_g^*$

(3) $\quad z=h$

(4) $\quad \theta(z) = \dfrac{gh^2}{2zR_g^*}\left[1+\left(\dfrac{z}{h}\right)^2\right]^{\kappa+1}$

$\quad N_B^2(z) = gz\left[\dfrac{2}{h^2}\dfrac{1}{\left[1+(z/h)^2\right]}-\dfrac{1}{z^2}\right]$

4.17 Derive Equation 4.77. (1) What is the effect on the static stability if the mean molecular weight decreases with altitude within a layer of the atmosphere? Radio-occultation measurements of Uranus' troposphere by Voyager 2 in 1986 showed that the layer between pressure levels 1 and 1.3 bar showed a decrease in the mixing ratio of methane $dx_{CH_4}/dz = -2.4\times10^{-3}$ km^{-1}. Taking a mean temperature for this layer 81 K and a mean molecular weight 2.3 g mol^{-1}, (2) what is the contribution to the static stability of this layer due to this methane abundance decrease?

Solution

(1) If $d\bar\mu/dz > 0$, the static stability S_T increases

(2) $(\bar T/\bar\mu)(d\bar\mu/dz) = -1.25$ K km^{-1}

APPENDIX 4.A

TABLE 4.A.1
Earth's Reference Vertical Temperature
Profile (Month June)

z (km)	P (mbar)	T (K)
0	1010	301
1	900	295
2	802	289.5
3	712	284
4	631	278
5	557	272.6
6	491	267
7	432	260
8	378	254
9	330	247
10	287	239
11	248	231
12	214	223
13	183	215
14	156	207
15	131	201
16	110	198
17	93	198
18	78	202
19	66	205
20	56	209
21	47	212
22	40.5	214
23	34.5	217
24	29.5	219
25	25	221
30	12	231
35	6	244
40	3	255
45	1.5	263
50	0.8	264
55	0.4	255
60	0.2	241
65	0.1	221
70	0.04	206
75	0.02	203
80	0.082	202
85	0.035	194
90	0.014	187

TABLE 4.A.1 (continued)
Earth's Reference Vertical Temperature Profile (Month June)

z (km)	P (mbar)	T (K)
95	0.006	186
100	0.0002	190

Source: Data taken from Houghton, J.T., *Physics of the Atmosphere*, Cambridge University Press, Cambridge, U.K., Appendix 5, p. 284, and British Atmospheric Data Center, COSPAR International Reference Atmosphere (CIRA-86), 2006, http://badc.nerc.ac.uk/data/cira

TABLE 4.A.2
Venus' Vertical Temperature Profile

P (mbar)	T (K)	z (km)
4.73E−010	125	180
7.2E−010	125	176
1.13E−009	125	172
1.84E−009	126	168
3E−009	126	164
5.017E−009	127	160
8.68E−009	129	156
1.53E−008	131	152
2.83E−008	134	148
5.65E−008	140	144
1.26E−007	147	140
3.3E−007	146	136
9.9E−007	142	132
3.3E−006	135	128
1.2E−005	125	124
5.17E−005	117	120
0.00021	129	116
0.0008	140	112
0.0027	150	108
0.0085	160	104
0.0275604	175	100
0.103352	182	95
0.373889	187	90
1.286828	192	85
4.123928	208	80

(continued)

TABLE 4.A.2 (continued)
Venus' Vertical Temperature Profile

P (mbar)	T (K)	z (km)
12.159	220	75
34.0452	232	70
50.155875	236	68
73.3593	241	66
106.39125	245	64
154.014	250	62
219.87525	261	60
309.04125	274	58
427.5915	289	56
580.59225	308	54
774.123	328	52
1,013.25	347	50
1,317.225	361	48
1,681.995	374	46
2,137.9575	387	44
2,705.3775	400	42
3,384.255	413	40
4,204.9875	428	38
5,187.84	444	36
6,353.0775	460	34
7,731.0975	476	32
9,352.2975	492	30
11,247.075	509	28
13,476.225	526	26
15,908.025	543	24
18,846.45	560	22
22,190.175	576	20
25,939.2	593	18
30,296.175	609	16
35,159.775	626	14
40,631.325	642	12
46,812.15	658	10
53,904.9	674	8
61,706.925	689	6
70,522.2	705	4
80,249.4	720	2
91,192.5	735	0

Sources: Data from Moroz, V.I., *Space Sci. Rev.*, 29, 1981 and Seiff, A. et al., *Adv. Space Res.*, 5, 1985 [from 0 to 100 km], complemented with data from Hunten, D.M. et al. (ed.), *Venus*, Arizona University Press, Tucson, AZ, Table AIII, p. 1048.

TABLE 4.A.3
Mars' Vertical Temperature Profile

P (mbar) HN	T (K)	z (km)
4.24212E−005	139	100
5.5332E−005	139	98
7.21224E−005	139	96
9.40644E−005	139	94
0.000122684	139	92
0.000160145	139	90
0.000209244	139	88
0.000272844	139	86
0.000356796	139	84
0.000465552	139	82
0.000608652	139	80
0.000796272	139	78
0.001041132	139	76
0.001361676	139	74
0.0017808	139	72
0.00233412	139.5	70
0.00304644	140	68
0.00398136	141	66
0.00519612	142	64
0.00676068	143	62
0.00878952	144.2	60
0.01139712	145.7	58
0.01474884	147.2	56
0.0190164	148	54
0.0245496	150.3	52
0.031482	152.2	50
0.0403224	154.1	48
0.051516	156	46
0.0656352	158	44
0.083316	160	42
0.105576	162.4	40
0.133242	164.8	38
0.1677132	167.5	36
0.210516	170	34
0.263304	172.5	32
0.328176	175	30
0.407676	177.5	28
0.50562	180	26
0.624552	182.5	24
0.770196	185.2	22
0.947004	188.2	20
1.1607	191.4	18

(continued)

TABLE 4.A.3 (continued)
Mars' Vertical Temperature Profile

P (mbar) HN	T (K)	z (km)
1.417644	194.6	16
1.72674	197.8	14
2.0988	201.4	12
2.53764	205	10
3.05916	209.2	8
3.68244	212.4	6
4.41384	213.4	4
5.299788	213.8	2
6.36	214	0

Source: Data from Seiff, A., *Adv. Space Res.*, 2, 3, 1982.

TABLE 4.A.4
Titan's Vertical Temperature Profile

z (km)	T (K)	P (mbar)
0	93.9	1440
0.5	93.3	1400
1	92.7	1370
1.5	92.1	1340
2	91.5	1300
3	90.5	1240
4	89.5	1180
5	88.5	1120
6	87.3	1060
8	85.5	957
10	83.6	859
12	82.1	770
14	80.6	688
16	79.2	614
18	78	547
20	76.7	487
22	75.7	432
24	74.7	383
26	73.9	340
28	73.1	300
30	72.3	265
32	71.8	235
34	71.5	207
36	71.4	183
38	71.2	161

TABLE 4.A.4 (continued)
Titan's Vertical Temperature Profile

z (km)	T (K)	P (mbar)
40	71.1	142
42	71.1	126
44	71.2	111
46	71.4	98.1
48	71.5	86.7
50	71.7	76.7
52	72.2	67.9
54	72.8	60.1
56	73.8	53.4
58	75.4	47.5
60	77.6	42.3
62	80.9	37.9
64	85.9	34.2
66	92.6	31
68	100.5	28.3
70	106.1	26
72	111.9	24
74	116.6	22.3
76	119.7	20.7
78	122.7	19.3
80	125.8	18
82	128.8	16.8
84	131.5	15.7
86	133.8	14.7
88	135.8	13.8
90	137.6	13
92	139.1	12.2
94	140.6	11.5
96	142.1	10.8
98	143.7	10.2
100	145.1	9.61
102	146.5	9.06
104	147.8	8.56
106	149	8.08
108	150.1	7.63
110	151	7.22
112	151.9	6.82
114	152.6	6.45
116	153.4	6.1
118	154.3	5.78
120	155.3	5.48
122	156.3	5.19
124	157.5	4.91

(continued)

TABLE 4.A.4 (continued)
Titan's Vertical Temperature Profile

z (km)	T (K)	P (mbar)
126	158.6	4.66
128	159.6	4.42
130	160.5	4.19
132	161.2	3.9
134	161.8	3.78
136	162.4	3.59
138	163	3.41
140	163.7	3.24
142	164.5	3.08
144	165.3	2.93
146	166	2.78
148	166.7	2.65
150	167.4	2.52
152	167.9	2.4
154	168.5	2.29
156	169	2.17
158	169.5	2.07
160	169.9	1.97
162	170.4	1.88
164	170.8	1.79
166	171	1.71
168	171.2	1.63
170	171.4	1.55
172	171.6	1.48
174	171.8	1.41
176	172	1.35
178	172.2	1.28
180	172.4	1.22
182	172.5	1.16
184	172.7	1.11
186	172.9	1.06
188	173	1.01
190	173.2	0.964
192	173.4	0.92
194	173.6	0.877
196	173.8	0.837
198	173.9	0.798
200	174	0.759

Source: Data from Lellouch, E. et al., *Icarus*, 79, 328, 1989.

TABLE 4.A.5
Jupiter's Vertical Temperature Profile

P (mbar) Voyager	T (K) Voyager	P (mbar) Galileo	T (K) Galileo
1	169.1	0.2824	157.2
1.12	168.8	0.6192	168.6
1.26	160.7	1.342	160.5
1.41	152.8	3.079	149.8
1.58	151.4	7.177	158.1
1.78	152.3	16.4	143.8
2	154.9	43.74	122.6
2.24	156	135.8	113.2
2.51	155.5	351.5	122.9
2.82	156.6	468	130
3.16	158.8	500	132.79
3.55	160.6	600	140.84
3.98	161.4	700	148.04
4.47	159	800	154.57
5.01	160.5	900	160.56
5.62	164	1,000	166.1
6.31	165.9	1,200	176.11
7.08	163	1,400	184.99
7.94	159.5	1,600	193
8.91	154.8	1,800	200.31
10	147	2,000	207.06
11.22	140.4	2,200	213.32
12.59	133.5	2,400	219.19
14.13	129.2	2,600	224.71
15.85	124.9	2,800	229.92
17.78	121.9	3,000	234.87
19.95	119.9	3,200	239.59
22.39	118.1	3,400	244.09
25.12	116.6	3,600	248.4
28.18	115.6	3,800	252.55
31.62	111.7	4,000	256.53
35.48	111.4	4,200	260.38
39.81	113	4,400	264.09
44.67	114.2	4,600	267.68
50.12	114.6	4,800	271.16
56.23	115.3	5,000	275.54
63.1	115.4	6,000	290.1
70.79	114.6	7,000	303.88
79.43	113.7	8,000	316.31
89.13	112.8	9,000	327.67
100	113.5	10,000	338.16
112.2	112.9	11,000	347.92
125.89	111.4	12,000	357.08

(*continued*)

TABLE 4.A.5 (continued)
Jupiter's Vertical Temperature Profile

P (mbar) Voyager	T (K) Voyager	P (mbar) Galileo	T (K) Galileo
141.25	110.9	13,000	365.71
158.49	111.7	14,000	373.87
177.83	112.9	15,000	381.64
199.53	114.1	16,000	389.04
223.87	115.7	17,000	396.13
251.19	117.4	18,000	402.93
281.84	119.5	19,000	409.46
316.23	121.2	20,000	415.75
354.81	123.8	21,000	421.83
398.11	127.2	22,000	427.71
446.68	130.1		
501.19	132.5		
562.34	136.9		
630.96	142.4		
707.95	146.9		
794.33	151.2		
891.25	157.5		
1000	165		

Source: Data from Voyager radio occultation profile from Lindal, G.F., *Astron. J.*, 103, 975, 1995, complemented with the Galileo probe profile measurements from Seiff, A. et al., *J. Geophys. Res.*, 103, 22857, 1998.

TABLE 4.A.6
Saturn's Vertical
Temperature Profile

P (mbar)	T (K)
0.56	143
0.63	142.2
0.71	141.3
0.79	141.2
0.89	141.3
1	141.6
1.12	142.1
1.26	142.5
1.41	141.5

TABLE 4.A.6 (continued)
Saturn's Vertical
Temperature Profile

P (mbar)	T (K)
1.58	140.3
1.78	139
2	137.7
2.24	137.3
2.51	136.9
2.82	135.9
3.16	133.4
3.55	131.2
3.98	129
4.47	126.2
5.01	123.5
5.62	120.6
6.31	117.7
7.08	114.8
7.94	112
8.91	108.4
10	104.8
11.22	102.2
12.59	99.5
14.13	97.3
15.85	95
17.78	92.7
19.95	90.5
22.39	88.7
25.12	87
28.18	85.6
31.62	84.2
35.48	83.8
39.81	83.3
44.67	82.8
50.12	82.3
56.23	82.1
63.1	82.1
70.79	82.4
79.43	82.6
89.13	82.8
100	83.4
112.2	84.1
125.89	84.9
141.25	85.7

(continued)

TABLE 4.A.6 (continued)
Saturn's Vertical
Temperature Profile

P (mbar)	T (K)
158.49	86.8
177.83	88.2
199.53	89.8
223.87	91.1
251.19	92.4
281.84	93.8
316.23	95.6
354.81	97.9
398.11	100.7
446.68	103.9
501.19	107.5
562.34	111.6
630.96	115.8
707.95	120.3
794.33	125
891.25	129.9
1000	134.8
1122.02	139.9
1258.93	145
1298.48	146.2

Source: Data from Voyager radio occultation measurements according to Lindal, G.F., *Astron. J.*, 103, 975, 1995.

TABLE 4.A.7
Uranus' Vertical
Temperature Profile

P (mbar)	T (K)
0.5	129
0.56	116.2
0.63	103.5
0.71	96.4
0.79	88.8

**TABLE 4.A.7 (continued)
Uranus' Vertical
Temperature Profile**

P (mbar)	T (K)
0.89	80.8
1	76.4
1.12	73.9
1.26	72.8
1.41	71.6
1.58	70.3
1.78	69.1
2	68.5
2.24	68.2
2.51	68
2.82	68.3
3.16	68.8
3.55	69.4
3.98	69
4.47	68.4
5.01	67.9
5.62	67
6.31	64.8
7.08	63.8
7.94	63.4
8.91	63.3
10	63.4
11.22	64.3
12.59	65.9
14.13	66.2
15.85	61.1
17.78	58.8
19.95	58.2
22.39	57.2
25.12	56.4
28.18	55.7
31.62	55.2
35.48	54.9
39.81	54.6
44.67	54.3
50.12	53.9
56.23	53.5
63.1	53.4
70.79	53.4

(continued)

TABLE 4.A.7 (continued)
Uranus' Vertical
Temperature Profile

P (mbar)	T (K)
79.43	53.3
89.13	53.2
100	53
112.2	53
125.89	53.1
141.25	53.4
158.49	53.5
177.83	53.9
199.53	54.3
223.87	54.7
251.19	55.2
281.84	55.7
316.23	56.3
354.81	57.2
398.11	58.3
446.68	59.6
501.19	61.3
562.34	63.2
630.96	65.3
707.95	67.6
794.33	70.3
891.25	73.2
1000	76.4
1122.02	80.1
1258.93	82.9
1298.48	83.5
1412.54	85.9
1584.89	89.2
1778.28	92.6
1995.26	96
2238.72	99.8
2308.81	100.9

Source: Data from Voyager radio occultation measurements according to Lindal, G.F., *Astron. J.*, 103, 975, 1995.

TABLE 4.A.8
Neptune's Vertical
Temperature Profile

P (mbar)	T (K)
0.35	130.6
0.4	130.1
0.45	130.2
0.5	129.8
0.56	128.7
0.63	127.8
0.71	127.1
0.79	126.2
0.89	125
1	123.9
1.12	122.9
1.26	121.5
1.41	119.6
1.58	117.1
1.78	114.6
2	112.3
2.24	109.9
2.51	106.9
2.82	103.2
3.16	100.4
3.55	97.6
3.98	94.4
4.47	90.9
5.01	87.4
5.62	84.1
6.31	80.9
7.08	77.7
7.94	74.9
8.91	72.3
10	69.7
11.22	67.6
12.59	65.9
14.13	64.3
15.85	62.8
17.78	61.5
19.95	60.4
22.39	59.2
25.12	58.3
28.18	57.5

(*continued*)

TABLE 4.A.8 (continued)
Neptune's Vertical
Temperature Profile

P (mbar)	T (K)
31.62	56.7
35.48	56
39.81	55.3
44.67	54.7
50.12	54.1
56.23	53.6
63.1	53.2
70.79	52.7
79.43	52.5
89.13	52.1
100	51.8
112.2	51.8
125.89	51.7
141.25	51.7
158.49	51.8
177.83	52.1
199.53	52.3
223.87	52.6
251.19	53.1
281.84	53.6
316.23	54.3
354.81	55.1
398.11	56
446.68	57.2
501.19	58.8
562.34	60.6
630.96	62.4
707.95	64.3
794.33	66.4
891.25	68.7
1000	71.5
1122.02	74.5
1258.93	77.9
1298.48	78.7
1412.54	81.4
1584.89	85.8
1778.28	90.3
1995.26	94.4
2238.72	98.4
2308.81	99.4
2511.89	102.2

**TABLE 4.A.8 (continued)
Neptune's Vertical
Temperature Profile**

P (mbar)	T (K)
2818.38	106.3
3162.28	110.4
3548.13	114.5
3981.07	118.7
4466.84	122.9
5011.87	127.2
5623.41	131.2
6267.92	135

Source: Data from Voyager radio occultation measurements according to Lindal, G.F., *Astron. J.*, 103, 975, 1995.

5 Clouds in Planets

All planets and satellites with substantial atmospheres have clouds and suspended particles. In most cases, the clouds and aerosols determine the visual appearance of these bodies. Clouds and aerosols fully cover Venus, Titan, and the giant and icy planets, and they partially cover the surfaces of Earth and Mars. Clouds are present in a variety of shapes and sizes (horizontal and vertical extents), in response to their origin serving as diagnostic of the dynamical mechanisms. Clouds are one of the ingredients of meteorological phenomena that occur in most planetary atmospheres. They play an important role in determining the atmospheric heat balance and thermal structure. Their formation mechanisms are not a simple issue; on the contrary they are complicated and involve many facets of atmospheric science such as cloud particle physics (microphysics) and thermodynamics. Suspended particles in the atmosphere can form from a variety of processes, including condensation, chemical and photochemical reactions, outgassing or lifting from the surface (dust, evaporation and sublimation, volcanic activity), and from particle bombardment of atoms and molecules in the upper atmosphere. The sizes of cloud particles may span several orders of magnitude and they can interact and grow through complicated mechanisms. Except for the Earth and in part for the terrestrial planets and Titan (where probes have measured their properties in situ), most of the fundamental aspects of cloud constitution and formation mechanisms on the planets are poorly known. However, since, in general, clouds in planetary atmospheres form from the condensation of minor atmospheric constituents, a simple approach can be followed: thermodynamic principles, chemistry basics, and the use of the Clausius–Clapeyron equation.

In this chapter, we present the fundamentals of cloud formation and microphysical processes. We also present the properties and nature of the clouds present in planetary atmospheres. We also extend our previous study of radiative transfer of Chapters 3 and 4 to include the role played by aerosols and cloud particles in the radiation field and its importance for the study of the structure and optical properties of clouds and hazes in planets.

5.1 CHEMISTRY AND SURFACE PROCESSES

Surfaces act on the terrestrial planets and satellites as the lower rigid boundary for the atmospheres. A variety of physical and chemical processes occur on them that impact the atmospheres, as already discussed in Chapter 2. On one hand, on bodies with important internal activity, outgassing from volcanic and similar escaping active areas can inject particulates and gases into the atmosphere, reacting there, and forming aerosols that add to the solid released compounds. Suspended particles can also be injected to the atmosphere by surface winds (lifted dust). Surfaces (including solid and liquid deposits) are also the place where condensation and evaporation, and sublimation of active species occur, giving rise to condensation cycles as occurs with water on Earth, carbon dioxide on Mars, methane on Titan, and similarly with these and other species in cold satellites and dwarf planets. We leave to be covered in Chapter 6 the study of photochemistry, surface, and the diffusion processes. In this section, we examine the fundaments of the chemical cycles in their role on compositional cyclical modifications.

5.1.1 CHEMICAL EQUILIBRIUM

In Chapter 4, we introduced the concept of local thermodynamic equilibrium (LTE) in the context of the radiation field. Here, we extend it to the case of the chemical reactions.

5.1.1.1 Basics of Chemical Equilibrium

At temperatures $\leq 2000\,\mathrm{K}$ most atoms are neutral, and since their thermal energy $k_B T$ is comparable to the strengths of some of the strongest chemical bonds, diatomic and polyatomic molecules can form. In a closed system where no matter or radiation enters or leaves, and where the temperature and pressure remain constant, the chemical species will reach equilibrium. The chemical reactions are such that the total rate of production of a compound equals the total rate of its destruction, so the composition remains the same. The relative amounts of the compounds involved that are present at equilibrium are given by the *reaction constant K*. This magnitude is a function of temperature but not of the total amount of the chemicals present. Assume that a reaction between two compounds X and Y produces the compound XY and has the generic form

$$aX + bY = cXY \tag{5.1}$$

After a very long time, we will have X, Y, and XY present in equilibrium amounts. Measuring their concentrations at equilibrium gives the constant

$$K = \frac{[XY]^c}{[X]^a [Y]^b} \tag{5.2}$$

where the square brackets denote the concentrations and the exponents (a, b, c) are the numerical coefficients fitting the reaction (5.1). For example, in the giant planets, reaction of carbon C and hydrogen H_2 give rise to methane CH_4 (hydrogen is much more abundant that carbon) and other related products such that ethane C_2H_6, C_2H_4, and C_2H_2.

5.1.1.2 More Detailed Thermochemical Equilibrium

The thermodynamic equilibrium of a homogeneous system requires thermal and mechanical equilibrium. In a heterogeneous system as the atmospheres, more than one phase can be involved, and for it to be in thermodynamic equilibrium requires it to be also in chemical equilibrium, with no conversion of mass occurring from one phase to the other. Without going into the details, the derivation of the chemical equilibrium constant requires the introduction of auxiliary thermodynamic energy variables. In our case, the most important is the *Gibbs function* (or *Gibbs free energy*):

$$G = U + PV - TS \tag{5.3}$$

where
   U is the internal energy of the system (see Equation 4.60)
   S is the entropy (see Equations 4.69)

Additional useful thermodynamic functions are the *enthalpy* ($H = U + PV$) and the *Helmholtz function* ($U - TS$).

Assume *thermochemical equilibrium* such that no irreversible reactions occur in the system, and G and the entropy S do not change in time. When homogeneous reactions occur between a number i of ideal gases at partial pressures P_i, the Gibbs function changes according to

$$dG = -S\,dT + V\,dP + \sum_i \left[\left(\frac{\partial G}{\partial n_i} \right)_{P=1\,\mathrm{atm}} + RT \ln P_i \right] dn_i = 0 \tag{5.4}$$

For a general gas-phase reaction of the type

$$a_1 X_1 + a_2 X_2 + \cdots = b_1 Y_1 + b_2 Y_2 + \cdots \tag{5.5}$$

integration of Equation 5.4 keeping T and P constant, gives

$$\Delta G(T, 1\,\mathrm{atm}) + RT \left[\frac{P(Y_1)^{b_1} P(Y_2)^{b_2} \cdots}{P(X_1)^{a_1} P(X_2)^{a_2} \cdots} \right] = 0 \tag{5.6}$$

Accordingly, only the temperature intervenes in the reaction. The equilibrium constant for the temperature T is now defined by the quantity in the brackets, which contains only pressure variables,

$$K_p(T) = \frac{P(Y_1)^{b_1} P(Y_2)^{b_2} \cdots}{P(X_1)^{a_1} P(X_2)^{a_2} \cdots} = \exp\left(-\frac{\Delta G(T, 1\,\mathrm{atm})}{RT} \right) \tag{5.7}$$

The equilibrium constant for the above reaction can be written as

$$K_p(T) = \sum_i b_i c_i(T) - \sum_i a_i c_i(T) \tag{5.8}$$

where coefficients $c_i(T)$ are known for the different chemical species.

In general, we expect chemical equilibrium in atmospheres as long as the reaction time rate is small in comparison to characteristic mixing times. However, disequilibrium processes occur in atmospheres and alter the predicted thermochemical equilibrium of the species. In such cases, the air parcel changes significantly its Gibbs free energy or the entropy with time. Chemical disequilibration in a parcel is due to one or both of two causes: irreversible chemical reactions and net exchange of material with surroundings. For example, high in the atmospheres, the action of solar radiation induces photochemical reactions (photolysis and ionization and dissociation, see Chapter 6). Mixing, diffusion, and particulate sedimentation also give rise to disequilibrium species. For example, the reactions that give rise to the ortho–para hydrogen conversion at low temperatures in the giant planets (commented at the end of Chapter 4), or the rapid vertical transport of compounds formed at high temperatures up to low temperature atmospheric levels as CO and N_2 and other exotic species like GeH_4 and AsH_3 in the giant planets. Measurements of the vertical profiles of these species provide constraints of the vertical motions and turbulence and diffusion in the atmosphere. Lighting within the clouds, magnetospheric particle bombardment in the upper atmosphere, and even micrometeoroids and particles dropping from the rings in the giant planets contribute to form disequilibrium species.

5.1.2 Surface–Troposphere Cycles of Condensable Gases and Aerosols

Chemical active gases are continually injected into the atmosphere from the surface and from photochemical transformed species in the upper levels, and removed from it again to the surface by chemical reactions and deposition through a variety of processes as rainout and by rock weathering. These mass flows into the bulk atmosphere give rise to "cycles" for some of the constituents that in some cases intervene decisively in the cloud and aerosol formation. We synthetically review what occurs in each planet.

5.1.2.1 Venus

The most important cycle in Venus involves sulfur compounds as SO_2 and SH_2 released from the surface (perhaps by volcanic or similar outgassing processes). They react with photodissociated oxygen in the upper atmosphere to produce SO_3 that subsequently reacts with H_2O to form the SO_4H_2 liquids droplets that constitute the cloud layers (Figure 5.1). The rainout of the sulfuric acid droplets ends as soon as they encounter the deep high temperatures of Venus and evaporate. Clouds cover completely the planet and remain at altitudes between ~30 and 75 km above the surface. This Venus cycle also involves reactions between different forms of sulfur (S_2, S_3, S_4, and S_8) and CO and COS. The basic photodissociation reactions and their products are given in Section 6.2.2.

5.1.2.2 Earth

On a short timescale, the gases that suffer major compositional changes in Earth's troposphere through cyclical interactions with the surface are water vapor and those

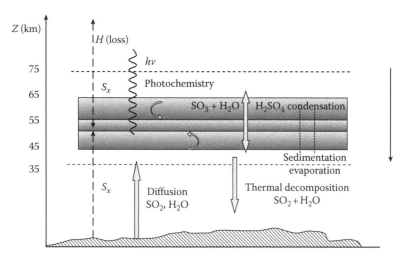

FIGURE 5.1   Scheme of Venus cloud—surface interactions and cycles.

resulting from biogenic and lightning activity as carbon dioxide, methane, hydrogen, ammonia, hydrogen sulfide, the nitrogen oxides, and nitric acid. The surface is also the source of an active exchange of particulate matter with the atmosphere by injection of suspended aerosols lifted by the winds.

Water vapor is highly variable in abundance in the atmosphere because of the large temperature variations on Earth's surface (at a fixed pressure of 1 bar), which causes it to condense into liquid and ice phases, and to sublimate and evaporate approximately following the diurnal and annual temperature cycles of our planet. In the coldest and driest areas, the water mole fraction may drop to low concentrations as 10^{-6}, whereas under the moist tropical monsoon conditions it can reach 4×10^{-2}. Water is the main source for clouds and this will be detailed in the next section.

The carbon dioxide mole fraction varies between 2 and 4×10^{-4} with diurnal oscillations due to removal by daytime photosynthesis and release by plant respiration, and with an annual cycle due to plant growth and decay. Added to this is the secular long-term increase produced by industrial processes and the burning of fossil fuels. The CO_2 largest reservoir is the crust where it is present in carbonate minerals in the form of sedimentary and metamorphic rocks. If all this carbonate content were released to the atmosphere, about 60 bar of equivalent pressure of CO_2 would be formed. The Earth's carbon cycle also involves methane whose mole fraction changes within the range ~$1–2 \times 10^{-6}$.

The nitrogen cycle involves NO_x species (NO, NO_2, NO_3, etc.) and ammonia (NH_3 and NH_4^+ ions) with the intervention of microbial and industrial processes. The ammonia mole fraction variability is ~2×10^{-8}.

The sulfur cycle involves SO_2 and SH_2 with mole fraction variability $2–20 \times 10^{-9}$ due to biogenic anaerobic decay and to volcanic and smelter eruptions. They participate in the formation of H_2SO_4 and $(NH_4)SO_4$ aerosols (sulfuric acid droplets) that form acid rain. Between an altitude of 18 and 25 km, there is a layer (the "Junge" layer)

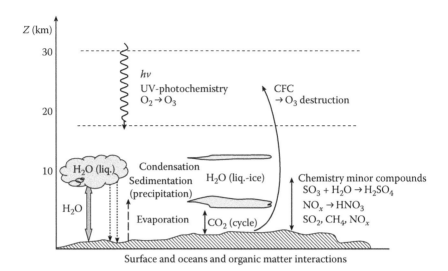

FIGURE 5.2   Scheme of Earth basic cloud—surface interactions, chemistry, and cycles.

that contains abundant sulfuric acid due to oxidation and hydration of SO_2, and where the chemistry remembers that occurring in Venus's cloud.

A number of other anthropogenic gases containing, for example, halogen compounds present variable mole fractions in the atmosphere ($<10^{-9}$). They participate in and induce chemical reactions in cloud chemistry and participate in modifications of the ozone layer as already presented in Chapter 2. In particular, the chlorine cycle involves chlorofluorocarbons (CFC) from biological and anthropogenic processes (see Section 6.2.2), but also includes marine NaCl and NH_4Cl, HCl, Cl, and ClO aerosols.

Insoluble minerals from continental areas are also injected in the air by wind erosion and volcanoes. This airborne dust includes particulates of $CaSiO_3$, SiO_2, Al_2O_3, etc. Organic aerosols (spores, pollen, and unsaturated compounds) are common over large forest areas. Figure 5.2 shows a basic scheme of these cycles.

5.1.2.3 Mars

Mars is the archetype of a planet with a strong surface–atmosphere aerosol and cloud interaction that involves three main seasonal cycles: CO_2, H_2O, and dust. The CO_2 cycle follows the seasonal condensation and sublimation of its large ice deposits in the polar regions. Up to a 30% of the atmospheric mass varies in response to the seasonal radiative heating and cooling as a function of latitude (Figure 4.13). Carbon dioxide condensation occurs during the polar nights sublimating during the spring and summer seasons. This huge change of the atmospheric mass produces important meridional transports of heat, momentum, and atmospheric trace constituents that intervene in cloud formation (water and CO_2 particulates).

The column abundance of water vapor on Mars varies spatially and seasonally due to exchange with ground reservoirs and to meridional transport in the atmosphere. Its condensation produces icy water clouds. During late spring and summer,

CO_2 sublimates in the northern pole leaving a residual water-ice cap that is one of the sources of water vapor. In the southern pole, a small permanent CO_2 ice cap acts as a cold trap for water vapor. The unfrosted ground (known as the Martian regolith) also acts as a source and sink for water vapor following the seasonal solar radiation cycle. Direct measurements indicate that near the surface, the water mass mixing ratios may vary from 10^{-4} to 10^{-8}. The maximum integrated northern hemisphere water vapor abundance during spring and summer reaches $\sim 7.5 \times 10^{11}$ kg (Figure 5.3).

Soil particles (dust) with a large size distribution cover the Martian surface, and when the winds reach a critical value they provide the atmosphere with small suspended dust particles. This is particularly notorious during the great dust storms when the injection of large amounts of dust fully cover the surface of the planet. The dust lifting initiates in the *saltation layer* next to the ground with a thickness of about 20 cm. There, sand-sized particles are temporarily raised and blown by winds. To inject particles with sizes $\sim 10–100\,\mu m$, a threshold value for the wind speed is $\sim 4\,m\,s^{-1}$.

Obviously the CO_2, water, and dust cycles are coupled. Figure 5.4 shows a basic scheme of these cycles. The injection of airborne dust represents a source of opacity to solar radiation and acts at the same time as condensation nuclei for cloud formation. This alters regionally the radiative balance, affecting the local dynamics or the general circulation, depending on the horizontal extent.

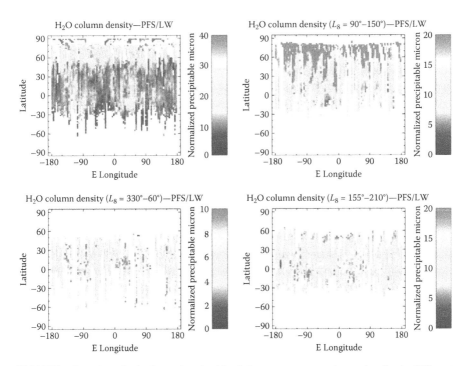

FIGURE 5.3 Maps (latitude vs. longitude) of the water vapor column density in different Martian seasons obtained with PFS/LW instrument onboard Mars Express. (From Fouchet, T. et al., *Icarus*, 190, 32, 2007. With permission.)

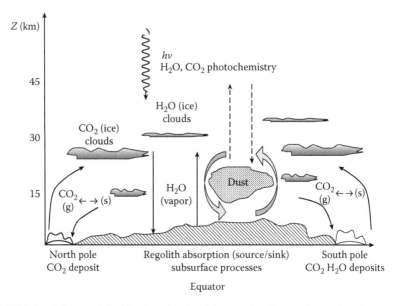

FIGURE 5.4   Scheme of the Martian cloud and dust surface interaction and cycles.

5.1.2.4 Titan

The hazes that cover the surface of Saturn's largest moon are the result of methane photochemical processes that occur at the top of the atmosphere (see Section 2.4.1). However, methane itself is apparently subjected to a cycle of condensation and sublimation–evaporation with origin at the surface. This has been called the *methalogical cycle*, for analogy with the hydrological cycle on Earth. Titan's surface contains icy and liquid deposits of methane and ethane (lakes) where the evaporation and vapor injection into the atmosphere leads to condensate clouds. Different models predict the precipitation of these compounds reaching the surface and closing the cycle (see Figure 5.5). It has also been proposed that outgassing of methane from *cryovolcanism* can be another source for this gas.

5.2 CLOUDS PROPERTIES

Generally speaking, those substances or their reaction products, whose bond strength is weak, tend to suffer *phase changes* in the pressure–temperature conditions in a planet playing a very significant role in atmospheric physics and chemistry. They are mostly responsible of cloud formation by condensing from the gas phase to the liquid or solid phases. Thermodynamic diagrams are used to represent the phase space of a substance. For every real substance, there exists a relation between the pressure, specific volume (or the density), and temperature (the equation of state). The $P–V_V–T$ diagram for a substance includes the gas, liquid, and solid phases. Where gas and liquid phases exist in equilibrium, the gas is usually called a vapor (a *saturated vapor*) and the liquid is *saturated liquid*. In pressure–temperature projected planes

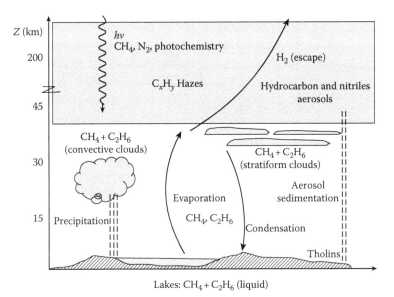

FIGURE 5.5 Scheme of the Titan methane and ethane cycles. Cloud and haze interaction with the surface and lakes.

(*P–T* diagram), the curve separating both phases is the saturation vapor pressure curve (the pressure exerted by a saturated vapor or liquid).

5.2.1 MOISTURE VARIABLES AND DEFINITIONS

In Section 3.5, we introduced the composition definitions for species in atmospheres. Here, we extend the definitions when moisture is present. A vapor in solution with the air obeys the equation of state

$$P_V V_V = R_V^* T \qquad (5.9)$$

where
P_V is the partial pressure of the condensable gas (vapor)
$V_V = V/m_V$ is the specific volume of vapor

and

$$R_V^* = \frac{R_g}{\mu_V} = \frac{R_d^*}{\varepsilon_\mu} \qquad (5.10)$$

where $\varepsilon_\mu = \mu_V/\mu$ is the ratio between the vapor molecular weight and the mean atmospheric molecular weight. The *absolute humidity* measures the absolute concentration of vapor

$$\rho_V = \frac{1}{V_V} = \frac{P_V}{R_V^* T} \qquad (5.11)$$

whereas the relative concentration is called the *specific humidity*

$$q = \frac{\rho_V}{\rho} = \frac{m_V}{m} \tag{5.12}$$

which is dimensionless and usually expressed in g kg^{-1} (grams of vapor per kilogram of atmosphere). The abundance of the condensable gases is measured in terms of the *vapor mass mixing ratio* x_{MV} (again in units of g kg^{-1}, see Equation 3.65). Defining the total gas mass as the sum of the dry and vapor $m = m_d + m_V$, we have

$$x_{MV} = \frac{\rho_V}{\rho_d} = \frac{q}{1-q} \approx q \tag{5.13}$$

since $m_d \gg m_V$. Therefore, q and x_{MV} are interchangeably and both are conserved quantities outside the regions where condensation occurs. It is convenient to express q and x_{MV} in terms of the partial pressure P_V since the chemical equilibrium of the condensable component is controlled by its absolute concentration. The equation of state for the dry component of the atmosphere is

$$P_d V_d = R_d^* T \tag{5.14}$$

Since $P = P_V + P_d \sim P_d$, using Dalton's law (Section 3.5), we get

$$x_{MV} = \varepsilon_\mu \frac{P_V}{P} \tag{5.15}$$

When managing a mixture of dry and vapor gases it is also convenient to incorporate the vapor variables into the dry air properties, so using the above definitions, the specific gas constant can be rewritten in terms of the dry specific constant and previous vapor definitions as (see Problem 5.2)

$$R_g^* = \left[1 + \left(\frac{1}{\varepsilon_\mu} - 1 \right) q \right] R_d^* = f_V R_d^* \tag{5.16}$$

so the equation of state for the mixture can be written as

$$P = \rho f_V R_d^* T \tag{5.17}$$

which also makes it convenient to introduce a *virtual temperature*

$$T_V = f_V T \tag{5.18}$$

that "redefines" or "re-scales" the temperature to take into account the humidity (vapor mixing ratio) and vapor molecular weight. The equation of state can be rewritten as

$$P = \rho R_d^* T_V \tag{5.19}$$

Let us now consider the situation of condensation. As stated above, when the vapor is in chemical equilibrium with the condensed phase it is said to be *saturated*. The *saturation vapor pressure* $P_{VS}(T)$ with respect to the liquid or solid phases is the equilibrium vapor pressure and is described by the *Clausius–Clapeyron* relationship (see below). Then the *saturated mass mixing ratio* is given by

$$x_S(T,P) = \varepsilon_\mu \frac{P_{VS}(T)}{P} \tag{5.20}$$

The *relative humidity RH* (in percent) is then defined as

$$RH(P,T) = 100 \frac{P_V}{P_{VS}} \approx 100 \frac{x_V}{x_S} \tag{5.21}$$

The determination of the vapor abundance in a gas mixture can also be done by means of the *vapor volume mixing ratio* (remember the definitions introduced in Section 3.5), also called the *vapor molar mixing ratio* or *vapor mole fraction*,

$$x_{VV} = \frac{V_V}{V} = \frac{P_V}{P} = \frac{x_{MV}}{\varepsilon_\mu} \tag{5.22}$$

However, because of the sharp dependence of the saturation vapor pressure with temperature, measurement of the moisture content using the RH can be misleading and care must be taken (Problems 5.4 and 5.5).

In a sample of moist air, the specific heats at constant volume (C_{vm}) and constant pressure (C_{pm}) are given by (see Equation 4.78)

$$C_{vm} = C_{vd} \left[\frac{1 + x_{MV} r_v}{1 + x_{MV}} \right] \tag{5.23a}$$

and

$$C_{pm} = C_{pd} \left[\frac{1 + x_{MV} r_p}{1 + x_{MV}} \right] \tag{5.23b}$$

where $r_v = C_{vv}/C_{vd}$ and $r_p = C_{pv}/C_{pd}$, where the sub-indexes V and d indicate the heat capacity of the vapor at constant volume and pressure (C_{vv}, C_{pv}) and of the dry air (C_{vd}, C_{pd}) (Problem 5.3). As a first approximation, they can be assumed to be temperature independent. In Tables 4.1 and 4.2, we gave the basic composition and thermodynamic properties of planetary atmospheres so they can be used for the cloud formation study.

5.2.2   LATENT HEAT AND THE CLAUSIUS–CLAPEYRON EQUATION

The energy released or absorbed during a phase change (vapor–liquid–solid) is called *latent heat*. Kirchhoff's equation gives its variation with temperature:

$$\left(\frac{\partial L_i}{\partial T} \right)_P = \Delta C_P \tag{5.24}$$

The specific heats are denoted as follows:

$$\text{vapor} \rightleftarrows \text{liquid (condensation, evaporation)} : L_v \text{(vaporization)}$$

$$\text{liquid} \rightleftarrows \text{solid (freezing, fusion)} : L_f \text{(fusion)}$$

$$\text{vapor} \rightleftarrows \text{solid (condensation, sublimation)} : L_s \text{(sublimation)}$$

and are related as

$$L_s = L_f + L_v \tag{5.25}$$

When saturation occurs and a phase change takes place, the temperature variation of the saturation vapor pressure, or, in other words, the slope of the vapor pressure curve $P_{VS}(T)$ in Equation 5.20 that marks the phase transition where the two phases are in equilibrium, is given by the Clausius–Clapeyron equation:

$$\frac{dP_{VS}}{dT} = \frac{L_i}{T(V_2 - V_1)} \tag{5.26}$$

where
  L_i is the latent heat of the phase transition (in J g^{-1})
  $V_i = 1/\rho_i$ is the specific volume (phase 1 = vapor phase, 2 = liquid or ice phase)

A useful approximation to Equation 5.26 can be obtained under the following assumptions: neglect the temperature variations in the latent heat, assume the vapor is an ideal gas, and the specific volume of the liquid or solid phases is also neglected compared to that of the vapor, which is $V_2 - V_1 \sim V_V = 1/\rho_V = R_V^* T/P$. Then Equation 5.26 can be rewritten as

$$\frac{dP_{VS}}{dT} = \frac{L_i P_{VS}}{R_V^* T^2} \tag{5.27}$$

Integration of Equation 5.27 with L_i=constant gives

$$P_{VS}(T) = P_{VS0} \exp\left[L_i \left(\frac{1}{R_V^* T_0} - \frac{1}{R_V^* T} \right) \right]$$
(5.28)

where P_{VS0} is the saturation vapor pressure at temperature T_0. A more accurate expression for the latent heat can be obtained by integrating the relationship (Equation 5.24). To do this, expanding the specific heat for each phase as $C_P(T) = \alpha + \beta T + \cdots$, we obtain

$$L = L_0 + \Delta\alpha T + \frac{\Delta\beta}{2} T^2 + O(T^3)$$
(5.29)

where
L_0 is an integration constant
α and β are empirically determined constants for each phase

$\Delta\alpha$ and $\Delta\beta$ indicates the change of the constants α and β between the two phases Combining Equations 5.24 and 5.27 we find the general form for the saturation vapor pressure curve:

$$\ln\left(P_{VS}\right) = \ln\left(P_{VS0}\right) + \frac{1}{R_V^*}\left[-\frac{L_0}{T} + \Delta\alpha \ln T + \frac{\Delta\beta}{2} T + O(T^2) \right]$$
(5.30)

In Table 5.1, we give the values of the coefficients in Equation 5.30 for the main species able to condense in planetary atmospheres. In Figure 5.6, we show as an example the pressure–temperature phase diagram for water, which is a usual condensable in planetary atmospheres. The saturation vapor pressure curves for each compound, for the vapor–liquid and vapor–solid equilibrium transitions, will be used in the range of temperatures found for each planet's tropospheres to search for cloud forming compounds.

5.2.3 CONDENSATION LEVEL AND CLOUD FORMATION

When atmospheric parcels are transported upward through the troposphere, they encounter lower temperatures and for some gases their partial pressure P_V becomes equal or exceeds the saturation vapor pressure $P_{VS}(T)$. If *condensation nuclei* are present in the atmosphere (i.e., small particles of micron or submicron size where the gas can condense), cloud formation takes place by a process known as *heterogeneous nucleation* whenever

$$P_V = x_{VV} P(T) \geq P_{VS}(T)$$
(5.31)

TABLE 5.1

Saturation Vapor Pressure and Latent Heat Coefficients for the Condensable Compounds of Interest in Planetary Atmospheres

Component	$\ln (P_{VS0})(P_{VS0}$ in bars)	L_0 (J g^{-1})	$\Delta\alpha$ (J g^{-1} K^{-1})	$\Delta\beta/2$ (J g^{-1} K^{-2})
SO_4H_2	16.256	865.8	—	—
H_2O	25.096	3148.2	—	-8.7×10^{-3}
CO_2	26.100	639.6	—	-1.7×10^{-3}
NH_3	27.863	2016	-0.888	—
NH_4SH	75.678	2915.7	-1.760	7.8×10^{-4}
CH_4	1.627	553.1	1.002	-4.1×10^{-3}
SH_2	17.064	747	—	-2.9×10^{-3}
C_2H_6	10.136	521.4	—	—
Fe	1.894	7097	—	—
$MgSiO_3$	11.554	4877.5	—	—

Source: Adapted from Sánchez-Lavega, A. et al., *Am. J. Phys.*, 72, 767, 2004 and references therein.

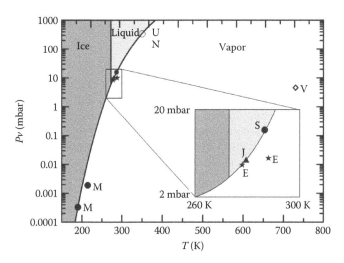

FIGURE 5.6 The phase diagram for water with the temperature–pressure range for the planets indicated to show the condensate phase that forms on them: V (Venus surface), E (Earth surface and cloud level), M (Mars surface and cloud level), J (Jupiter), S (Saturn), U (Uranus), and N (Neptune). Details are shown in the inset for some of them. Note that for convenience, the inset's vertical scale is linear. (From Sánchez-Lavega, A. et al., *Am. J. Phys.*, 72, 767, 2004. With permission.)

Cloud formation can also occur under homogeneous condensation, which is when supersaturation takes place and the condensate growths by nucleation. Equation 5.31 is equivalent to state that the *RH* of the condensing layer is greater than 100%. Thus, for a given atmosphere, a comparison of its vertical temperature profile $T(P)$, as given in Chapter 4 with the saturation vapor pressure curve $P_{VS}(T)/x_{VV}$, tells us which gases can condense to form clouds and at what pressure (altitude) these clouds form. The abundance of the condensing molecules above the condensation level (the locus of the $T(P)$ and $P_{VS}(T)/x_{VV}$ curves) is determined by the saturation vapor pressure.

For single liquid–vapor or solid–vapor transitions, the cloud base is located at a pressure P_{cl} and temperature T_{cl} such that $P_{cl}(T_{cl}) = P_{VS}(T_{cl})/x_{VV}$. In Figure 5.7, we show the vertical temperature profiles for each planetary atmosphere in the altitude range of interest (upper troposphere–lower stratosphere) as well as the saturation vapor pressure curves for the different condensates. The crossing point between the temperature profile and the saturation pressure vapor curve marks the altitude for the cloud base formation. In some cases, different condensation lines have been plotted for the same species (molecule) because they depend on the values of the concentration of the gas (that are not well known) or simply because they vary at different locations.

In Table 5.2, we list the values of the measured molar fractions of the putative condensates and then we present averaged cloud properties. It is evident that the condensed phase (liquid or solid) depends on the atmospheric temperature–pressure relation and on the condensate phase diagram. As an example, we add to the phase diagram in Figure 5.7 the pressure–temperature ranges for the different planets where water clouds form. In the first column of Table 5.2, we give the expected phase of the corresponding condensate.

Additional to this single one phase change, a two-component reaction can occur to produce clouds. One classical example is the formation of ammonium hydrosulfide clouds in the atmospheres of the giant planets according to the reaction

$$NH_3(g) + H_2S(g) \rightleftarrows NH_4SH(s)$$

For such a case, an approximate equation similar to Equation 5.30 can be found for the vapor pressure curve by considering the partial pressures of both NH_3 and H_2S.

5.2.4 CLOUD DENSITY AND VERTICAL EXTENT

A simple approach to the vertical extent of a condensing cloud can be derived using Equation 5.28 and the assumption of hydrostatic equilibrium. Defining in Equation 5.28 $T_0 = T_{cl}$ for the temperature at the cloud base, we have

$$P_{VS}(T) = P_{VS,cl} \exp\left[-\frac{L_i(T_{cl} - T)}{R_V^* T T_{cl}} \right] \tag{5.32}$$

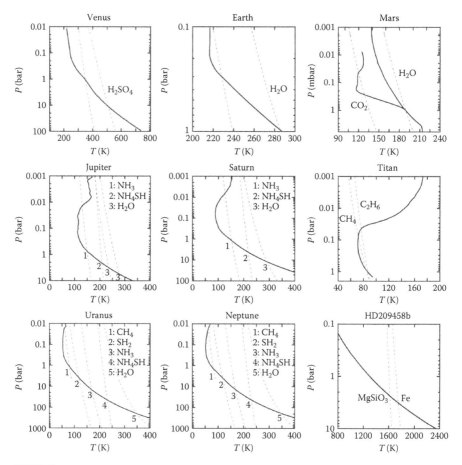

FIGURE 5.7   The vertical temperature profile in the atmospheres of the planets is shown by a continuous line and is compared to the saturation vapor pressure curves for the condensates (dashed lines) for each planet. The points where the two curves cross mark the cloud base for the specific condensate. For Venus' sulfuric acid clouds and the Earth's and Jupiter's water clouds, two saturation curves are given corresponding to two limiting abundance cases. Because of Mars' tenuous atmosphere and seasonal variability, two vertical temperature profiles are given, one (right) is a yearly average profile and the other (left) is a cold profile introduced to illustrate the CO_2 atmospheric condensation. (From Sánchez-Lavega, A. et al., *Am. J. Phys.*, 72, 767, 2004. With permission.)

where $P_{VS,cl}$ is the saturated vapor pressure curve at cloud level. Assume now a simple adiabatic temperature profile for the atmosphere (Section 4.5.1). Then $T_{cl} - T = (g/c_p)(z - z_{cl})$ and since, in general, $(T - T_{cl})/T \ll 1$, we take $T \sim T_{cl}$, and write

$$P_{VS}(T) = P_{VS,cl} \exp\left[-\frac{L_i(g/c_p)(z - z_{cl})}{R_V^* T_{cl}^2}\right]$$

(5.33)

TABLE 5.2
Predicted Planetary Atmosphere Condensates and Cloud Characteristics

Object and Cloud	x_{VV}	μ_v	P (bar)	T (K)	H (km)	H_d (km)	H_d/H	ρ_d (g cm^{-3})	Γ_s (K km^{-1})	Γ_s/Γ_d
Venus										
SO_4H_2 (l)	2.0×10^{-6}	98.08	1.0	348	7.4	1.1	0.15	4.5×10^{-8}	10.5	1.000
SO_4H_2 (l)	2.0×10^{-3}	98.08	11.3	510	11.0	2.4	0.22	2.4×10^{-4}	9.3	0.880
Earth										
H_2O (s)	2.5×10^{-4}	18.02	0.30	229	6.8	0.9	0.13	5.3×10^{-7}	9.4	0.957
H_2O (l)	0.015	18.02	0.96	285	8.4	1.5	0.18	5.9×10^{-5}	5.0	0.508
Mars										
CO_2 (s)	0.95	44.01	2.0×10^{-4}	127	6.4	1.1	0.17	4.6×10^{-6}	0.77	0.17
H_2O (s)	3.0×10^{-4}	18.02	1.0×10^{-3}	190	9.6	1.3	0.14	2.5×10^{-9}	3.90	0.87
Jupiter										
NH_3 (s)	2.0×10^{-4}	17.00	0.75	150	23	3.0	0.13	1.6×10^{-6}	1.93	0.967
NH_4SH (s)	3.6×10^{-5}	50.00	2.20	210	32	1.2	0.04	6.1×10^{-6}	1.90	0.948
H_2O (Galileo)(s)	5.0×10^{-5}	18.02	3.20	228	35	4.3	0.12	1.2×10^{-6}	1.98	0.990
H_2O (Solar) (l)	1.7×10^{-3}	18.02	5.7	280	43	7.4	0.17	4.4×10^{-5}	1.73	0.867
Saturn										
NH_3 (s)	2.0×10^{-4}	17.00	1.2	150	58	8.4	0.14	2.3×10^{-6}	0.68	0.970
NH_4SH (s)	3.6×10^{-5}	50.00	4.0	215	84	3.4	0.04	1.0×10^{-5}	0.67	0.954
H_2O (Solar) (s)	1.7×10^{-3}	18.02	9.3	285	111	21.0	0.19	6.3×10^{-5}	0.62	0.883
Titan										
CH_4 (s)	0.05	16.04	0.90	84	18	4.6	0.26	4.0×10^{-4}	0.59	0.456
C_2H_6 (s)	10^{-5}	30.00	1.04	87	19	3.1	0.16	2.6×10^{-7}	1.3	0.995

(continued)

TABLE 5.2 (continued)
Predicted Planetary Atmosphere Condensates and Cloud Characteristics

Object and Cloud	x_{vv}	μ_v	P (bar)	T (K)	H (km)	H_d (km)	H_d/H	ρ_{cl} (g cm^{-3})	Γ_s (K km^{-1})	Γ_s/Γ_d
Uranus										
CH$_4$ (s)	0.02	16.04	0.7	68	28	5.9	0.21	1.9×10^{-4}	0.36	0.511
SH$_2$ (s)	3.7×10^{-5}	34.06	4.0	117	48	7.0	0.15	3.6×10^{-6}	0.70	0.994
NH$_3$ (s)	2.0×10^{-4}	17.00	15.0	169	69	11.1	0.16	2.3×10^{-5}	0.68	0.977
NH$_4$SH (s)	3.6×10^{-5}	50.00	50.0	240	99	4.5	0.05	9.9×10^{-5}	0.67	0.963
H$_2$O (s)	1.7×10^{-3}	18.02	200.0	350	144	36.0	0.25	8.4×10^{-4}	0.65	0.934
Neptune										
CH$_4$ (s)	0.02	16.04	0.9	69	22	4.8	0.22	2.4×10^{-4}	0.440	0.520
SH$_2$ (s)	3.7×10^{-5}	34.06	3.8	117	38	5.5	0.15	3.4×10^{-6}	0.845	0.994
NH$_3$ (s)	2.0×10^{-4}	17.00	14.0	168	55	8.7	0.16	2.1×10^{-5}	0.830	0.977
NH$_4$SH (s)	3.6×10^{-5}	50.00	50.0	240	78	3.6	0.05	9.8×10^{-5}	0.820	0.963
H$_2$O (s)	1.7×10^{-3}	18.02	200.0	350	114	29.0	0.25	8.3×10^{-4}	0.790	0.933
HD209458b										
MgSiO$_3$ (s)	7.52×10^{-5}	100.4	2.4	1620	840	78	0.09	1.5×10^{-6}	0.58	0.974
Fe (s)	6.77×10^{-5}	55.84	3.2	1750	909	112	0.12	6.8×10^{-7}	0.59	0.987

Source: Adapted from Sánchez-Lavega, A. et al., *Am. J. Phys.*, 72, 767, 2004 and references therein.
Notes: H is the gas scale height at the cloud formation level. (l, Liquid; s, Solid)

where the units for L_i are in J kg^{-1} and for c_p are in J K^{-1} kg^{-1}, and z_{cl} is the altitude level of cloud formation. We define a vertical scale height for the cloud as

$$H_{cl} = \frac{R_V^* T_{cl}^2 c_p}{g L_i} \qquad (5.34)$$

A useful parameter for comparing the cloud vertical extent among the different planets is the ratio between scale heights of the cloud and of the atmosphere:

$$\frac{H_{cl}}{H} = \frac{c_p T_{cl}}{L_i} \frac{R_V^*}{R_g^*} \qquad (5.35)$$

If we compare the data for the different planets given in Table 5.2, we see that in general $H_{cl}/H \sim 0.05-0.2$, and thus the condensed cloud vertical extents are expected to be thin relative to the atmospheric scale height. This obviously corresponds to the case of formation of stratified clouds without any dynamical process involved. The cloud density can be obtained from the ratio of the condensate mass (mass mixing ratio times the atmospheric mass) to the total volume:

$$\rho_{cl} = x_{MV} \frac{P_{cl}/g}{H_{cl}} \qquad (5.36)$$

Note that ρ_{cl} will give the maximum cloud density because we have assumed that all the vapor in the atmosphere condenses with no precipitation occurring. In general, Equation 5.36 must be multiplied by a factor ranging from 0.01 to 1 to treat a more realistic cloud density distribution (particle size distribution), with the cloud density decreasing with altitude from the cloud base as the condensate is removed upon condensation. In Figure 5.8, we present the predicted vertical structure in Jupiter and Saturn according to a simple thermodynamic scheme.

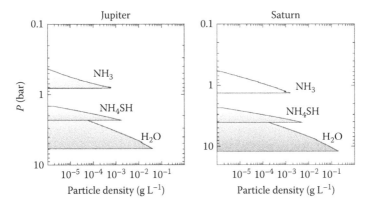

FIGURE 5.8 Scheme of the main cloud expected to form and their vertical structure in Jupiter and Saturn for a solar abundance of the reference elements. (Courtesy of Ricardo Hueso, UPV-EHU, Bilbao, Spain.)

5.2.5 WET ADIABATIC LAPSE RATE

The latent heat released during cloud formation heats the atmosphere locally, changing the vertical temperature lapse rate from a dry adiabatic (Section 4.5.1) to a *wet adiabatic lapse rate*, Γ_S. It can be obtained from the Clausius–Clapeyron equation. We start with the first law of thermodynamics and the hydrostatic equation and assuming from simplicity that only a single component condenses. We also assume that the amount of heat removed from an ascending parcel by the condensed phase is small compared with that remaining in the parcel, a process referred to as *pseudoadiabatic*. At saturation, the amount of heat deposited in the condensing layer is $\delta Q = -L_i\, dx_S$, and the first law of thermodynamics (Section 4.5.1) can be written as

$$c_P\, dT = -g\, dz - L_i\, dx_S. \tag{5.37}$$

Here, dx_S denotes the change in the saturation mixing ratio of the condensable gas, that is, the mass of the vapor that condenses out per unit mass (kilogram) of dry gas (formed by the noncondensable species). We take logarithms in Equation 5.20, differentiate (ε_μ is constant), to obtain

$$\frac{dx_S}{x_S} = \frac{dP_{VS}}{P_{VS}} - \frac{dP}{P} \tag{5.38}$$

If we use Equation 5.27 for dP_{VS}/P_{VS} and the hydrostatic relationship (4.1) for dP, and introduce both in Equations 5.37 and 5.38, we obtain the *pseudoadiabatic gradient*:

$$\Gamma_S = -\left(\frac{dT}{dz}\right)_S = \frac{g}{c_P} \cdot \frac{\left(1 + (L_i x_S / R_g^* T)\right)}{\left(1 + (L_i^2 x_S / c_P R_V^* T^2)\right)} \tag{5.39a}$$

Inside the cloud, the vertical temperature gradient becomes Γ_S. Due to the latent heat release at condensation, $\Gamma_S \le \Gamma_d$. In Table 5.2, we give the value of the pseudoadiabatic gradient for the condensing clouds for each planet. The ratio Γ_S/Γ_a can be used to estimate the influence that condensation and cloud formation have on the thermal structure of the planetary atmospheres.

It is straightforward to extend Equation 5.39a to the case of a multicomponent (j-index) saturation (e.g., as with the NH$_4$SH cloud):

$$\Gamma_S = -\left(\frac{dT}{dz}\right)_S = \frac{g}{c_P} \cdot \frac{\left(1 + \left(\sum_j L_{i,j} x_{S,j} \big/ R_g^* T\right)\right)}{\left(1 + (1/c_P T^2)\sum_j \left(L_{i,j}^2 x_{S,j} / R_{V,j}^*\right)\right)} \tag{5.39b}$$

5.2.6 EQUIVALENT POTENTIAL TEMPERATURE AND CONDITIONAL INSTABILITY

When saturation and condensation occurs, it is convenient to introduce an *equivalent potential temperature* in a way similar to that followed in Section 4.5.2, adding to Equation 4.69a the contribution to the entropy due to the latent heat:

$$dS = C_P \frac{dT}{T} - nR_g \frac{dP}{P} + \frac{d(L_i x_S)}{T} \tag{5.40}$$

For a pseudoadiabatic process, no entropy change occurs ($dS=0$), and integration of (5.40) after dividing by C_p gives

$$\theta_e(T,P) = T \left(\frac{P}{P_0} \right)^{-\kappa} \exp\left(\frac{L_i x_S}{C_p T} \right) = \text{constant} \tag{5.41a}$$

a quantity that is called the equivalent potential temperature that combined with (4.68) gives

$$\theta_e(T,P) = \theta(T,P) \exp\left(\frac{L_i x_S}{C_p T} \right) \tag{5.41b}$$

Accordingly, θ_e remains constant during a pseudoadiabatic process, and this requires that during condensation, a reduction of x_S must be accompanied by an increase in θ to preserve θ_e (=constant).

5.2.6.1 Conditional Instability

The arguments presented in Section 4.5.3 can now be extended to include the atmospheric stability of atmospheric parcels to vertical displacements in terms of the saturation processes that define the wet adiabatic gradient Γ_S. Since $\Gamma_S < \Gamma_d$ (see Table 5.2), the following possible situations occur:

$$
\begin{aligned}
&\Gamma < \Gamma_S && \text{absolutely stable} \\
&\Gamma = \Gamma_S && \text{neutral} \\
&\Gamma_S < \Gamma < \Gamma_d && \text{conditionally unstable} \\
&\Gamma = \Gamma_d && \text{neutral} \\
&\Gamma > \Gamma_d && \text{absolutely unstable}
\end{aligned}
\tag{5.42}
$$

Figure 5.9 shows the different possibilities in term of the actual vertical temperature gradient Γ. Similarly, we can redefine the Brunt–Väisälä frequency for the wet atmosphere by changing in Equation 4.74 the temperature by the virtual temperature, as given by Equation 5.18; then

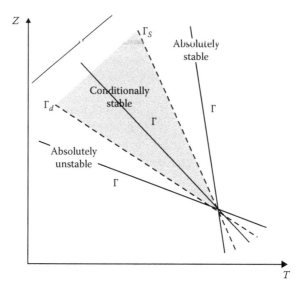

FIGURE 5.9   Stability conditions to vertical motions for a dry or moist air parcel. The actual vertical temperature gradient is Γ (Γ_d is the dry lapse rate and Γ_s is the moist lapse rate).

$$N_B^2 = \frac{g}{T_V}\left(\frac{dT_V}{dz} + \frac{g}{C_{p_d}}\right)$$ (5.43)

5.3   MICROPHYSICS

If we want to understand the life of a cloud (its formation, evolution, and destruction mechanisms), we must immerse in the processes that occur at the micron or submicron scales; all them are known collectively as *microphysical processes*. Some occur in isolated form and can be treated individually, but in most cases they occur simultaneously, complicating their formulation. The ensemble behavior and their role in the overall cloud structure is not a simple issue and a parameterization approach of the processes is necessary. Microphysical processes can be grouped in the following categories: nucleation and particle growth under condensation–evaporation, vapor diffusion, collection (coalescence), drop breakup, fallout (precipitation), ice nucleation, and melting. They involve the transformations between the vapor, liquid, and solid phases according to the phase diagram of the condensing substances. In a closed atmospheric parcel, the mass of a substance is preserved if no chemical transformations occur within it, so the total mixing ratio of the substance is given by the sum of the mixing ratios of the different phases considered.

The equations describing the microphysical processes depend on the atmospheric density that vary over orders of magnitude with altitude and, as we have seen, from one to other planet or satellite considered. This implies the existence of different physical regimes that depend, apart from the atmospheric density, on the particle

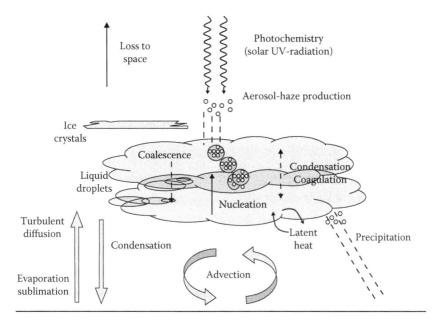

FIGURE 5.10 Typical processes that intervene in cloud and aerosol formation in the terrestrial planets and Titan.

size and on the vertical velocities the particles can encounter. Two nondimensional numbers are useful to characterize the situation and make a classification: (1) the *Knudsen number* $Kn = \ell_{fp}/a$, which is the ratio between the gas molecular mean free path ℓ and the particle size a, and (2) the *Reynolds number* $Re = 2a\rho v/\eta_D$, where ρ and η_D are the gas density and dynamic viscosity and v the gas velocity. In the tropospheres where most particles form, $Kn \ll 1$ and $Re \ll 1$ (called *laminar regime*), and this is the situation we will present. In the opposite side, under a *turbulent regime*, intense vertical motions modify the particle growth (the coalescence and sedimentation processes). In Figure 5.10, we summarize schematically the main microphysical processes in the clouds of the terrestrial planets and Titan.

   In the following, we describe the fundamentals of the different microphysical processes. The incorporation of microphysics to cloud dynamics requires the use of the continuity equation for the moisture content that combines all the categories of the condensable species and of the temporal evolution of the different processes. This equation and the associated source/sink terms are coupled to the dynamical equations, and this problem must be solved numerically. A detailed treatment of these processes is obviously beyond the scope of this introductory book and the reader is referred to specific books in the literature.

5.3.1 GROWTH OF CLOUD PARTICLES: NUCLEATION

There are two ways for particles to nucleate and grow within a cloud. When *homogeneous nucleation* occurs, the vapor molecules may come together by collisions to condense forming an embryonic particle around which the vapor deposits and the particle grows. In *heterogeneous nucleation*, the vapor molecules add on an existing substance, foreign in composition to the vapor. Typically, they are small particles (micron and submicron in size) known as *cloud condensation nuclei* (CCN). For a cloud particle to grow by a nucleation process, a "free energy barrier" must be overcome, and this is best described in terms of the Gibbs free energy introduced in Section 5.1.1. For a heterogeneous system formed by a droplet and the surrounding vapor, Equations 5.3 and 5.4 can be used to calculate the Gibbs energy, incorporating the work done by the surface tension force σ_T (units of surface energy per unit area) at the interface of the growing particle to form the incremental area, dS, between the vapor and the condensed phase, $\sigma_T \, dS$. The nucleation of a particle of radius a requires a change in Gibbs free energy:

$$\Delta G = 4\pi a^2 \sigma_T - \frac{4}{3}\pi a^3 \rho_c R_V^* T \ln\left(\frac{P_V}{P_{VS}(T)}\right) \tag{5.44}$$

The first term on the right corresponds to the work necessary to create the interface, and the second term is the energy change associated with the vapor molecules going into the condensed phase, ρ_c being the density of the condensed phase. This equation shows that the Gibbs free energy depends on the humidity in the atmosphere surrounding the surface of the particle (see Equation 5.21). For subsaturated ($P_V/P_{VS} < 1$) and saturated ($P_V/P_{VS} = 1$) conditions, the logarithm in (5.44) is negative or zero and ΔG is an increasing function with the radius. However, for supersaturation conditions ($P_V/P_{VS} > 1$), the logarithm is negative and the Gibbs function has a maximum. Considering $\partial \Delta G/\partial a = 0$, we get the particle critical radius

$$a_c = \frac{2\sigma_T}{\rho_c R_V^* T \ln(P_V/P_{VS})} \tag{5.45}$$

which is known as the *Kelvin formula*. Note that this radius has sense for supersaturated conditions since $a_c \to \infty$ when $P_V \to P_{VS}$. For a droplet in equilibrium at the radius a_c, condensation of a small quantity of vapor increases the radius slightly. On the contrary if evaporation occurs, then the droplet shrinks and it can eventually disappear. The critical radius is a strong function of the RH, and changes orders of magnitudes for a small change in the RH (Problem 5.7). The stability of small particles requires high supersaturation conditions. As an example, the growth of water droplets by homogeneous nucleation will require a RH of 300%–400%. For a RH of 101%, droplets with size below 0.12 μm are unstable and tend to evaporate. On Earth, typical supersaturation conditions rarely exceed 101%, so heterogeneous nucleation on CCN aerosols is the dominant mechanism in forming liquid cloud droplets. If ice particles are nucleated either from the liquid or the vapor phase to shapes different from the spherical, then a geometrical correction factor must be multiplied to the right part of Equation 5.45.

5.3.2 Diffusion: Condensation–Evaporation–Sublimation

A cloud particle (liquid droplet or ice crystal) may continue to grow as vapor diffuses from the neighboring to the cloud particles and condenses. On the contrary, when evaporation occurs, the particle decreases its size as the vapor diffuses away. Assuming there is a vapor density gradient in the region surrounding the particle, the condensation/evaporation processes may be represented by the Fick diffusion law:

$$\frac{\partial \rho_V}{\partial t} = D_V \nabla^2 \rho_V \tag{5.46}$$

where
  ρ_V is the vapor density
  D_V is the diffusion coefficient for the vapor in the atmosphere, which we assume to be constant

For steady state conditions ($\partial \rho_V/\partial t=0$) and spherical symmetry, the above equation reduces to

$$\nabla^2 \rho_V = \frac{1}{r^2}\frac{\partial}{\partial r}\left(r^2 \frac{\partial \rho_V}{\partial r}\right) = 0 \tag{5.47}$$

If the vapor density at the surface is $\rho_V(a)$ and far away from it is $\rho_V(\infty)$ (as $r \to \infty$, the free atmosphere value), the solution to (5.47) can be written as

$$\rho_V(r) = \rho_V(\infty) - \frac{a}{r}\left[\rho_V(\infty) - \rho_V(a)\right] \tag{5.48}$$

The rate of variation of the mass of the cloud particle m_p is then given by

$$\frac{dm_p}{dt} = 4\pi a^2 D_V \left.\frac{d\rho_V}{dt}\right|_a \tag{5.49}$$

where $D_V(d\rho_V/dt)$ is the flux of vapor in the radial direction across the surface at radius a of the cloud particle. From (5.48) and (5.49), we get

$$\frac{dm_p}{dt} = 4\pi a D_V \left[\rho_V(\infty) - \rho_V(a)\right] \tag{5.50}$$

If $\rho_V(\infty) > \rho_V(a)$, vapor diffuses and condenses on the particle surface. If ice crystals are being formed, then electrostatic forces play a role in the particle growth and the radius a in (5.50) must be substituted by the electrical capacity at the interface expressed in length units that represent the shape of the particle (for a sphere is a). The growth of ice particles by diffusion of ambient vapor toward the particle is called *deposition* and the inverse situation is known as *sublimation*.

Using for the vapor density the ideal gas law $\rho_V = P_V / R_V^* T$, we get from (5.50)

$$\frac{dm_p}{dt} = \frac{4\pi a D_V}{R_V^* T} \left[P_V(\infty) - P_V(a) \right] \tag{5.51}$$

If $\rho_V(\infty) < \rho_V(a)$, the condensate evaporates from the particle surface and diffuses away from it. In addition, when condensation occurs at the surface, latent heat is released, and a temperature gradient is established in the droplet. The heat is transported by conduction according to the equation

$$\frac{dQ}{dt} = -4\pi r^2 K_T \frac{dT}{dr} \tag{5.52}$$

where

$(dQ/dt) = L_i \, (dm_p/dt)$ is the conductive heating rate at the particle surface (J s^{-1})
L_i is the latent heat of evaporation (sublimation for ice)
K_T is the thermal conductivity of the atmosphere (J cm^{-1} s^{-1} K^{-1})

Integration of (5.52) from the particle surface to infinity gives

$$L_i \frac{dm_p}{dt} = 4\pi a K_T \left[T(a) - T(\infty) \right] \tag{5.53}$$

Combining Equations 5.51 and 5.53 allows evaluating the rate of particle growth for different atmospheric conditions. The radius growth rate can be obtained making $dm_p/dt = 4\pi \rho_p \, a^2 (da/dt)$, see Problem 5.8. Note that in the case of ice crystal growth, a variety of shapes, such as plate-like, columnar, prism-like, and dendrites can occur, with significant deviations from the simple spherical shape assumed in the previous formulation.

5.3.3   COLLISION AND COALESCENCE

A variety of particle sizes are encountered within a cloud. Large particles fall faster than the smaller ones, collide among themselves and, depending on the interaction they have (which depends on the sizes and trajectories of the particles, and on electrical forces between both, etc.), coalesce and grow. For large particles, the growing rate is a function of a parameter called the *collision efficiency* ε_c that represents the probability that collision and coalescence occurs as the large drop falls. The particle size a grows through this process according to the equation

$$\frac{dm_p}{dt} = \pi a^2 \varepsilon_c \rho_c (w_l - w_s) \tag{5.54}$$

where

w_l and w_s are the fall speed of the large and small particles (in general $w_l \gg w_s$)
$\rho_c \sim \rho_p$ is the density of the condensed substance (particle)

The *collision efficiency* ε_c increases with the particles size and can be calculated as $\varepsilon_c = (a-a')^2/(a+a')^2$ a and a' being the radius of the large and small particle, respectively.

When treating with ice particles, the collision process is called *aggregation*. The ice particles can collect liquid drops (a process called *rimming*) or other smaller ice particles, a mechanism that depends strongly on temperature. On Earth, this produces aggregates with a variety of shapes, representing the *hail* and *hailstones* with typical sizes of 1 cm that can reach in extreme cases up to 10–15 cm.

5.3.4 SEDIMENTATION

As the cloud particle grows, it also falls in the atmosphere. Its motion equation is given according to Newton's second law by

$$m_p \frac{d^2 z}{dt^2} = (m_p - m_f)g - F_f \tag{5.55}$$

where $(m_p - m_f)g$ represents the weight force minus the buoyancy. Assuming for simplicity that the particles are spherical, the buoyancy force can be written as $m_f g = (4/3)\pi a^3 g(\rho_c - \rho_{atmos}) \approx (4/3)\pi a^3 g \rho_c$ since the density of the condensed phase is much greater than that of the atmosphere. The frictional force exerted on the falling particle by the atmosphere is $F_f = K_f \eta_D w$. Here η_D is the dynamical viscosity of the atmosphere (unit kg m^{-1} s^{-1}), $K_f = 6\pi a(C_D Re/24)$ is a friction coefficient that depends on a *drag coefficient* C_D and on the *Reynolds number Re* that characterizes the motion regime in the atmosphere, and $w = dz/dt$. The particle reaches a *terminal fall speed* when both forces in the right part of (5.55) equal so the limit speed is given by

$$w_L = \frac{2}{9} \frac{a^2 g \rho_c}{(C_D Re/24)\eta_D} \tag{5.56}$$

If the descending particle moves in a laminar regime (in general small speeds that apply to small particles), $C_D Re/24 = 1$, and this equation reduces to

$$w_L = \frac{2}{9} \frac{a^2 g \rho_c}{\eta_D} \tag{5.57}$$

known as the *Stokes law* (see Problem 5.10). The fall speed can be parameterized in terms of the size to include nonspherical shapes. As an example, for large ice particles on Earth, the following law has been empirically obtained:

$$w_L(\text{m s}^{-1}) = k_{ice}(2a)^{k'_{ice}} \tag{5.58}$$

where k_{ice} and k'_{ice} are shape-dependent constants. In practice, there is a limit in the terminal speed on raindrops because dynamic friction with the atmosphere tends to break apart the larger raindrops. This limit is of importance in the rainfall of water droplets on Earth and methane on Titan.

5.3.5 SIZE AND SHAPE DISTRIBUTION

The above mechanisms for particle growth will produce a variety of size distributions for liquid particles and of size and shape distributions for solid particles within the cloud (Figure 5.11). In addition, the above equations show that the size distribution will be time- and space-dependent. To have an idea of what kind of size distribution we could expect within a cloud, stationary conditions are usually assumed. Let us denote as $\eta(r)$ the *size distribution function* of particles within a cloud that represents the number of particles within the range of sizes between r and $r+dr$ per unit volume (unit m^{-3}). The total number of particles per unit volume is then

$$N_p = \int_{r_1}^{r_2} \eta(r)\,dr \qquad (5.59)$$

where (r_1, r_2) are the maximum and minimum particle sizes. The *condensable content* within the cloud or cloud density (see Equation 5.36, Table 5.2), that is, grams of the condensable for unit volume of the atmosphere (unit g m^{-3}), is given by

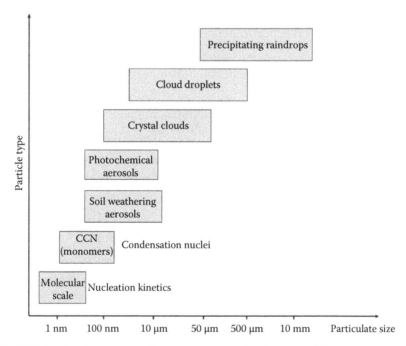

FIGURE 5.11 Particle types and size ranges resulting from the different microphysical processes.

$$\rho_{cl} = \frac{4}{3}\pi\rho_c \int_0^\infty r^3 \eta(r)\, dr \tag{5.60}$$

In Earth's liquid clouds, this is usually also termed as the *liquid water content* (LWC) of the cloud. The size distribution allows defining a *mean particle size* $\langle r \rangle$ as

$$\langle r \rangle = \frac{\int_{r_1}^{r_2} r\eta(r)\, dr}{\int_{r_1}^{r_2} \eta(r)\, dr} = \frac{1}{N_p}\int_{r_1}^{r_2} r\eta(r)\, dr \tag{5.61}$$

and an effective radius r_{eff}

$$r_{eff} = \frac{\int_{r_1}^{r_2} r^3 \eta(r)\, dr}{\int_{r_1}^{r_2} r^2 \eta(r)\, dr} = \frac{1}{G}\int_{r_1}^{r_2} \pi r^3 \eta(r)\, dr \tag{5.62}$$

where $G_p = \int_{r_1}^{r_2} \pi r^2 \eta(r)\, dr$ is called the geometrical cross-sectional area of particles per unit volume. The *effective variance* determines the width of the size distribution

$$\upsilon_{eff} = \frac{1}{G_p r_{eff}^2} \int_{r_1}^{r_2} \pi r^2 (r - r_{eff})^2 \eta(r)\, dr \tag{5.63}$$

which is a dimensionless parameter. For Earth clouds, some commonly used size distributions are (1) the *Hansen distribution*,

$$\eta(r) = C_g r^{(1-3b)/b} e^{-r/ab} \tag{5.64a}$$

where $a = r_{eff}$, $b = \upsilon_{eff}$, and C_g is a normalization constant, and (2) the *log-normal distribution*,

$$\eta(r) = \frac{1}{(2\pi)^{1/2}\sigma_g}\frac{1}{r}\exp\left[-\frac{(\ln r - \ln r_g)^2}{2\sigma_g^2}\right] \tag{5.64b}$$

where $\ln r_g = \int_0^\infty \ln r\eta(r)\, dr$ and $\sigma_g^2 = \int_0^\infty (\ln r - \ln r_g)^2 \eta(r)\, dr$.

5.4 RADIATIVE TRANSFER IN CLOUDS

As already presented in Chapters 3 and 4, the radiative transfer, that is, the propagation of electromagnetic waves through the atmosphere, is essential to understand the physics and chemistry of atmospheres. We previously presented radiative transfer in a purely gaseous medium, considering essentially the absorption component. To complete the scheme, we present in this section the fundamentals of the electromagnetic radiation propagation in an atmosphere containing suspended particles (aerosols and condensate clouds) mixed with the gas, including the effects of the scattering of electromagnetic radiation by the gas.

5.4.1 DEFINITIONS

A medium is optically characterized by the *refraction index n*, which has two components: a real n_r, which is defined as the ratio of the speed of light in the vacuum to the speed of light in the medium, and an imaginary part n_i, which characterizes the absorption effect by the medium on the radiation beam. *Refraction* occurs when a beam of light changes its direction when passing from one medium to another. If their refractive indexes are n_1 and n_2, and the angles of incidence and refraction are φ_1 and φ_2 relative to the normal to the surface, *Snell's law* follows: $n_1 \sin \varphi_1 = n_2 \sin \varphi_2$. For mirror *reflection* in the surface, we have $\varphi_1 = -\varphi_2$ (same angle changing only the direction). *Dispersion* refers to the separation of a beam of light into colors (wavelengths) due to the dependence of the refraction index $n(\nu)$ with frequency or wavelength. According to Snell's law, each wavelength will be dispersed with a different angle. *Diffraction* is the process by which the direction of propagation of a wave changes when the wave encounters an obstacle. For the visible light, it represents the bending it suffers as it passes the obstacle. *Scattering* usually refers to the sum of the effects of reflection and refraction for a particle larger than wavelength. Therefore, it is a process that has a strong angular dependence.

Usually, scattering is separated in the *forward* and *backward* directions relative to the incident angle of the beam on the particle. Charged particles, atoms and molecules, and solid and liquid particulates in atmospheres all scatter radiation. A useful classification of these processes in atmospheric physics can be done in terms of "the size" a of the scatter particle. In order of size magnitude, we have from gas molecules ($a \sim 10^{-4}\,\mu$m) to aerosols ($a \sim 0.1$–$1\,\mu$m), to liquid droplets ($a \sim 2$–$10\,\mu$m), ice crystals ($a \sim 10$–$100\,\mu$m) and raindrops and hail with ~ 1 mm–1 cm. It is very useful to introduce the *size parameter* relative to the wavelength of radiation $\lambda x = 2\pi a/\lambda$ that serves to make the following classification of the scattering processes: (1) When $x \ll 1$, we have *Rayleigh scattering*, responsible of the blue color sky and polarization produced by Earth's molecules. (2) When $x \gtrsim 1$, the scattering is referred to as *Mie* (or also *Lorentz–Mie*) and is typically produced by small spherical particles. (3) When $x \gg 1$, the asymptotic approach to the theory of electromagnetic propagation allows to treat the light as a beam of parallel rays obeying the reflection and refraction Snell's law (*geometric optics*) but including the diffraction theory produced by the object boundary shape. *Rainbows* and *glory* on the Earth's atmosphere are phenomena that can be explained by this theory. The description of these processes relies on the application of the electromagnetic theory whose base is the Maxwell equations (Section 1.6.1). *Raman scattering* refers to the case when a frequency shift occurs in the incident radiation following the scattering process. In atmospheric physics this situation is encountered in some particular cases as in the ultraviolet in Uranus and Neptune, but we do not consider it here. In atmospheres, the particles are considered to be widely separated so each one scatters light independently of the others, which is referred to as *independent scattering*. For a given volume (or layer) of the atmosphere, we say that *multiple scattering* occurs when the beam of light is scattered subsequently by different particles. The reader is referred to the radiative transfer books in the bibliography to find the full details of the scattering of light in atmospheres.

Let us now consider an atmospheric layer that contains a mixture of gases and particles that absorb and scatter the incoming radiation. A fundamental parameter that serves to describe the interaction of radiation with the layer is the *optical depth* introduced in Section 3.3 and discussed in Chapter 4 to describe the gas scattering. Extending its definition to the different contributing processes, we define the total optical depth τ of the layer (which is obviously a function of frequency or wavelength) as the sum

$$\tau = \tau_{s,p} + \tau_{a,p} + \tau_{s,g} + \tau_{a,g} \tag{5.65}$$

in which each term represents, respectively, the particle's scattering and absorption optical depths and the gas scattering and absorption optical depths. The meaning of the last term was already discussed in detail in Section 3.3.2, and the combination of gas scattering and absorption optical depths were introduced in Section 4.2.2 (see Equation 4.15). It can now be decomposed in the two contributions to the gas optical depth as

$$\tau_g(v) = \tau_{g,s} + \tau_{g,a} = \int_0^\infty \alpha_v(z)\rho(z)\,dz = \int_0^\infty \sigma_v(z)\rho(z)\,dz + \int_0^\infty k_v(z)\rho(z)\,dz \tag{5.66}$$

where absorption k_v, scattering σ_v, and mass extinction coefficient $\alpha_v = k_v + \sigma_v$ are given in cm^2 g^{-1}, and for simplicity we have taken the layer orientation perpendicular to the vertical direction z. Combining these coefficients with the density, we can define a *volume coefficient* β (units of cm^{-1}) for absorption ($\beta_a = k_v\rho$), scattering ($\beta_s = \sigma_v\rho$), and extinction ($\beta_e = \alpha_v\rho$). It is also customary to refer to or use the coefficients in (5.66) as *cross sections* for absorption, scattering, or extinction, respectively, to denote the amount of fractional energy removed from the incident beam by the particle, given them simply as α, σ, and k (area units, cm^2). In such a case, the volume scattering coefficient is obtained by summing the scattering cross sections of all the particles in a unit volume. For identical particles with a concentration N_p per unit volume, we have as before $\beta_a = N_p k$, $\beta_s = N_p\sigma$, and $\beta_e = N_p\alpha$. When the gas abundance is given in cm-amagat (Section 3.5), the coefficients are given in units of (cm-amagat)$^{-1}$ and are denoted as $\tilde{\alpha}, \tilde{\sigma}, \tilde{k}$. The following relationships between these definitions are useful for what follows:

$$\alpha(cm^2) = \alpha_v(cm^{-2}g^{-1}) \frac{\langle \mu \rangle}{x_{VV} N_A}$$

$$\alpha_v(cm^{-2}g^{-1}) = \tilde{\alpha}[(cm-amagat)^{-1}] \frac{x_{VV} V_0}{\langle \mu \rangle} \tag{5.67}$$

$$\tilde{\alpha}[(cm-amagat)^{-1}] = \alpha(cm^2) N_0$$

As before, x_{VV} is the volume mixing ratio of the absorbing gas, V_0 is the volume occupied by 1 mol of gas at S.T. P., 22.4×10^3 cm³; $\langle \mu \rangle$ is the mean molecular weight of the atmosphere (g mol⁻¹); N_A is the Avogadro number (6.02×10^{23} particles mol⁻¹); and N_0 is Loschmidt's number (2.68×10^{19} cm⁻³).

5.4.2 GAS ABSORPTION AND SCATTERING

We consider first the effects of a pure gaseous atmosphere (free of particulates) on the incident radiation, following the theory presented in Sections 3.3, 3.4, and 4.2, to calculate the different terms in (5.66).

5.4.2.1 Gas Absorption: A Useful Simple Approach

To simplify, let us assume that for a given gas and absorption band, we introduce a mean effective absorption coefficient $\langle \tilde{k} \rangle$ integrated over the absorption band (Section 3.4). Then, if Z_A denotes the *column abundance* of the absorbing gas above a given level (Equation 3.75), we can simply write (Equation 3.33) $\tau_{g,a} = \langle \tilde{k} \rangle Z_A$. According to the definitions introduced in Section 3.5, we have

$$\tau_{a,g} = \langle \tilde{k} \rangle Z_A = \langle \tilde{k} \rangle H_0 P = \langle \tilde{k} \rangle \frac{x_{VV} V_0}{\langle \mu \rangle g} P \tag{5.68}$$

where
  g is the acceleration of gravity (cm s⁻²)
  P is the pressure in dynes cm⁻² (1 bar = 10^6 dynes cm⁻²)
  Z_A is given in cm-amagat and $\langle \tilde{k} \rangle$ in (cm-amagat)⁻¹

See Problems 5.12 and 5.13.

According to the Beer–Lambert law (also called Bougert law) (see Equation 3.28), for an incident beam with a flux F_0 (W m⁻²) at the top of the atmosphere, the transmitted flux F at a pressure level P along the spectral band is simply given by

$$F(P) = F_0 \exp(-\tau_{g,a}) = F_0 \exp\left(-\langle \tilde{k} \rangle \frac{x_{VV} V_0}{\langle \mu \rangle g} P\right) \tag{5.69}$$

5.4.2.2 Rayleigh Scattering

The scattering of electromagnetic radiation produced by gas molecules and very small particles is well explained by the theory developed by Rayleigh. It assumes that the small particle (or gas molecule) acts as an elementary dipole in interaction with the incident electric field of the electromagnetic wave. If we do not consider polarization, the intensity of radiation $I_v(\Theta)$ scattered by the particle in the angular direction Θ (considered from the direction of propagation), for an incident beam with flux density F_{v0}, is given by

$$I_v(\Theta) = \sigma_R(\Theta)F_{v0} \tag{5.70}$$

where the Rayleigh scattering cross section is given by

$$\sigma_R(\Theta) = \frac{9\pi^2 V^2}{2\lambda^4}\left(\frac{n^2-1}{n^2+2}\right)^2 (1+\cos^2\Theta) \tag{5.71}$$

in units of cm^2 sr^{-1}, where V is the volume of the particle, n is the refractive index, and λ is the wavelength of the radiation. The total scattering cross section (unit cm^2) produced by the particle is obtained upon integration of (5.71) over the 4π steradians

$$\sigma_R = \int_{4\pi} \sigma_R(\Theta)\,d\Omega = \frac{24\pi^3 V^2}{\lambda^4}\left(\frac{n^2-1}{n^2+2}\right)^2 \tag{5.72a}$$

and for a spherical particle with radius a, we get

$$\sigma_R = \frac{128\pi^5 a^6}{3\lambda^4}\left(\frac{n^2-1}{n^2+2}\right)^2 \tag{5.72b}$$

In a medium with identical particles and concentration N_p, the volume coefficient for Rayleigh scattering is given by

$$\beta_R = N_p\sigma_R = \frac{24N_p\pi^3 V^2}{\lambda^4}\left(\frac{n^2-1}{n^2+2}\right)^2 \tag{5.73a}$$

For gas molecules with refractive index n_g this expression can be rewritten as

$$\beta_R = \frac{32\pi^3}{3N_p\lambda^4}(n_g-1)^2 \tag{5.73b}$$

and finally the *Rayleigh optical depth* for a gaseous medium can be obtained using the hydrostatic relationship and definitions in Section 3.5 as

$$\tau_{s,g} = \tau_R = \int \beta_R \, dz = \sigma_R \frac{N_A}{\langle\mu\rangle g} P = \sigma_R \frac{x_{VV} V_0}{\langle\mu\rangle g} P \tag{5.74}$$

with the same units as in (5.68). The scattering cross section for gas molecules can be described to a good approximation in the general form

$$\sigma_R = 4.58 \times 10^{-25} \frac{A_R}{\lambda^4} \left(1 + \frac{B_R}{\lambda^2}\right) \tag{5.75}$$

A_R and B_R being constants that depend on the scattering atom or molecule, and where σ_R is in cm^2 for the wavelength is expressed in microns. In Table 5.3, we give their values for the planets and their major atmospheric components. Combining (5.74) and (5.75) it is straightforward to note that the Rayleigh optical depth can be expressed for each planetary atmosphere as a function of wavelength and pressure (i.e., altitude), $\tau_R(P, \lambda)$, see Problem 5.14.

In the above description, it has been assumed that the scattering particles are isotropic, that is, the induced dipole moment is aligned with the incident electric vector. However, real molecules are weakly anisotropic and a small correction factor, called the *depolarization factor* δ, must be introduced, which translates in a multiplier factor of the order of 1.01–1.06 in the above equation. Rayleigh theory also explains the *polarization* that gas molecules produce on the incident electromagnetic waves and then on the scattered intensity. Without going into the details, assuming linear polarization, the intensity of polarized light I_{Pol} is related to the total intensity scattered in the direction Θ by the equation

$$I_{Pol}(\Theta) = I(\Theta) \frac{\sin^2\Theta}{1 + \cos^2\Theta + 2\delta/(1-\delta)} \tag{5.76}$$

For a beam of light suffering single scattering (i.e., only a primary scattering), Equations 5.70 and 5.76 allow the calculation of the intensity distribution and

---

TABLE 5.3

Rayleigh Scattering Parameters for Atmospheres

Planet (Gas)	$A_R \times 10^{-4}$	$B_R \times 10^{-2}$
Jupiter ($H_2 + He$)	1.6	1.4
Saturn ($H_2 + He$)	1.8	1.6
Earth ($N_2 + O_2$)	9	1.1
Mars, Venus (CO_2)	23	1.3
CH_4	4.4	—

polarization of the radiation from any direction relative to the Sun's position. The real situation usually requires multiple scattering calculations and the full radiative transfer theory must be applied. Rayleigh theory explains well-known phenomena on Earth's atmosphere as the blue color of the sky due to the λ^{-4} dependence of the scattering. For example, blue light with 420 nm is scattered five times more than red light of 650 nm. It also explains the almost full polarization of the sky near $\Theta = 90°$ and the almost zero polarization at $\Theta = 0°$ and 180°.

5.4.3 Particle Absorption and Scattering (Mie Scattering)

Repeating the definition of Equation 5.70 now for a particle, we have

$$I_\nu(\Theta) = \sigma_p(\Theta)F_{\nu 0} \tag{5.77}$$

where now $\sigma_p(\Theta)$ is the angular cross section for the particle. Figure 5.12 shows examples of the angular dependence of the scattering intensity. The Lorenz–Mie

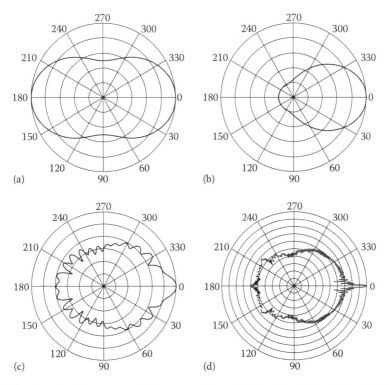

FIGURE 5.12 Angular patterns of the scattered intensity for spherical particles of different sizes corresponding to an incident radiation with a wavelength of 500 nm. The spheres are formed by water (refractive index 1.337724) immersed on air (refractive index 1.0002882). Particle size: (a) molecule (Rayleigh scattering), (b) 0.1 μm, (c) 1 μm, (d) 10 μm. Light is directed from left to right. The radial axes are not to scale (logarithmic representation for (c) and (d)). (Courtesy of Santiago Pérez-Hoyos, UPV-EHU, Bilbao, Spain.)

theory for the scattering of electromagnetic waves by particulates relies on Maxwell equations and appropriate boundary conditions to explain the interaction of the incident electromagnetic wave on the particle. Its derivation is beyond the scope of this book and they can be found in a number of textbooks and reviews (listed in the bibliography). Here we use the basic ideas and formulations that are relevant to the study of planetary atmospheres.

Assume that the particle is a homogeneous sphere. The Mie solution to the above problem gives for the scattering cross section

$$\sigma_P(\Theta) = \left(\frac{\lambda^2}{8\pi^2}\right)\left(|S_1(\Theta)|^2 + |S_2(\Theta)|^2\right) \tag{5.78}$$

where S_1 and S_2 are dimensionless complex amplitudes that can be expressed as infinite series in terms of the scattering angle and size and particle composition. The size a is incorporated in these amplitudes through the parameter $x = 2\pi a/\lambda$ (of interest here for $x \geq 1$) and the composition through the refractive index $n = n_r - in_i$. The real and imaginary refractive indexes n_r and n_i are related to the phase speed of the electromagnetic wave in the medium relative to that in the vacuum and to its absorption, respectively.

The total scattering cross section is obtained by integration over all directions:

$$\sigma_P = \int_{4\pi} \sigma_P(\Theta)\,d\Omega = 2\pi \int_0^{2\pi} \sigma_P(\Theta)\sin\Theta\,d\Theta \tag{5.79}$$

Mie scattering theory uses dimensionless *efficiency factors* to compare the scattering cross section area to the geometric cross section of the particle. They are denoted as Q_i with sub-indexes $i = e, s, a$ running for extinction, scattering, and absorption as

$$Q_i = \frac{\sigma_i}{\pi a^2} \tag{5.80}$$

and where for extinction we have

$$Q_e = Q_s + Q_a \tag{5.81}$$

Mie theory allows to calculate $Q_e(x, n)$ from the complex amplitudes (S_1 or S_2) evaluated at $\Theta = 0°$. $Q_e(x, n)$ is an oscillating function of x as shown in Figure 5.13. For $x < 1$ the slope in $Q_e(x, n)$ is approximately 4, consistent with Rayleigh scattering, whereas for large particles $x \gg 1$ one obtains the results expected from a simple ray tracing (the geometric optics) with a limiting value for $Q_e(x, n) \sim 2$. The largest oscillations in $Q_e(x, n)$ occur in the case of nonabsorbing particles ($n_i = 0$) for which $Q_e = Q_s$, and are smoothed as n_i increases with $Q_s \sim 1$ for large particles. The volume extinction, scattering, and absorption coefficients (β_i, $i = e, s, a$) are obtained by integrating the

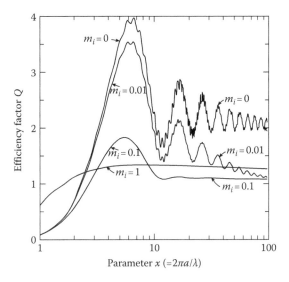

$$\text{Parameter } x \ (=2\pi a/\lambda)$$

FIGURE 5.13 Scattering efficiency factors as a function of the parameter $x=2\pi a/\lambda$ for different cases of dielectric spheres: absorbing particles characterized by the imaginary refractive index (denoted here as m_i) and nonabsorbing ($m_i=0$). (From Hansen, J.E. and Travis, L.D., *Space Sci. Rev.*, 16, 527, 1974. With permission.)

cross sections of the individual particles in a unit volume, using the efficiency factors, and taking into account the size distribution $\eta(r)$ (see Section 5.3.5):

$$\beta_i = \int_0^\infty \pi r^2 Q_i(x,n)\eta(r)\,dr \quad (i=e,s,a) \tag{5.82}$$

The volume coefficients serve to define two other important parameters in the scattering theory. The first is the *albedo for single scattering* $\tilde{\omega}_0$ (or single-scattering albedo) defined as

$$\tilde{\omega}_0 = \frac{\beta_s}{\beta_e} = \frac{\beta_s}{(\beta_s+\beta_a)} \tag{5.83}$$

which represents the fraction of a light beam that undergoes the scattering event. It gives information on the scattering versus absorption in a polydisperse medium. The single scattering albedo value ranges between 0 and 1, such that the extremes are for non-scattering particles ($\tilde{\omega}_0 = 0$) and for nonabsorbing particles ($\tilde{\omega}_0 = 1$). Care must be taken in the use of $\tilde{\omega}_0(\lambda)$ since it is a nonlinear function and becomes significant for three and more decimals (e.g., $\tilde{\omega}_0 = 0.995$ is significant against 0.999).

The other parameter is the *phase function* $p(\Theta)$ that serves to evaluate the directional dependence of the scattering processes, defined as

$$p(\Theta) = \frac{4\pi\beta_s(\Theta)}{\beta_s} \qquad (5.84)$$

The normalization to the solid angle is given by

$$\left(\frac{1}{4\pi}\right)\int_{4\pi} p(\Theta)\,d\Omega = 1 \qquad (5.85)$$

Common scattering phase functions are the *Rayleigh scattering* (representative for molecules)

$$p_R(\Theta) = \frac{3}{4}(1+\cos^2\Theta) \qquad (5.86a)$$

and the *isotropic scattering* medium (constant phase function)

$$p(\Theta) = \tilde{\omega}_0 \qquad (5.86b)$$

The Mie scattering phase functions for any particle type must be fully calculated from (5.83) with the use of (5.78). However, it is customary to use synthetic phase functions that represent the measured scattered intensity reasonably well without going into detailed calculations, and that at the same time cover other aspects not taken into account by the Mie theory as the effects introduced by the deviations from the spherical shape of the particles. Two common cases in planetary atmospheres are the *Henyey–Greenstein phase function* given by

$$p_{HG}(\Theta) = \frac{(1-g_s^2)}{(1+g_s^2-2g_s\cos\Theta)^{3/2}} \qquad (5.86c)$$

and the double Henyey–Greenstein phase function, a variant of the previous one

$$p_{2HG}(\Theta) = f_s p_{HG}(g_{s1},\Theta)+(1-f_s)p_{HG}(g_{s2},\Theta) \qquad (5.86d)$$

The parameter g_s is called the *asymmetry factor* (similarly its two terms g_{s1} and g_{s2} in 5.86d) and is calculated according to $g_s = (1/4\pi)\int_{4\pi} p(\Theta)\cos\Theta\,d\Omega$. It serves to evaluate the asymmetry degree in the angular distribution of the phase function, with g_s increasing toward unity as more energy is scattered in the forward direction and g_s tending to -1 for strong backward scattering. The other parameter f_s is the ratio between the forward-to-backward scattering with g_s and f_s both being wavelength dependent (see Problems 5.16 and 5.17). In Figure 5.14, we show some representative examples of the atmospheric phase functions and scattered intensity distribution. The two-term Henyey–Greenstein phase function can simulate

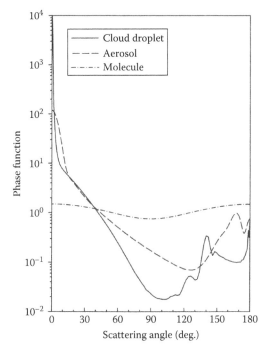

FIGURE 5.14 Phase functions (normalized) versus scattering angle computed with the Mie theory for different sizes of spheres when illuminated with a wavelength of 500 nm: cloud droplets (~10 μm), aerosols (~1 μm), molecules (~10^{-4} μm). (Adapted from Hansen, J.E. and Travis, L.D., *Space Sci. Rev.*, 16, 527, 1974. With permission.)

particles with forward and backward peaks. A simplified first-order representation of the forward and backward scattering distribution is given by the *anisotropic scattering* phase function

$$p_a(\Theta) = \omega_0(1 + g_s\cos\Theta) \tag{5.86e}$$

which reduces to the isotropic case for $g_s = 0$, with backscattering for $g_s > 0$ and forward scattering for $g_s < 0$.

The optical depth due to particles (including absorption and scattering) in an atmospheric layer located between levels z_0 and z_1 (or using the hydrostatic condition between pressure levels P_1 and P_2) is given by

$$\tau_P(z) = \int_{z_0}^{z_1} \beta_e(z)\,dz = -\int_{P_0}^{P_1} \frac{\beta_e(P)}{\rho g}\,dP \tag{5.87}$$

As stated above, for large particles ($x \gg 1$), the Mie theory gives $Q_e \sim 2$ and assuming that they have a single size $a \sim r_{eff}$ and that the number density N_p distributes homogeneously in the vertical within a layer with thickness H_{cl} (i.e., the cloud scale height), we can calculate the optical depth simply as

$$\tau_{cl} = 2\pi a^2 N_p H_{cl} \qquad (5.88)$$

A derived useful parameter is the mass per unit area within the cloud (unit kg m^{-2}), $(4/3)a\rho_P(\tau_{cl}/Q_e)$ with ρ_P being the particle density.

5.4.4 THE RADIATIVE TRANSFER EQUATION INCLUDING PARTICULATES

In Sections 4.2 and 4.3, we introduced the radiative transfer equation and found solutions for the propagation of a light beam in a pure gaseous atmosphere. The scattering behavior was not considered explicitly, nor was the presence of suspended particles and their effect in absorption and scattering on the incident light. The explicit inclusion of these processes into the radiative transfer equation makes it a complicated integro-differential equation that must be solved in most cases using numerical methods (in particular, when including multiple scattering). Going into the details of the methodologies is beyond the objective of this book and the reader can find several books (listed in the bibliography) that are explicitly dedicated to present different approaches and solution methods for this equation and a variety of situations. Therefore, our objective is to present the form and structure of the equation for a planetary atmosphere that contains condensate clouds and aerosols, and solutions for simplified cases.

We search for the diffusely reflected, emitted, and transmitted intensity of the solar radiation that enters into the atmosphere and suffers absorption and multiple scattering by the gases and suspended particles. The reflected (mainly in the visible part of the spectrum) and emitted (mainly in the thermal infrared part of the spectrum) and transmitted radiation can be measured from above the atmosphere as emergent intensity, from a satellite or using ground-based or space-based telescopes. It can also be measured within the atmosphere (e.g., from a descending probe) or from the ground (except in the case obviously of the giant and icy planets). When the atmosphere is thin enough, the emergent intensity from a planet or satellite includes the contribution from the atmosphere (and its clouds) and ground, as on Earth and Mars but also, at some specific wavelengths, from the atmosphere, clouds, and ground in Venus and Titan.

The emergent intensity depends basically on wavelength (as extensively shown in Chapter 3) and geometry (viewing angles). The wavelength dependence can be split into two parts, one corresponding to the reflected spectrum (typically from ~200 nm to 5–10 μm) and the other to the thermal emitted spectrum (3–10 μm to the microwaves). Different treatments to the radiative transfer problem can be performed when studying one part of the spectrum or the other. The geometry dependence can also be separated into two parts. First, the variation of the radiation field with altitude that occurs as the radiation penetrates the atmosphere and the density and optical thickness increase with depth. Second, the information of the direction of the incident radiation (location of the Sun relative to the planet and observer) and of the emergent intensity from a point of the planetary disk relative to the observer position. In summary, the radiation field depends on wavelength and geometry (altitude and angular direction).

For most situations of practical interest, the atmosphere can be considered plane-parallel (thickness much smaller than the planetary radius), allowing for a simplified form of the radiative transfer equation. Figure 5.15 shows a scheme of the geometrical behavior we use. The direction of the incident direct solar flux is specified by $(-\mu_{\varphi 0}, \phi_0)$, which are the cosine of the incident angle (relative to the normal of the plane parallel atmosphere, vertical direction) and the azimuth angle, respectively. The multiple scattered radiation is specified by the direction (μ'_φ, ϕ) and the emergent radiation by (μ_φ, ϕ), where μ_φ is the cosine of the emission angle. The radiative transfer equation (4.16) can be rewritten for the radiation incident at the top of the atmosphere using a source function that explicitly includes the following three main parameters of the scattering and absorption processes: the optical depth, the single scattering albedo, and the phase function (all are functions of wavelength),

$$\mu_\varphi \frac{dI(-\mu_{\varphi 0}, \phi_0; \mu_\varphi, \phi)}{d\tau}$$

$$= I(-\mu_{\varphi 0}, \phi_0; \mu_\varphi, \phi) - \frac{\tilde{\omega}}{4\pi} \int_0^{2\pi} \int_{-1}^1 I(-\mu_{\varphi 0}, \phi_0; \mu'_\varphi, \phi') p(\mu_\varphi, \phi; \mu'_\varphi, \phi') d\mu'_\varphi \, d\phi'$$

$$- \frac{\tilde{\omega}}{4\pi} F_\odot p(\mu_\varphi, \phi; -\mu_{\varphi 0}, \phi_0) \exp\left(-\frac{\tau}{\mu_{\varphi 0}}\right) - (1 - \tilde{\omega}) B[T(\tau)] \qquad (5.89)$$

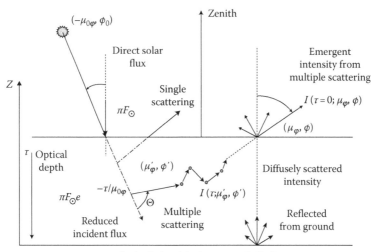

FIGURE 5.15 Geometry employed for the study of the transfer of diffuse solar intensity in a plane parallel atmosphere.

where the first term accounts for the energy lost by absorption and scattering out of the beam, the second and third terms on the right are source terms describing the addition into the beam from scattering, and the fourth source term is that due to the thermal radiation (emission of infrared radiation into the beam) with $B[T(\tau)]$ the Planck function. Inclusion of polarization will make this equation more complicated since the intensity I is replaced by the four-element Stokes vector making the phase function a 4×4 matrix. We remember that in this equation the intensity I is given in units (W m^{-2} μm^{-1} sr^{-1} Hz^{-1}) and $\pi F_\odot$ is the incident solar flux in units (W m^{-2} μm^{-1} Hz^{-1}) (see Chapter 1). From the spherical geometry, the scattering angle introduced previously is related to the incoming and outgoing directions by the expression

$$\cos\Theta = \mu_\varphi\mu_{\varphi 0} + (1-\mu_\varphi^2)^{1/2}(1-\mu_{\varphi 0}^2)^{1/2}\cos(\phi-\phi_0) \tag{5.90a}$$

When observing planetary atmospheres, it is convenient to express the incident, emission, and azimuth angles in terms of the coordinates of the point in the disk: its longitude from the central meridian ΔL and its planetographic latitude φ_g, and of the Sun—observer phase angle α (angle between the Sun, center of the body, and observer, then $\Theta = \pi - \alpha$)

$$\mu_\varphi = \sin B_\oplus \cdot \sin\varphi_g + \cos B_\oplus \cos\varphi_g \cos(\Delta L)$$
$$\mu_{\varphi 0} = \sin B'_\odot \cdot \sin\varphi_g + \cos B'_\odot \cos\varphi_g \cos(\Delta L - \alpha) \tag{5.90b}$$

where we have introduced as $B_\oplus$ the sub-observer planetocentric latitude and $B'_\odot$ the sub-solar planetocentric latitude. To gain insight on the solutions to Equation 5.89, we consider in the following some simple situations.

5.4.4.1 Purely Absorbing Atmosphere

Neglecting scattering and thermal emission, Equation 5.89 reduces to the simple absorption case (Beer's law) previously treated, which for the upward and downward directions can be written as

$$\mu_\varphi \frac{dI(\mu_{\varphi 0},\phi_0;\mu_\varphi,\phi)}{d\tau} = I(\mu_{\varphi 0},\phi_0;\mu_\varphi,\phi)$$
$$-\mu_\varphi \frac{dI(-\mu_{\varphi 0},\phi_0;\mu_\varphi,\phi)}{d\tau} = I(-\mu_{\varphi 0},\phi_0;\mu_\varphi,\phi) \tag{5.91}$$

with simple exponential solutions for the intensity in terms of the optical depth. A simple but practical and illustrative case of interest would correspond to that of a purely absorbing atmosphere above a reflecting layer. This could occur, for example, at the wavelength of an absorption band in the near infrared where Rayleigh scattering and thermal emission can be neglected, and where the reflecting layer could be the ground or a thick cloud deck (Problem 5.21). The solution to (5.91) can be written as

$$I = I_0 \exp\left[-\tau_{a,g}\left(\frac{1}{\mu_\varphi} + \frac{1}{\mu_{\varphi 0}}\right)\right] \tag{5.92}$$

where
  I_0 is the incident intensity
  $\tau_{a,g}$ is the total optical depth of the atmosphere that could be calculated, for example, using (5.68)

The trivial case of a very thin atmosphere without any kind of scattering or absorption inside it reduces to an emergent intensity $I = I_0$.

5.4.4.2 Infrared Intensity: Thermal Radiation and the Temperature Profile

Now consider the case where absorption and infrared emissions are considered but scattering is neglected (see Section 4.3.1). Equation 5.89 reduces to the following equation for the upward thermal infrared radiation:

$$\mu_\varphi \frac{dI(\mu_{\varphi 0}, \phi_0; \mu_\varphi, \phi)}{d\tau} = I(\mu_{\varphi 0}, \phi_0; \mu_\varphi, \phi) - B\left[T(\tau)\right] \tag{5.93}$$

In applications to planetary atmospheres, one deals with the emergent intensity at the top of the atmosphere, which in terms of the total optical depth τ at the surface level, is given by

$$I(0; \mu_\varphi, \phi) = I(\tau; \mu_\varphi, \phi)e^{-\tau/\mu_\varphi} + \int_0^\tau B\left[T(\tau')\right]e^{-\tau'/\mu_\varphi}\frac{d\tau'}{\mu_\varphi} \tag{5.94}$$

The first term on the right-hand side is the radiance emitted by the surface $I(\tau; \mu_\varphi, \phi)$ (which can be considered isotropic) attenuated at the top by the atmospheric absorption, and the second term denotes the contribution from the emission by the atmosphere. Note that the Planck function varies with temperature and it must be integrated in the optical depth that it is itself a function of altitude (pressure). Moreover, note that although it is not indicated, it must be remembered that in these expressions there is an explicit dependence with wavelength.

As an application, if we assume that we observe in the vertical direction $\mu_\varphi \sim 1$, the emitted radiance from the surface or from an optically thick cloud deck can be expressed as $I(\tau_T) = \varepsilon_S B(T_S)$, the emissivity $\varepsilon_S \sim 1$ and T_S being the surface or bottom cloud deck temperature. The monochromatic transmittance $T(\tau) = e^{-\tau}$ (see Equation 3.30) and the *weighting function* $\partial T(\tau)/\partial \tau = -e^{-\tau}$ are useful parameters for remote sensing studies. Then it follows that at the top of the atmosphere the intensity given by (5.94) can be rewritten as in terms of the altitude (expressed in pressure coordinates) as

$$I(0) = B(T_S)T(\tau) + \int_\tau^0 B(\tau)\frac{\partial T(\tau)}{\partial \tau}d\tau = B(T_S)T(P_S) + \int_{P_S}^0 B[T(P)]\frac{\partial T(P)}{\partial P}dP \quad (5.95)$$

According to this equation, the emerging radiance is a result of the product of the Planck function (which includes the temperature information), the spectral transmittance (which includes the density profiles of the absorbing gases, see Equation 3.31), and the weighting functions. Measurement of the radiances emergent from a body at selected spectral wavelengths corresponding to very different opacities with weighting function peaks at different heights can be used to mathematically invert (5.95) and retrieve the temperature as a function of altitude (see Section 4.3.1).

5.4.4.3 Single-Scattering Approximation

When the optical depth of the layer is small (typically $\tau < 0.1$), the single scattering of the direct solar radiation dominates the intensity field. This occurs for thin cirrus clouds and aerosols on Earth, for thinner clouds of Mars, and for some high-altitude thin hazes in giant planets. In this situation, we retain the first and the third terms on the right-hand side of Equation 5.89. The solution for the intensity at the top of the atmosphere is given by

$$I(0; \mu_\varphi, \phi) = I(\tau; \mu_\varphi, \phi)e^{-\tau/\mu_\varphi} + \int_0^\tau \frac{\tilde{\omega}_0}{4\pi\mu\varphi}F_{\odot p}p(\mu_\varphi, \phi; -\mu_{\varphi 0}, \phi_0)e^{-(\tau+\tau')/\mu_\varphi}d\tau'$$

$$= I(\tau; \mu_\varphi, \phi)e^{-\tau/\mu_\varphi} + \frac{\mu_{\varphi 0}F_{\odot p}}{\pi}\frac{\tilde{\omega}_0}{4(\mu_\varphi + \mu_{\varphi 0})}p(\mu_\varphi, \phi; -\mu_{\varphi 0}, \phi_0)$$

$$\times\left\{1 - \exp\left[-\tau\left(\frac{1}{\mu_\varphi} + \frac{1}{\mu_{\varphi 0}}\right)\right]\right\} \quad (5.96)$$

where the solar flux on the planet $F_{\odot p}$ was introduced in Section 1.4.3. For a black surface such that the upward reflected intensity is zero ($I(\tau; \mu_\varphi, \phi)=0$) and for very small τ we obtain the so-called *bidirectional reflectance*:

$$R(\mu_\varphi, \phi; -\mu_{\varphi 0}, \phi_0) = \frac{\pi I(0; \mu_\varphi, \phi)}{\mu_{\varphi 0}F_{\odot p}} = \tau\frac{\tilde{\omega}_0}{4\mu_\varphi\mu_{\varphi 0}}p(\mu_\varphi, \phi; -\mu_{\varphi 0}, \phi_0) \quad (5.97)$$

This formula can be employed, for example, to retrieve the optical depth of aerosols on Earth's atmosphere from satellite remote sensing measurements. The case $R=1$ corresponds to the perfect Lambert reflector.

5.4.4.4   Isotropic Scattering

In some situations, the phase function can be assumed to be isotropic, that is, independent of the scattering angle. Then, neglecting the thermal contribution, Equation 5.89 can be written as

$$\mu_\varphi \frac{dI(\mu_\varphi)}{d\tau} = I(\mu_\varphi) - \frac{\tilde{\omega}_0}{2} \int_{-1}^{1} I(\mu_\varphi) d\mu_\varphi - \frac{\tilde{\omega}_0}{4\pi} F_{\odot p} e^{-\tau/\mu_{\varphi 0}} \tag{5.98}$$

where the intensity is obviously a function of the optical depth. For a semi-infinite atmospheres such that $\tau \to \infty$, a solution to (5.98) can be expressed in terms of the so-called *Chandrashekar H-functions*:

$$I(0; \mu_\varphi) = \frac{\tilde{\omega}_0 F_{\odot p}}{4\pi} \frac{\mu_{\varphi 0}}{\mu_\varphi + \mu_{\varphi 0}} H(\tilde{\omega}_0, \mu_{\varphi 0}) H(\tilde{\omega}_0, \mu_\varphi) \tag{5.99}$$

These functions are tabulated in several books (see list in the bibliography), but for the purposes of this book, a reasonable first order analytical approach (accurate to 10%) can be used:

$$H(\tilde{\omega}, \mu_\varphi) \approx \frac{1 + \sqrt{3}\mu_\varphi}{1 + \mu_\varphi \left[3(1 - \tilde{\omega}_0) \right]^{1/2}} \tag{5.100}$$

5.4.4.5   Thick Clouds

The total albedo of a cloud depends on its optical depth, phase function, and single scattering albedo. When τ is large and $\tilde{\omega}_0 \to 1$ (thick nonabsorbing clouds), solving the radiative transfer equation (5.89) by a method called the two-stream approximation gives the *Lacis and Hansen albedo of a cloud*:

$$A_{cl} = \frac{(1 - g_s)\tau\sqrt{3}}{2 + (1 - g_s)\tau\sqrt{3}} \tag{5.101}$$

which is only dependent on the asymmetry parameter and optical depth.

5.4.4.6   Reflectivity Laws

A useful empirical formula that reproduces the limb-darkening behavior of the reflectivity of planetary surfaces and cloudy atmospheres reasonably well was introduced by Minnaert in 1941:

$$\left(\frac{I}{F} \right) = \left(\frac{I}{F} \right)_0 \cdot \mu_\varphi^{K_M - 1} \cdot \mu_{\varphi 0}^{K_M} \tag{5.102}$$

where (I/F) is the reflectivity at a point of the disk observed at scattering angles (μ_φ, $\mu_{\varphi 0}$), $(I/F)_0$ is the reflectivity for $\mu_\varphi = \mu_{\varphi 0} = 1$, and K_M is the limb-darkening coefficient that depends on wavelength. The limb-darkening coefficient K_M is usually retrieved from the observations. Expressing the Minnaert law logarithmically

$$\ln\left[\mu_\varphi\left(\frac{I}{F}\right)\right] = K_M \cdot \ln(\mu_\varphi \mu_{\varphi 0}) + \ln\left(\frac{I}{F}\right)_0$$

the measured reflectivity for different values of the scattering angles (different points of the disk) gives in a linear fit the slope K_M. The *Minnaert reflectivity law* is related to the geometric albedo p_0 (Section 1.4.3) by

$$\left(\frac{I}{F}\right)_0 = p_0 \frac{2K_M + 1}{2} \tag{5.103}$$

One of the most common procedures used to retrieve the optical properties of the clouds in an atmosphere is to compare the reflectivity measured at different points on the disk (e.g., the reflectivity curves from the center of the disk to the limb at selected wavelengths) with those resulting from a model (obtained numerically solving the radiative transfer equation or, if possible, by simple analytical solutions as presented above) (see Figure 5.16). However, in general, one confronts a multi-parametric problem in which the number of free parameters in the model is large, and this requires the use of as many wavelengths as possible and scattering geometries (angles of illumination and view) to be able to constrain the model parameters.

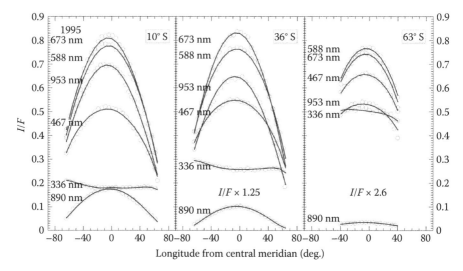

FIGURE 5.16 Comparison between observations of the reflectivity curves (from limb to limb) and the results produced by different types of haze and cloud models of Saturn diffuse reflection. Results are given for different latitude bands and observing wavelengths. (Courtesy of Santiago Pérez-Hoyos, UPV-EHU, Bilbao, Spain.)

5.4.5 CLOUD FORCING AND AEROSOL HEATING RATES

In Section 4.3.3, we presented the formalism used to calculate the heating and cooling rates produced by gaseous absorption and emission in a planetary atmosphere. Suspended aerosols and clouds in atmospheres also act as a thermal source for heating (by sunlight absorption in a layer) or cooling the atmosphere (reflecting sunlight and cooling the atmosphere below the layer in an anti-greenhouse effect). Clouds can also block the thermal infrared radiation coming from lower levels, enhancing the temperatures (greenhouse effect). The thermal forcing mechanisms by clouds are an important part of dynamics and climate in planetary atmospheres. However, it is a complicated issue to model due to the coupling between thermal changes, microphysics, and dynamics. From a known or assumed vertical distribution of aerosols (particle density variation with altitude) and their properties (size, size distribution, shape, optical properties), the penetration level of the incoming solar radiation and the corresponding heating rates (Section 4.3.3) can be calculated once the radiative transfer equation is solved.

An approximate estimate of the aerosol heating rate ($K\ s^{-1}$) can be obtained from a simple parameterization of the type

$$\left(\frac{Q}{\rho C_p}\right)_a = \left(\frac{g}{C_p}\frac{dF_v}{dP}\right)_a \approx \frac{g}{C_p}\frac{f_a F_\odot}{2P_{cl}}\langle\xi_P(\varphi_c)\rangle \qquad (5.104)$$

where

f_a is a parameter that gives the fraction of sunlight absorbed by aerosols (a number between 0 and 1)

P_{cl} is the pressure reference level above which the aerosol heating is located

$\langle\xi_P(\varphi_c)\rangle$ is the fractional length of daytime (a number between 0 and 1), which depends on the planet orbital tilt and latitude (see Section 1.4.3)

5.5 OBSERVED CLOUD PROPERTIES: OVERVIEW

We have seen in previous sections that a large variety of cloud compositions and particles properties are present in planetary atmospheres. A summary is given in Tables 5.2, 5.4, and 5.5. Images of planetary clouds, their spatial structure and coverage, and aspect, are shown in Figures 5.17 through 5.27. More images are available in Chapters 8 and 9.

There are two cases where the major clouds come from chemical reactions. One is Venus' clouds composed of sulfuric acid (H_2SO_4) resulting from the reaction cycle between SO_2 and H_2O as the main ingredients. The other is ammonium hydrosulfide (NH_4SH) clouds, which are present in giant and icy planets, resulting from a reaction between NH_3 and SH_2, as shown in Section 5.2.3. In addition, the main production processes of the hydrocarbon and nitrile hazes in Titan's atmosphere and in the stratospheres of giant and icy planets result from photochemical reactions (Tables 5.4 and 5.5).

TABLE 5.4

Measured/Inferred Cloud Properties in Terrestrial Planets, and Satellites Titan and Triton

Planet or Satellite	Cloud Type	Altitude (km)	Particle Size (μm)	Number Density (cm⁻³)	Composition
Venus	*Upper haze*	79–90	0.4	500	SO_4H_2 (aqueous)+contaminant
	Upper cloud	56–70	1 (0.4–2)	1–1500	Same
	Middle cloud	50–56	1.5 (0.3–7)	1–300	Same
	Lower cloud	47–50	2 (0.4–8)	1–1200	Same
	Lower haze	31–47	0.3 (0.3–2)	50–150	Same

Total optical depth τ (600 nm)=20–40; Surface coverage=100%.
Mostly of stratiform type (equator: probably convective-"dry cumulus" like).
Contaminant: UV-absorber of unknown origin mixed with upper cloud particles (droplets and crystals).

Earth	*Polar mesospheric* (*)	79–90	0.05		H_2O ice
	Polar stratospheric	15–25	1–10		H_2O ice-HNO_3
	Stratospheric aerosols	10–30	0.1		Sulfates
	Cirrus cloud	7–16	10–100	1000	H_2O ice
	Stratus, Cumulus, Nimbus clouds	1–18	10–1000	0.001–1	H_2O liquid, H_2O ice
	Fogs	0–1	0.5–10	1000	H_2O liquid

(*) "Noctilucent" clouds
Total optical depth τ (600 nm)=5–10, maximum (dense clouds) τ (600 nm)=300–400
Surface coverage=40%

					Composition
	Volcanic particles	5–35	0.1–10		Minerals, sulfates, ash
	Smoke aerosols	0–10	0.1–1		Soot, ash, tars
	Dust (storms)	0–3	1–10		Silicates, clays
	Tropospheric aerosols (CCN)	0–10	0.1–1		Sulfates, nitrates, minerals
Mars	Stratiform	0–60	0.1–10	10^{-2} to 10^{-4}	CO_2 ice
	Isolated clouds—Fogs	0–50	1–4	10^{-2} to 10^{-4}	H_2O ice
	Winter poles	0–20	0.1–10	10^{-2} to 10^{-4}	CO_2, H_2O, ice
	Dust storms	0–50	0.4–2.5	1–100	Minerals

Dust optical depth $\tau = 0.3$–6, clouds optical depth τ (600 nm) = 0.01–1
Cloud surface coverage = 5%; dust surface coverage = 0%–100%

					Composition
Titan	*High hazes*	150–350	Aggregates (0.05)		HC_3N, C_2H_5CN, C_4N_2, C_2H_2
	Main haze (maximum)	0–150 (50–150)	0.5–0.9 (aggregates)	1–60	Same
	Upper clouds (isolated)	30–60	1–3	0.1–1	C_2H_6
	Lower clouds	10–30	10–5000	0.1–1	CH_4–C_2H_6

Haze optical depth τ (600 nm) = 6–7
Haze surface coverage = 100%; Clouds surface coverage = 5%–10%

					Composition
Triton	*High hazes*	0–30	0.15		Photochemical
	Cloud	0–8	0.25		N_2 ice

Haze optical depth τ (470 nm) = 3–8×10^{-3}
Geyser-like plume activity also injects particulates in this tenuous atmosphere (see Chapter 6)

TABLE 5.5

Measured/Inferred Cloud Properties in Giant and Icy Planets

Planet	Cloud Type	Pressure (bar or mbar)	Particle Size (μm)	Number Density (cm⁻³)	Composition
Jupiter	Stratospheric haze	1–70 mbar	0.1–1		Hydrocarbons
	Tropospheric haze	0.1–0.5 bar	1	1	Hydrocarbons
	Upper cloud (Galileo)	0.5–1 bar (0.46–0.53)	3–100 (0.5–0.9)	1 (4–7)	NH_3 ice + chromophore
	Middle cloud (Galileo)	2–3 bar (0.75–1.3)	100 (0.8–1.1)	0.1 (33)	NH_4SH
	Lower cloud (Galileo)	4–6 bar (2.4–3.6)	60 (1–4)	2 (0.5)	H_2O

Galileo probe measurements between brackets

Stratospheric haze optical depth τ (600 nm) = 0.1–0.2

Tropospheric haze + upper cloud: τ (600 nm) = 3–10; coverage = 100%

Saturn	Stratospheric haze	1–50 mbar	0.1–0.2		Hydrocarbons
	Tropospheric haze	0.1–0.5 bar	1–2	1	Hydrocarbons
	Upper cloud	1.5 bar	100	0.08	NH_3 ice + chromo
	Middle cloud	4 bar	500	0.0004	NH_4SH
	Lower cloud	9–11 bar	100	1	H_2O

Stratospheric haze optical depth τ (600 nm) = 0.2–1

Tropospheric haze + upper cloud: τ (600 nm) = 5–30; coverage = 100%

Uranus	Upper haze	0.2–2.5 mbar	0.02	25	C_2H_4
	Middle haze	2.5–14 mbar	0.05	7	C_2H_2
	Lower haze	14–100	0.26	3	C_2H_6
	Haze cloud	0.1–1.2 bar	0.5	2	CH_4 ice
	Middle cloud	3 bar	300	0.01	SH_2
	Lower cloud	6–7 bar	300	0.04	NH_3
	Deep clouds	>20–30 bar	200–1000	1–0.0003	NH_4SH–H_2O–SH_2–NH_3
Hazes optical depth τ (600 nm) = 0.4; coverage = 100%					
Neptune	Upper haze	1.4–6 mbar	0.2	0.06	C_2H_2–HCN
	Middle haze	6–10 mbar	0.2	1	C_2H_2–C_3H_6–C_3H_8
	Lower haze	10–20	0.2	5	C_2H_6
	Haze cloud	0.3–1.5 bar	5000 (cloud)	0.0001	CH_4 ice
	Middle cloud	2–7 bar	300	0.01	SH_2
	Lower cloud	3 bar	1000	4×10^{-4}	NH_3
	Deep clouds	>20–30 bar	200–3000	10^{-5} to 10	NH_4SH–H_2O–SH_2–NH_3
Hazes optical depth τ (600 nm) = 0.1; coverage = 100%					

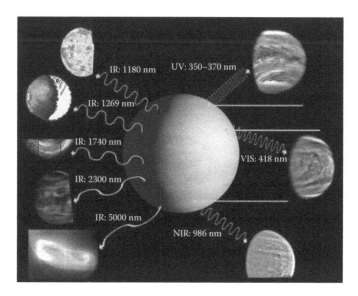

FIGURE 5.17 Observations of Venus's clouds (and other features) from space as a function of the wavelength employed. Nighttime (left part): surface (1180 nm), oxygen airglow emission (1269 nm), lower clouds from opacity to thermal radiation from deeper levels (1740 and 2300 nm), dipole from emission of the upper cloud (5000 nm). Daytime (right part): sunlight reflection by the upper cloud (350–370, 418, and 986 nm). Composed from images obtained by Pioneer-Venus and Venus Express spacecrafts. (Courtesy of NASA and ESA, Plymouth, U.K.).

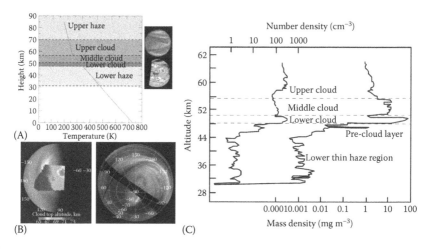

FIGURE 5.18 Structure of Venus's cloud layers. (A) Location in altitude as compared with the temperature profile. Upper layer in image in ultraviolet (upper inset, Pioneer-Venus) and lower cloud layer in image at 2.3 μm (lower inset, Galileo NIMS). (Courtesy of NASA-JPL, Los Angeles, CA.) (B) Altitude of the south polar upper clouds in Venus observed in the ultraviolet (descending clouds toward the pole as indicated by the gray scale inset in the image) from Venus-Express VMC. (Courtesy of ESA, Paris, France.) (C) Vertical structure of Venus's clouds: number density, extinction coefficient, mass loading measured by Pioneer-Venus probes. (Adapted from Knollenberg, R.G. and Hunten, D.M., *J. Geophys. Res.*, 85, 8039, 1980.)

FIGURE 5.19 Earth's map projection showing the cloud coverage. (Courtesy of NASA Earth Observatory, Greenbelt, MD.)

FIGURE 5.20 Thin CO_2 condensate clouds in Mars. Image from Mars Reconnaissance Orbiter (MRO). (Courtesy of NASA/JPL-Caltech, Los Angeles, CA/University of Arizona, Tucson, AZ.)

Some clouds are expected to be really exotic for solar system standards, as those predicted in giant extrasolar planets of refractory materials like iron and enstatite ($MgSiO_3$). These planets are nicknamed "hot Jupiters" because they are gas giants at high temperature (~1500 K) due to the proximity to their parent star and the intense stellar radiation they suffer. In such hot environments, only these "clouds" are able to form.

Mars' main cloud condensate is a special case since the main atmospheric constituent (CO_2) condenses massively in the surface of the polar areas when the temperatures become low enough during winter season forming great deposits of carbonic snow. Water-ice clouds are not so common but form as fogs and in particular over the poles in winter season. In addition, the atmosphere usually contains important quantities of dust lifted from the ground by blowing winds. Rare, but important event in this context are the large-scale dust storms that develop when winds are intense enough to lift and spread the dust along the entire planet.

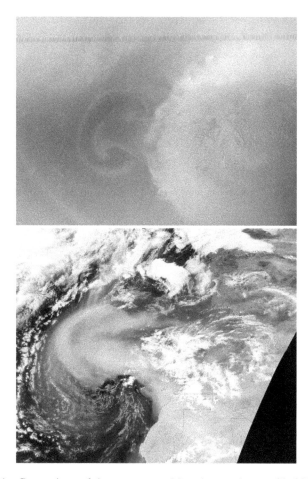

FIGURE 5.21 Comparison of dust storms on Mars (upper, August 29, 2000) and Earth over Canary I. (West of Africa) (lower, February 26, 2000). From: Mars Orbiter Camera (MOC) onboard Mars Global Surveyor and SeaWiFS spacecrafts. (Courtesy of NASA-JPL and Malin S.S.S, Los Angeles, CA.)

The clouds predicted by thermochemical studies are essentially colorless (white). However, the visible clouds of Venus and those of giant planets show a faint color. The agent that makes Venus's clouds yellowish and the Jovian clouds brown or reddish is unknown. Most surely, these contaminant molecules must be the result of nonequilibrium processes in the atmosphere. In addition, the formation of high-altitude hazes by photochemical processes in the giant and icy planets, together with the absorption and Rayleigh scattering of the solar radiation by the gases, cause the subtle colored appearance of cloudy planetary atmospheres. This also occurs in the satellite Titan, which has a dense and vertically extended reddish haze layer made of hydrocarbons.

Stratified clouds are apparently the dominant type of clouds in planets with their vertical extent being a fraction of the scale-height. For them, the dominant heat exchange with the atmosphere is radiation. Venus's cloud layers are of this type.

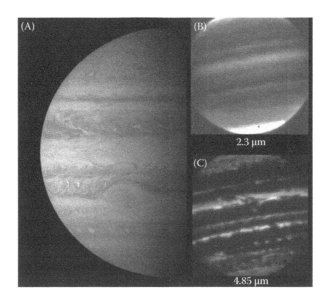

FIGURE 5.22 Jupiter's clouds: (A) Visual range (400–900 nm): Tropospheric cloud (ammonia and chromophores) and hazes from reflected sunlight (left: Courtesy of Cassini ISS image in December 2000 NASA-ESA.). (B) Upper aerosol layers in reflected light at 2.3 μm wavelength of the methane absorption band (upper right: Courtesy of NASA-JPL University of Hawaii (July 27, 1995).). (C) Lower cloud (ammonia and NH_4SH) at 4.85 μm wavelength from opacity to thermal infrared radiation. (Courtesy of NASA-JPL, Los Angeles, CA and University of Hawaii, Honolulu, HI (July 27, 1995).)

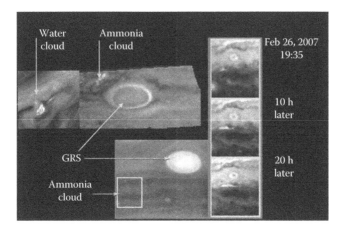

FIGURE 5.23 Detection of isolated pure "fresh" ammonia and water clouds in Jupiter (marked by arrows). The water cloud is formed by a convective storm near the Great Red Spot (cumulus-like clouds) detected at visual wavelengths. The ammonia clouds are shown in a single Galileo NIMS image and in a New Horizons image, showing its temporal evolution. Mosaic from Galileo and New Horizons spacecraft images. (Courtesy of NASA-JPL, Los Angeles, CA.)

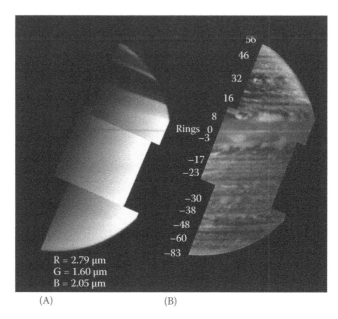

FIGURE 5.24 Saturn's hazes and clouds from Cassini VIMS instrument: (A) Upper aerosol layers in reflected light at near infrared wavelengths; (B) middle clouds (ammonia and NH_4SH) at $4.85\,\mu m$ wavelength from opacity to thermal infrared radiation. (Courtesy of NASA-JPL, Los Angeles, CA.)

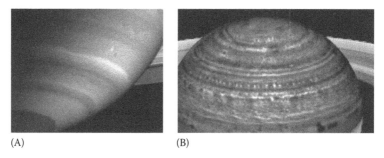

(A)   (B)

FIGURE 5.25 Saturn's hazes and upper and middle clouds from Cassini ISS (A) and VIMS (B) instruments: (A) Tropospheric haze and upper cloud layers in reflected light (visual wavelength); (B) middle cloud (ammonia and NH_4SH) at $4.85\,\mu m$ wavelength from opacity to thermal infrared radiation. (Courtesy of NASA-JPL, Los Angeles, CA.)

However, vertical mixing between particles initially in different cloud layers occur due to dynamical processes making the vertical extent of the clouds a scale-height or more. Cumulus clouds resulting from intense vertical motions related to wet convection are observed in localized areas of the Earth and on the giant planets, Jupiter and Saturn, and in Titan. For them, the dominant heat exchange with the atmosphere is latent heat. Venus's transient cellular cloud fields are convective but not related to moisture. Orographic clouds form on Earth and Mars but not in Venus (and apparently not in Titan) since they locate high far away from the surface.

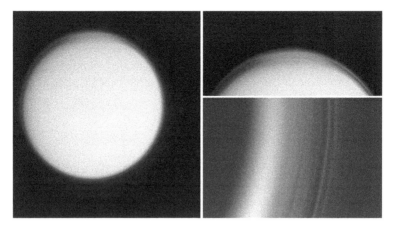

FIGURE 5.26 Titan hazes observed in the visual range in full disk (100% coverage) and their upper altitude stratigraphy in the limb obtained with the Cassini ISS instrument. Image: PIA 06122-06123-06160. (Courtesy of NASA-JPL, Los Angeles, CA.)

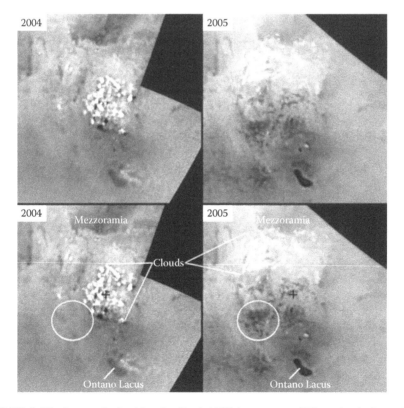

FIGURE 5.27 Image obtained by the Cassini ISS instrument of Titan clouds formed by methane (probably mixed with ethane) located close to the surface in the south pole over different years. The position of a possible lake of liquid hydrocarbons (informally called Ontario Lacus) is indicated. Image PIA11147. (Courtesy of NASA-JPL, Los Angeles, CA.)

Sedimentation of the cloud and haze particulates occurs between the layers where they reside and in some cases between the clouds and the surface. This last is obviously the case of the Earth, where precipitation of liquid and solid water is abundant and occurring practically in all the latitudes and epochs. Apparently this also occurs, perhaps more sporadically, in Titan (methane and ethane) being perhaps the source that replenishes the lakes.

PROBLEMS

5.1 Consider the following two reactions: $CO + 3H_2 \rightleftarrows CH_4 + H_2O$ (1), $N_2 + 3H_2 \rightleftarrows 2NH_3$ (2). Assuming thermochemical equilibrium, find the dependence of the mole fractions of methane and carbon monoxide in reaction (1), and of ammonia and nitrogen in reaction (2), with atmospheric pressure.

Solution

$$x_{V,CH_4} = \frac{x_{V,CO_2}(x_{V,H_2})^3}{x_{V,H_2O}} K_p(1)P^2, \quad x_{V,CO} = \frac{x_{V,CH_4} x_{V,H_2O}}{x_{V,H_2}} \frac{1}{K_p(1)P^2}$$

$$x_{V,NH_3} = \left[K_p(2) x_{V,N_2}(x_{V,H_2})^3 \right]^{1/2} P, \quad x_{V,N_2} = \frac{(x_{V,NH_3})^2}{(x_{V,H_2})^3} \frac{1}{K_p(2)P^2}$$

5.2 Derive Equations 5.15 and 5.16.

5.3 Derive Equations 5.23a, b and show that an approximate expression for the specific heats in a moist water atmosphere is a linear function of the water vapor mixing ratio with $C_{pd} = 1000$ J kg^{-1} K^{-1}, $C_{pv} = 1810$ J kg^{-1} K^{-1}, $C_{vd} = 717$ J kg^{-1} K^{-1}, $C_{vv} = 1350$ J kg^{-1} K^{-1}.

Solution

$$C_{vm} \sim C_{vd}(1 + 0.9x_V); \quad C_{pm} \sim C_{pd}(1 + 0.8x_V)$$

5.4 A sample of moist air in the atmospheres of Earth (containing H_2O), Jupiter (NH_3), and Titan (CH_4) is at a temperature of 280, 165, and 84 K, respectively, and in all cases at 1 bar of pressure. The measured mole fractions are 8×10^{-3} (H_2O), 10^{-4} (NH_3), and 3×10^{-2} (CH_4). Calculate for each case: (a) mass mixing ratio, (b) virtual temperature, (c) absolute humidity, (d) relative humidity.

Solution

Case	x_{MV} (g kg^{-1})	T_V (K)	ρ_V (kg m^{-3})	RH (%)
Earth (H_2O)	5	280.8	6.1×10^{-3}	29.6
Jupiter (NH_3)	0.76	164.9	1.2×10^{-4}	5
Titan (CH_4)	16	87.08	6.4×10^{-2}	70

5.5 (a) What is the specific and relative humidity of a parcel of air on Earth's surface ($P=1$ bar) that has a mole fraction content of water of 7×10^{-3} at a temperature of 30°C and 0°C? (b) What should be the specific humidity under the same two previous conditions on Earth's surface for a parcel to have a relative humidity of 50%? (c) Galileo probe measured a mole fraction of water of 5×10^{-5} in Jupiter's atmosphere at $P=5$ bar and $T=0$°C. What is the specific and relative humidity at this level of the Jovian atmosphere?

Solution

(a) $q=x_{MV}=4.3$ g kg^{-1}, $RH=77.7\%$ (273 K), 12% (303 K)
(b) $q=x_{MV}=4.5$ g kg^{-1} (273 K) and $q=x_{MV}=29$ g kg^{-1} (303 K)
(c) $q=x_{MV}=0.4$ g kg^{-1}, $RH=2.8\%$

5.6 Temperature measurements in a region of the Earth's lower troposphere give a vertical temperature gradient $\Gamma=6.5$ K km^{-1}. Find the empirical $P(T)$ profile in the troposphere assuming it follows an adiabatic if the surface temperature and pressure are T_0, P_0 given in Table 4.2. Find the saturation vapor pressure relationship and the altitude in kilometers where you should expect the clouds to form for two different values of the water molar mixing ratio: (a) $x_{VV}=0.015$ (wet conditions), (b) $x_{VV}=2.5\times10^{-4}$ (dry conditions).

Solution

$$P(T)=1.013\left(\frac{T(\mathrm{K})}{288}\right)^{5.2}$$

$$P_{VS}(T)=\frac{1}{0.015}\exp\left[25.096-\frac{6823.15}{T(\mathrm{K})}-0.019T(\mathrm{K})\right]$$

(a) $T=285$ K, $P=0.96$ bar, $z\sim500$ m
(b) $T=230$ K, $P=0.3$ bar $z\sim10$ km

5.7 A liquid water droplet formed by homogeneous nucleation on Earth's atmosphere has a radius $a_c=0.1\,\mu$m at $T=278$ K when the relative humidity $RH=101\%$. (1) What would be the radius for $RH=112\%$. (2) Find an expression for the number of molecules n_c required to give the critical radius a_c (Equation 5.45). (3) Calculate the critical radius and number of molecules of water for $RH=110\%$ for the following conditions: $T=288$ K, $\sigma_T=7.5\times10^{-6}$ J cm^{-2}, $\rho_c=1$ g cm^{-3}, $\mu_c=18$ g mol^{-1}, N_A Avogadro number.

Solution

(1) $a_c=0.009\,\mu$m

(2) $n_c=\dfrac{32\pi\sigma_T^3\mu_c N_A}{3\rho_c^2\left[R_V^*T\ln(P_V/P_{VS})\right]^3}$

(3) $a_c=0.011\,\mu$m, $n_c=2.32\times10^5$

5.8 (1) Calculate the time taken by a water droplet (density 1000k gm^{-3}) to grow by condensation and diffusion from 1 to $100\,\mu\text{m}$ on Earth's atmosphere given that $(P_V(\infty) - P(a)) = 20$ Pa at $10°\text{C}$. Take $D_V = 1.8 \times 10^{-5}\,\text{m}^2\,\text{s}^{-1}$. (2) What is the temperature difference between the surface of the droplet and a distance point in the atmosphere? Take $K_p = 2.4 \times 10^{-2}\,\text{W m}^{-1}\,\text{K}^{-1}$. (3) Repeat this calculation for an ice crystal of ammonia (870k gm^{-3}) in Jupiter's atmosphere assuming the same pressure vapor difference when $T = 150\,\text{K}$. Take for the hydrogen atmosphere $D_V = 12 \times 10^{-5}\,\text{m}^2\,\text{s}^{-1}$ and $K_p = 16 \times 10^{-2}\,\text{W m}^{-1}\,\text{K}^{-1}$.

Solution

(1) 30.2 min
(2) 0.28 K
(3) 2.2 min, 0.38 K

5.9 Calculate the time taken by a water droplet with a radius of $20\,\mu\text{m}$ to grow when colliding and coalescing with particles of size $1\,\mu\text{m}$ on Earth's atmosphere. Take the cloud density from Table 5.3 ($5.9 \times 10^{-5}\,\text{g cm}^{-3}$) and a fall speed assuming a laminar regime if $\eta_D = 1.7 \times 10^{-5}\,\text{kg m}^{-1}\,\text{s}^{-1}$.

Solution
32.3 s.

5.10 Calculate the terminal fall speed of a particle with a radius $20\,\mu\text{m}$ in the following atmospheres: (1) Water drop (density $1000\,\text{kg m}^{-3}$) on Earth's atmosphere ($293\,\text{K}$); (2) ice ammonia crystal (density $860\,\text{kg m}^{-3}$) on Jupiter's atmosphere (temperature $140\,\text{K}$); (3) ice methane crystal (density $490\,\text{kg m}^{-3}$) on Titan's atmosphere ($85\,\text{K}$); (4) ice methane crystal (density $490\,\text{kg m}^{-3}$) on Uranus's atmosphere ($65\,\text{K}$). Use for the hydrogen atmospheres $\eta_0 = 0.84 \times 10^{-5}\,\text{kg m}^{-1}\,\text{s}^{-1}$ and for nitrogen atmospheres $\eta_{D0} = 1.66 \times 10^{-5}\,\text{kg m}^{-1}\,\text{s}^{-1}$ at $T_0 = 293\,\text{K}$, and assume an empirical temperature dependence $\eta_D = \eta_{D0}(T/T_0)^{0.75}$.

Solution

(1) Earth: $w_L = 5\,\text{cm s}^{-1}$
(2) Jupiter: $w_L = 38\,\text{cm s}^{-1}$
(3) Titan: $w_L = 0.9\,\text{cm s}^{-1}$
(4) Uranus: $w_L = 14\,\text{cm s}^{-1}$

5.11 Calculate the sedimentation time for a water droplet to fall a cloud scale height (H_{cl}) in the Earth's atmosphere for the following cases: (a) stratus cloud under a laminar regime with $a = 10\,\mu\text{m}$, $H_{cl} = 1\,\text{km}$; (b) cumulus cloud under a laminar regime with $a = 1\,\text{mm}$, $H_{cl} = 5\,\text{km}$; (c) cumulus cloud under a turbulent regime with $a = 1\,\text{mm}$, $H_{cl} = 5\,\text{km}$, for a drag coefficient $C_D = 0.2$ and velocity $w = 30\,\text{m}\,\text{s}^{-1}$.

Solution

(a) 21.2 h
(b) 38.2 s
(c) 1014.5 s

5.12 The effective absorption coefficient for the methane bands centered at 619, 725, and 890 nm are 6×10^{-6}, 4×10^{-5}, and 4×10^{-4} (cm-amagat)$^{-1}$ at the temperatures of Saturn's equatorial troposphere. Calculate the pressure level at which the optical depth due to the absorption by methane reaches unity for all three cases. Take $x_{VV}(CH_4) = 4.5 \times 10^{-3}$, $g = 9$ m s^{-2}, $\langle \mu \rangle = 2.12$ g mol^{-1}.

Solution
3.15 bar (619 nm), 470 mbar (725 nm), 47 mbar (890 nm).

5.13 The absorption cross section of the 420 nm band of the pollutant agent NO_2 in Earth's atmosphere is 5.4×10^{-19} cm^2. If its mixing ratio is 0.01 ppmv, what is the optical absorption depth at this wavelength on Earth's surface? Compare it with the Rayleigh optical depth at the Earth's surface due to the entire atmosphere at this wavelength.

Solution
$\tau_a = 0.11$, $\tau_R = 0.3$.

5.14 Find the dependence of the Rayleigh optical depth with wavelength (in microns) and pressure (in bars) on Venus, Earth, Jupiter, and Saturn at their equatorial atmospheres using the parameters in Table 5.4 and g and $\langle \mu \rangle$ from the tables in the book. Then find the atmospheric pressure level in each planet at which the Rayleigh optical depth is unity for a wavelength of 350 nm.

Solution
Venus $\tau_R = 0.0162\lambda^{-4} (1 + 0.013\lambda^{-2})P$, Earth $\tau_R = 0.00857\lambda^{-4} (1 + 0.011\lambda^{-2})P$ Jupiter $\tau_R = 0.0083\lambda^{-4} (1 + 0.014\lambda^{-2})P$, Saturn $\tau_R = 0.0244\lambda^{-4} (1 + 0.016\lambda^{-2})P$. Venus (0.83 bar), Earth (1.6 bar, non sense), Jupiter (1.62 bar), Saturn (0.54 bar).

5.15 The aerosol size distribution in polluted urban air can be represented by the *Junge distribution* $\eta(r) = cr^{-\upsilon}$, where c and υ are adjustable parameters. Find an expression for the extinction coefficient in terms of an integral over the size parameter x and that for a typical value for the distribution $\upsilon = 4$, we have $\beta_e \approx \lambda^{-1}$.

Solution

$$\beta_e = \pi c \left(\frac{\lambda}{2\pi} \right)^{3-\upsilon} \int x^{-\upsilon} Q_e(x) dx \approx \lambda^{3-\upsilon}$$

5.16 Tropospheric aerosols in Jupiter respond to a phase function p_{2HG}, which, for blue wavelength (440 nm), has the following values for the parameters: $g_{s1} = 0.80$, $g_{s2} = -0.75$, $f_s = 0.97$. The single scattering albedo at this wavelength is 0.990. Compare the value of this phase function with that predicted by isotropic and Rayleigh scattering for the scattering angles $\Theta=0°$, $90°$, $180°$. Where they differ significantly so isotropic and Rayleigh scattering must be disregarded as representatives for Jupiter?

Solution

$$\Theta = 0° \quad (p_{2HG} = 46.1,\ p_{istr} = 0.99,\ p_R = 1.5)$$

$$\Theta = 90° \quad (p_{2HG} = 1.66,\ p_{istr} = 0.99,\ p_R = 0.75)$$

$$\Theta = 180° \quad (p_{2HG} = 0.94,\ p_{istr} = 0.99,\ p_R = 1.5)$$

They differ at low scattering angles.

5.17 Calculate the asymmetry factor for isotropic and Rayleigh scattering phase functions, and for the anisotropic phase function $p(\Theta)=1+a\cos\Theta$.

Solution

$g_s=0$ (isotropic and Rayleigh); g_s (anisotropic)$=a/3$.

5.18 Measurements of the dependence of the optical depth with altitude at a wavelength of 550 nm during the descent of the Huygens probe in Titan's hazy atmosphere gave $\tau=\tau_0 - \tau_1 z$ ($\tau_0=4.5$, $\tau_1=0.03\,\mathrm{km}^{-1}$) with z ranging from 150 km altitude to the surface $z=0$ km. Assume that the haze particles fall with the Stokes speed with a constant mass flux $\Phi=5\times 10^{-12}$ kg m^{-2} s^{-1} at each altitude level. If the atmospheric viscosity is given by

$$\eta_D(T) = 1.8\times 10^{-5}\left(\frac{416.2}{T+120}\right)\left(\frac{T}{296.2}\right)^{3/2}\ \mathrm{kg\,m^{-1}\,s^{-1}}$$

find an expression for the particle size a and number density N_p in terms of η_D, Q_e, Φ, ρ_P, and τ_1, ρ_P being the particle's density. Calculate the expected value for a and N_p at the surface where $T=94$ K if the particles are composed of methane ($\rho_P=420$ kg m^{-3}) and $Q_e=1.2$.

Solution

$$a=\left(\frac{27\Phi\eta_D Q_e}{8\rho_P^2 g\tau_1}\right)^{1/3};\quad N_p=\left(\frac{8\rho_P^2 g}{27\pi\Phi\eta_D}\right)^{2/3}\left(\frac{\tau_1}{Q_e}\right)^{5/3};\quad a=1.5\,\mu\mathrm{m},\ N_p=13\,\mathrm{cm}^{-3}$$

5.19 For small particles ($x \ll 1$), the Mie efficiency factors can be approximated by the expressions

$$Q_s = \frac{8}{3}x^4\left|\frac{n^2-1}{n^2+1}\right|^2,\quad Q_a = -4x\,\mathrm{Im}\left(\frac{n^2-1}{n^2+1}\right)^2$$

Consider the upper hydrocarbon haze in Uranus' atmosphere being formed by particles with $a=0.01\,\mu\mathrm{m}$, $n_r=1.5$, and $n_i=0.05$ at a wavelength of 600 nm, and extending vertically 40 km in altitude. What are the values of the efficiency factors? What is the optical depth of the haze at this wavelength if the number density is 25 particles cm^{-3}?

Solution
$Q_a = 10^{-2}$, $Q_s = 2 \times 10^{-5}$, τ_H (600 nm) = 1.9×10^{-6}.

5.20 Find Equation 5.88. Then consider the case of a water cloud on Earth formed by droplets with an effective size $r_{eff} \sim a = 10\,\mu$m and number density $1\,$cm^{-3} extending vertically 5 km. What is the cloud optical depth at 600 nm? What is the cloud mass per unit area?

Solution
$\tau_{cl} = 3.14$; $2\,$kg m^{-2}.

5.21 Measurements of the reflectivity of a bright dense cloud in Saturn's atmosphere, the "Great White Spot," when it was in a position in the disk such that $\mu_\varphi = \mu_{\varphi0} = 0.95$, gave 0.9 at a wavelength of 750 nm (no gas absorption) and 0.68 at 725 nm where an absorption band due to methane has an absorption coefficient $\langle \tilde{k} \rangle = 4 \times 10^{-5}$(cm-amagat)$^{-1}$. Assuming that the cloud reflectivity behaves like a Lambert surface (*) (the so-called "reflecting layer model"): (1) Calculate its Lambert albedo A_L neglecting Rayleigh scattering by the atmosphere. (2) What is the altitude level (pressure) in the atmosphere of the cloud top? (3) What is its expected reflectivity in the deeper 890 nm methane absorption band that has $\langle \tilde{k} \rangle = 4 \times 10^{-4}$ (cm-amagat)$^{-1}$. (4) Justify neglecting Rayleigh scattering.
(*) A *Lambert* surface has a reflected intensity $\mu_{\varphi0}A_L$, where A_L is the surface albedo for normal incidence, which can obviously be a function of wavelength.

Solution

(1) $A_L = 0.95$
(2) $P = 100$ mbar
(3) R (890 nm) = 0.053
(4) $\tau_R = 0.008$ (law)

5.22 Assume that Mars' surface follows a Minnaert reflection law, so measurements at a wavelength of 1 μm give $K = 0.9$ and $(I/F) = 0.27$ when $\mu_\varphi = 0.85$ and $\mu_{\varphi0} = 0.75$. A thin dust layer with optical depth 0.08 and single-scattering albedo 0.95 develops in the area. If the dust is represented by a Henyey–Greenstein phase function with asymmetry parameter $g_s = 0.8$, (1) calculate the reflectivity $(\pi I(0, \mu_\varphi)/\mu_{\varphi0}F_\odot)$ of the surface and dust layer assuming that the surface alone has a reflectivity $(I/F)_0 = \pi I(\tau, \mu_\varphi)/\mu_{\varphi0}F_\odot$ for the following values of the azimuth scattering angle $\Delta\phi = 20°$, 45°, 90°, and 180° using the single scattering approach; (2) quantify if Rayleigh scattering will play a role in the observed reflectivity calculating its optical depth at the observed wavelength.

Solution

(1) $\left(\dfrac{I}{F}\right)_0 = \dfrac{\pi I(0, \mu_\varphi)}{\mu_{\varphi0}F_\odot} = 0.48(20°), 0.34(45°), 0.32(90°), 0.32(180°)$
(2) $\tau_R = 2.6 \times 10^{-4} \ll \tau_{dust}$

5.23 Consider a simplified model of Venus's three layer densest clouds as a unique cloud layer extending from 47 to 67 km altitude above the surface, formed by isotropic scattering particles with a size of $2\,\mu m$ and a number density $50\,cm^{-3}$. (1) Calculate their total optical depth at a wavelength of $0.9\,\mu m$. (2) Calculate their reflectivity $R = (\pi I(0,\mu_\varphi)/\mu_{\varphi 0}F_\odot)$ at the center of the disk when $\mu_\varphi = \mu_{\varphi 0} = 1$ if the first order approach to the H-function is employed for the following single-scattering albedos: 1, 0.999, 0.990. (3) Compare it with the cloud albedo predicted by the simple approximate formula (5.101).

Solution

(1) $\tau_{cl}\,(900\,nm) = 25$

(2) $R = \dfrac{\pi I(0,\mu_\varphi)}{\mu_{\varphi 0}F_\odot} = 0.93(\tilde{\omega}_0 = 1), 0.83(\tilde{\omega}_0 = 0.999), 0.67(\tilde{\omega}_0 = 0.990)$

(3) $A_{cl} = 0.96$

5.24 What is the contribution to the heating rate by aerosols in Saturn's equatorial upper troposphere if the fractional length of the day is 0.5 and the flux deposition fraction 0.16 for aerosols located above 0.43 bar?

Solution
$0.017\,K\,day^{-1}$.

5.25 During a global dust storm in Mars, the dust optical depth varies according to $\tau(z) = \tau_0 \exp(-z/H)$, τ_0 being its value at the surface and H the atmospheric scale height. Assuming the dust to be a good absorber, we neglect multiple scattering effects. Find an expression for the heating rate $Q/\rho C_p$ following Equation 5.104 if the sunlight absorption fraction of the incoming solar radiation $F_\odot$ is f_a, the surface pressure P_0, and the scattering angles represented by $\mu_\varphi, \mu_{\varphi 0}$. Apply this for $\tau_0 = 2, f_a = 0.1$ and $\mu_\varphi = 1, \mu_{\varphi 0} = 1$ for altitudes $z = 1$ and $5\,km$.

Solution

$$\frac{Q}{\rho C_p} = \frac{g}{C_p} f_a F_{\odot p} \frac{\tau_0}{P_0}\left(\frac{1}{\mu_\varphi} + \frac{1}{\mu_{\varphi 0}}\right)\exp\left[-\tau\left(\frac{1}{\mu_\varphi} + \frac{1}{\mu_{\varphi 0}}\right)\right];$$
$$8.4\times10^{-4}\,K\,s^{-1}(z = 1\,km); 1.6\times10^{-4}\,K\,s^{-1}(z = 5\,km)$$

6 Upper and Tenuous Atmospheres

This chapter is dedicated to presenting the fundamentals of the structure of low-density atmospheres that correspond to two situations, the upper part of the atmospheres of planets and satellites so far discussed, and those bodies with tenuous atmospheres. Since its physical behavior is dictated in part by their common low-density nature and low pressure, we treat them in a unique chapter. However, we separate it in two parts to differentiate the physical and chemical properties of the upper part of a massive atmosphere from the tenuous atmosphere case that is related to the presence of the solid surface.

The frontier between both types of situations is not strict, and neither is the definition of upper atmosphere. Planets and satellites with a solid surface and substantial atmospheres are Venus, Earth, and Titan, whereas planets without surface but with dense atmospheres are Jupiter, Saturn, Uranus, and Neptune, and the family of extrasolar giant planets so far discovered. Mars represents a transition planet with an atmospheric density midway between those of Earth and Titan on the one hand, and the low-density cases as Triton and Pluto on the other. I will consider Mars for most situations within the group of dense atmospheres due to the similarity in the dynamical phenomena occurring on it (e.g., a rich meteorology in its troposphere). Bodies with tenuous atmospheres are then Mercury and the Moon, the Galilean satellites, the small satellite of Saturn Enceladus, the dwarf planet Pluto, and Neptune's satellite Triton. In Mars, Io, Triton, and Pluto, the main atmospheric constituent exists both as a vapor in the atmosphere and as a condensate ice on the surface. Other large bodies in the Kuiper Belt region (KBO), the "transneptunians" (TNO), could have atmospheres close in their properties to those of Pluto and Triton. We will not consider in the book the case of the atmospheres of very small bodies as the comets since this represents a very special case of extreme temporal variability due to the orbital characteristics, small body mass and size (typically between 5 and 50 km), and low mean density ($\sim$1 g m^{-3}).

In order to classify the "upper" and "tenuous" character of the atmospheres, we present in Table 6.1 a comparative set of data (the mass per unit area and mean atmospheric density) at particular levels of the atmospheres of these bodies. For the very tenuous atmospheres (Mercury and Moon, and the Galilean satellites Europa, Ganymede, and Callisto), one must go to the upper thermospheres of massive atmospheres to find similar density conditions.

TABLE 6.1
Comparison of Upper and Tenuous Atmospheric Densities

Planet/Satellite	P/g (kg m^{-2}) Tropopause[a]	$\langle \rho \rangle$ (kg m^{-3}) Tropopause[a]	P/g (kg m^{-2}) Stratopause[b]	$\langle \rho \rangle$ (kg m^{-3}) Stratopause[b]
Venus	1124	0.21	562	6.6×10^{-2}
Earth	1020	0.16	10	1.2×10^{-3}
Mars	0.26	3.8×10^{-5}	0.13	1.9×10^{-5}
Titan	560	0.49	12	2×10^{-3}
Jupiter	1000	0.03	20	4.9×10^{-4}
Saturn	1250	0.03	11	3.5×10^{-4}
Uranus	1800	0.06	18	2.3×10^{-4}
Neptune	7400	0.1	74	4.2×10^{-4}

Planet/Satellite	P/g (kg m^{-2}) Surface	$\langle \rho \rangle$ (kg m^{-3}) Surface
Io	5.6×10^{-3}	6.4×10^{-7}
Triton	1.3	9×10^{-5}
Pluto	2.3	1.1×10^{-4}
Others[c]	$\sim 10^{-7}$ to 10^{-11}	$\sim 10^{-11}$ to 10^{-13}

[a] Tropopause altitude taken from Table 4.2.
[b] Stratopause location is not well-defined in Venus, Mars, and Titan (see Chapter 4). Here, for the stratopause, we used the tabulated data in *Encyclopedia of Standard Planetary Information, Formulae and Constants* (NASA PDS Atmospheric Node), at: http://atmos. nmsu.edu/education_and_outreach/encyclopedia/encyclopedia.htm
[c] Mercury, Moon, Europa, Ganymede, Callisto.

6.1 UPPER ATMOSPHERES: INTRODUCTION

In Section 4.1.2, we introduced the vertical structure of planetary atmospheres and its layered classification (Figure 4.2). The upper parts of massive atmospheres (the exosphere) is a rarefied gas at very low density and pressure ($\sim 10^{-6}$ or 10^{-10} relative to the troposphere values) in which the mean-free path of the atoms is greater than the atmospheric scale height, and the condition of local thermodynamic equilibrium does not hold (Sections 4.3 and 5.1). The solar radiation, in particular, the most energetic parts of its spectrum (i.e., at wavelengths shorter than UV-visible), and energetic particles from the solar wind and from the magnetic environment of the planet, if present, control the physical and chemical properties of the upper atmosphere. Photochemical reactions produce new compounds from those present in the atmosphere, introducing composition changes and giving rise to a layer of ionized gases (*the ionosphere*) formed by a plasma of electrons and positive and negative ions. It shows a very complex dynamics due to the additional action of magnetic and electric forces.

In terms of its composition, the atmosphere can be divided in two broad regions. On one hand is the *homosphere*, a layer where the composition is homogeneous, not varying with altitude due to the mixing produced by the different dynamical processes. On the other hand is the *heterosphere*, where the diffusion mechanism separates compounds according to their molecular weight. Both regions are separated by the *homopause* (sometimes also called the *turbopause*). It is defined as the level where the molecular and eddy diffusion processes become equally important. The whole region (homosphere and heterosphere) is extremely sensitive to the temperature changes induced by the day to night and by the annual solar irradiation cycles, and is called the *thermosphere*. In the extreme idealized situation, the temperature of the whole layer changes according to these temporal periods, with small temperature gradient in the vertical, and with the deposited heat transported only by conduction. The atmosphere ends in the *exosphere*, which is the escape region of the compounds to the outer space (see Section 2.4.1 and Table 2.2).

The study of the upper atmospheres where the atomic and molecular dissociation and ionization processes dominate is called *aeronomy*, a term introduced by S. Chapman and widely accepted since 1954. At the low densities of the upper atmospheres, the optical emission of light by discrete atomic and molecular transitions occurs, a phenomenon known as *airglow*, sometimes distinguished as *dayglow* or *nightglow* depending on which hemisphere (day or night) the emission takes place. Similarly, the *aurora* is also an upper atmosphere emission of light from atoms and molecules, but the excitation source being the bombarding electrons and ions from the solar wind and magnetosphere. Aurora phenomena occur mainly in the polar region where the planetary magnetic field lines converge.

Figure 6.1 is an extended version of Figure 4.2 (see also Figure 3.28), representing the vertical structure of a planetary atmosphere and its layered classification in response to a variety of processes.

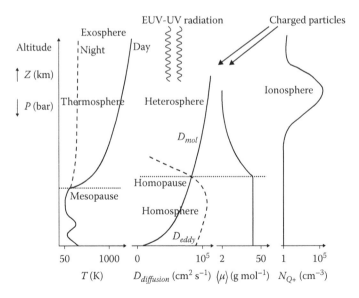

FIGURE 6.1   Scheme of the vertical structure of dense atmospheres in terms of the temperature (see details in Figure 4.2), diffusion mechanisms, and mean molecular weight (see text) and electrical charge.

6.2 PHOTOCHEMISTRY

6.2.1 FUNDAMENTALS

As presented in Chapter 4, the absorption of solar radiation heats the atmosphere but also changes its chemical state (i.e., its composition). Photochemistry refers to a chemical reaction that is induced by the absorption of electromagnetic waves (radiation) and whose products are new compounds or the formation of aggregates (aerosols, see Figure 3.28). The most important energy source is obviously the solar photons that reach all planetary atmospheres, most importantly, their upper part that represents the boundary with the outer space. As already discussed, solar photons are characterized by their frequency ν, wavelength λ ($\nu = c/\lambda$), or energy $E = h\nu$. In Chapter 3, we showed that radio and far-infrared wavelengths ($\lambda \gtrsim 100\,\mu m$) excite electrons in the lowest quantum states of molecules (i.e., the rotational levels). The vibrational levels are excited by infrared photons ($\lambda \sim 2\text{--}20\,\mu m$), while visible and ultraviolet (UV) radiation excite the higher quantum electronic levels in atoms and molecules. As briefly commented in Section 2.4.1, the higher energetic photons may break up molecules in atoms or ions, a process known as *photodissociation* or *photolysis* (this occurs typically for $\lambda \le 1\,\mu m$), or extract electrons from atoms and molecules by *photoionization*, a process that typically occurs for $\lambda \le 1000\,\text{Å}$ for the outer shell electrons and for $\lambda \le 100\,\text{Å}$ (UV- and x-ray radiation) for the inner shell electrons in the atom. Table 6.2 gives the energy thresholds for ionization and dissociation

TABLE 6.2
Dissociation and Ionization Thresholds of Fundamental Molecules[a]

Species	Energy (eV) Dissociation	Wavelength (Å) Ionization	Energy (eV) Ionization
H_2	4.48	804	15.41
N_2	9.76	796	15.58
O_2	5.12	1026	12.1
CO	11.11	885	14
H_2O	5.12 ($\rightarrow H + OH$)	985	12.6
CH_4	4.45 ($\rightarrow CH_3 + H$)	954	13
CO_2	5.46 ($\rightarrow CO + O$)	899	13.79
NH_3	3.9 ($\rightarrow NH_2 + H$)	1221	10.15

Sources: Data from Chamberlain, J.W. and Hunten, D.M., *Theory of Planetary Atmospheres: An Introduction to Their Physics and Chemistry*, Academic Press, New York, 1987; Bauer, S.J. and Lammer, H., *Planetary Aeronomy: Atmosphere Environments in Planetary Systems*, Springer, Berlin, Germany, 2004.

[a] For converting the energy to wavelength use λ (Å) $= 12{,}395/Energy$ (eV), but note that for photodissociation, these are not necessarily the required wavelength thresholds because the cross section may be much smaller than the corresponding wavelength.

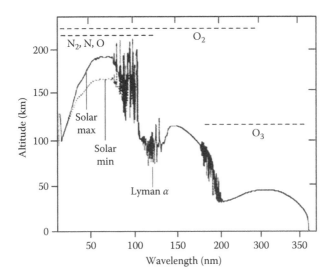

FIGURE 6.2 Penetration level for radiation in the Earth's atmosphere at short wavelengths (up to 300 nm). (Adapted from Meier, R.R., *Space Sci. Rev.*, 58, 1, 1999. With permission.)

of most common molecules in planetary atmospheres, and Figure 6.2 shows as an example the penetration level (barrier level) of high-energy solar photons in Earth's atmosphere. The reverse process to photodissociation is called *recombination*. A typical photochemical reaction can be written as

$$X + h\nu \rightarrow X^* \tag{6.1}$$

where the molecule X absorbs the photon energy $h\nu$ and excites it to the molecular state X^* that can be a higher electronic, rotational, or vibration state, or a dissociated or ionized product. The new molecule can react with other species and produce secondary and other compounds that cannot exist in thermodynamic equilibrium. The photochemical processes are nonthermal processes. The instantaneous rate of production of X^* is given by the equation as follows:

$$\frac{d[X^*]}{dt} = J \cdot [X] \tag{6.2}$$

where

$[X^*]$ denotes the concentration (molecules cm^{-3}) (Section 5.1.1)
J is the *photolysis* or *photodissociation* rate given by

$$J(z) = \int_{\nu_0}^{\infty} k_\nu \pi F_\nu e^{-\tau_\nu(z)/\mu_\varphi} \, d\nu \tag{6.3}$$

where

k_ν is the monochromatic photon absorption cross section
μ_φ is the cosine of the angle between the solar direction and the local vertical

F_ν is the monochromatic photon flux outside the Earth's atmosphere (units photons $cm^{-1} s^{-1} Hz^{-1}$)

$h\nu_0$ is the threshold energy

$J(z)$ in s^{-1} defines the production rate of X* per molecule of X

Its inverse defines a characteristic *lifetime* for species X:

$$\tau_{Ch} = \frac{1}{J} = \frac{[X]}{|d[X]/dt|} \qquad (6.4)$$

In the absence of other processes, [X] would decrease with time exponentially $\sim\exp(-Jt)$.

Consider now a two-body or *bimolecular* reaction:

$$X + Y \rightarrow Z \qquad (6.5)$$

the conversion rate of species Z is expressed as

$$\frac{d[Z]}{dt} = k_{XY}[X][Y] \qquad (6.6)$$

where the *bimolecular reaction rate* k_{XY} ($cm^3\ s^{-1}$) is usually expressed as

$$k_{XY} = c_1 \left(\frac{T}{300}\right)^{c_2} e^{-E_0/kT} \qquad (6.7)$$

where

E_0 is the *activation energy* needed to overcome the potential barrier for the reaction to occur

c_1 and c_2 are constants

For planetary atmospheres, a crude upper limit to this coefficient can be obtained from the collision theory that assumes an elastic collision rate between molecules in the gas kinetic regime:

$$k_{g.k.} = \sigma_x \langle v \rangle \approx 2 \times 10^{-10} \left(\frac{T}{300}\right)^{1/2} \qquad (6.8)$$

where

σ_x is the collisional cross section

$\langle v \rangle = \sqrt{2kT/m}$ is the thermal velocity of the particle

Similarly, for a three-body reaction we have

$$X + Y + M \rightarrow Z + M \qquad (6.9)$$

in which M is a species that helps to balance the kinetic energy and momentum in the reaction. The rate for the production of Z is given by

$$\frac{d[Z]}{dt} = k_{XYM}[X][Y][M] \tag{6.10}$$

the *three-molecular reaction rate* k_{XYM} ($cm^6 \ s^{-1}$) can be written as

$$k_{XYM} = 10^{-12} k_{XY}^2 \approx 4 \times 10^{-32} \left(\frac{T}{300} \right)^{1/2} \tag{6.11}$$

In the following, we apply this formulation and definitions to planetary atmospheres according to their main composition.

6.2.2 APPLICATIONS TO TERRESTRIAL AND GIANT PLANETS, TITAN, TRITON, AND PLUTO

In Figures 6.3 and 6.4, we summarize the known measured and modeled vertical composition of the terrestrial planets and Titan, and the giant planets resulting from the various processes occurring in the atmosphere. The main photochemical reactions that took place in the upper atmosphere are presented in the following sections.

6.2.2.1 Earth

In the case of our planet, we concentrate on the most interesting photochemical product, the stratospheric ozone. The formation of the ozone stratospheric layer for a pure oxygen atmosphere on Earth is described by the *Chapman reactions*. The photolysis of molecular oxygen takes place due to UV photons with $\lambda \leq 240$ nm (Figure 6.2)

$$O_2 + h\nu \rightarrow O + O \quad (J_2) \tag{6.12a}$$

followed by a recombination due to a three-body process:

$$O + O_2 + M \rightarrow O_3 + M \quad (k_{12}) \tag{6.12b}$$

and photolysis of ozone by photons of wavelength less than 1100 nm:

$$O_3 + h\nu \rightarrow O + O_2 \quad (J_3) \tag{6.12c}$$

as well as by collisions with oxygen atoms:

$$O + O_3 \rightarrow O_2 + O_2 \quad (k_{13}) \tag{6.12d}$$

These reactions take place simultaneously and when an equilibrium state is reached, the number of ozone molecules formed per unit volume and time equals those

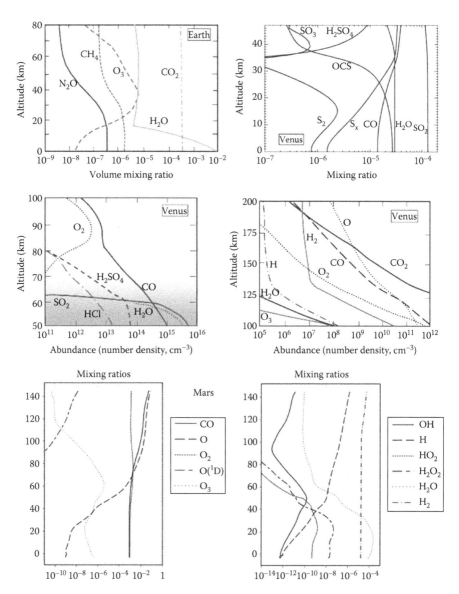

FIGURE 6.3 The vertical distribution of the main compounds in the atmospheres of the terrestrial planets according to current models. The following sources have been used. Earth. (Adapted from Goody, R.M. and Yung, Y.L., *Atmospheric Radiation: Theoretical Basis*, Oxford University Press, New York, 1995. With permission.) Venus. (Adapted from Krasnopolsky, V.A. and Parshev, V.A., in *Venus II*, 1997, p. 431; Krasnopolsky, V.A., *Icarus*, 191, 25, 2007. With permission.) Mars. (From Moudden, Y. and Mc Connell, J.C., *Icarus*, 188, 18, 2007. With permission.)

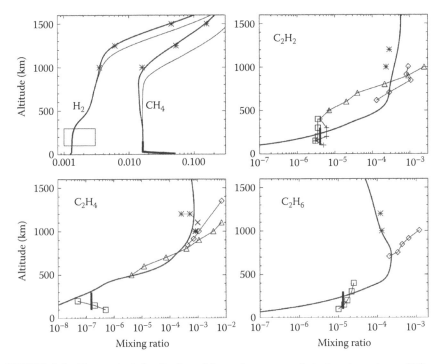

FIGURE 6.4 The vertical distribution of the main compounds in the atmosphere of Titan. (From Krasnopolsky, V.A., *Icarus*, 201, 226, 2009. With permission.)

destroyed. Using the above definitions, the net change in the oxygen and ozone number densities are given by

$$\frac{d[O]}{dt} = 2J_2(z)[O_2] + J_3(z)[O_3] - k_{13}[O][O_3] - k_{12}[O][O_2][M] \quad (6.13a)$$

$$\frac{d[O_3]}{dt} = k_{12}[O][O_2][M] - k_{13}[O][O_3] - J_3(z)[O_3] \quad (6.13b)$$

In chemical equilibrium, the net changes $d[O]/dt$ and $d[O_3]/dt$ are equal to zero. This gives the following vertical profiles for oxygen and ozone:

$$[O] = \frac{J_2(z)[O_2]}{k_{13}[O_3]} \quad (6.14a)$$

$$[O_3] = \frac{k_{12}[O][O_2][M]}{k_{13}[O] + J_3(z)} \quad (6.14b)$$

The observed vertical profiles follow these laws with O increasing with altitude, O_2 decreasing with altitude, and O_3 peaking at about 30 km altitude.

In the real Earth atmosphere, ozone is also destroyed by *catalytic cycles* that involve typically a pair reaction:

$$X + O_3 \rightarrow XO + O_2 \tag{6.15}$$

$$XO + O \rightarrow X + O_2 \tag{6.16}$$

with the net effect of destroying ozone:

$$O + O_3 \rightarrow 2O_2 \tag{6.17}$$

The molecule X is a catalyst compound that is not consumed in the reactions. Examples are atomic hydrogen and OH, nitrogen oxides (NO and NO_2), and chlorine (Cl and ClO).

One of the major problems with Earth's ozone is its destruction in the polar areas, most importantly on the Antarctic (*the ozone hole*), as a consequence of dynamical and chemical processes (discussed previously in Section 2.5.7). The ozone is measured by its *column abundance* defined by the total number of molecules of ozone in a vertical column of atmosphere of unit cross section:

$$N(O_3) = \int_0^\infty [O_3]\,dz \tag{6.18}$$

The column ozone is measured in *Dobson units* (DU), defined as the height of the column (in hundredths of millimeter) that will be reached if all the ozone molecules in it were brought to a pressure of 1 atm and a temperature of 0°C. Then, 100 DU means a height column of 1 mm. One DU is equivalent to 2.66867×10^{20} ozone molecules per square meter or 0.447 mmol of ozone per square meter. The ozone layer has typically 300 DU above 25 km altitude. Another way to measure it is simply by its partial pressure.

The Antarctic ozone hole occurs in the winter night when an intense circumpolar vortex rotating eastward forms in the stratosphere above the Antarctic continent. Air within the vortex gets cold by the absence of solar heating, and heterogeneous reactions of chlorine species on the surface of particles (most chlorofluorocarbons [CFC] of anthropogenic origin transported from industrial regions) destroys O_3. Simplifying, free Cl atoms are produced by photolysis from CFCs according to the reactions

$$CFCl_3 + h\nu \rightarrow CFCl_2 + Cl \quad (\lambda < 226 \text{ nm})$$
$$CF_2Cl_3 + h\nu \rightarrow CF_2Cl + Cl \quad (\lambda < 214 \text{ nm}) \tag{6.19a}$$

then reacting with O_3:

$$Cl + O_3 \rightarrow ClO + O_2$$
$$ClO + O \rightarrow Cl + O_2 \tag{6.19b}$$

The net result is to produce the ozone destruction. One atom of chlorine destroys many atoms of ozone. For historical reasons, a reference number of 220 DU is usually taken as the starting point to refer to the ozone hole initiation. Figure 6.3 shows the averaged vertical composition of minor species in Earth's atmosphere.

6.2.2.2 Venus

The most abundant compound of Venus and Mars atmospheres is CO_2. It is photodissociated for $\lambda < 1700$ Å (see Table 6.2) according to the reaction as follows:

$$CO_2 + hv \rightarrow CO + O \tag{6.20}$$

which is followed by oxygen formation by a two-body reaction:

$$O + O \rightarrow O_2 \tag{6.21a}$$

Thus, O, CO, and O_2 in these two planets are of photolytic origin. This last reaction is slow but can be accelerated by the intervention of a catalytic compound in a three-body collision:

$$O + O + M \rightarrow O_2 + M \tag{6.21b}$$

Recombination of CO_2 is also favored by a three-body process:

$$CO + O + M \rightarrow CO_2 + M \tag{6.22}$$

On Venus, chlorine can act as the catalytic agent for this reaction. It also intervenes in the cloud top and atmospheric chemistry upon its photodissociation:

$$HCl + hv \rightarrow H + Cl \tag{6.23}$$

The other important photodissociation reaction that takes place at cloud level is

$$SO_2 + hv \rightarrow SO + O \tag{6.24a}$$

and then the oxygen readily reacts to form

$$SO_2 + O \rightarrow SO_3 \tag{6.24b}$$

which is the source for the sulfuric acid vapor that condenses to form the main component of Venus clouds as shown in Chapter 5:

$$SO_3 + H_2O \rightarrow H_2SO_4 \tag{6.24c}$$

Figure 6.3 shows the vertical distribution of species at different levels in the atmosphere of Venus.

6.2.2.3 Mars

The same CO_2 photolysis scheme (Equations 6.20 through 6.23) occurs on Mars. On this planet, photodissociation of water vapor that occurs for $\lambda < 1900$ Å is another source of oxygen:

$$H_2O + h\nu \rightarrow OH + H \tag{6.25a}$$

The hydroxyl radical can react with H atoms to recombine and form water via the three-body reaction:

$$OH + H + M \rightarrow H_2O + M \tag{6.25b}$$

The oxygen atoms formed as in Equation 6.20 can react with molecular oxygen to form ozone (which is preserved over the winter poles) or with OH to form hydroperoxyl radical HO_2. In addition OH oxidize CO by

$$OH + CO \rightarrow H + CO_2 \tag{6.25c}$$

It is still an open question of how the photochemical equilibrium of H_2O and CO_2 is maintained, in Mars since their sources are at different levels and dynamical transport intervenes. Figure 6.3 shows the vertical distribution of species in the atmosphere of Mars.

6.2.2.4 Titan

The most active photochemical species on Titan is methane. The photochemical hydrocarbon products are the same as those indicated for the giant planets (Equations 6.27), but include also higher order alkanes as C_3H_8 and C_4H_{10}.

As on Earth, nitrogen, the most abundant gas in the atmosphere, is highly inert and does not react directly with the hydrocarbons. However, nitrogen atoms and ion N^+ produced during the dissociation of N_2 by energetic magnetospheric electrons, solar photons with $\lambda < 1000$ Å, and electrons produced by cosmic rays react with methane to produce nitriles as HCN (hydrogen cyanide), HC_3N (cyanoacetylene), C_2N_2 (cyanogen), CH_3C_2H (methylacetylene), and high-order polymers.

The presence of oxygen in form of CO, CO_2, and H_2O in a reducing ambient, as is Titan's atmosphere, leads to an interesting photochemistry with intermediate photolytic compounds as H_2CO and HCO. Figure 6.4 shows the vertical distribution of the main species in the atmosphere of Titan.

6.2.2.5 Giant Planets

Photochemistry on the giant planets involves a variety of molecules containing H (H_2, CH_4, NH_3, SH_2, and PH_3). Most reactions occur above the ~50–100 mbar level due to limitations by Rayleigh scattering on UV photons whose unit optical depth at 1500 Å is reached at this level (Equation 5.74).

1. Molecular hydrogen suffers direct photodissociation for $\lambda < 1000–1100$ Å in the upper exospheric layer:

$$H_2 + h\nu \rightarrow H + H \qquad (6.26a)$$

followed by a three-body recombination reaction:

$$H + H + M \rightarrow H_2 + M \qquad (6.26b)$$

2. For methane, CH_4 photolysis occurs for $\lambda < 1600$ Å to produce methylene CH_2 and methyl radical CH_3:

$$CH_4 + h\nu \rightarrow CH_2 + H_2 \qquad (6.27a)$$

$$CH_4 + CH_2 \rightarrow 2CH_3 \qquad (6.27b)$$

Further reactions of these products, including CH leads to the formation of ethylene C_2H_4:

$$CH + CH_4 \rightarrow C_2H_4 + H \qquad (6.27c)$$

and subsequently diacetylene (C_4H_2), and ethane C_2H_6, acetylene C_2H_2, etc. All this gives rise to a rich hydrocarbon chemistry with catalytic reactions of acetylene, producing polyacetylenes $C_{2n}H_2$ and small quantities of other more complex organics as propane C_3H_8, butane C_4H_{10}, etc., and whose mixing ratios rapidly decrease with increasing depth (scheme in Figure 3.28).

3. Ammonia NH_3 suffers photolysis for $\lambda < 2300$ Å:

$$NH_3 + h\nu \rightarrow NH_2 + H \qquad (6.28a)$$

which can undergo a three-body reaction to reconstitute ammonia or form hydrazine N_2H_4:

$$NH_2 + NH_2 + M \rightarrow N_2H_4 + M \qquad (6.28b)$$

a compound that can condense in the upper atmosphere or suffer photolysis that converts it to nitrogen. The efficient destruction of ammonia in the upper atmospheres of the giant planets and its condensation in the tropospheres makes it only abundant deep in the giant planet atmospheres where its absorption can be only inferred from radio observations.

4. Phosphine PH_3 suffers photolysis by photons with wavelengths in the range 1600 Å $< \lambda < 2300$ Å:

$$PH_3 + h\nu \rightarrow PH_2 + H \qquad (6.29)$$

with PH_2 suffering a reaction series very similar to that of ammonia. It forms diphosphine P_2H_4 and, through photolysis, triclinic P_4, also known as red phosphorous, a compound that condenses to form solid red aerosols. They have been invoked as a possible chromophore agent to tint the reddish cloud features observed particularly in Jupiter, as the Great Red Spot. Phosphine is oxidized by water reaction at $300 < T < 800\,K$, producing P_4O_6. Therefore, phosphine detection occurs in an altitude range that is limited at high altitudes by photolysis and at lower levels by its oxidization.

5. Hydrogen sulfide H_2S suffers photolysis by photons with $\lambda < 3170$ Å with reaction as follows:

$$SH_2 + h\nu \rightarrow HS + H \tag{6.30}$$

with the hydrosulfuryl radicals forming sulfur S and S_2, and subsequently S_8, hydrogen polysulfides H_2S_x, and ammonium polysulfides $((NH_4)_x)S_y$, compounds with yellow, brown, and orange colorations that have been proposed as chromophore agents for Jupiter and Saturn. As shown in Section 5.2.3, deeper in the atmosphere of the giant planets, ammonia reacts with H_2S to produce ammonium hydrosulfide $(NH_4)SH$ that condenses and forms clouds.

Figure 6.5 shows the vertical distribution of species from the upper tropospheres upward in the giant planets.

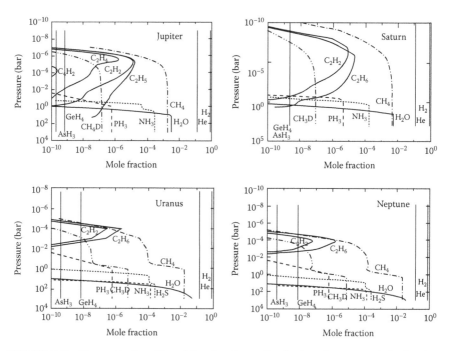

FIGURE 6.5 The vertical profiles of the abundance distribution of compounds in the giant planets from models and measurements. (Reproduced from Irwin, P.G.J., *Giant Planets of Our Solar System: An Introduction*, Springer Verlag-Praxis, Chichester, U.K., 2009. With permission.)

TABLE 6.3
Main Species and Their Principal Photochemical Products

Species	Principal Products	Planet/Satellite Dominant
H_2 ($\lambda < 1000$ Å)	H	J, S, U, N
CO_2 ($\lambda < 1700$ Å)	CO, O, O_2	V, M
O_2 ($\lambda < 2400$ Å)	O_3	E
H_2O ($\lambda < 1900$ Å)	OH, H	M
SO_2 ($\lambda < 2320$ Å)	SO, SO_3, S_x, H_2SO_4 ($+H_2O$)	V, Io
CH_4 ($\lambda < 1600$ Å)	C_2H_2, C_2H_6	J, S, Ti, U, N
NH_3 ($\lambda < 2300$ Å)	N_2, N_2H_4	J, S
PH_3 ($\lambda < 2300$ Å)	P_4	J, S
SH_2 ($\lambda < 2700$ Å)	S_x	J, S

6.2.2.6 Triton and Pluto

In the tenuous and frigid bulk nitrogen atmosphere of these two similar bodies where methane is a minor species (see Section 6.8), we expect the photolytic reactions to produce hydrocarbons and nitriles similar to those produced on Titan. A summary of the photochemical products resulting from the main species in planetary atmospheres is presented in Table 6.3.

6.3 PHOTOIONIZATION: IONOSPHERES

In Table 6.2, we gave the ionization thresholds for the most common molecular species in the atmospheres. It can be seen that ionization occurs at solar radiation UV-, EUV-, and x-ray wavelengths ($\lambda < 1300$ Å). Additionally, ionization is also produced by the energetic particles penetrating and impinging at different levels of the atmospheres coming from the galactic cosmic rays, solar particles (protons plus some alpha particles), and from particles trapped in the magnetospheres of the planets (e.g., electrons associated with auroras).

The recombination of atomic ions is a very slow process compared to molecular recombination. Therefore, a population of free electrons and ions results where ionization takes place, forming a charged layer called the *ionosphere* (Figure 6.1). The electron and ion density distribution with altitude is determined by a balance between the ionization and the recombination rates of ions and electrons, and by their reaction with other species that leads to other ions. Ionospheres are highly variable since ionization is a process that depends on the solar radiation flux that is subjected to the day–night and seasonal cycles, showing variability with latitude. Moreover, since at short wavelengths the solar flux varies significantly, the ionosphere structure follows the 11 year solar cycle as well as the 27 day rotation period of the Sun, and a random variability due to the intrinsic solar changes related to the presence of magnetic disturbances as the "solar flares." Here we consider the idealized but instructive ionization model originating solely by solar photons, which is sufficient for the introductory purposes of this book.

The properties of planetary ionospheres are performed during occultation experiments (stellar and radio) by spacecrafts and in situ, during close spacecraft encounters with the planets, or during the atmospheric descent of planetary probes.

6.3.1 IONIZATION BY PHOTONS

When the energy of the photons is above the ionization potential threshold of a neutral compound X, an ion–electron pair is produced (recombination in the other direction), so according to the "law of mass action," we have

$$X + h\nu \rightleftarrows X^+ + e^- \tag{6.31}$$

In equilibrium, the electron density concentration N_e (cm^{-3}) obeys a continuity equation:

$$\frac{dN_e}{dt} = P_e - L_e = P_e - \alpha_r (N_e)^p \tag{6.32}$$

where P_e is the production rate (cm^3 s^{-1}), and the loss rate by recombination L_e has been set equal to the p-power of the electron density times a recombination coefficient α_r (cm^{-3} s^{-1}). This relationship is valid when there is only one type of ionized molecules and we set $p = 2$ for the dissociative and radiative recombination reaction (6.31). To derive the electron–ion production rates, we assume the following simplifications: (a) the ionization radiation is monochromatic (constant absorption cross sections), (b) the atmosphere is isothermal (scale-height H constant), (c) a plane-parallel atmosphere, and (d) the atmosphere is formed by only one species. Then we can write

$$P(z) = \sigma_i N_i(z) \pi F(z) \tag{6.33}$$

where

    σ_i is the ionization cross section
    N_i is the density of the ionizable constituent and πF is the solar flux penetration with altitude

For radiation incident with an angle i_z with respect to the vertical (the solar zenith angle), the altitude dependence of the solar flux can be written as

$$\mu_\varphi \pi \frac{dF}{dz} = N_i \sigma_a F_\odot \tag{6.34}$$

where

    μ_φ is the cosine of the zenith angle
    σ_a is the absorption cross section (usually above σ_i)
    $\pi F_\odot$ is the solar flux outside the atmosphere

Integration of (6.34) gives

$$F(z) = F_\odot \exp\left[-\left(\frac{\sigma_a}{\mu_\varphi}\right)\int_z^\infty N_i(z)\,dz\right]$$
(6.35)

Since the ionizing photon flux decreases with depth due to absorption in the atmosphere, above the altitude reference level z_0, we can express

$$N_i(z) = N_i(z_0)\exp\left[-\frac{z-z_0}{H}\right]$$
(6.36)

and then the ion production rate is given by

$$P(z,\mu_\varphi) = \pi F \sigma_i N_i(z_0)\exp\left[-\frac{z-z_0}{H} - \frac{\sigma_a N_i(z_0)H}{\mu_\varphi}e^{-(z-z_0)/H}\right]$$
(6.37)

where H is the scale height of the absorber. The ionizing photon flux here is normalized relative to the required ionization energy E_i integrated over a given wavelength bandwidth, $F = \int (F_\lambda/E_i)\,d\lambda$ (unit cm^{-2} s^{-1}). The optical depth (3.33) can be expressed as

$$\tau(z) = N_i(z_0)H\sigma_a e^{-(z-z_0)/H}$$
(6.38)

The maximum production rate occurs for $\tau/\mu_\varphi = 1$, and after differentiation of (6.37), we get

$$P_{max}(z_{max},\mu_\varphi) = \pi F N_i(z_0)\sigma_i \exp\left[-\frac{z_{max}-z_0}{H}-1\right] = \mu_\varphi \frac{\sigma_i}{\sigma_a}\frac{\pi F}{eH}$$
(6.39)

where σ_i/σ_a is called the *ionization efficiency*. Accordingly, (6.37) can finally be rewritten as

$$P(z,\mu_\varphi) = P_{max}\exp\left[1 - \frac{z-z_M}{H} - \frac{1}{\mu_\varphi}e^{-(z-z_M)/H}\right]$$
(6.40)

where z_M is the value of z_{max} for $\mu_\varphi = 1$. The above expression is known as the *Chapman production function*. The ionization efficiency is 1 for atomic species, but as indicated above is <1 for molecular species. Introducing this expression into (6.32) and neglecting mass motion (implicit in the term d/dt), we find that the electron density under equilibrium situation is given by

$$N_e(z,\mu_\varphi) = \sqrt{\frac{P_{max}}{\alpha_r}}\exp\left\{\frac{1}{2}\left[1 - \frac{z-z_M}{H} - \frac{1}{\mu_\varphi}e^{-(z-z_M)/H}\right]\right\}$$
(6.41)

This defines the *Chapman layer*. See Problem 6.2 for an approximate *parabolic expression* for (6.11) and for a numerical estimation for the electron density.

6.3.2 PLANETARY IONOSPHERES

An ionosphere is present in all planets with substantial atmospheres. The main source for ionization is the major atmospheric component for each planet: CO_2 for Venus and Mars, O_2 and N_2 for the Earth, H_2 for the giant planets, and N_2 for Titan. Peak electron densities range from ~10^5 cm^{-3} for the terrestrial planets to ~10^4 cm^{-3} for the icy distant planets. We briefly describe in what follows, the basic properties of these ionospheres.

6.3.2.1 Earth

The ionosphere of our planet is complex and shows a high variability related to the diurnal and seasonal cycles and to solar activity. It is divided in three regions (Figure 6.6), from bottom to top called D ($z \sim 60$–90 km, $N_e = 10^2$–10^4 cm^{-3}); E ($z \sim 105$–160 km, $N_e = 10^5$ cm^{-3}); and F that is broad and extends from this level up to more than 1200 km, itself divided into two layers F_1 ($z \sim 160$–180 km, $N_e = 10^5$–10^6 cm^{-3}), and F_2 (peak at $z \sim 300$ km, $N_e = 10^5$–10^6 cm^{-3}).

The D-layer has its origin on x-ray ionization of O_2 and N_2, and Lyman α ionization of NO. Since there is a substantial atmospheric density at this level ($z = 75$–90 km), a variety of ionization and photoionization reactions occur, resulting in a predominance of negative ions O_2^- formed by the *three-body attachment* reaction:

$$O_2 + e^- + O_2 \rightarrow O_2^- + O_2 \qquad (6.42a)$$

This layer is absent in nighttime.

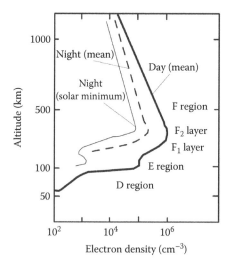

FIGURE 6.6 Earth ionosphere: vertical profile of the number density of electrons (from theory and observations) and its diurnal and solar cycle variability.

The E-layer and F_1 regions are of the Chapman type. The E-region results from direct photoionization of O_2:

$$O_2 + h\nu \rightarrow O_2^+ + e^- \quad \lambda < 1027 \text{ Å} \tag{6.42b}$$

X-rays coming from the solar corona photoionize O, O_2, and N_2 and produce ions by *change–exchange* reactions:

$$O_2 + N_2^+ \rightarrow N_2 + O_2^+$$
$$O + N_2^+ \rightarrow N + NO^+ \tag{6.42c}$$

These ions recombine with electrons to give O and N.

The F_1 layer forms from photoionization of atomic O and molecular N_2:

$$O + h\nu \rightarrow O^+ + e^- \qquad \lambda < 911 \text{ Å}$$
$$N_2 + h\nu \rightarrow N_2^+ + e^- \qquad \lambda < 796 \text{ Å} \tag{6.42d}$$

The F_2 layer has oxygen atom photoionization as the main production mechanism of oxygen ions. The upper part of the terrestrial ionosphere has H^+ as the predominant constituent and is sometimes called *protonosphere*.

6.3.2.2 Venus and Mars

The ionospheres of Venus and Mars are of the Chapman E- and F_1-type. They have peak electron densities, due to the UV photoionization of CO_2, with values $\sim 10^5$ cm^{-3} (Mars) and $\sim 3-5 \times 10^5$ cm^{-3} (Venus) at an altitude $\sim 110-150$ km above the surface (Figure 6.7). CO_2 ionization produces electrons:

$$CO_2 + h\nu \rightarrow CO_2^+ + e^- \quad \lambda < 900 \text{ Å} \tag{6.43a}$$

whose reaction with oxygen produces the principal ion constituent O_2^+:

$$O + CO_2^+ \rightarrow O_2^+ + CO$$
$$O + CO_2^+ \rightarrow O^+ + CO_2 \tag{6.43b}$$
$$O^+ + CO_2 \rightarrow O_2^+ + CO$$

The dissociative recombination of these ions with electrons produces CO and O.

The proximity of Venus to the Sun makes its ionosphere strongly influenced by the interaction with the solar wind that compresses, accelerates, heats, and removes the ionospheric plasma. Since Venus lacks an intrinsic internal planetary magnetic field, an *ionopause* barrier to solar wind particles forms at altitudes of 290 km at the subsolar point and at 1000 km at the terminator, expanding further on the nightside. A night ionosphere is replenished from the transport of O^+ ions from day to night side by fast high-altitude horizontal winds, and by chemical reactions producing O_2^+.

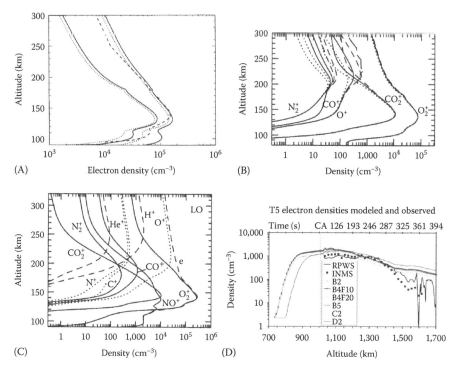

FIGURE 6.7 Terrestrial planet ionospheres: vertical profiles of the electron density in Mars (A) and ions in Venus (B) and (C). (From Fox, J.L., *Adv. Space Res.*, 33, 132, 2004. With permission.) Vertical profile of electron density in Titan (D). (From Cravens, T.E. et al., *Icarus*, 199, 174, 2009. With permission.)

6.3.2.3 Giant Planets

The giant planets have Chapman-like ionospheres formed by the ionization of hydrogen by solar XUV (X and ultraviolet) radiation and, since all of them have powerful magnetospheres, also by the impact of trapped particles. The direct photoionization yields hydrogen ions:

$$H_2 + h\nu \rightarrow H_2^+ + e^- \quad \lambda < 804 \text{ Å}$$

$$H_2 + h\nu \rightarrow H + H^+ + e^- \qquad\qquad (6.44a)$$

$$H + h\nu \rightarrow H^+ + e^-$$

The ion H_2^+ reacts rapidly to form the ion H_3^+:

$$H_2^+ + H_2 \rightarrow H_3^+ + H \qquad\qquad (6.44b)$$

Other ions that intervene in these ionospheres are He^+ and CH_3^+, produced by photolysis of helium and methane. Figure 6.8 shows the structure of the ionospheres

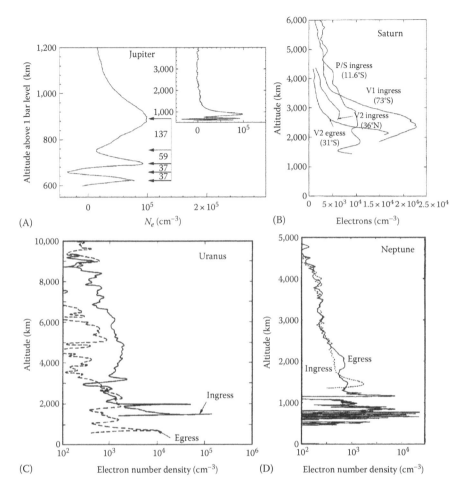

FIGURE 6.8 Giant planet ionospheres: vertical profiles of the electron density in Jupiter, Saturn, Uranus, and Neptune. Jupiter from Galileo measurements showing oscillations (probably associated to the propagation of gravity waves) with the distances between peaks maximum and minimum indicated (in km). (From Matcheva, K.I. et al., *Icarus*, 152, 347, 2001 and references therein. With permission.) Saturn, Uranus, and Neptune from Voyager radio-occultation experiments. (From Majeed, T. et al., *Adv. Space Res.*, 33, 197, 2004 and references therein. With permission.)

of the giant planets. They extend from altitudes ranging from $z \sim 500$ to $10,000\,km$ above the 1 bar level, with electron density reaching peaks of $\sim 10^4–10^5\,cm^{-3}$ but with considerable vertical structure.

6.3.2.4 Titan

Solar X- and EUV radiation and energetic plasma particles from Saturn's magnetosphere ionize the neutral molecules of Titan's massive nitrogen and hydrocarbon atmosphere, creating a main ionosphere at altitudes between about 600 and

2000 km. The primary ions CH_4^+ and N_2^+ produced by electron impact have rather low densities. The main ions are due to hydrocarbons (CH_5^+, H_2CN^+, $HCNH^+$, $C_2H_5^+$, $C_3H_5^+$, $C_4H_3^+$, and others). The total ion density (including all important ion species) has peak values $\sim 10^3$ cm^{-3} between altitudes $z \sim 1000$ and 1400 km, and the peak electron density is $\sim 5 \times 10^3$ cm^{-3} at $z \sim 1000$ km (Figure 6.7).

6.3.3 ELECTRICAL PROPERTIES

The ionosphere is a plasma flow formed by electrons and ions (mixed with neutral species) sensitive to the presence of electric and magnetic fields. Ions and electrons move in ionic and electronic currents in response to the forces originated by these fields. Added to these motions is the wind that acts on the neutral atmosphere leading to collisions between the charged and neutral particles. The collision frequency is an important parameter for the conductivity calculations in the ionosphere. Also relevant is the *gyrofrequency* associated to the gyre of electrons and ions around the magnetic field lines. Within the plasma flow, electric fields can be created by charge separation (e.g., electrons and ions) due to diffusion. In addition, magnetic fields of external or intrinsic origin can also be present. If we neglect the collisions, the plasma velocity can be calculated from the forces due to the electric $\vec{E}$ and magnetic $\vec{B}$ fields as

$$\vec{v} = \frac{\vec{E} \times \vec{B}}{B^2} \tag{6.45}$$

The effect of the wind on the neutral particles can be introduced as a parameterized drag force. As a result, the plasma is carried to some degree with the wind. The region where the wind is most effective in producing an electric current is called the *dynamo region*. The relative drift motion of electrons and ions induces an electric current $\vec{J}$ (unit A m^{-2}) that according to the generalized *Ohm's law* (Equation 1.140) can be written as

$$\vec{J} = \sigma_c \left(\vec{E} + \vec{v} \times \vec{B} \right) \tag{6.46}$$

where σ_c is the electrical conductivity (unit S m^{-1}, where S (siemens) = A V^{-1}). In the general case where collisions are present, the magnetic field makes the medium anisotropic to the applied electric field, and the electrical conductivity is a tensor magnitude (a 3×3 matrix). In the reference frame moving with the neutral gas and taking the magnetic field aligned along the z-axis, the current vector is given by

$$\vec{J} = \begin{pmatrix} \sigma_1 & \sigma_2 & 0 \\ -\sigma_2 & \sigma_1 & 0 \\ 0 & 0 & \sigma_0 \end{pmatrix} \begin{pmatrix} E_x \\ E_y \\ E_z \end{pmatrix} \tag{6.47}$$

The conductivity elements of this tensor are the following:

1. *Direct or longitudinal* conductivity

$$\sigma_0 = \left(\frac{N_e}{m_e \nu_e} + \frac{N_i}{m_i \nu_i} \right) q_e^2 \qquad (6.48a)$$

2. *Pedersen* conductivity

$$\sigma_1 = \left(\frac{N_e}{m_e} \frac{\nu_e}{(\nu_e^2 + \omega_e^2)} + \frac{N_i}{m_i} \frac{\nu_i}{(\nu_i^2 + \omega_i^2)} \right) q_e^2 \qquad (6.48b)$$

3. *Hall* conductivity

$$\sigma_2 = \left(\frac{N_e}{m_e} \frac{\omega_e}{(\nu_e^2 + \omega_e^2)} - \frac{N_i}{m_i} \frac{\omega_i}{(\nu_i^2 + \omega_i^2)} \right) q_e^2 \qquad (6.48c)$$

where
  subindexes e and i stand for electron and ion, respectively
  N_e and N_i are the particles number density
  m_e and m_i are their masses
  ν_e and ν_i are the collision frequencies
  ω_e and ω_i are the cyclotron frequencies ($\omega_e = q_e B/m_e$, $\omega_i = q_e B/m_i$)

For example, at 100 km above Earth's surface, the conductivity elements range from $1\,\text{S m}^{-1}$ (σ_0) to $10^{-3}\,\text{S m}^{-1}$ (σ_1 and σ_2).

Although in terms of the atmospheric abundance, the particle number in the ionosphere is low, the plasma has a profound effect on the propagation of electromagnetic waves. They suffer refraction or reflection with attenuation due to the complex nature of the index of refraction of the plasma, which is a function of the electric conductivity. Below a *critical frequency* $\nu_{critical}(\text{MHz}) \approx 0.009\sqrt{N_e}$ (for N_e given in electrons per cm^{-3}), the radio waves suffer refraction. As it is well known, this is a fundamental aspect for Earth-based long-distance radio communications and for spacecraft radio-occultation experiments.

6.4 UPPER ATMOSPHERES HEATING BALANCE

The temperature structure of the low-density upper atmospheres results from a balance between the heating sources and heat loss processes, and the heat transport mechanisms. Radiative transfer (Section 4.2) is important in the lower and densest part of the upper atmospheres. However, heat conduction becomes important in the thermosphere, which is the subject of this chapter.

Neglecting the effect of the mass motion of the neutral gas and for a plane-parallel layer heated above the altitude level z, the heat diffusion equation (including sources) can be written as (see Equation 1.126)

$$\rho C_p \frac{\partial T(z,t)}{\partial t} - \frac{\partial}{\partial z} \left(K_T(z) \frac{\partial T(z,t)}{\partial z} \right) = \sum_i Q_i(z,t) \qquad (6.49)$$

where

Q_i is the net volume heating rate (W m^{-3}) due to the i-sources

$K_T(z)$ is the thermal conductivity that is a function of altitude (or equivalently temperature): $K_T(T)$

The thermal conductivity in atmospheres results from particle collisions, and according to the kinetic theory, the molecular heat conduction is proportional to the mean molecular velocity that varies as $T^{1/2}$ and is independent of pressure. More precisely, experimental data and quantum mechanical calculations indicate that a more general expression is $K_T(T)=A_T T^s$ where the coefficient A_T has units W m^{-1} K$^{-(s+1)}$ and s is a number that depends on the atmospheric composition (see Section 6.5 for more details).

The main heat sources contributing to Q_i are

1. The absorption of solar x-ray and EUV radiation that is in the origin of the ionization and photochemistry processes and, in the subsequent reactions, that liberate heat.
2. The penetration (precipitation) of energetic charged particles from the solar wind and magnetospheres.
3. Joule heating in the ionosphere produced by electric currents.
4. Dissipative processes resulting from dynamics (tides, gravity waves, turbulent motions, and molecular viscosity).

The first mechanism dominates in general. Photons with $\lambda \leq 2000\text{--}3000$ Å are absorbed by atomic and molecular species and the net heat is given by $Q_{XUV}(z, t) - Q_{IR}(z, t)$ where Q_{XUV} is the total heat volume production rate and Q_{IR} is the radiative heat loss in the infrared. For a given species j we can write

$$Q_{XUV}(z,t) - Q_{IR} = \varepsilon_j N_j \sigma_{aj} F_{XUV} \exp\left(-\frac{\tau}{\mu_\varphi}\right) \tag{6.50}$$

where

N_j is the species density (cm^{-3})

ε_j is the fraction of absorbed XUV radiation transformed to thermal energy (the *heating efficiency*, $\varepsilon_j \sim 0.3\text{--}0.6$)

σ_{aj} is the XUV absorption cross section ($\sim10^{-18}$ to 10^{-16} cm^2)

F_{XUV} is the energy flux at the particular wavelength (W m^{-2})

The optical depth is given by $\tau = \displaystyle\int_z^\infty \sigma_{aj} N_j\, dz$

The energy flux and photon flux (cm^{-2} s^{-1}) are related by

$$F_{XUV} = \frac{hc}{\lambda}\Phi_{XUV} \tag{6.51}$$

The radiative loss term Q_{IR} depends on the number density of emitting species and is a function of temperature. Its exact form must be obtained from radiative transfer calculations. The term ε_j allows for an approximate account of the heat losses, mostly by infrared radiation.

The second mechanism, the energetic particles bombardment, usually becomes important in the polar region of magnetized planets where they are associated to aurora phenomena. The third mechanism, Joule heating, is a friction term produced by the collisions of charged particles dissipating electrical energy. For electrical currents that are transverse to the magnetic field, the Joule heating rate is given by

$$Q_J = J_x^2 \left(\frac{\sigma_1}{\sigma_1^2 + \sigma_2^2} \right) = \frac{J_x^2}{\sigma_{Cowling}} \tag{6.52}$$

where the term between brackets is the Cowling conductivity.

Let us illustrate some characteristics of the heating of the planetary thermospheres by considering a simple case in which we assume steady state conditions with the daily absorbed solar radiation being equal to the conduction of heat, so Equation 6.49 can be written as

$$-\frac{d}{dz}\left(K_T \frac{dT}{dz} \right) = Q_{XUV} - Q_{IR} \tag{6.53}$$

On the other hand, upon integration of (6.50) taking $\mu_\varphi = 1$, we have

$$\int_z^\infty (Q_{XUV} - Q_{IR}) dz = -\varepsilon_j F_{XUV} \left(1 - e^{-\tau} \right) \tag{6.54}$$

since $dz = d\tau/N_j k_j$. Accordingly, the vertical temperature gradient is obtained from (6.53)

$$\frac{dT}{dz} = \frac{\varepsilon_j F_{XUV}(1 - e^{-\tau})}{K_T(T)} \tag{6.55}$$

In thermospheres, the thermal conductivity is large. In its upper part where $\tau \to 0$, we have that $dT/dz \sim 0$ and the outer layer becomes isothermal with $T = T_\infty$ (the exospheric temperature). In its lower part where $\tau \to \infty$, the maximum temperature gradient is given by

$$\left(\frac{dT}{dz} \right)_{max} \simeq \frac{\varepsilon_j F_{XUV}}{K_T(T)} \tag{6.56}$$

More generally, we can integrate (6.55) to derive a relationship between the temperatures at the thermosphere boundaries: the upper exospheric temperature (T_∞) and the temperature at the lower, mesopause boundary (T_0). Using $K_T = A_T T^s$ and assuming that $\tau \to 0$ at z_∞ and that $\tau \to \infty$ at z_0, and a single compound for simplicity, we have

$$T_\infty^{s+1} = T_0^{s+1} + \frac{\varepsilon F_{XUV}}{A_T}(s+1)\left[(z_\infty - z_0) + \frac{1}{N(z_\infty)k}\right] \qquad (6.57)$$

Comparison between observations and this first-order estimate of the exospheric temperatures in the giant planets indicate an underestimation by a factor of 3–6, indicating that the other sources of heating (energetic particle precipitation and dissipation by inertia–gravity waves) are important in these planets (see Problems 6.3 and 6.4). The calculation of the conductive time constant in the thermosphere is proposed in Problem 6.5.

6.5 DIFFUSION PROCESSES

The vertical distribution of atmospheric constituents under the low densities in the upper atmospheres is basically controlled by *diffusion processes*. The diffusion of species in a medium is a transport mechanism that arises from a gradient in the concentration. Two types of particle diffusion occur in the atmospheres, one due to the vertical movement of molecules called *molecular diffusion*, and the other due to the vertical transport of air parcels, collectively known as *turbulent* or *eddy diffusion*.

The diffusion in the vertical direction for a certain gas species with number density N_i is described by the one-dimensional continuity equation as follows:

$$\frac{\partial N_i}{\partial t} + \frac{\partial \Phi_i}{\partial z} = P_{Ni} - L_{Ni} \qquad (6.58)$$

where Φ_i, P_i, L_i are the vertical flux (cm^{-2} s^{-1}), and the production and loss rate (cm^{-3} s^{-1}), respectively, of the ith constituent. Let us consider the two contributions to the particle flux separately, since molecular diffusion dominates at very low densities above the homopause, and the eddy diffusion is predominant below the homopause.

6.5.1 MOLECULAR DIFFUSION

Assume that a minor species has a vertical distribution $N_i(z)$ in an atmosphere where its density distribution in equilibrium is $N_{iE}(z)$. The diffusion process is described by the *first Fick's law* that relates the upward flux (i.e., the upward mean diffusion velocity for this species w_i and concentration) and the gradient in the concentration according to the equation

$$\Phi_i = N_i w_i = -D_i N_{iE} \frac{\partial\left(N_i / N_{iE}\right)}{\partial z} \qquad (6.59)$$

where D_i (m² s⁻¹) is the molecular diffusion coefficient. From the kinetic theory of gases, the diffusion coefficient can be written as

$$D_i = \frac{1}{3}\ell_{fp}\langle w_i\rangle = \frac{1}{3}\ell_{fp}\left(\frac{8R_gT}{\pi\mu_i}\right)^{1/2} \tag{6.60}$$

where the mean free path $\ell_{fp} = 1/\sqrt{2}N_i\sigma_i$, σ_i being the collision cross section and μ_i the molecular weight for the species (see Chapter 2). The diffusion coefficient can then be written $D_i = \langle w_i\rangle/3\sqrt{2}N_i\sigma_i \propto N_i^{-1}T^{1/2}$, although for atmospheric calculations, is more practical than the use of a semiempirical form (generic, suppressing the species subindex) that is obtained from fittings of the concentration distributions to experimental data (Equation 2.27):

$$D = \frac{b_c}{N} = \frac{A_D T^s}{N} \tag{6.61}$$

where b_c is the *binary collision parameter*, with the value of the constant A_D and exponent s depending on the diffusing gases. Note that this expression has the same form as the one used for the thermal conductivity in the previous section, as diffusion and thermal conduction in gases are similar phenomena. The exponent s is the same in both cases, but the constant A_D in units of m² s⁻¹ K⁻ˢ differs from the thermal one A_T. In Table 6.4, we give the data for these constants for several mixtures of gases to calculate the diffusion coefficient.

At the low densities in the upper atmospheres, the rapid response to heating and cooling means that near isothermal conditions met, so large variations in the vertical scale height H can be ruled out (see Section 4.1.1). In addition, hydrostatic equilibrium can be assumed. Under such situation, the diffusion equation (6.59) can be rewritten in a more useful form. Since the vertical variation of pressure with altitude is given by $P(z) = P_0\exp(-z/H)$, the number density distribution (in equilibrium) has a similar exponential dependence:

$$N_{iE}(z) = \frac{P(z)}{k_B T} = N_0 e^{-z/H^*} \tag{6.62}$$

where N_0 is the concentration at pressure level P_0 and H^* is rewritten from its form (Equation 4.5) as

$$\frac{1}{H^*} = \frac{1}{H} + \frac{1+\alpha_{TD}}{T}\frac{dT}{dz} \tag{6.63}$$

A *thermal diffusion factor* term α_{TD} has been added to take into account the molecular diffusion produced by the temperature gradient when deviations from the isothermal condition exist. This coefficient depends on the species molecular weight: for light minor constituents (H, D, He), we can take $\alpha_{TD} = -0.25$ and for other species $\alpha_{TD} = -5/13$. Now, substituting these expressions in (6.59), we get

TABLE 6.4
Diffusion Parameters

Gas 1	Gas 2	A_D (×10⁻¹⁶)ᵃ	s
H	H_2	145	1.61
	Air	65	1.7
	CO_2	84	1.6
H_2	Air	26.7	0.75
	CO_2	22.3	0.75
	N_2 (100 K)	18.8	0.82
H_2O	Air	1.37	1.07
CH_4	N_2	7.34	0.75
Ne	N_2	11.7	0.743
Ar	Air	6.73	0.75

Source: Data from Chamberlain, J.W. and Hunten, D.M., *Theory of Planetary Atmospheres: An Introduction to Their Physics and Chemistry*, Academic Press, New York, 1987, Appendix VII.

Diffusion parameters applicable to Equation 6.61.

ᵃ A_D unit is cm² s⁻¹ K⁻ˢ, and is the column value ×10¹⁶.

Gas H_2	A_D (×10⁻¹⁶)ᵃ	s	Gas He	A_D (×10⁻¹⁶)	s
CH_4	23	0.765	CH_4	23	0.75
CH_3	23	0.765	CH_3	23	0.75
C_2H_2	13.8	0.834	C_2H_2	87	0.5
C_2H_4	16.2	0.791	C_2H_4	40.1	0.618
C_2H_6	20.1	0.738	C_2H_6	33.5	0.633
C_4H_2	14.4	0.75	C_4H_2	12.4	0.75
H	83	0.728	H	10.4	0.732

Source: Data from Atreya, S.K. et al., in *Uranus*, 1991, p. 129.

$$\Phi_i = -N_i D_i \left(\frac{1}{N_i} \frac{dN_i}{dz} + \frac{1}{H_i} + \frac{1+\alpha_{TD}}{T} \frac{dT}{dz} \right) \qquad (6.64a)$$

or

$$\Phi_i = N_i w_i = N_i D_i \left(\frac{1}{H_i^*} - \frac{1}{H_{iE}^*} \right) \qquad (6.64b)$$

where we have introduced the scale height for the equilibrium distribution H_{iE}^*. The time needed to reach the diffusive equilibrium can be obtained from (6.58), with the source and sink terms null. Then $\partial N_i / \partial t = -\partial \Phi_i / \partial z$, and using (6.64), the molecular diffusion of the ith species has a time scale $\tau_D = \left[(1/N_i)(\partial N_i / \partial t) \right]^{-1} \sim H_i^* / w_i \sim H^2 / D_i$ since $w_i \sim D_i / H$.

Consider now that this ith compound has a mixing ratio or mole fraction $x_{VVi} = N_i/N$, so differentiating with respect to altitude, we have

$$\frac{1}{x_{VVi}} \frac{dx_{VVi}}{dz} = \frac{1}{N_i} \frac{dN_i}{dz} - \frac{1}{N} \frac{dN}{dz} \tag{6.65}$$

Substituting in (6.64a), we get

$$\Phi_i = \Phi_L - ND_i \frac{dx_{VVi}}{dz} \tag{6.66}$$

where we have introduced a *limiting flux* Φ_L:

$$\Phi_L = -N_i D_i \left(\frac{1}{N} \frac{dN}{dz} + \frac{1}{H_i} + \frac{1 + \alpha_{TD}}{T} \frac{dT}{dz} \right) \tag{6.67}$$

For very small or a null temperature gradient (pure isothermal conditions), the limiting flux can be expressed using Equation 6.61 as

$$\Phi_L = -N_i D_i \left(\frac{1}{N} \frac{dN}{dz} + \frac{1}{H_i} \right) = \frac{N_i D_i}{H} \left(1 - \frac{\mu_i}{\langle \mu \rangle} \right) \approx \frac{N_i D_i}{H} \approx \frac{b_{Di} (N_i/N)}{H} = \frac{b_{Di} x_{VVi}}{H} \tag{6.68}$$

This expression gives the maximum rate of molecular diffusion for this element. It corresponds in other words to the case when the light element is completely mixed.

6.5.2 EDDY DIFFUSION

Diffusion due to macroscopic mixing by turbulent motions or due to other dynamical processes can be characterized in a similar way to molecular diffusion by means of an equivalent parameter, the *eddy diffusion coefficient* D_{eddy} (m^2 s^{-1}). Unfortunately, the relationship between D_{eddy} and actual atmospheric motions is complex. There is no general way for the derivation of this coefficient, although for some processes, as for example, when mixing by gravity waves dominates (as occurs in the outer planets), a description in terms of the wave parameters can be obtained and it can be shown that $D_{eddy} \sim N^{-0.5}$.

Prandtl's mixing length theory can be used to describe the mixing of parcels in terms of a characteristic distance ℓ_{eddy} equivalent to the free path length. A crude estimate of the coefficient is $D_{eddy} = u' \ell'$ where u' is the velocity fluctuation about a mean value in the characteristic length scale $\ell_{eddy} = \ell'$. For vertical motions, we can take $\ell' \sim H$ and $u' = w'$. We will discuss this topic in detail in Chapter 9 within the dynamical context. For the moment, it is sufficient according to (6.59) to write the particle flux of the ith species due to turbulent mixing as

$$\Phi_i = -D_{eddy} N \frac{d}{dz} \left(\frac{N_i}{N} \right) \tag{6.69a}$$

Then from (6.67), we get

$$\Phi_i = N_i w_i = -D_{eddy} N_i \left(\frac{1}{N_i} \frac{dN_i}{dz} + \frac{1}{H} + \frac{1}{T} \frac{dT}{dz} \right) \qquad (6.69b)$$

As before, the characteristic time constant for eddy diffusivity can be similarly obtained as $\tau_{eddy} = H^2/D_{eddy}$. As already indicated, the altitude where $\tau_{eddy} = \tau_D$ defines the *homopause*.

Combining molecular and eddy diffusion processes, using (6.66), the flux for the *i*th compound can be expressed as

$$\Phi_i = \Phi_L - N(D_i + D_{eddy}) \frac{dx_{Vvi}}{dz} \qquad (6.70)$$

If the minor compound occurs in many molecular combinations, its total flux can be obtained by summing terms like the second one in (6.70) for each combination, weighting by the number of atoms in each molecule.

6.5.3 DIFFUSION PROCESSES IN PLANETARY ATMOSPHERES

The vertical distribution of minor constituents in the upper atmospheres of the planets can be reproduced from the coupling of photochemistry and diffusion models. When diffusion or other transport mechanisms dominate over chemical transformations (photochemical or thermochemical) or phase changes, that is, when the characteristic times for the processes are $\tau_{Chem} > \tau_D$, τ_{eddy}, *disequilibrium* occurs, and at a given atmospheric level, one can find species with abundances above or below their expected chemical equilibrium values. Diffusion processes can explain the vertical mixing ratio profiles of some species when compared with those predicted by photochemical models alone. From these comparisons, one can empirically retrieve the diffusion coefficients. Table 6.5 gives the measured values for the D_{eddy} in the upper atmospheres of the planets and Table 6.6 gives the location and atmospheric properties of the homopauses.

1. On Earth, molecular diffusion makes the O/O_2 mixing ratio to be less than expected from photochemical models at $z \sim 100$ km but greater at altitudes $z \geq 100$ km. A downward flux of oxygen atoms occurs between 120 and 70 km with $\Phi = 10^{12}-10^{10}$ cm^2 s^{-1}, whereas an upward flux occurs in the lower thermosphere. In the mesosphere and lower thermosphere, diffusion controls the CO, CO_2, and Ar profiles. For the stratosphere, suitable tracers of diffusion are CH_4, N_2O, and the chlorofluoromethane compounds. Figure 6.9 shows the derived eddy coefficient profile for the Earth's atmosphere.

2. On Venus and Mars, CO_2 flows up, but CO, O, and O_2 flow down below the homopause. In Mars, at 50 km above the surface, the turbulent diffusion is high with $D_{eddy} \sim 10^8$ cm^2 s^{-1}, but in the lower atmosphere ($z \sim 0-20$ km), it drops to $D_{eddy} \sim 10^5$ cm^2 s^{-1}. In Venus, $D_{eddy} \sim 10^5-10^6$ cm^2 s^{-1} high above the

TABLE 6.5
Eddy Diffusion Coefficient in Planetary Atmospheres

Planet	D_{eddy} (cm² s⁻¹)	Altitude Range
Venus	$1-4 \times 10^6$	$z = 80-110$ km
Earth	$0.3-1 \times 10^6$	$z = 55-100$ km
Mars	10^5	$z < 40-50$ km
	10^8	$z > 50$ km
Titan	2×10^2	$z = 50$ km
	10^5	$z = 190$ km
	8×10^6	$z = 500$ km
	7×10^7	$z = 1000$ km
Jupiter	4.5×10^2	$P = 0.14$ bar
	1.5×10^6	$P = 10^{-6}$ bar
Saturn	5×10^3	$P = 0.13$ bar
	2×10^8	$P = 10^{-9}$ bar
Uranus	10^4	$P = 5 \times 10^{-3}$ to 10^{-5} bar
Neptune	8×10^3	$P = 2 \times 10^{-3}$ bar
	10^5	$P = 5 \times 10^{-5}$ bar
	$5 \times 10^6 - 5 \times 10^7$	$P = 10^{-6}$ bar

Note: See also Figures 6.9 through 6.11.

TABLE 6.6
Homopause Characteristics

Planet	Pressure (bar)	$D_{eddy} = D_i$ (cm² s⁻¹)	Number Density (cm⁻³)
Venus	2×10^{-8}	10^7	7.5×10^{11}
Earth	3×10^{-7}	$0.3-1 \times 10^6$	10^{13}
Mars	2×10^{-10}	$1.3-4.4 \times 10^8$	10^{10}
Titan	6×10^{-10}	10^8	2.7×10^{10}
Jupiter	10^{-6}	1.5×10^6	1.4×10^{13}
Saturn	10^{-9}	10^8	1.2×10^{11}
Uranus	2×10^{-5}	10^4	1.1×10^{15}
Neptune	10^{-8}	3×10^7	3×10^{11}

Source: Data from Atreya, S.K. et al., in *Uranus*, 1991, p. 145 and references therein.

surface (at $z \sim 70-110$ km). A full profile for Venus and Mars is shown in Figure 6.10.

3. On Titan, where the hazes extend to great altitudes, D_{eddy} increases strongly with altitude from the lower atmosphere where $D_{eddy} \sim 10^2$ cm² s⁻¹ to the high atmosphere where $D_{eddy} \sim 10^8$ cm² s⁻¹ at 1000 km (Figure 6.9).

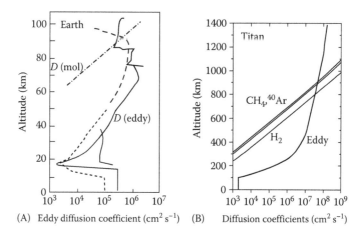

FIGURE 6.9 The vertical profiles of the eddy diffusion coefficient in Earth (A) and Titan (B) atmospheres. For the Earth, the eddy profiles are based on different studies of the mixing ratio distributions of N_2O and CFC compounds. The molecular diffusion coefficient is also indicated. (Based on data taken from Patra, P.K. and Lal, S., *J. Atmos. Sol-Terr. Phys.*, 59, 1149, 1997.) The vertical profiles of the molecular diffusion coefficient (with molecules indicated) and of the eddy diffusion coefficient in Titan's atmosphere. (From Lavvas, P.P. et al., *Planet. Space Sci.*, 56, 68, 2008. With permission.)

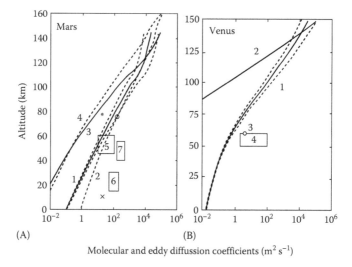

FIGURE 6.10 The vertical profiles of the molecular and eddy diffusion coefficients in the atmospheres of Venus and Mars. (A) Curves 1,3 are the eddy and molecular diffusion coefficients under conditions during "Viking" descent (northern hemisphere summer, latitude 22.5°). Curves 2,4 are the same but during a winter dust storm. Symbols 5,6,7 data from clouds, dust and ozone, measurements using "Fobos-2" spacecraft. Symbols 8,9,10 are from a numerical model. The dotted lines near curve 1 represent the probable error in the eddy diffusion coefficient. (B) Curves 1,2 are the eddy and molecular diffusion coefficients. Symbols 3,4 come from radio measurements. The dotted lines near curve 1 represent the probable error in the eddy diffusion coefficient. (From Izakov, M.N., *Planet. Space Sci.*, 49, 47, 2001. With permission.)

4. On the giant planets, D_{eddy} is calculated using different methods, although the most employed are the vertical distribution of CH_4 and the abundance of disequilibrium species such as GeH_4, PH_3, CO, and HCN not expected in the upper atmospheres of Jupiter and Saturn. These compounds are expected to be present deep in the atmospheres at $T > 1000\,K$ but not at lower temperatures, so rapid vertical transport by dynamics from the deep atmosphere is required. On the other hand, the spread of the debris left by the impact of comet SL-9 in 1994 in Jupiter's atmosphere above the altitude level $P < 10\,mbar$ implied $D_{eddy} \sim 10^9 - 10^{10}\,cm^2\,s^{-1}$.

The great differences in the methane and hydrocarbon abundances in Uranus and Neptune upper atmospheres can be explained by much greater eddy diffusion in these planets than in Jupiter and Saturn. This is also supported by the detection of CO and HCN in Neptune that requires rapid vertical transport.

Figure 6.11 shows the vertical profiles for the molecular and eddy diffusion coefficients in the giant and icy planets. A reasonable analytical approximation for the vertical dependence of D_{eddy} from the tropopause to the homopause in the giant planets is given by $D_{eddy}(z) = D_H\,N_H/N(z)^{0.5}$, where D_H and N_H are the molecular diffusivity and the number density at the homopause (see Table 6.6), respectively.

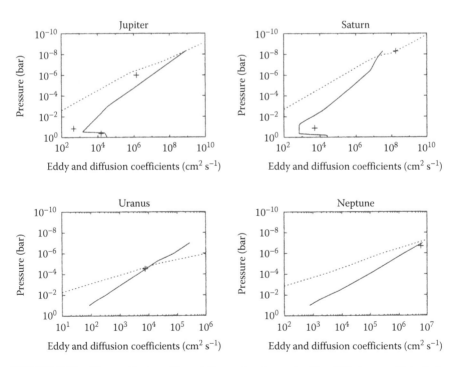

FIGURE 6.11 The vertical profiles of the molecular and eddy diffusion coefficients in the giant planets. (Reproduced from Irwin, P.G.J., *Giant Planets of Our Solar System: An Introduction*, Springer Verlag-Praxis, Chichester, U.K., 2009. With permission.)

Problems 6.9 through 6.13 deal with the diffusion mechanism in different planetary atmospheric environments.

6.6 AIRGLOW AND AURORA

Airglow and *aurora* are radiation emission processes occurring in the upper atmospheres of the planets. Similar phenomena also occur in the tenuous atmospheres of some satellites. They are distinguished from other emission mechanisms from their trigger sources and their spatial and temporal properties. Airglow and aurora represent a broad subject of research whose detailed coverage is out of the aim of this book. Here, we just outline their basic properties and make a comparison among the different atmospheres.

6.6.1 AIRGLOW

Airglow emission occurs continuously, covering virtually all latitudes. When observed during the night, the phenomenon is sometimes called *nightglow*. Airglow arises mainly from discrete atomic and molecular emissions, and its spectrum is formed by lines and bands, although a continuum component is observed in the Earth case. The main emission mechanisms are

1. Direct scattering of sunlight radiation
2. Emissions associated with the formation and destruction of compounds in the ionosphere
3. Emission resulting from photochemistry of neutral components

They involve the atomic and molecular excitation and ionization by solar EUV photons and photoelectrons, followed by deexcitation and radiative recombination emissions (Sections 2.4.1, and Sections 6.2 and 6.3). One important mechanism is the *resonant* and *fluorescent scattering* of sunlight. Consider a three-level atom (or ion, radical, molecule) whose ground-based state J_0 is excited to an upper level J' when it absorbs a solar photon with energy $h\nu_0$ (Figure 6.12). Deexcitation can be directly to the level J_0 by emitting a photon with the same wavelength (*resonance scattering*) or through a cascade along a series of intermediate energy levels, in this example, through a single level J_1 (*fluorescent scattering*), emitting photons with longer wavelengths. The level splitting can be originated by the atomic fine or hyperfine structure or by the molecular vibrational and rotational sublevel structure, as shown in Chapter 3. The intensity of the emission line can be expressed as

$$I = g_{ph}Np(\Theta) \tag{6.71}$$

where
g_{ph} is the photon scattering coefficient (in photons/s molecule) that depends on the atomic parameters and on the intensity of the incident solar flux
$p(\Theta)$ is the phase function for single scattering (normalized to unity, Section 5.4.2)
N is the number of atoms (ions, molecules) in the line of sight (in a plane-parallel atmosphere corrected from the vertical by the factor $1/\mu_\varphi$)

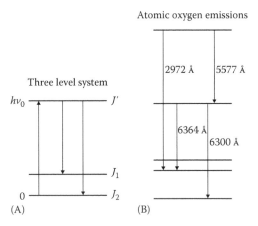

FIGURE 6.12 (A) Excitation mechanism by resonant and fluorescence scattering in a three-level system (one upper state and two ground states) and (B) transitions in atomic oxygen emission observed in terrestrial airglow or aurora.

For a tangentially viewed atmosphere, we can express $N = N(r)(2\pi r H)^{1/2}$, where r is the radius of the tangentially viewed spherical shell, H is the local scale height, and $N(r)$ is the local density of scattering molecules. The airglow emission rate is usually measured in *rayleigh* unit (R; *kilo-rayleighs* kR, and *mega-rayleigh* MR), where $1\,R = 10^6$ photons cm^{-2} s^{-1}.

In the Earth, the "visible" optical emissions from 300 to 1000 nm are dominated by O_2 transitions (named as the *Herzberg* and *Chamberlain* UV blue lines, see Figure 6.12) and by hydroxyl OH (*Meinel* bands), including also single emissions from atomic O I, Li I, Na I, K I, and Ca II. The emission comes from altitude levels $z \sim 90$–100 km. A prominent intense characteristic emission known as the *green oxygen line* occurs at 557.7 nm in the E-layer and results from the double reaction known as *Barth mechanism*:

$$O + O + M \rightarrow O_2^* + M$$

$$O_2^* + O \rightarrow O_2 + O$$

This emission is used to study the atmospheric motions (velocity) of the neutral air by Doppler shift measurements.

Other interesting emissions are the following. The resonance triplet from oxygen O I at around 130.4 nm has an intensity of 7.5 kR and comes from altitudes ~190 km, which together with the 630 nm emission by O I (viewed at limb) is used to retrieve the vertical atomic oxygen profile. Another peculiar line occurs at 391.4 nm from N_2^+ emission, and since its intensity is directly proportional to the ionization rate of the molecule, it becomes useful to measure the solar energy input in the upper atmosphere. Nitrogen airglow emissions by N^+ and N_2 (in the short UV at 91.6–108.5 nm) have also been observed in the atmospheres of the satellites Titan and Triton.

In Venus and Mars, airglow optical emissions at wavelengths 300–800 nm are dominated by O_2 with the participation of CO_2 as a third-body reaction element. In

Venus, emission profiles along the limb (i.e., intensity as a function of altitude) are produced by prominent O_2 bands (peak at altitudes $z \sim 100–130\,km$). Other detected nightglow emissions are due to NO (an UV emission probing the 115 km altitude level with intensity ~1.9 kR) and OH. The O_2 nightglow emission in the near infrared (1.27 μm) is also prominent (intensities ~4–6 MR), and its morphology distribution has been used to detect motions in Venus mesosphere, in particular, to capture the solar–antisolar circulation at altitudes of $z \sim 95–110\,km$ (Figure 6.13). UV emissions from O I at 130.4–135.6 nm have been observed in both planets, with intensities depending on CO_2 dissociation and CO formation. In Mars, emissions due to CO, O, and CO_2^+ in UV (130–400 nm) come from altitudes ~120–200 km with limb intensities ranging from 1 to 800 kR. As in Earth and Venus, the most intense infrared nightglow emission is produced by O_2 at 1.27 μm, reaching intensities ~3–26 MR.

Resonant scattering by atomic hydrogen of the Lyman α emission (121.5 nm) is present in all planets (including Mercury) and in Titan's tenuous torus. This emission is known as the *planetary corona*, and it's an extended emission glow that surrounds the planet. Hydrogen is present in the upper atmospheres due to molecular dissociation and, because of its lightest mass, by its diffusive separation. The emission intensity reaches ~40 kR on Venus, ~8 kR on Earth, and ~5–6 kR on Mars. In the hydrogen dominated atmospheres of the giant planets, a rich dayglow emission spectrum due to H occurs around the Lyman α emission in the wavelength range

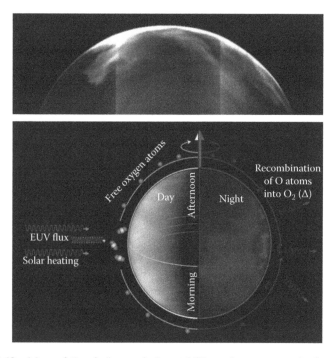

FIGURE 6.13 Maps of the airglow emission at 1.27 μm due to oxygen in the atmosphere of Venus as obtained with the VIRTIS instrument onboard Venus Express. Scheme of the oxygen photochemistry and recombination processes in Venus. (Courtesy of R. Hueso, Universidad del País Vasco UPV-EHU, Spain.)

60–160 nm (the Lyman α emission reaches in Jupiter intensities ~12–24 kR). It is accompanied by a band of electronic transitions occurring in the H_2 molecule (the *Lyman and Werner* band systems).

The helium resonance lines, in particular He I (58.4 nm) has been observed in Mercury with intensity ~70 kR, in Venus with ~600 kR, in the Earth, and in the giant planets (in Jupiter it reaches an intensity ~5 kR). The study of this emission line is important since it allows inferring by remote sensing the helium abundance in Jupiter.

6.6.2 AURORA

Aurora is also a radiation emission phenomenon from the upper atmosphere that is distinguished from airglow in the excitation source. In the aurora, the species excitation is produced by accelerated charged particles (electrons and ions from the solar wind and magnetosphere) when they precipitate into the atmosphere from the outer space. The emission depends on the incident particle energy whose range is typically between 1 and 10 keV (in Jupiter's powerful environment they can reach up to 100 keV). The most intense and spectacular auroras are related to the presence of a planetary magnetic field that acts as a focusing mechanism for the particles that impact into the upper atmosphere. In this case, the charged particles coming from the magnetosphere are accelerated and guided along the magnetic field lines (Section 1.6.3), producing a bombardment of the upper atmospheric gases. Some species can be excited by these collisions, emitting radiation when deexciting in a broad wavelength range from x-rays to radio wavelengths, depending on the planet magnetic field properties and particle energy, and on the atmospheric species involved. The emission occurs in oval and ringlike shape areas that form around the magnetic poles (typically at magnetic latitudes ~60°–80°), known as the *auroral zone*. In addition, secondary photoelectrons and atmospheric ions produced by the primary collisions contribute to the aurora emission.

Table 6.7 summarizes the main properties of observed aurora phenomena in planetary atmospheres. For the planetary magnetic field characteristics, refer to Table 1.7. Particles in the magnetosphere could have their origin in the solar wind and, therefore, the aurora emission shows variability related to the solar activity cycle. Averaged over the whole atmosphere, the energies involved in aurora processes is low (~10^{-4} W m^{-2}), although locally it can be very high. For example, in the foot of the magnetic tube connecting Io with Jupiter (a region of 100 km^2 in Jupiter's upper atmosphere), the energies involved are ~10 W m^{-2}. Table 6.8 summarizes the total power emitted in the different spectral ranges by the three most intense auroras observed in the solar system. The aurora radio emissions occur at the *local electron cyclotron frequency* $f_{ce} = q_e B/2\pi m_e = 2.8 \times B$, where q_e is the electron charge, m_e the electron mass, and B the intensity of the local magnetic field, and the frequency is in MHz and the magnetic field in Gauss units.

Although Venus lacks of an intrinsic magnetic field, its proximity to the Sun generates an x-ray aurora emission by the excitation due to the solar wind ions, as well as of atomic oxygen in the night side of the planet. Auroras on Earth are seen from the ground, predominantly at high latitudes during nighttime, as diffuse variable bands

TABLE 6.7

Aurora in Planetary Atmospheres

Planet Body	Excitation Source	Atmospheric Emitting Species	Dominant Wavelength	Auroral Power (W)
Venus	e^-, H^+, He^{++} (sw) H^+, O^+, He^+, C^+ (atmos)	CO_2, N_2, O, CO	UV (OI)	—
Earth	e^-, H^+, He^+ (sw) H^+/H, O^+/O (atmos)	N_2, O_2, O	γ, X, UV, optical, IR	10^{10}
Jupiter	e^-, H^+ (sw, atmos) O, S, Na ions (satellites)	H_2, H, ions (H_3^+), hydrocarbons	X, UV, optical, IR	10^{14}
Io	Io torus particles Jupiter magnetosphere	SO_2, SO, S, O, and ions, Na, Cl	UV, optical	—
Ganymede	Ganymede magnetosphere Jupiter magnetosphere	O_2, O_3, O	UV, optical	—
Europa	Io torus particles Jupiter magnetosphere	O_2, O_3, Na, O (ion)	UV	—
Saturn	e^-, H^+ (sw, atmos) O^+, N^+, N_2^+, H^+ (satellites, rings)	H_2, H, ions (H_3^+), hydrocarbons	X, UV, IR	10^{11}
Uranus	sw, atmos	H_2, H, CH_4	X(?), UV	10^{11}
Neptune	N^+, H^+ (Triton)	H_2, H, CH_4	X(?), UV	$<10^8$

TABLE 6.8

Emitted Power of the Most Intense Aurora Emissions

Wavelength Range	Earth	Saturn	Jupiter
X-rays (0.1–0 keV)	$10–30 \times 10^6$	$<5–15 \times 10^9$ W	$1–4 \times 10^9$ W
FUV (80–180 nm)	$\sim 50 \times 10^9$ W	$<50 \times 10^9$ W	$2–10 \times 10^{12}$ W Up to 10^{14} W
IR H_3^+ (3–4 mm)	—	$150–300 \times 10^9$ W	$4–8 \times 10^{12}$ W
Radio	4×10^7 W	4×10^8 W	4×10^{10} W

of arcs (Figure 6.14). A blue emission comes from nitrogen at wavelengths 3914 and at 4278 Å. A green–red color originates from the green (forbidden) and red oxygen lines at 5577 Å (altitudes ~ 200 km) and 6300–6364 Å. From space, they are registered as full ovals or as arcs with a ringlike shape surrounding the pole, in particular, in UV wavelengths (Figure 6.15). Earth auroras are also detected at other wavelengths from space (x-rays) to radio (frequencies < 1 MHz). X-rays from the aurora come from line (by neutral species) and continuum emission due to high-speed

FIGURE 6.14 Images of the aurora emission on Earth as observed from the ground in the visual range over Alaska (USA). (Courtesy of J. Strang, USAF, Alaska; Wikipedia; B. Kuenzli, in Astronomy Picture of the Day, October 9, 2007. Available at: http://en.wikipedia.org/wiki/Aurora(astronomy).)

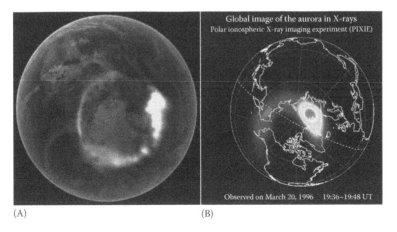

FIGURE 6.15 Oval arc aurora emission (overlaid around Earth's poles) as observed from space. (A) UV emission capture by MIME on September 11, 2005; (B) X-ray emission captured by satellite Polar (PIXIE experiment) on March 1996. (Courtesy of NASA, Washington, DC.)

impinging electrons. The aurora oval region increases its meridional width with geomagnetic activity from ~10° to 20°.

The four giant planets show intense aurora emissions from x-rays to radio wavelengths (Tables 6.7 and 6.8) due to the presence of magnetic fields, particularly powerful in Jupiter (Figure 6.16). Aurora emissions from a number of compounds have been detected: NH_3, CH_4, C_2H_2, C_2H_6, ion H_3^+ and trace gases as C_2H_4, C_3H_4, and C_6H_6. The UV emission in Jupiter is seen as an oval shape (Figure 6.16), rapidly variable, and correlated to the H_3^+ infrared emission in the wavelength range ~3.5–4.2 μm. Jupiter shows localized spots of enhanced aurora emissions associated to the footprints of the satellites Europa, Io, and Ganymede, where the charged particles enter Jupiter's ionosphere following the magnetic field lines that connect the planet with the satellites.

Saturn shows also a prominent oval shape aurora confined to latitudes 70°–75° around both poles (Figure 6.17). The aurora processes are driven by a mixture of solar wind particles and magnetosphere particles injected by water products originating in the massive rings that surround the planet and in the material expelled from the geysers in the moon Enceladus.

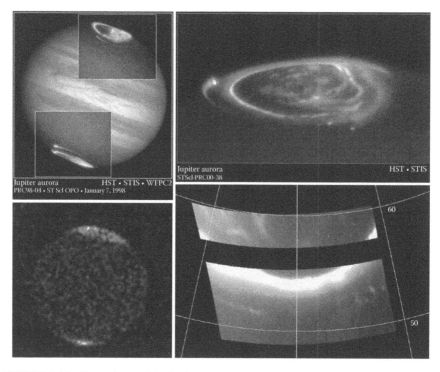

FIGURE 6.16 Four views of Jupiter's aurora rings and oval aurora. Upper left: over the two poles in the UV wavelength (overlain a visible Jupiter image, Hubble Space Telescope, HST); Upper right: details of the arcs (HST); Lower left: emission in x-rays (from the satellite Chandra); Lower right: the arc seen by the Galileo spacecraft. (Courtesy of NASA/ESA (HST), Plymouth, U.K., NASA-JPL (Chandra, Galileo), Los Angeles, CA, J. Clarke, University of Michigan, Ann Arbor, MI, and NASA, Washington, DC.)

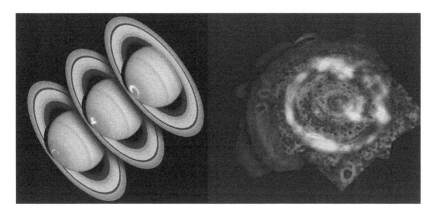

FIGURE 6.17 Saturn's aurora. Left: Variability at UV wavelengths of Saturn's dynamic oval aurora (overlain a visible Saturn image) for 24, 26, and 28 January 2004. (Courtesy of J. Clarke, Boston University, Boston, MA and Z. Levay, STScI, Baltimore, MD.) Right: Saturn's north pole aurora emission in the infrared at 4 µm, overlain over an infrared image of Saturn's polar hexagon, both obtained by Cassini VIMS instrument on November 10, 2006. (Courtesy of NASA/ESA, Plymouth, U.K. and ASI Tempe, AZ.)

Due to its proximity to the Sun, the Earth aurora is highly variable with the solar 11-year activity cycle, in particular is very sensitive to magnetic storms (flares, coronal mass ejections) that disturb the geomagnetic field. In addition, the sequential effect in a matter of few days produced by a coronal mass ejection in the aurora enhanced emission in Earth, Jupiter, and Saturn, triggered by the passage of the solar wind particles was observed at the end of year 2000.

6.7 GAS PRODUCTION MECHANISMS IN TENUOUS ATMOSPHERES

Tenuous atmospheres surround the other planets (Mercury and dwarf planet Pluto) and some of the most massive satellites. In Table 6.1, we compared the density of these atmospheres with the upper parts of the planets with dense atmospheres, and in Table 6.9, we give their basic properties. Some of these atmospheres are so tenuous that they have densities lower than those produced with vacuum pumps in the laboratory. They are so rarefied as to have nothing but an exosphere (surface boundary exosphere) with gases escaping at the speed of sound and atmospheric density falling off as $1/r^2$. Their origin is multiple, produced by a variety of processes (Figure 6.18). On one hand, the atmospheres of Mercury and the Moon and some icy satellites are produced by surface sputtering from the bombardment with particles from the solar wind, magnetospheres, and micrometeoroids. On the other hand, the atmospheres of Pluto and of Neptune's satellite Triton (and probably those of Kuiper Belt Objects) result from the sublimation of surface ices when these bodies are heated along their eccentric orbits by solar radiation. Finally, there are bodies whose atmospheres result from the outgassing due to internal heat in the body, which melts material escaping via volcanoes (as occurs in Jupiter's satellite Io) and by geyser-like activity (as in

TABLE 6.9

Main Properties of Tenuous Atmospheres

Planet/Satellite	Main Composition (Ranges, %)	Surface P_s (bar), T_s (K)	Aerosols	Optical Depth
Mercury	O (42–52), Na (29–39)	5×10^{-15}	—	—
	H (22), He (6–8), K (0.6)	725–100 (day/night)		
	N, Ar, Ne, Ca			
Moon	He (50), Ar (50), Na	3×10^{-15}	—	—
	K	277		
Io (J)	SO_2 (99) SO, O, S_2	10^{-7} to 10^{-9}	Volcanic aerosols	0.1
	Na, K, Cl, NaCl, O_2	110–120 at noon		
	CS_2, CO, H_2S, KCl	300–600 (volcano)		
Europa (J)	O_2 (99), Na, K	2×10^{-12}	—	—
		85–135		
Ganymedes (J)	O_2 (99), H	2×10^{-12}	—	—
		~100		
Callisto (J)	CO_2, O_2 (?)	~10^{-12}	—	—
		~100–150		
Enceladus (S)	H_2O (91), N_2 or CO (4),	Inhomogeneous	Geyser	
	CO_2 (3), CH_4	65–85	Water-ice	
		(120–155 plumes)	Dust	
Triton (N)	N_2 (99), CH_4	10^{-5}	N_2 ice (?)	0.1–0.3
		38	CH_4 hazes	
Pluto	N_2 (99), CH_4	1.5×10^{-5}	CH_4	>0.15
		45		

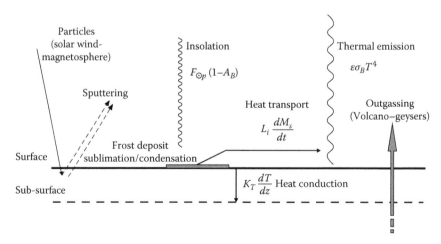

FIGURE 6.18 Scheme of the physical processes and main production mechanisms of gases in tenuous planetary atmospheres.

Triton and Saturn's small satellite Enceladus). The gases produced or liberated by these mechanisms become partially trapped in the lower gravity field of these bodies, expanding in some cases to their surroundings where they follow the orbit, as exemplified in the case of Io.

6.7.1 SPUTTERING

Sputtering occurs effectively on the surfaces of planetary bodies with very tenuous atmospheres so the bombarding particles can reach the surface at high velocity (expressed in terms of energy, in electron volts [eV]). The collisions liberate surface particles, a process that depends on two aspects: (1) on one hand, on the incident particle mass (and particle number), their charge and speed (and velocity distribution and incident angle), and on the environment conditions (presence of a magnetic field or a thin atmosphere); (2) and on the other hand, on the surface properties, its composition and texture (e.g., grainy or powder, icy or solid refractory). The surface particles can be liberated directly from the collision or following internal collisions within the thin upper surface layer, whose thickness is measured by a parameter called the *effective penetration depth* or *stopping depth* for the impinging particles. The bombarding particles can in addition alter chemically this thin surface layer, affecting the sputtering process.

The impact of charged particles from the solar wind (effective in Mercury and the Moon) and from particles trapped by the magnetospheres of the giant planet on the surfaces is controlled by the magnetic field itself, which can shield the surface or on the contrary accelerate them into it. The particle energy must take into account its orbital and plasma velocity relative to the surface of the body, that is itself moving along its orbit and in rotation. The sputtering process involves the calculation of the energy transferred by the impinging particles to the atoms in the stopping depth layer, and then the number of them that acquire enough energy to overcome the surface binding forces and abandon it to the atmosphere. The averaged transferred energy is represented by the energy loss per unit path length of the particle in the stopping depth, dE/dz. The energy is transferred by elastic nuclear collisions and by electronic excitations and ionizations of the surface atoms, so it can be written as the summation of two terms $dE/dz \approx (dE/dz)_n + (dE/dz)_e$. The process is characterized by a parameter called the *sputtering yield Y* that is defined as the number of atoms or molecules ejected per incident particle, accordingly $Y_{sput} \approx Y_n + Y_e \cdot Y_n$ dominates in bodies with surfaces formed by refractory materials, but Y_e becomes most important for icy surfaces. According to the standard linear collision cascade theory, these parameters can be written as

$$Y_n \approx C \left[\frac{\ell (dE/dz)_n}{U_b} \right]^p$$

$$Y_e \approx C \left[\frac{\ell f_e (dE/dz)_e}{U_b} \right]^p$$

(6.72)

where
   ℓ is the mean space between the atoms or molecules
   C is a constant
   U_b is the average binding energy of the surface atoms
   f_e is the fraction of electronic energy converted into energetic atomic motions (about 0.1–0.2)
   the exponent $p \sim 1$–3 depending on the energy of the bombarding particles

For a density N_p of impinging particles, moving with a velocity distribution $\eta(\bar{v})$, the surface averaged particle sputtered flux can be written as

$$\Phi_s = \iint Y_{sput}\left[-\hat{n}(\bar{u}+\bar{v})\right]N_p\eta(\bar{v})\,d^3v\,\frac{d\Omega_s}{4\pi} \tag{6.73}$$

Here
   $\bar{u}$ is the average flow velocity of the particles relative to the body
   $\hat{n}$ is the local surface normal

The flux is obtained integrating over the solid angle and velocity distribution. If we denote by Φ_i the incident particle flux on the surface, then the average yield per incident particle is given by $\langle Y_{sput}\rangle = \Phi_s/\Phi_i$. See Problem 6.14.

The collisions of micrometeoroids or larger bodies with the surface can also inject dust or other material to the atmosphere and represent a source of energy to drive sublimation.

6.7.2 Surface Ice Sublimation

As already indicated, the sublimation and condensation of surface ices (whole surface or localized deposits) that follows the seasonal insolation cycle is an important mechanism for replenishing or depleting the atmosphere in bodies with tenuous atmospheres: from Mars to icy satellites (Galileans and Triton), and dwarf planet Pluto. In general, atmospheres formed by the sublimation of a body uniformly covered by frost are centered at the subsolar point expanding from this area to the rest of the body. The motion of the sublimating vapor forms a "sublimating wind" that transports the mass across the body tending to produce isothermal atmospheres.

When the surface temperature reaches the critical value in the phase diagram for the sublimation of the frost, a vapor is released into the atmosphere leaving behind the ground substrate. The sublimation of surface ices into a vacuum is a rapid, nonequilibrium process. The process is described by the thermodynamic equation including solar (or stellar) heating, radiative losses, latent vapor heating, and conductive transport across the surface layer (Section 1.5.4, and Equations 2.10 and 6.49) expressed as

$$M_S C_p \frac{dT}{dt} = F_{\odot p}(1-A_B) - \varepsilon\sigma_B T^4 + L_i \frac{dM_f}{dt} + K_T \frac{dT}{dz} \tag{6.74}$$

where

M_S is the mass of the frost deposited or sublimated per unit area (kg m^{-2})

C_p is the specific heat capacity of the frost deposit

dT/dt is the rate of change of the temperature of the frost

$F_{\odot p}$ is the incident solar flux (W m^{-2}, a function of latitude and longitude, Section 1.4.3)

A_B is the albedo of the frost or of the substrate

ε is the emissivity of the frost or substrate

T is the frost or ground temperature

L_i is the latent heat of the frost for sublimation or condensation

dM_S/dt is the rate of frost sublimation/condensation per unit area

K_T is the thermal conductivity of the substrate

dT/dz is the thermal gradient in the substrate

Table 6.10 presents the thermophysical properties of the most common sublimating ices in the planetary bodies at low temperatures.

In one extreme situation, if the energy goes into changing the frost deposit temperature and no energy goes into sublimation/condensation of the frost, then $L_i\, dM_S/dt = 0$ and no vapor is injected into the atmosphere. In the other extreme, if all the energy goes into subliming/condensing frost and no energy goes into changing the frost deposit temperature, $M_S C_p\, dT/dz = 0$, then

$$F_{\odot p}(1 - A_B) = \varepsilon \sigma_B T^4 + L_i \frac{dM_S}{dt} + K_T \frac{dT}{dz} \tag{6.75}$$

To further simplify, if we assume that no heat is conducted to the substrate, this energy balance reduces to heating by absorption, and cooling by radiation and latent heat release:

TABLE 6.10
Thermophysical Properties of Sublimating Ices at Low Temperatures

Ice	C_p (J kg^{-1} K^{-1})	K_T (J m^{-1} s^{-1} K^{-1})	ρ (g cm^{-3})	Γ (J m^{-2} s$^{-1/2}$ K^{-1})
H_2O (40 K)	350	2×10^{-3}	930	26
N_2 (40 K)	1300	2.1×10^{-4}	1000	5.3
CO_2 (40 K)	450	2.4×10^{-3}	1700	14
CH_4 (40 K)	180	5×10^{-4}	520	26
Rhea regolith (90 K)	350	2×10^{-7}	500	0.3

Source: Adapted from Spencer, J.R. and Moore, J.M., *Icarus*, 99, 261, 1992.

For CH_4 from Lellouch, E. et al., *Icarus*, 147, 220, 2000.

For CO_2 from Palmer, E.E. and Brown, R.H., *Icarus*, 195, 434, 2008.

Γ is the thermal inertia defined as: $\Gamma = \sqrt{K_T \rho C_p}$.

$$F_{\odot p}(1 - A_B) = \varepsilon \sigma_B T^4 + L_i \frac{dM_S}{dt} \tag{6.76}$$

For sublimation occurring in free vacuum, the mass and latent heat transport can be related to the vapor pressure equation (for sublimation) and temperature through an expression given by Estermann (1955):

$$\frac{dM_S}{dt} = P_{vsat}(T)\left[\frac{1}{2\pi R_g^* T}\right]^{1/2} \tag{6.77a}$$

where dM_S/dt is given in SI units in kg m^{-2} s^{-1}. The time integration of Equations 6.76 or 6.77a over a given area gives the total amount of gas release into the atmosphere (Problems 6.15 through 6.17). If a balance between sublimation and condensation is considered, then for a surface pressure P_s, the above equation can be written as

$$\frac{dM_S}{dt} = \left[P_{vsat}(T) - P_s\right]\left[\frac{1}{2\pi R_g^* T}\right]^{1/2} \tag{6.77b}$$

For $P_s > P_{vsat}$ condensation occurs, while if $P_s < P_{vsat}$ sublimation takes place. A first order estimate of the atmospheric pressure distribution across a body produced by the frost sublimation can be obtained assuming equilibrium with the local saturation vapor pressure at the equilibrium temperature. This is sometimes called a *"buffered"* atmosphere. The surface temperature can be written according to Equation 6.76 in terms of the angle φ from the subsolar point, including the temperature T_N of the nonilluminated side (the night) as

$$T_s(\varphi) = \left[\frac{F_{\odot p}(1 - A_B)}{\varepsilon \sigma_B}\right]^{1/4}(\cos\varphi)^{1/4} + T_N \tag{6.78}$$

for $\varphi < 90°$ and $T_s = T_N$ for $\varphi > 90°$. It follows that the atmospheric pressure will be simply for this atmosphere:

$$P_s = P_{vsat}(T_s) \tag{6.79}$$

See Problem 6.18.

6.7.3 OUTGASSING FROM THE INTERIOR: VOLCANIC AND GEYSER ACTIVITY

Volcanic and geyser-like outgassing activity occurs currently on the following known geologically active bodies: the Earth, Io, Enceladus, Titan (suspected), and Triton. The gases are releases from sporadic localized areas at the surface (calderas, rifts, vents, etc.). In contrast to sublimating atmospheres whose source is around the subsolar region, volcanic atmospheres can be centered at any position on the body and can have source areas at different locations acting simultaneously. As a consequence of the gas injection, the atmospheric pressure increases locally around these sources. We do not enter on the energy sources, heat transport mechanisms, and origin of this activity, for example, hot volcanism as on the Earth and Io or *cryovolcanism* in the other bodies (see Chapter 1 for some details, in particular Section 1.5). The mass of the gases released per unit area will simply be given by

$$\left(\frac{dM}{dt}\right)_e = \rho v_e \tag{6.80}$$

where
   ρ is the gas density
   v_e its ejection speed
   $(dM/dt)_e$ is given in $kg\,m^{-2}\,s^{-1}$

If the diameter D of the vent or caldera is known, then the ejected gas mass per unit time is $dM/dt_e = \pi(D/2)^2 \rho v_e$ $(kg\,s^{-1})$. See Problem 6.10.

6.8   OVERVIEW OF THE TENUOUS ATMOSPHERES

The main properties, atmospheric composition, and surface temperature and pressure of planets, dwarf planets, and satellites with tenuous atmospheres are summarized in Table 6.9. In the following sections, we add some additional comments on their characteristics.

6.8.1   MERCURY AND MOON

These two bodies have extremely rarefied (i.e., low collision rate between atoms) and low-density atmospheres. Their low mass atomic composition (H and He) comes from the capture of the solar wind particles, whereas the higher mass atoms (Na, Ca, K on Mercury) have their origin in the surface sputtering produced by micrometeorites and energetic particle bombardment. Surface *photodesorption* from solar photons (photodesorption also called photosputtering is a phenomenon whereby species are released from or through a surface by background energetic radiation), and direct evaporation from the hot Mercury surface can also intervene in creating the atmosphere.

On Mercury, the most abundant elements are O and Na with atomic densities $\sim 10^4$ atoms cm^{-3} and He with densities 6×10^3 atoms cm^{-3}. Due to Mercury's low gravity, the sodium atoms form a tail that leaves the planet and extends away by ~ 16 planetary radii. On Mercury, the exobase is in fact placed in the surface where the temperature $\sim 600\,K$. The magnetic field of Mercury plays a role not only in protecting the atmosphere from erosion and escape (Section 2.4.1) but also in moving the ionized atoms from the intense UV solar radiation areas from one place to another.

Helium and argon are the most abundant components of the lunar atmosphere with day–night variable densities in the range $\sim 10^3$–10^4 atoms cm^{-3}. Their lifetime above the surface is low. The scale height for these tenuous atmospheres is for

Mercury ~13–95 km and for the Moon is about 65 km. In the cold polar areas of both bodies (where $T \sim 50$–60 K), there could be water–ice deposits (in the base of craters protected by the solar radiation), and molecules could escape and survive just above the surface. In fact, the fly-by of Mercury by the Messenger spacecraft in 2008 discovered water-related ions like O^+, OH^-, and H_2O^+ (in trace quantities) above the surface. It is not yet established if their origin is in the hypothetic polar deposits or is due to comet material or produced by sputtering.

6.8.2 Io

The satellite Io is unique in the solar system due to its high volcanic activity (200 volcanic calderas, most with sizes above 20 km) and the presence of a complex atmosphere and interaction with Jupiter's magnetosphere. The primary sources for its atmosphere are the volcanic gaseous emissions (typically 10 sources are simultaneously active) and the sublimation of the surface SO_2 frost (Problems 6.18 through 6.20) (Figure 6.19). The SO_2 released by volcanic eruption freezes and deposits onto the surface when the temperature descends below the condensation value as during

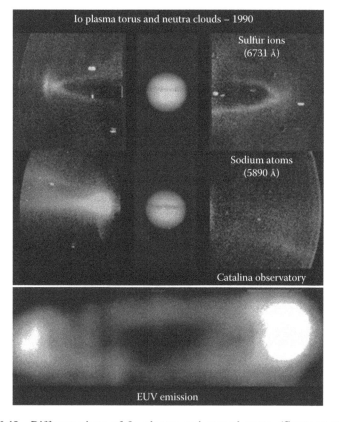

FIGURE 6.19 Different views of Io plasma and neutral torus. (Courtesy of Cassini, Laboratory for Atmospheric and Space Physics (LASP), University of Arizona, Boulder, Colorado and N. M. Schneider and J. T. Trauger, Catalina Observatory, Tucson, Arizona.)

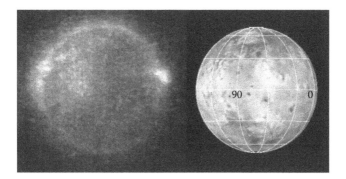

FIGURE 6.20 Io aurora emission (left) with a satellite image as a reference (right), as observed by the Galileo spacecraft. (Courtesy of NASA-JPL, Los Angeles, CA.)

the night hemisphere or during the eclipses by Jupiter. Temperatures at noon reach on Io a value of 110–120 K, but on the volcano calderas they are as high as 300–600 K (with hot spots of 900–1200 K).

The atmosphere is inhomogeneous in its spatial distribution and has additional components as SO, S_2, S, O and traces of Na, K, Cl, NaCl, and H (Table 6.9). On the average, the column abundance of SO_2 is $N \sim 10^{16}$–10^{17} cm^{-2} with patches that cover 11%–15% of the surface of a higher density $N \sim 10^{18}$ cm^{-2}. Above some volcanoes, the S_2 abundance reaches $N \sim 10^{16}$ cm^{-2}. The column abundances for the other species are $\sim 10^{12}$–10^{14} cm^{-2}. An isothermal exosphere is predicted above the exobase ~ 100–300 km from the computed heating–cooling rates and the use of Equation 6.53. An ionosphere with peak electron densities $N \sim 10^4$–10^5 cm^{-3} extends between altitudes ~ 200 and 400 km above the surface.

Around Io, a plasma torus (donut shape) and a neutral gaseous tail (banana shape) extend along its orbit. The neutral "cloud" has O and S atoms as main constituents with traces of Na and K that are more easily detected by resonance scattering. The cloud extends forward, with the elements following a keplerian orbit. The plasma is mostly composed of ions of O, Cl, S, and SO_2 mixed with protons, and concentrated in the center of the torus, and is spatially and temporally variable. The excitation of the ions by electron collision produces emission that makes the torus visible at optical wavelengths.

Aurora emissions also occur as a consequence of the SO_2 atmosphere when the electrons excite oxygen and sulfur atoms, and dissociate and excite SO_2. The aurora emissions are concentrated in the equatorial area and are detected in the visible and far-ultraviolet wavelengths (Figure 6.20).

6.8.3 EUROPA, GANYMEDE, CALLISTO

The three satellites of Jupiter, Europa, Ganymede, and Callisto, have tenuous atmospheres that form above a water–ice rich crust mixed with other icy components as CO_2 (Table 1.6). The Europa water–ice surface and crust contain impurities such as hydrated salts and a reddish component. Its temperature ranges between 85 and 133 K depending on the albedo of the markings, but at the poles, it can descent to 50 K. Due to its close proximity to Jupiter, immersed in its intense magnetic field, Europa surface is bombarded with particles and heavy ions with energies up to 10 MeV, producing sputtering that together with radiolysis (the dissociation of surface molecules by solar UV radiation) creates an exosphere basically made of oxygen (O_2) with few sodium atoms. For oxygen, the disk averaged vertical column density in the trailing hemisphere ranges from 10^{14} to 10^{15} cm^{-2}. For Na atoms, the column abundance is 10^9 cm^{-2} spreading in a tail out of the satellite more than 25 satellite radius.

Ganymede is the largest satellite and possesses an intrinsic dipolar magnetic field and a surrounding magnetosphere. The sputtering and sublimation of the surface ices produce an atmosphere of molecular oxygen similar to that of Europa. The oxygen atmosphere is seen by the airglow or aurora emission at UV wavelengths of 130.4 and 135.6 nm produced by the dissociation excitation of the molecule from the electron impact. A thin veil of hydrogen also surrounds the satellite. The vertical column density for O_2 is in the range 10^{14}–10^{15} cm^{-2} and that of H is 2×10^{12} cm^{-2}. Models predict that the sublimation of H_2O at the subsolar point should dominate the atmospheric mass production. The models also predict the existence of an ionosphere, but its existence and properties are a controversial matter.

Callisto has a low-density atmosphere formed by CO_2 detected by its airglow emission 100 km above the surface, with a total column abundance of ~8×10^{14} molecules cm^{-2}. The models suggest that an ionosphere and oxygen atmosphere with a similar density should also be present. Callisto does not have an intrinsic magnetic field but instead a field induced by the Jovian electromagnetic environment. A strong electrodynamic interaction with this atmosphere and ionosphere is also predicted.

6.8.4 ENCELADUS

Enceladus, a small satellite of Saturn, has giant active plumes of gas and water–ice particles that are a source for Saturn's E ring. The plume emanates from multiple fractures called as the "tiger stripes," a tectonically active area in the south polar region. The mean frigid surface temperatures on Enceladus range from ~37 K in the north polar night to a diurnal range of 50–75 K at low latitudes to perhaps 145 K or more in the warmest surface area of the tiger stripes where temperatures can elevate to 180 K. The plume is composed of 91% H_2O, 3.2% CO_2, 4% N_2 or CO, and 1.6% CH_4T. The water vapor column abundances are ~10^{16} cm^{-2} with a flux of 10^{27}–10^{28} molecules s^{-1}. Due to the weak satellite gravity ($g = 0.12$ m s^{-2}), most of the gases and particles ejected by the plumes get lost in space, with some plumes extending 1000–3000 km above the surface (Figure 6.21).

6.8.5 TRITON

Triton is the largest satellite of Neptune that has a very cold surface temperature of 38 K due to great distance from the Sun. It possesses a tenuous atmosphere of nitrogen (integrated column density ~9×10^{22} cm^{-2}) with trace amounts of methane (mixing ratio 10^{-4}). The surface is covered with a thin layer of nitrogen and methane ice that sublimates and condenses following the insolation cycle so the atmosphere is in equilibrium with surface nitrogen ice. Triton shows also cryovolcanic activity

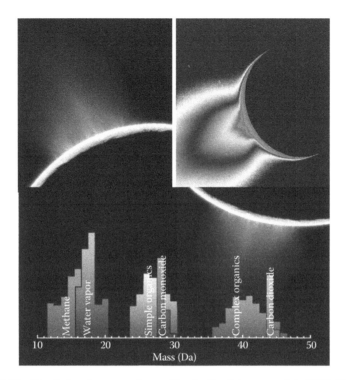

FIGURE 6.21 Enceladus plumes as observed by the Cassini spacecraft in 2005–2008 (upper two images). Eight source locations for these geysers have been identified along surface fractures in the south polar region. The lower panel is a mass spectrum that shows the chemical constituents sampled in Enceladus' plume by Cassini's ion and neutral mass spectrometer during its fly-through of the plume on March 12, 2008. The amounts are given in atomic mass per elementary charge (Daltons [Da]) of water vapor, methane, carbon monoxide, carbon dioxide, simple organics, and complex organics identified in the plume. (Courtesy of NASA/ESA, Plymouth, U.K. and ASI, Tempe, AZ.)

with eruptions of geyser-like plumes that reach altitudes of ~8 km, injecting nitrogen to the atmosphere that is spread by the winds (Problem 6.22).

The temperature lapse rate is $dT/dz = -g/C_p = -0.75$ K km^{-1} if dry convection is assumed to dominate the heat transport in the lower atmosphere. However, the latent heat released during the sublimation at the surface contributes to the thermal structure. Then, the temperature structure can be assumed to be that of a N_2 saturated adiabat, decreasing with altitude from the surface up to a minimum at the tropopause at ~8 km altitude. From the hydrostatic equation, the gas perfect law and the Clausius–Clapeyron equation, the wet adiabatic lapse rate (5.42) is given by

$$\frac{dT}{dz} = -\frac{gT}{L_{N_2}} \tag{6.81}$$

where the latent heat of sublimation of nitrogen per unit mass is at 38 K, $L_{N_2} = 2.5 \times 10^5$ J kg^{-1}. This predicts a temperature decrease with altitude at a rate 0.1 K km^{-1}.

However, dynamical transport by winds of heat and mass influences the temperature structure of the troposphere (in Section 7.2.9 we give the mass condensation flow). Above tropopause, heat conduction becomes important and a heated thermosphere (see Section 6.4) develops with a temperature of ~100 K (Figure 6.22; Problem 6.22).

A diffuse haze formed by small particles of 0.1 μm covers the satellite (Table 5.4), being most probably composed of hydrocarbons and nitriles formed by photochemical reactions induced by sunlight (Figure 6.23). The haze extends vertically up to 20–30 km with maximum particulate concentration in the lower altitude 5 km.

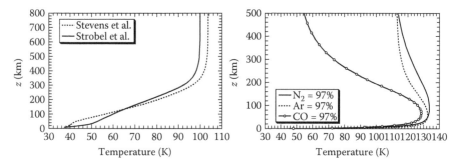

FIGURE 6.22 Models of the vertical temperature profiles in the atmospheres of the satellite Triton (left) and of the dwarf planet Pluto (right). (From Strobel, D.F. et al., *Icarus*, 120, 266, 1996. With permission.)

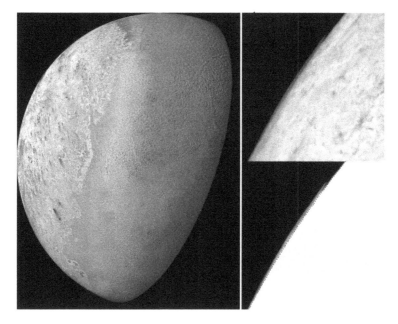

FIGURE 6.23 Triton geysers are visible as dark elongated patches in the frozen surface of the satellite (left). The thin cloud and haze layers are seen above the limb (right) detached in the lower disk saturated image. Images obtained by the spacecraft Voyager 2 during its fly-by in 1989. (Courtesy of NASA-JPL, Los Angeles, CA.)

Discrete densest clouds 10 km wide and few hundredth km long are seen in the troposphere (altitudes 1–3 km). They probably consist of nitrogen ice suspended particles (sizes ranging from 0.2 to 2 μm) formed from the condensation of the main atmospheric constituent (Table 5.4). A third aerosol component is released by the geyser-like plumes formed by dark particulates made probably of complex organic material.

6.8.6 PLUTO

Pluto has large similarities with Triton in that it possesses a nitrogen ice covered surface (mixed with traces of methane and carbon monoxide) at a low temperature (~40–50 K) that sublimates and freezes with the seasonal cycle, forming a nitrogen atmosphere with traces of the other two compounds (total bulk density at surface $\sim 10^{15}–10^{16}$ cm^{-3}). The atmosphere mass varies seasonally, significantly (Section 6.7.2) due to its considerable eccentric orbit around the Sun (perihelion at 29.7 AU and aphelion at 49.54 AU on a 248-year orbital period), with additional variations at specific locations with season due to Pluto's high obliquity, as observed from stellar occultation data sets in time. Observations and models to date support the interpretation that Pluto's atmosphere represents equilibrium sublimation of surface ices (N_2, CH_4, and CO) due to solar heating.

The radiative–conductive models of the vertical thermal structure constrained by the stellar occultation observations indicate that the temperature increases in the atmosphere at a rate of 2.2 K km^{-1} from the surface at 40 K up to a maximum of 100 K and then should decrease slowly with altitude (Figure 6.22). The occultation results also indicate the existence of mass motion (winds) in the tenuous atmosphere, perhaps originating in the day-to-night temperature contrasts.

Pluto has a large moon when compared to the planet size, Charon (radii 603.5 km) that orbits with a period synchronized to Pluto's spin of 6.387 days. No atmosphere has been detected on it but occultation observations indicate an upper limit of 110 and 15 nbar, assuming respectively a pure nitrogen or pure methane atmosphere.

PROBLEMS

6.1 On Earth, below 75 km in height, it holds in Equation 6.13 that $k_{12}[O][O_2]$ $[M] \gg k_{13}[O][O_3]$. Then combine Equations 6.14 to find a simplified expression relating the equilibrium ozone concentration $[O_3]$ as a function of the $[O_2]$ concentration in Earth's atmosphere. Calculate the equilibrium ozone concentration at the altitude $z = 25$ km where the mean photolysis rate for oxygen J_2 ($z = 25$ km) $= 3 \times 10^{-12}$ s^{-1} and ozone production by reaction (6.12b) has $k_{12} = 6 \times 10^{-46}$ $T/300^{-2.3}$ (m^6 s^{-1}), and photodissociated by reaction (6.12c) has J_3 ($z = 25$ km) $= 5.5 \times 10^{-4}$ s^{-1}, and reaction (6.12d) has $k_{13} = 8 \times 10^{-18}$ $e^{-2060/T}$ (m^3 s^{-1}). Take at $z = 25$ km the air density $= 40$ g m^{-3}, $T = 221.6$ K.

Solution

$$[O_3] = [O_2] \left(\frac{J_2 K_{12}}{J_3 K_{13}} [M] \right)^{1/2}, \quad [O_3] = 1.5 \times 10^{13} \text{ cm}^{-3}$$

6.2 Find an approximate form to Equation 6.41 by expanding the exponential about the altitude of maximum electron density z_M for small vertical distances compared to H and for an overhead Sun ($\mu_\varphi = 1$) (*parabolic approximation*). Consider the case of Earth's ionosphere at $z = 200$ km with electrons resulting from oxygen ionization. For $\tau = 1$, the number density of oxygen atoms is $N_O = 10^{10}$ cm^{-3}, $\sigma_O = k = 10^{-17}$ cm^2, $\alpha_r = 10^{-7}$ cm^3 s^{-1}, the ionization energy 13.6 eV at 0.1 μm where the energy flux is 3×10^{-3} W m^{-2} μm^{-1} in the bandwidth 0.05–0.1 μm. Calculate at $z = z_M$ the electron density using the parabolic approximation.

Solution

$$N_e(z,1) \approx \sqrt{\frac{P_{max}}{\alpha_r}} \left[1 - \left(\frac{z - z_M}{2H} \right)^2 \right], \quad N_e = 1.07 \times 10^5 \text{ cm}^{-3}$$

6.3 Derive Equation 6.57. Calling Z the altitude term in the right part of this equation, calculate the exospheric temperature of the Earth ($T_0 = 300$ K, $Z = 100$ km, $\epsilon F_{XUV} = 1.1 \times 10^{-3}$ W m^{-2}, $A_T = 3.6 \times 10^{-3}$ W m^{-1} K$^{-(s+1)}$, and $s = 0.8$) and in Jupiter ($T_0 = 180$ K, $Z = 100$ km, $A_T = 2.52 \times 10^{-3}$ W m^{-1} K$^{-(s+1)}$ and $s = 0.75$). Compare this first order calculation with the observed values T_∞ (Earth) and T_∞ (Jupiter) ~1000 K. What conclusion can you derive?

Solution
(1) T_∞ (Earth) = 448 K, T_∞ (Jupiter) = 357 K
Other important heating sources are in action.

6.4 From Equation 6.53 find an expression for the column integrated heating rate Q_H (in W m^{-2}) required to produce the boundary thermospheric temperatures T_∞ (exosphere) and T_0 (base of thermosphere) (cf. also Equation 6.57) if the thermal conductivity is $K_T = A_T T^s$ and the energy flux is deposited in a layer of thickness Δz above the base of the thermosphere at level z_0, and conducted downward where it is radiated away. Apply to calculate Q_H in Uranus thermosphere assuming $T_\infty = 800$ K, $T_0 = 100$ K, and $\Delta z = 2000$ km. Take $A_T = 2.5 \times 10^{-3}$ W m^{-1} K$^{-(s+1)}$, and $s = 0.75$.

Solution

$$Q_H = \frac{A_T}{(s+1)\Delta z} (T_\infty^{s+1} - T_0^{s+1}), \quad Q_H(\text{Uranus}) = 8.5 \times 10^{-5} \text{ W m}^{-2}$$

6.5 Find an expression for the molecular thermal diffusivity D_T as a function of pressure and temperature if $K_T = A_T T^s$. Calculate its value for helium ($A_T = 2.1 \times 10^{-3}$ W m^{-1} K$^{-(s+1)}$ and $s = 0.75$, $C_p/R_g = 2.5$) at 500 K and 1 atm. Find an expression for D_T for a mixture of two gases with mole fractions ν_1 and ν_2 if $C_{pi} = (\alpha_i + \beta_i T)/\mu_i$.

Solution

$$D_T = \frac{A_T T^{s+1}}{P(C_p/R_g)}, \quad D_T(\text{He}) = 4.4 \times 10^{-4} \ \text{m}^2 \text{s}^{-1} \text{ at } 500 \text{ K and 1 atm};$$

$$D_T(T,P) = \frac{A_T T^{s+1}}{(P/R_g)\left[v_1^2(\alpha_1 + \beta_1 T) + v_2^2(\alpha_2 + \beta_2 T)\right]}$$

6.6 Find an expression for the time constant for conductive heat transport in the thermosphere in terms of the mass per unit area (m_S), temperature contrast across it (ΔT), temperature gradient, and thermal conductivity. Apply to the thermosphere of (1) Venus ($m_S = 1 \text{ kg m}^{-2}$, $C_p = 670 \text{ J kg}^{-1} \text{K}^{-1}$, $T = \Delta T = 100 \text{ K}$, $dT/dz = 1.1 \times 10^{-3} \text{ K m}^{-1}$, $K_T = A_T T^s$ with $A_T = 0.82 \times 10^{-5} \text{ W m}^{-1} \text{K}^{-(s+1)}$, and $s = 1.82$), (2) Neptune ($m_S = 0.9 \text{ kg m}^{-2}$, $C_p = 13{,}000 \text{ J kg}^{-1}$ K^{-1}, $T = 160 \text{ K}$, $\Delta T = 25 \text{ K}$, $dT/dz = 8.3 \times 10^{-5}$ K m^{-1}, $A_T = 2.5 \times 10^{-3}$ W m^{-1} K$^{-(s+1)}$, and $s = 0.74$).

Solution

$$\tau_{cond} = m_S C_p \Delta T \left(K_T \frac{\partial T}{\partial Z} \right)^{-1}; \quad \tau_{cond}(\text{Venus}) = 195 \text{ days}$$

$$\tau_{cond}(\text{Neptune}) = 347 \text{ years}$$

6.7 Assume an equilibrium situation for the species distribution on Earth's thermosphere such that at the altitude $z_0 = 150 \text{ km}$, the abundances of the following components are: $N(\text{Ar}) = 5 \times 10^{13} \text{ m}^{-3}$, $N(\text{N}_2) = 5 \times 10^{16} \text{ m}^{-3}$, $N(\text{O}) = 5 \times 10^{16} \text{ m}^{-3}$, and $N(\text{He}) = 2 \times 10^{13} \text{ m}^{-3}$. If the thermospheric temperature remains constant at $T = 1000 \text{ K}$ up to $z = 500 \text{ km}$, calculate the relative abundances of argon, atomic oxygen, and helium, relative to molecular nitrogen between these two altitude levels.

Solution

$$\left(\frac{N_E}{N_0} \right)_i \left(\frac{N_0}{N_E} \right)_{N_2} = 1(\text{N}_2), 0.06(\text{Ar}), 174(\text{O}), 22{,}620(\text{He})$$

6.8 Venus upper atmosphere shows strong temperature changes with local time. During day time above the reference level $z_0 = 160 \text{ km}$, the temperature in the thermosphere reaches 300 K, descending during nighttime to 150 K so the layer is called the *cryosphere*. Find the level above which the abundance of helium relative to the predominant CO_2 atmosphere dominates both in day and nighttime.

Solution
Thermosphere at $z = 163.5 \text{ km}$, cryosphere at $z = 167 \text{ km}$.

6.9 Calculate the vertical number density distribution for a minor constituent $N_1(z)$ in an isothermal atmosphere for a constant flux and for a total number density distribution for the eddy diffusion coefficient varying with altitude according to: (1) $D_{eddy}=D_0$(constant); (2) D_{eddy} (z) exp $(z - z_0)/H$.

Solution

$$(1) \ N_1(z) = \left[N_1(z_0) + \frac{\Phi_1 H}{D_0} \right] \exp\left[-\frac{(z-z_0)}{H} \right] - \frac{\Phi_1 H}{D_0}$$

$$(2) \ N_1(z) = \left[N_1(z_0) - \frac{\Phi_1}{D_0}(z-z_0) \right] \exp\left[-\frac{(z-z_0)}{H} \right]$$

6.10 Consider the diffusion of methane on nitrogen in the cold atmosphere of Titan. (1) Calculate the methane flux due to molecular and eddy diffusion at altitudes $z=500$ km $(T=140$ K, $N=10^{14}$ cm^{-3}, $N(CH_4)=3\times10^{10}$ cm^{-3}) and at $z=1000$ km $(T=180$ K, $N=10^{10}$ cm^{-3}, $N(CH_4)=4\times10^8$ cm^{-3}) assuming a simple linear interpolation between these two levels and finding other suitable values for the other quantities from tables or diagrams in the text. (2) Compare the chemical time constant at 1000 km if the methane photolysis rate is 6×10^{-8} s^{-1} with the molecular and eddy diffusion times at this level. (3) What are the diffusion time scales at 500 km?

Solution

(1) Φ_{CH_4} (500 km) $\sim 5\times10^{10}$ cm^{-2} s^{-1}, Φ_{CH_4} (1000 km) $\sim -2.2\times10^{11}$ cm^{-2} s^{-1}
(2) At $z=1000$ km : $\tau_{Chem}=1.6\times10^7$ s, $\tau_{eddy}=8.7\times10^4$ s, $\tau_{CH_4}=4.1\times10^5$ s
(3) At $z=500$ km : $\tau_{eddy}=7\times10^8$ s, $\tau_{CH_4}=8\times10^6$ s

6.11 Compare the limiting flux and the Jeans escape flux (from Chapter 2) to evaluate the escape rate of hydrogen atoms in Earth's atmosphere. For the limiting flux, take a hydrogen mixing ratio (for all hydrogen bearing molecules H_2O, CH_4 and H_2) of 10^{-5} below the homopause at $z \sim 100$ km $(T=210$ K) and use the diffusion coefficients for H_2 on air. For the Jeans escape flux of hydrogen atoms, take as typical exospheric parameters $N_{exo}=10^5$ cm^{-3} and $T_{exo}=900$ K, and values for the magnitudes from tables in the book.

Solution
$\Phi_L=2.4\times10^8$ cm^2 s^{-1}; $\Phi_J=3.3\times10^7$ cm^2 s^{-1}.

6.12 Consider an isothermal atmosphere in a steady state that is irreversibly losing a compound due to photodissociation at a rate J (see Equation 6.4) directly proportional to the number density N and no production term in Equation 6.58. Assuming that J and D_{eddy} are constant, find the differential equation for $N(z)$ and its solution for the boundary conditions that require N to be finite for $z \to \infty$ and $N=N_0$ at the surface in terms of the scale height h for this compound and the mean atmospheric scale height H. Find the asymptotic values for $N(z)$ for the cases $(J/K)/(1/4H^2) \gg 1$ and $\ll 1$.

Solution

$$\frac{d^2N}{dz^2} + \frac{1}{H}\frac{dN}{dz} = \frac{J}{K}N; \quad N(z) = N_0 e^{-z/h} \quad \text{with } h = \left[-\frac{1}{2H} + \sqrt{\frac{1}{4H^2} + \frac{J}{K}} \right]^{-1}$$

For $(J/K)/(1/4H^2) \ll 1$, $h=0$ and $N(z)=N_0$.

For $(J/K)/(1/4H^2) \gg 1$, $h=(K/J)^{1/2}$ and $N(z)=N_0 \exp[z(K/J)^{-1/2}]$

6.13 Apply the results of Problem 6.12 for the following two situations: (1) Calculate h in Jupiter's troposphere ($H=17$ km) for ammonia that is irreversible converted to N_2 and N_2H_4 (hydrazine) in 38 days assuming that $D_{eddy}=2\times10^4$ cm^2 s^{-1}. (2) Calculate the eddy diffusion coefficient in Triton tenuous atmosphere ($H=20$ km) if methane, whose scale height is 8 km, is photodissociated in 3.17 years.

Solution

(1) $h=2.8$ km
(2) $D_{eddy}=4.4\times10^3$ cm^2 s^{-1}

6.14 Protons trapped in the Jovian magnetosphere moving at a speed of 13,800 km s^{-1} impinge on the surface of the satellite Ganymede. (1) How this speed compares with the orbital speed of the satellite? (2) What is their kinetic energy? (3) What is the sputtering yield Y_{sput} for O_2 if from laboratory data impacting protons give $\ell(dE/dz)_e/U_b=800$ and $Cf_e=0.125$? (4) If the ion flux for this proton energy is $\Phi_i=100$ protons/(cm^2 s sr keV)$^{-1}$, what is the sputtered O_2 flux?

Solution

(1) $v_{orb}=10.8$ km s^{-1}
(2) E_K (protons)$=1000$ keV
(3) $Y_{sput}=100$ molecules O_2 per proton
(4) $\Phi_s=10^7$ O_2 molecules (cm^2 s sr)$^{-1}$

6.15 (1) Calculate the sublimation mass rate for CO_2 frost on Mars for the temperatures 145 and 180 K using Equation 6.77. (2) Compare the value at 145 K with that expected from the thermal balance (Equation 6.76) for CO_2 ice in Mars poles (take an average $F_{op}=78$ W m^{-2}, $A_B=0.65$, $\varepsilon=0.95$, $L_{CO_2}=6\times10^5$ J kg^{-1}). Data

$$P_{vsat}(\text{bar}) = \frac{1}{760}10^{\left(-\frac{1275.62}{T}+0.006833T+8.3071\right)}$$

Solution

(1) (a) $T=145$ K, $dM_s/dt=10^{-5}$ kg m^{-2} s^{-1}; (b) $T=180$ K, $dM_s/dt=8\times10^{-4}$ kg m^{-2} s^{-1}
(2) $T=145$ K, $dM_s/dt=5.8\times10^{-6}$ kg m^{-2} s^{-1}

6.16 Using the data from question (1) (a) in the previous Problem 6.15, estimate the total CO_2 mass sublimated in a year from a Martian pole that extends up to latitude 55°. Then estimate how much the atmospheric pressure will increase on Mars due solely to this sublimation if the gas is distributed over the whole Mars surface.

Solution

$$M_{CO_2} = 4.1 \times 10^{15} \text{ kg}, \quad \Delta P = 1.08 \text{ mbar}$$

6.17 Assume that the satellite Triton is fully covered with N_2 frost ($A_B = 0.8$). (1) Determine the flow of sublimated nitrogen, vertically integrated Φ (kg m^{-1} s^{-1}), from the summer to the winter hemisphere if half of the deposited solar energy is employed to sublimate the ice (take an average $F_{\odot p} = 1.5$ W m^{-2}, $R_T = 1353$ km, $L_{N_2} = 2.5 \times 10^5$ J kg^{-1}). (2) Calculate the time constant for the interhemispheric mass flow that is approximately the ratio between the integrated column mass of the atmosphere and the flux of mass subliming from the surface as obtained above. (3) Compare this value with Triton day and year, and with the radiative time constant at the surface. Take suitable values for the parameters from tables in the book.

Solution

(1) $\Phi = \dfrac{F_{\odot p}(1 - A_B)R_T}{4L_{N_2}} = 0.4 \text{ kg m}^{-1}\text{s}^{-1}$

(2) $\tau_\Phi = \dfrac{P/g}{[F_{\odot p}(1 - A_B)/4L_{N_2}]} = 77 \text{ days}$

(3) $\tau_\Phi/\tau_{day} = 13$, $\tau_\Phi/\tau_{year} = 1.3 \times 10^{-3}$ (frost on Triton is in instantaneous equilibrium with solar radiation), $\tau_{rad} = 1.32$ Triton days.

6.18 A model of Io atmosphere assumes that is buffered due to sublimation/condensation of SO_2 frost. Calculate the atmospheric pressure at the subsolar point if the saturation vapor pressure for SO_2 in bars is given by $P_{vsat}(T) = 1.516 \times 10^8 \exp(-4510/T)$ and $T_N = 50$ K, $A_B = 0.96$, $\varepsilon = 0.9$.

Solution
$P_s = 1.1 \times 10^{-7}$ bar.

6.19 Episodic volcanic activity in Io (of silicate and sulfur origin) and geyser-like activity (frozen N_2) in Triton produce a local injection of gases into the atmosphere. Assume that gas ejection occurs in Io volcanoes through a section with a diameter of 1 km and in Triton through a hole with a diameter 350 m. Measurements of the altitude reached by the gaseous plume give 400 km on Io and 8 km on Triton. (1) Calculate the vertical velocity at the ejection point assuming that the gas moves in the free space (no atmospheric pressure present). (2) Calculate the maximum ejected mass per unit time if the gas density at the ejection point in Io is 1.5 kg m^{-3} and in Triton geyser is 10^{-4} kg m^{-3}.

Solution

(1) (a) Triton: $v_e = 110\,\mathrm{m\ s^{-1}}$, (b) Io: $v_e = 1200\,\mathrm{m\ s^{-1}}$
(2) (a) Io: $(dM/dt)_e = 7.4 \times 10^7\,\mathrm{kg\ s^{-1}}$; (b) Triton: $(dM/dt)_e = 1.06 \times 10^3\,\mathrm{kg\ s^{-1}}$

6.20 Consider a model of a satellite where sublimation and a volcanic source inject the atmospheric mass. (1) Find an expression for the atmospheric pressure at an angular distance φ from the subsolar point assuming that sublimation is controlled by the buffered model (Equations 6.78 and 6.79) and that the mass flux injected from the volcanic source is $(dM/dt)_e$ and occurs during a time Δt after which it is distributed homogeneously above the satellite surface. (2) Apply to the satellite Io for $\varphi = 30°$ using the data of Problem 6.18 for a volcano that has ejected a mass flux of $10^4\,\mathrm{kg\ s^{-1}}$ during 30 days.

Solution

(1) $P_s(\varphi) = P_{vsat}(T_s) + \left(\dfrac{dM}{dt}\right)_e \left(\dfrac{\Delta t\, g}{4\pi R_p^2}\right)$ for $T_s(\varphi)$ given by Equation 6.78

(2) Io: $P_s(30°) = 6.2 \times 10^{-8}$ bar

6.21 Spectral measurements in Mercury's atmosphere of the resonance scattering lines by sodium D_1 and D_2 at 589 and 589.6 nm, respectively, give an integrated intensity $4\pi I = 400\,\mathrm{kR}$, and for the resonance emission line by calcium Ca I at 422.67 nm give $4\pi I = 422$ R. If the photon scattering coefficient is $g(\mathrm{Na}) = 0.8$ and $g(\mathrm{Ca}) = 2.3$ photons $\mathrm{s^{-1}\ molecule^{-1}}$, calculate the integrated column density of Na and Ca atoms and their partial pressure on Mercury surface. Take in (6.71) the phase function $p(\Theta) = 1/4\pi$.

Solution

$$N(\mathrm{Na}) = 5 \times 10^{12}\,\mathrm{atoms\ cm^{-2}}, \quad P_s(\mathrm{Na}) = 7 \times 10^{-14}\,\mathrm{bar}$$
$$N(\mathrm{Ca}) = 1.8 \times 10^8\,\mathrm{atoms\ cm^{-2}}, \quad P_s(\mathrm{Ca}) = 4.5 \times 10^{-18}\,\mathrm{bar}$$

6.22 Assuming steady state conditions, find the expression for the heat transport equation in spherical coordinates if there is only a radial dependence and the volume heating rate is Q. Give the integral solution for the radial temperature dependence if the thermal conductivity is given by $K_T = A_T T^s$. Then find the integrated energy flux F_0 in terms of T and r and of the temperature T_0 at a reference level r_0. From the conductive temperature lapse rate and using the hydrostatic equation, find the pressure versus temperature dependence. Apply these equations to the case of the atmosphere of Triton to find F_0 and the surface pressure assuming that at the tropopause $r_0 = 8\,\mathrm{km}$ ($T_0 = 37\,\mathrm{K}$, $P_0 = 6 \times 10^{-6}$ bar), and that in the thermosphere $r = 400\,\mathrm{km}$, $T = 95\,\mathrm{K}$ (for $s = 1.35$, $A_T = 2.6 \times 10^{-5}$ W $\mathrm{m^{-1}\ K^{-(1+s)}}$).

Solution

$$-\frac{1}{r^2}\frac{d}{dr}\left(r^2 K_T \frac{dT}{dr}\right) = Q(r); \quad T(r)^{s+1} = T_0^{s+1} + \frac{(s+1)}{A_T}\int_{r_0}^{r}\frac{dr}{r}\int_{x}^{\infty} x^2 Q(x)\,dx;$$

$$F_0 = \frac{A_T}{s+1}\frac{T^{s+1} - T_0^{s+1}}{\left(\left(r_0^2/r\right) - r_0\right)}; \quad P(T) = P_s \exp\left[\frac{g}{R_g^*}\frac{A_T}{F_0}\frac{T^s - T_0^s}{s}\right]$$

$$F_0 = -1.15\times10^{-6}\,\mathrm{W\,m^{-2}}; \quad P_s = 10^{-5}\,\mathrm{bar}$$

7 Global Atmospheric Motions

Motions in planetary atmospheres obey the laws of hydrodynamics that are valid for any fluid subjected to the general restrictions that apply to a planet: (1) spherical geometry (or spheroid) and (2) planetary rotation (atmospheric motions are considered relative to a non-inertial reference frame). Planetary atmospheres have an open boundary with the outer space (the exosphere, Chapters 1 and 6) and a rigid surface as a lower boundary in some cases (terrestrial planets, dwarf planets, and satellites), or a deep, not well-determined, lower boundary in the case of giant fluid planets. Atmospheres have external and internal energy sources to drive the motions (Chapter 1). They are by definition gaseous, and therefore are compressible fluids (obeying the perfect gas law as a first, in general good approach except in some particular cases). A large part of their mass is electrically neutral (called neutral atmosphere), and the motions can be described without the inclusion of electromagnetic forces. Therefore, motions in ionospheres are not considered here.

Since planets are large bodies ranging in two-scale orders (radii from ~10^5 km in large extrasolar planets to ~10^3 km), many different spatial scales are involved in the atmospheric motions. This chapter is devoted to the large-scale motions that globally affect the planet. In Chapters 8 and 9, we will extend the study of atmospheric dynamics to the formation of waves and instabilities within the global flow.

7.1 EQUATIONS OF ATMOSPHERIC MOTIONS

Motions are described by Newton's second law (in continuous fluid motions called *Navier–Stokes equation*) and the laws of thermodynamics. To these, we add the *continuity equation* that expresses the principle of mass conservation and the *equation of state* of the system (as stated above, the perfect gas law). Spherical (or spheroidal) coordinates are used in general (see Chapter 1) but Cartesian plane geometry description will be introduced when appropriate.

7.1.1 TOTAL AND PARTIAL DERIVATES

Atmospheric motions can be described according to the two classical views. In the Eulerian description, the flow is studied (measured) in a fixed point in space $\vec{r}$ relative to the reference frame, as is the situation with fixed stations on it (e.g., a Martian lander that is measuring wind speeds, temperature, pressure, etc.). In the Lagrangian description, the motion follows the atmospheric flow, that is, that of a small fluid parcel or "blob" particle, as marked, for example, with a dye. Although, most probably, the blob will be distorted in a short time, this view of the flow evolution is visually and conceptually important.

The rate of change of a field variable (say scalar A) following the motion (Lagrangian view) and its rate of change at a fixed point (Eulerian view) are related by means of the *total* derivative (*substantive, advective, or material* derivative) d/dt:

$$\frac{dA}{dt} = \frac{\partial A}{\partial t} + \left(u\frac{\partial A}{\partial x} + v\frac{\partial A}{\partial y} + w\frac{\partial A}{\partial z} \right) = \frac{\partial A}{\partial t} + \vec{u}\cdot\nabla A \qquad (7.1a)$$

$\vec{u} = u\hat{i} + v\hat{j} + w\hat{k}$ being the velocity vector that has components $u = dx/dt$, $v = dy/dt$, $w = dz/dt$ (expressed in Cartesian coordinates). Note that some textbooks use D/Dt instead of d/dt for the total derivative. The last term $\vec{u}\cdot\nabla A$ is called the *advection term* of the field variable A. The total derivative d/dt represents the rate of change with respect to time *following the motion* of the magnitude A (Lagrangian description), as compared to $\partial A/\partial t$ that represents the local rate of change with respect to time at a *fixed point* (Eulerian description). This relationship is symbolically written as

$$\frac{d}{dt} = \frac{\partial}{\partial t} + \vec{u}\cdot\nabla \qquad (7.1b)$$

(see Problem 7.1 for an application).

7.1.2 EQUATIONS

7.1.2.1 Continuity
The continuity equation expresses the principle of mass conservation in a control volume within the flow. If a flow element has a velocity $\vec{u}$ and density ρ, the continuity equation can be written as

$$\frac{\partial\rho}{\partial t} + \nabla\cdot(\rho\vec{u}) = 0 \qquad (7.2a)$$

or using the differentiation rule

$$\frac{\partial\rho}{\partial t} + \vec{u}\cdot\nabla\rho + \rho\nabla\cdot\vec{u} = 0 \qquad (7.2b)$$

and using the definition of the material derivative (7.1), we can write it as

$$\frac{d\rho}{dt} + \rho\nabla\cdot\vec{u} = 0 \qquad (7.2c)$$

For an incompressible fluid (constant density), the continuity equation reduces to

$$\nabla \cdot \vec{u} = 0 \qquad (7.3)$$

that is, $\mathrm{div}\,\vec{u} = 0$ (but see Pedlosky (1987) Chapters 1 and 6 for the use of this equation as a definition of incompressibility).

7.1.2.2 Equation of State

In general, we assume that atmospheres behave as ideal gases, as so far assumed:

$$P = \rho R_g^* T \qquad (7.4)$$

See Section 1.5.1 for other expressions for the equation of state (e.g., polytropic fluids).

7.1.2.3 Navier–Stokes Equation

The Navier–Stokes equation is the expression for the momentum conservation (Newton's second law) for fluid motions. We describe it in the following three situations.

7.1.2.3.1 Inertial Frame

In an inertial reference frame, Newton's second law for a fluid (Navier–Stokes equation) includes the action of the following three forces on a mass element: (1) the pressure gradient force ∇P (acting in the direction of the most rapid increase of pressure); (2) the gravity force (see Section 1.3), which accelerates the fluid parcel toward the center of a planet; (3) the frictional force $\vec{F}_f$ (interior to the fluid and from interaction with its boundaries). The momentum equation then becomes

$$\frac{d\vec{u}}{dt} = -\frac{\nabla P}{\rho} - g\hat{k} + \vec{F}_f \qquad (7.5)$$

The form of the frictional force is in general complicated and depends on the scale of the motion and on the intensity of the winds and wind shears acting on the fluid. We will discuss it later in this chapter, but initially to estimate an order of magnitude of its value, we introduce it in terms of a stress tensor τ_F that, for friction acting on the fluid in the x-direction, is written as $\tau_F = \eta_D (du/dz)$, η_D being the dynamic viscosity coefficient (the fluid is then said to be Newtonian). In three dimensions, it becomes $\vec{F}_f = \nabla \tau_F / \rho = (\eta_D / \rho) \nabla^2 \vec{u}$, $\upsilon_V = \eta_D / \rho$ being the kinematic viscosity (in units of m^2 s^{-1}).

7.1.2.3.2 Rotating Frame

Since all planets rotate about an axis with angular velocity $\vec{\Omega}$ (which is assumed to be constant), atmospheric motions are referred to relative to a non-inertial frame. For terrestrial and dwarf planets and satellites, this is the surface that rotates with the same angular velocity as the whole body. For giant fluid planets (Jupiter, Saturn, Uranus, and Neptune) that lack a solid surface, the atmospheric motions are generally referred to relative to a rotation frame, denoted as System III, as determined by the radio and magnetic field observations with a special note for Saturn (see Table 1.3).

In this frame, the Navier–Stokes equation must include the action of the "apparent or fictitious forces" resulting from the changes in the vectors and their time derivatives due to the planetary rotation. For a generic vector $\vec{A}$, its first and second derivatives are related in the *inertial frame I* and *rotating frame R* by the following relationships:

$$\left(\frac{d\vec{A}}{dt}\right)_I = \left(\frac{d\vec{A}}{dt}\right)_R + \vec{\Omega} \times \vec{A} \tag{7.6a}$$

$$\left(\frac{d^2\vec{A}}{dt^2}\right)_I = \left(\frac{d^2\vec{A}}{dt^2}\right)_R + 2\vec{\Omega} \times \left(\frac{d\vec{A}}{dt}\right)_R + \vec{\Omega} \times (\vec{\Omega} \times \vec{A}) \tag{7.6b}$$

In particular for $\vec{A} = \vec{r}$, the position or radius vector of a point, the following relationships for velocity and acceleration hold:

$$\vec{u}_I = \vec{u}_R + \vec{\Omega} \times \vec{r} \tag{7.7a}$$

$$\vec{a}_I = \vec{a}_R + 2\vec{\Omega} \times \vec{u}_R + \vec{\Omega} \times (\vec{\Omega} \times \vec{r}) \tag{7.7b}$$

The term $2\vec{\Omega} \times \vec{u}_R$ represents the *Coriolis acceleration* and is a vector perpendicular to the plane determined by $\vec{\Omega}$ and $\vec{u}_R$. The term $\vec{\Omega} \times (\vec{\Omega} \times \vec{r})$ is the *centripetal acceleration* and acts perpendicular to the rotation axis. Applying these relationships to Equation 7.5 and replacing $d\vec{u}/dt$ by $\vec{a}_I$ in (7.7b), the Navier–Stokes equation relative to the rotating frame (dropping the subindex R) can be written as

$$\frac{d\vec{u}}{dt} = -\frac{\nabla P}{\rho} - g\hat{k} - 2\vec{\Omega} \times \vec{u} - \vec{\Omega} \times (\vec{\Omega} \times \vec{r}) + \vec{F}_f \tag{7.8a}$$

The acceleration terms arising from the rotating frame are interpreted in terms of fictitious forces (per unit mass): the *Coriolis force* $-2\vec{\Omega} \times \vec{u}$ and the *centrifugal force* $-\vec{\Omega} \times (\vec{\Omega} \times \vec{r})$. The Coriolis force acts on a moving parcel if its direction is nonparallel to the rotating axis. The centrifugal force is usually added to the gravitational force (Section 1.3.4) to define an *effective gravity* $\vec{g}_{eff} = -g\hat{k} - \vec{\Omega} \times (\vec{\Omega} \times \vec{r})$, with the centrifugal term being in a different direction to gravity and much smaller in magnitude to it (see full development in Section 1.3.1). To simplify the notation, we drop in what follows the subindex *eff* and use the full vector $\vec{g}$ with a positive sign in the Navier–Stokes equation (it is therefore understood that the centrifugal term is included); so finally we get

$$\frac{d\vec{u}}{dt} = -\frac{\nabla P}{\rho} + \vec{g} - 2\vec{\Omega} \times \vec{u} + \vec{F}_f \tag{7.8b}$$

It must be noted that when making such a transformation from the inertial to the rotating system, the total time rate of change of any scalar magnitude remains unchanged. However, it must be stressed that the individual components of the substantial derivative are not invariant (see Pedlosky (1987) Chapter 1). The frictional force is assumed to remain invariant when changing from the inertial to the rotation frame, that is, $\vec{F}_f(\vec{u}_I) = \vec{F}_f(\vec{u}_R)$, which is the case for its expression given above for this force. The continuity and thermodynamic equations are not affected by the choice of the reference frame.

7.1.2.3.3 Spherical Coordinates (r, φ, ϕ)

In Figure 7.1, we show the nomenclature we use to describe the atmospheric motions in spherical coordinates: vector radius (r), longitude (ϕ), and latitude (φ). In Section 1.3.4, we have discussed the management of some quantities when deviations from the pure spherical shape (spheroid) are considered, and this will be incorporated later. A reference system linked to the "surface" (real surface or an altitude reference level in the giant planets) has coordinates (x, y, z): x-positive eastward (tangent to a parallel circle), y-positive northward (tangent to the meridional circle), and z-positive upward (radial direction), whose unit vectors are ($\hat{\imath}$, $\hat{\jmath}$, $\hat{k}$). This system is not strictly Cartesian because the directions of the axes are functions of the position on the spherical planet or, in other words, the unit vectors ($\hat{\imath}$, $\hat{\jmath}$, $\hat{k}$) are not constant. The

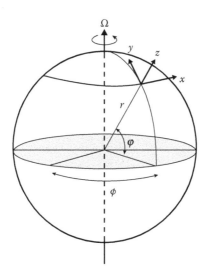

FIGURE 7.1 Local Cartesian coordinate systems for a spherical planet.

respective components of the velocity vector in these directions are $\bar{u}(u, v, w)$. A small incremental distance along these axes is related to the spherical coordinates by $dx = r \cos\varphi\, d\phi$ (eastward or *zonal* direction), $dy = r\, d\varphi$ (northward or *meridional* direction), $dz = dr$ (upward), and then

$$u = r\cos\varphi \frac{d\phi}{dt}, \quad v = r\frac{d\varphi}{dt}, \quad w = \frac{dz}{dt} \tag{7.9}$$

Using as a reference the "surface" level where $r = R_p$ (the planetary radius), we have in general that $r = R_p + z$. In terrestrial meteorology, the "weather layer" is taken as the troposphere (height 8–10 km), and then since $z \ll R_p$, it is customary to use the approach $r \sim R_p$ for the motions in the "weather layer" (which becomes a "shallow layer"). Here, we keep r as a variable instead of R_p for completeness in relation to the deep atmospheres of the giant planets. In this coordinate system, the three components of the Navier–Stokes equation can be written as (see Table 7.1)

$$\frac{du}{dt} = \left(2\Omega + \frac{u}{r\cos\varphi}\right)(v\sin\varphi - w\cos\varphi) - \frac{1}{\rho}\frac{\partial P}{\partial x} + F_{fx} \tag{7.10a}$$

$$\frac{dv}{dt} = -\frac{wv}{r} - \left(2\Omega + \frac{u}{r\cos\varphi}\right)u\sin\varphi - \frac{1}{\rho}\frac{\partial P}{\partial y} + F_{fy} \tag{7.10b}$$

$$\frac{dw}{dt} = \frac{u^2 + v^2}{r} + 2\Omega u\cos\varphi - \frac{1}{\rho}\frac{\partial P}{\partial z} - g + F_{fz} \tag{7.10c}$$

(F_{fx}, F_{fy}, F_{fz}) being the frictional force components per unit mass in this reference system (again in the eastward, northward, and upward directions). The total derivative (7.1b) can be written as

$$\frac{d}{dt} = \frac{\partial}{\partial t} + u\frac{\partial}{\partial x} + v\frac{\partial}{\partial y} + w\frac{\partial}{\partial z} \tag{7.11}$$

7.1.2.4 Thermodynamic Energy Equation

The thermodynamic energy equation is the expression of the conservation of energy applied to a moving fluid element. It is derived from the first law of thermodynamics that states that "the change in the internal energy of a system is equal to the difference between the heat added to the system and the work done by the system." A detailed derivation of the thermodynamic energy equation can be found in many

TABLE 7.1
Vector Relationships in the Sphere

Velocity field

Longitude: ϕ (zonal velocity $u\hat{i}$)

Latitude: φ (meridional velocity $v\hat{j}$)

Radial: r (vertical velocity $w\hat{k}$)

$$\vec{u} = u\hat{i} + v\hat{j} + w\hat{k}$$

$$\frac{d\vec{u}}{dt} = \hat{i}\frac{du}{dt} + \hat{j}\frac{dv}{dt} + \hat{k}\frac{dw}{dt} + u\frac{d\hat{i}}{dt} + v\frac{d\hat{j}}{dt} + w\frac{d\hat{k}}{dt}$$

$$\frac{d\hat{i}}{dt} = u\frac{\partial \hat{i}}{\partial x} = \frac{u}{r\cos\varphi}\left(\hat{j}\sin\varphi - \hat{k}\cos\varphi\right)$$

$$\frac{d\hat{j}}{dt} = u\frac{\partial \hat{j}}{\partial x} + v\frac{\partial \hat{j}}{\partial y} = -\frac{u\tan\varphi}{r}\hat{i} - \frac{v}{r}\hat{k}$$

$$\frac{d\hat{k}}{dt} = u\frac{\partial \hat{k}}{\partial x} + v\frac{\partial \hat{k}}{\partial y} = \frac{u}{r}\hat{i} + \frac{v}{r}\hat{j}$$

Vector operations in spherical coordinates:

$$\nabla\Phi = \frac{1}{r\cos\varphi}\frac{\partial\Phi}{\partial\phi}\hat{i} + \frac{1}{r}\frac{\partial\Phi}{\partial\varphi}\hat{j} + \frac{\partial\Phi}{\partial r}\hat{k}$$

$$\nabla\cdot\vec{u} = \frac{1}{r\cos\varphi}\left[\frac{\partial u}{\partial\phi} + \frac{\partial(v\cos\varphi)}{\partial\phi}\right] + \frac{1}{r^2}\frac{\partial}{\partial r}(r^2 w)$$

$$\nabla\times\vec{u} = \frac{1}{r^2\cos\varphi}\begin{vmatrix} r\cos\varphi\hat{i} & r\hat{j} & \hat{k} \\ \dfrac{\partial}{\partial\phi} & \dfrac{\partial}{\partial\varphi} & \dfrac{\partial}{\partial r} \\ ur\cos\varphi & vr & w \end{vmatrix}$$

$$\nabla^2\Phi = \frac{1}{r^2\cos^2\varphi}\frac{\partial^2\Phi}{\partial\phi^2} + \frac{1}{r^2\cos\varphi}\left[\cos\varphi\frac{\partial\Phi}{\partial\varphi}\right] + \frac{1}{r^2}\frac{\partial}{\partial r}\left(r^2\frac{\partial\Phi}{\partial r}\right)$$

where Φ is an scalar function. Here $w = dz/dt = dr/dt$ and the height $z = r - R_p$. For the *shallow atmosphere approximation*, we have $z \ll R_p$ and $r = R_p$, and the geometric divergence associated with the vertical displacements can be ignored. Then, these equations write

$$\nabla\Phi = \frac{1}{R_p\cos\varphi}\frac{\partial\Phi}{\partial\phi}\hat{i} + \frac{1}{R_p}\frac{\partial\Phi}{\partial\varphi}\hat{j} + \frac{\partial\Phi}{\partial z}\hat{k}$$

$$\nabla\cdot\vec{u} = \frac{1}{R_p\cos\varphi}\left[\frac{\partial u}{\partial\phi} + \frac{\partial(v\cos\varphi)}{\partial\phi}\right] + \frac{\partial w}{\partial z}$$

$$\nabla\times\vec{u} = \frac{1}{R_p^2\cos\varphi}\begin{vmatrix} R_p\cos\varphi\hat{i} & R_p\hat{j} & \hat{k} \\ \dfrac{\partial}{\partial\phi} & \dfrac{\partial}{\partial\varphi} & \dfrac{\partial}{\partial z} \\ uR_p\cos\varphi & vR_p & w \end{vmatrix}$$

$$\nabla^2\Phi = \frac{1}{R_p^2\cos^2\varphi}\frac{\partial^2\Phi}{\partial\phi^2} + \frac{1}{R_p^2\cos\varphi}\left[\cos\varphi\frac{\partial\Phi}{\partial\varphi}\right] + \frac{\partial^2\Phi}{\partial z^2}$$

meteorological textbooks (e.g., in Holton (2004) and Salby (1996) for two different approaches). In Section 1.4, we have presented the different types of energy sources that intervene in a planetary body, and in Section 1.5.4, the fundamentals of the energy transport mechanisms in a planetary body are presented. In previous chapters, we have already dealt, for convenience, with different forms of the energy equation (see, e.g., Equations 1.126, 2.3, 6.49, and in particular Chapter 4). In all these cases, instead of the energy, we have used the temperature as the variable representing the thermal or energy state evolution of the fluid element. The thermodynamic equation is then written as

$$\rho C_v \frac{dT}{dt} + P \nabla \cdot \vec{u} = \nabla(K_T \nabla T) - \nabla \vec{F}_{rad} + Q \tag{7.12a}$$

This equation expresses the rate at which the internal energy of a fluid element changes as work is performed on it (second term in the left part), and it exchanges heat by conduction (first term on the right), radiation (second term is the divergence of heat flux radiation), and other internal energy sources (Q, namely, various diabatic effects, latent heating, and frictional heating). Each term has units of power per unit volume (W m^{-3}).

From the continuity equation (7.3), the ideal gas equation (7.4), and the *Mayer relationship* ($C_p = C_v + R_g^*$), we get another useful form for this equation (Problem 7.2):

$$\rho C_p \frac{dT}{dt} - \frac{dP}{dt} = \nabla(K_T \nabla T) - \nabla \vec{F}_{rad} + Q \tag{7.12b}$$

Remember that (d/dt) on the left-hand side represents the total derivatives, so the relationship (7.1b) or (7.11) applies.

7.2   MOMENTUM EQUATION: BALANCES

The momentum equations (7.10) are quite complicated since they form a nonlinear coupled system. But one can expect that, based on the ranges of magnitude of the different terms in a planetary atmosphere (e.g., rotation period and friction forces), some of the terms can dominate over others. To have a first-order estimate of the importance of each term in the equations, it is customary to perform a *scale analysis*, sometimes called a *filtering*, of the equations of motions (7.10). When performing a filtering, it must be remembered that atmospheric dynamics involve different scales of motion (see Chapter 8) and, depending on what phenomena we are studying, one term can dominate over others. Each term must be appropriately evaluated depending on the motion scale under study. In general, the following order of magnitude scale applies to the different terms in these equations:

$$\frac{du}{dt}, \frac{dv}{dt} \rightarrow \frac{U^2}{L}$$

$$\frac{dw}{dt} \rightarrow \frac{UW}{L}, \frac{W^2}{H}$$

$$-2\Omega v \sin\varphi, 2\Omega u \sin\varphi, -2\Omega u \cos\varphi \rightarrow f_0 U$$

$$2\Omega w \cos\varphi \rightarrow f_0 W$$

$$\frac{uw}{r}, \frac{vw}{r} \rightarrow \frac{UW}{R_p}$$

$$-\frac{uw\tan\varphi}{r}, \frac{u^2\tan\varphi}{r}, -\frac{u^2+v^2}{r} \rightarrow \frac{U^2}{R_p}$$

$$-\frac{1}{\rho}\frac{\partial P}{\partial x}, -\frac{1}{\rho}\frac{\partial P}{\partial y} \rightarrow \frac{\Delta P}{\rho L}$$

$$-\frac{1}{\rho}\frac{\partial P}{\partial z} \rightarrow \frac{P_0}{\rho H}$$

$$-g \rightarrow g$$

$$F_{fx}, F_{fy} \rightarrow \frac{\eta_D U}{H^2}$$

$$F_{fz} \rightarrow \frac{\eta_D W}{H^2}$$

Here

U and W are characteristic horizontal and vertical velocities

L and H are the horizontal and vertical scale of motions

R_p is the planetary radius

$f_0 = 2\Omega \sin\varphi_0$ is the local *Coriolis parameter* evaluated at a fixed latitude (e.g., $\varphi_0 = 45°$ for a large range of midlatitude motions)

ΔP is the horizontal pressure fluctuation scale

P_0 is the mean pressure at the considered atmospheric level

There are two natural dimensionless numbers that arise from the scale analysis that are very important to characterize atmospheric motions. The *Rossby number* gives an estimate of the magnitude of the flow acceleration as compared to the *Coriolis force*:

$$Ro = \frac{(U^2/L)}{f_0 U} = \frac{U}{f_0 L} \tag{7.13}$$

The other is the *aspect ratio* that compares the vertical to the horizontal scales of motion:

$$\delta = \frac{D}{L} \tag{7.14}$$

Here, we use D as vertical scale instead of H (used for the density or pressure scale height, see Section 4.4.1) to cover a large range of possible vertical motions. In geophysical fluid dynamics, when studying large-scale motions whose aspect ratio $\delta = D/L \ll 1$, the flow tends to be essentially two dimensional with (u, v) dependent basically on (x, y).

7.2.1 TANGENT-PLANE GEOMETRY

The dependence of the Coriolis parameter $f = 2\Omega \sin \varphi$ upon latitude means that it increases toward the poles ($f \rightarrow 2\Omega$) but decreases toward the equator $f \rightarrow 0$. For large-scale meridional motions across the whole sphere, the Coriolis parameter must be fully retained. But for motions restricted to a latitude band where the variation of the Coriolis parameter (that is, the Coriolis force) with latitude is small, one of two simplifications are usually introduced to the momentum equations (7.10).

7.2.1.1 f-Plane Approximation

Here, the meridional displacements of atmospheric parcels are assumed to be small enough so f can be considered constant, that is, we take its value at the reference latitude φ_0:

$$f_0 = 2\Omega \sin \varphi_0 \tag{7.15}$$

This is equivalent to consider the motions taking place in a plane tangent to the spherical surface at the latitude φ_0.

7.2.1.2 β-Plane Approximation

For large-scale meridional motions but such that the variation of f with latitude is still small, we can make a further approximation expanding in a Taylor series $f(\varphi)$ about the reference latitude φ_0:

$$f(\varphi) = 2\Omega \sin \varphi = 2\Omega \left[\sin \varphi_0 + (\varphi - \varphi_0) \cos \varphi_0 + \cdots \right]$$

so that on the tangent plane where $\varphi - \varphi_0 \approx y/R_p$, we have

$$f(y) \approx f_0 + \beta y \tag{7.16a}$$

being the parameter β

$$\beta = \frac{2\Omega \cos \varphi_0}{R_p} = \left(\frac{df}{dy} \right)_{y=0} \tag{7.16b}$$

(7.16a) is called the beta plane approximation in which the Coriolis parameter is simplified by allowing it to vary linearly with the northward distance y.

7.2.2 GEOSTROPHIC BALANCE

In rapidly rotating planets, the Coriolis and the pressure gradient forces dominate in the momentum equation, $Ro \ll 1$ in (7.13) (away from the equator) and Equations 7.10a and b reduce to a simplified balance between these two terms, resulting in the so-called *geostrophic approximation* (Figure 7.2):

$$fu = -\frac{1}{\rho}\frac{\partial P}{\partial y} \tag{7.17a}$$

$$fv = \frac{1}{\rho}\frac{\partial P}{\partial x} \tag{7.17b}$$

According to Equations 7.17a and b, when this balance holds, the wind speed is directly proportional to the horizontal gradient of pressure and the wind intensifies as the isobars (at a given altitude level) come close together. Note that as $f \to 0$ at low latitudes, the geostrophic balance breaks around equatorial latitudes. Combining both velocity components we introduce the *geostrophic wind* as

$$\vec{V}_g = u_g\hat{i} + v_g\hat{j} = \hat{k}\times\frac{1}{\rho f}\nabla P \tag{7.18}$$

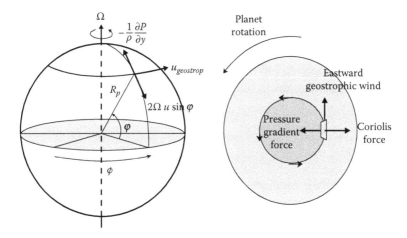

FIGURE 7.2 Diagram of main force balances under geostrophic conditions in planetary atmospheres.

The relationships (7.17a and b) are diagnostic equations for the flow balance. For prognostic equations, so the motion of the parcel can be described in time, we must retain the acceleration terms in Equation 7.10, and the horizontal momentum equation can be written in terms of the geostrophic components:

$$\frac{du}{dt} = fv - \frac{1}{\rho}\frac{\partial P}{\partial x} = f(v - v_g)$$
(7.19a)

$$\frac{dv}{dt} = -fu - \frac{1}{\rho}\frac{\partial P}{\partial y} = -f(u - u_g)$$
(7.19b)

Application of this balance to Earth, Mars, and Jupiter is illustrated in Problem 7.3.

7.2.3 CYCLOSTROPHIC BALANCE

Slowly rotating bodies have f small (also as for rapidly rotating planets in equatorial latitudes), so the Rossby number tends to be $Ro > 1$. The Coriolis force may be neglected in Equation 7.10a and b when compared to the pressure gradient force and to the centrifugal force. The dominant terms in Equation 7.10a and b reduce to a single balance between the latter two forces, called *cyclostrophic balance*:

$$\frac{u^2 \tan\varphi}{R_p} = -\frac{1}{\rho}\frac{\partial P}{\partial y}$$
(7.20)

from which the *cyclostrophic wind* speed is obtained:

$$u = \sqrt{-\frac{R_p}{\rho \tan\varphi}\frac{\partial P}{\partial y}}$$
(7.21)

In Figure 7.3, we schematically show this balance, and its application to Venus is illustrated in Problems 7.4 and 7.5.

7.2.4 GRADIENT WIND BALANCE

When a triple balance occurs among the Coriolis force, the centrifugal force, and the pressure gradient force, the flow is said to be in gradient balance and the resulting motion is called the *gradient wind*. Equation 7.10a and b can be simplified into a single equation:

$$\frac{u^2 \tan\varphi}{R_p} + fu = -\frac{1}{\rho}\frac{\partial P}{\partial y}$$
(7.22)

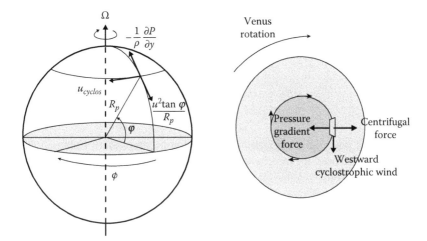

FIGURE 7.3 Cyclostrophic balance in a slowly rotating planet, illustrated for the case of Venus (planetary westward rotation).

which gives the *gradient wind speed*

$$u = -\frac{fR_p}{\tan\varphi} \pm \left(\frac{f^2 R_p^2}{2\tan^2\varphi} - \frac{2R_p}{\rho\tan\varphi}\frac{\partial P}{\partial y} \right)^{1/2} \tag{7.23}$$

Application of this balance to Titan is illustrated in Problem 7.4.

7.2.5 HYDROSTATIC EQUILIBRIUM AND GEOPOTENTIAL

Gravity is a very strong force acting in the radial direction, and the vertical component of the momentum equation (7.10c) for a parcel is dominated, on a global scale in planetary atmospheres, by a balance between it and the vertical pressure gradient force, as already discussed in Section 4.1.1. This is known as the *hydrostatic balance*:

$$\frac{\partial P}{\partial z} = -\rho g \tag{7.24}$$

Pressure and altitude are directly related, since $dP = -\rho g\, dz$, allowing for the use of pressure as vertical coordinate, with pressure decreasing monotonically with height. The total pressure at a given altitude z can be obtained upon integration of (7.24)

$$P(z) = \int_z^\infty \rho g\, dz \tag{7.25}$$

and the pressure at a point is the average weight per unit area of the total atmospheric column. We introduce the *geopotential function* in differential and integrated form as

$$d\Phi_G = g\,dz, \quad \Phi_G = \int_0^z g\,dz = gz. \tag{7.26}$$

It represents the potential energy per unit mass for an atmospheric column. This is the same definition used for the gravitational potential in Section 1.3.1 in the general planetary context, but renamed as geopotential in atmospheric physics (extending the prefix "geo" to any planet). Rearranging (7.24) and combining with the ideal gas relationship (7.4), one gets

$$g\left(\frac{\partial z}{\partial P}\right) = -\left(\frac{1}{\rho}\right) = -\frac{R_g^* T}{P} \tag{7.27}$$

where, it must be noted, the partial derivative implies obviously that no variability occurs with respect to the other three variables (x, y, t). The hydrostatic relationship in pressure coordinates becomes

$$\frac{\partial \Phi_G}{\partial P} = -\frac{R_g^* T}{P} \tag{7.28}$$

with the vertical velocity given in terms of the pressure coordinate as

$$\omega_P = \frac{dP}{dt} = \frac{\partial P}{\partial t} + \vec{V}_h \cdot \nabla_h P + w\left(\frac{\partial P}{\partial z}\right) \tag{7.29}$$

in units of mbar s^{-1}, where the subindex h means partial differentiation on the horizontal components. See Problems 7.6 and 7.7 for the application of the geopotential function to the geostrophic balance and for the calculation of the vertical velocity in pressure coordinates in the case of Earth's synoptic-scale motions.

7.2.6 BOUSSINESQ AND ANELASTIC APPROXIMATIONS

If one assumes that large-scale atmospheric motions are nearly in hydrostatic balance, other common simplifications are useful. The density and pressure can each be represented as the sum of two terms, a background reference (density ρ_0, pressure P_0), which depends only on the vertical coordinate, plus a small deviation (density ρ', pressure P') from the hydrostatically balanced reference state. Then $\rho(x,y,z,t) = \rho_0(z) + \rho'(x,y,z,t)$ and $P(x,y,z,t) = P_0(z) + P'(x,y,z,t)$. Thus, in the horizontal momentum equations (7.8b), we replace $\rho \rightarrow \rho_0$ and $P \rightarrow P_0$ and introduce a *buoyancy force* (dry atmosphere) to obtain

$$\frac{d\vec{u}}{dt} = -\frac{\nabla P'}{\rho_0} - g\frac{\rho'}{\rho_0}\hat{k} - 2\vec{\Omega}\times\vec{u} + \vec{F}_f \qquad (7.30)$$

To a high degree of approximation, the time derivative of the density in (7.2a) can be neglected and the continuity equation becomes

$$\nabla \cdot \rho_0\vec{u} = 0 \qquad (7.31)$$

Note that density variations with height are retained. The use of Equations 7.30 and 7.31 instead of (7.8b) and (7.2a) is usually referred to as the *anelastic approximation*. More strictly, the pure anelastic condition requires the flow to be not only hydrostatic but also isentropic, that is, $\theta=$constant (although in practice it is often taken to be nonisentropic). Anelastic means that elastic energy is not considered and, since the derivative of density is zero, sound waves are filtered out (see Chapter 8).

When the vertical motions occur on a distance smaller than the atmospheric scale height H, that is, in a shallow layer, ρ_0 can be considered constant in Equations 7.30 and 7.31. In such a case, the fluid can be treated as incompressible ($\nabla \cdot \vec{u} = 0,(7.3)$). The use of Equation 7.30 with $\rho_0=$constant, and (7.3) instead of (7.31) is usually referred to as the *Boussinesq approximation* that only retains density variations if coupled with gravity.

7.2.7 THERMAL-WINDS

For the above balances in the momentum equations (Sections 7.2.2 through 7.2.4), together with the ideal gas law and after some cross-differentiation, we can obtain the following relationships between the vertical shear of the horizontal velocity components and the horizontal temperature gradients.

7.2.7.1 Rapidly Rotating Planet

Assuming that the atmosphere is in hydrostatic and geostrophic balances, the variation with altitude of the horizontal wind velocity is directly proportional to the corresponding components of the horizontal temperature gradient according to the following relationships (Problem 7.8):

$$\frac{\partial u}{\partial z} = -\frac{g}{fT}\frac{\partial T}{\partial y} \qquad (7.32a)$$

$$\frac{\partial v}{\partial z} = \frac{g}{fT}\frac{\partial T}{\partial x} \qquad (7.32b)$$

Both equations can be combined into a single vector form for the geostrophic velocity:

$$\frac{\partial \vec{V}_g}{\partial z} = \frac{g}{fT}\hat{k}\times\nabla_h T \qquad (7.32c)$$

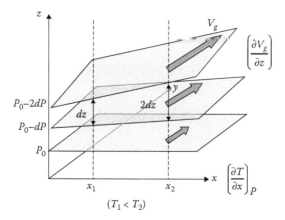

FIGURE 7.4 The thermal wind for geostrophic conditions: relationship between the horizontal temperature gradient (in the x-direction) and the vertical shear of the geostrophic wind.

Figure 7.4 illustrates the thermal wind concept. As an application, the dependence of the zonal wind velocity on altitude and latitude can be calculated if the velocity at a reference level (pressure P_0) is known by simply integrating Equation 7.32a:

$$u(z,\varphi) = u_0(z_0,\varphi) - \frac{g}{R_p f}\int_{z_0}^{z}\frac{1}{T}\left(\frac{\partial T}{\partial \varphi}\right)dz' \approx u_0(z_0,\varphi) - \frac{g}{fL}\frac{\Delta T}{\langle T\rangle}H\ln\left(\frac{P_0}{P}\right) \qquad (7.33)$$

with the last term being a first-order estimate using an average value for the vertical and horizontal variations of temperature $\langle T\rangle$. Here $dz = -H(dP/P)$ with $H = R_g^* T/g$ the vertical scale height, L is a characteristic meridional length (e.g., the zonal current width), and ΔT is the mean temperature variation across this length (along the corresponding altitude level). See Problem 7.9 for an application of this expression to the atmospheres of Jupiter and Saturn.

7.2.7.2 Slowly Rotating Planet

A similar relationship for the vertical wind shear can be found for a slowly rotating planet, assuming the atmosphere is under cyclostrophic and hydrostatic balances. Following similar steps, the thermal wind shear is now given by (Problem 7.10)

$$2u\tan\varphi\,\frac{\partial u}{\partial z} = -\frac{R_g^* R_p}{H}\frac{\partial T}{\partial y} \qquad (7.34)$$

As before, if the meridional temperature gradient and the wind speed at a given altitude are known, then the velocity as a function of altitude and latitude is given upon integration by

$$u^2(z,\varphi) = u_0^2(z_0,\varphi) - \frac{R_g^*}{\tan\varphi}\int_{z_0}^{z}\frac{1}{H}\frac{\partial T}{\partial \varphi}dz' \approx u_0^2(z_0,\varphi) - \frac{R_g^* R_p}{\tan\varphi}\frac{\Delta T}{L}\ln\left(\frac{P_0}{P}\right) \qquad (7.35)$$

where, instead of y as meridional coordinate, we have used the latitude as before. Again, the last term represents an order of magnitude estimate of the wind speed using averaged values for the vertical and horizontal directions. See Problems 7.11 and 7.12 for an application of this expression to the atmospheres of Venus and that of an extrasolar planet.

7.2.7.3 Intermediate Case of a Moderately Rotating Planet

From the gradient wind relationship (7.22) and hydrostatic balance, one can get the gradient wind version for the thermal wind shear (Problems 7.13 and 7.14):

$$\frac{\partial u}{\partial z}\left[\frac{2u\tan\varphi}{R_p}+f\right]=-\frac{R_g^*}{H}\frac{\partial T}{\partial y} \tag{7.36}$$

which can be integrated to get the wind speed at an altitude level if its value is known at another level and if the meridional temperature contrast is also known:

$$u^2(z,\varphi)\frac{\tan\varphi}{R_p}+u(z,\varphi)f=u_0^2(z_0,\varphi)\frac{\tan\varphi}{R_p}+u_0(z_0,\varphi)f-\frac{g}{R_p}\int_{z_0}^{z}\frac{1}{T}\frac{\partial T}{\partial\varphi}dz' \tag{7.37}$$

where the last term is of the order of $(R_g^*\Delta T/L)\ln(P_0/P)$. See Problem 7.15 for an application of this expression to the atmosphere of Titan.

7.2.8 MERIDIONAL CIRCULATION—HADLEY CIRCULATION

Differential solar heating between parallel circles can induce meridional motions. The atmosphere is heated at the subsolar point (the equator for a planet with spin axis perpendicular to the orbital plane) and the parcels of gas rise and move toward colder regions (North and South) where it becomes cold, descending and returning back toward the low latitudes. This overturning circulation forms meridional cells, that is, along the meridians, and the motion crosses latitude circles.

The best known example of this type of circulation occurs in Earth's tropics where it is called the *Hadley cell*. Similar meridional circulations are known to be present in bodies with a surface where solar radiation is deposited and becomes the main source for driving the motions (Venus, Earth, Mars, and Titan). On the giant planets, meridional circulations are predicted to occur at the stratospheric level between the cloud bands (adjacent belts and zones) or around the core of the zonal jets (clockwise or anticlockwise depending on the eastward or westward direction of the zonal wind and hemisphere).

A first-order estimation of the velocities involved in meridional circulations can be obtained from the thermodynamic equation (7.12b). For a steady state flow neglecting heat conduction, the equation reduces to

$$u\frac{\partial T}{\partial x}+v\frac{\partial T}{\partial y}+w\left(\frac{\partial T}{\partial z}+\frac{g}{C_p}\right)=\left(-\nabla\vec{F}_{rad}+Q\right)\left(\rho C_p\right)^{-1} \tag{7.38a}$$

Then neglecting the horizontal advection terms, spatially averaging, and with the radiative flux divergence parameterized as a linear temperature relaxation controlled by the radiative time constant (τ_{rad}) (Section 4.3), we can write

$$\overline{w}_{u,d}\left(\frac{\partial \overline{T}}{\partial z} + \frac{g}{C_p}\right) = -\left(\frac{(T - T_e)_{u,d}}{\tau_{rad}}\right), \tag{7.38b}$$

where the subindices (u, d) refer to the upward and downward motions in the cell, and the temperature difference is calculated between that observed (T) and that expected from radiative equilibrium $(T_e$, see, e.g., Chapter 4) in each branch (u, d). The order of magnitude meridional velocity can be estimated by a scaling analysis of the continuity equation (7.3) in the vertical–meridional plane:

$$\frac{\overline{v}}{Y_H} \approx \frac{\overline{w}}{H} \tag{7.39}$$

where Y_H and H (the scale height) are the meridional and vertical scales of the cell. See Problem 7.16 for an application of this meridional motion to Venus and Titan (more details can be found in Section 7.5.2).

The above is a very simplistic calculation, and the explanation of the observed meridional circulations must involve the full equations of motion with the appropriate boundary conditions (in particular on planets with a surface). This makes use of the relationship between the horizontal wind components and temperature (usually through the thermal wind equation and simplified forms of the momentum equation). Insight on axially symmetric circulations is provided by the *Held–Hou model* (see, e.g., Vallis (2006) and Read and Lewis (2004)), which assumes that the angular momentum per unit mass (see Section 7.6.3) is conserved when a parcel moves poleward along the upper branch of the cell (Figure 7.5). Near the ground, the surface friction keeps the zonal velocity close to zero (at least for Earth and Mars). This model predicts the width of the Hadley cell (when considering small latitude angles) to be

$$Y_H \approx \sqrt{\frac{5gH\Delta\theta}{3\Omega^2\theta_0}} \tag{7.40}$$

where
$\Delta\theta$ is the equilibrium equator–pole potential temperature difference
θ_0 is a reference temperature in the midaltitude of the cell

For Earth, this predicts $Y_H \sim 2000\,\text{km}$ or extending ~18° latitude in each hemisphere (for $\Omega = 7.3 \times 10^{-5}\,\text{s}^{-1}$, $H = 8\,\text{km}$, $\Delta\theta = 40\,\text{K}$, $\theta_0 = 255\,\text{K}$) and for Mars, $Y_H \sim 2000\,\text{km}$ or extending 35° latitude in each hemisphere (for $\Omega = 7.3 \times 10^{-5}\,\text{s}^{-1}$, $H = 11\,\text{km}$, $\Delta\theta = 65\,\text{K}$, $\theta_0 = 210\,\text{K}$). The Hadley cells break down at midlatitudes as the Coriolis force becomes

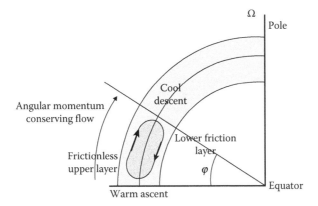

FIGURE 7.5 Schematic illustration of a meridional (Hadley cell) mechanism. Rising air near the equator moves poleward in the upper frictionless layer, descending at latitude φ and returning in a lower friction layer. The poleward-moving atmospheric parcels conserve the axial angular momentum, leading to a zonal flow that increases away from the equator.

large (geostrophic effect). On Earth, the Hadley cell is in fact limited to motions between the equator and midlatitudes, and then two weak cells, the *Ferrell* and *polar* cells, with opposed circulations, extend up to the poles. The Held–Hou model also predicts the maximum strength of the zonal wind speed at the top of the Hadley cell:

$$u_H \approx \frac{\Omega Y_H^2}{R_p} \qquad (7.41)$$

which gives for Earth $u_H \sim 55\,\text{m s}^{-1}$ and for Mars $u_H \sim 85\,\text{m s}^{-1}$.

In the slowly rotating bodies as Venus and Titan, (7.40) predicts the Hadley cell to expand poleward. We can expect the existence of at least one Hadley cell per hemisphere (at one altitude level) but, it is possible that there exist other cells above and below the main one. The condition for a dominant Hadley cell mechanism in a planetary atmosphere can be expressed as

$$Ro_T = \frac{gH\Delta\theta}{\Omega^2 R_p^2 \theta_0} \geq 1 \qquad (7.42)$$

where Ro_T is a Rossby number (7.13) in which the velocity is estimated from the thermal wind balance, using the planetary radius as the characteristic length.

7.2.9 CONDENSATION FLOWS

The condensation flow is the motion of air masses that result when the surface ices sublimate at a heated location due to insolation and then condensation in another distant place at a lower temperature. This mechanism is important in the tenuous atmospheres of Mars (due to sublimation and condensation of CO_2), and the satellites

Io and Triton (due to sublimation of ices in their frozen surfaces), as presented in Section 6.7.2. If all the sublimating mass occurring below a latitude circle φ is transferred across this circle with a meridional wind speed v, then mass continuity requires that

$$\frac{dM}{dt} = 2\pi R_p \cos\varphi \int_0^h \rho v \, dz \approx 2\pi R_p \cos\varphi \rho_0 \langle v \rangle h \qquad (7.43)$$

where dM/dt is the sublimating mass rate (in kg s^{-1}) that can be obtained as described in Section 6.7.2. This expression allows the retrieval of a mean meridional speed $\langle v \rangle$ where the last term in (7.43) is obtained assuming a constant density ρ_0 within a layer of altitude h. Its application to Mars is proposed in Problem 7.17.

7.3 VORTICITY

Formally, knowing the boundary and initial conditions for a particular atmospheric problem, a solution to the evolution of the different magnitudes describing the flow could be pursued from the set of equations of motion. However, this is a complicated problem since these equations are coupled and nonlinear, making them extremely sensitive to the initial conditions. But there are different aspects of geophysical fluid motions that are generally broad and sufficiently relevant to be described by one of the family of quantities that derive from the rotational proper-ties of the fluid.

The concepts of *circulation*, *vorticity*, and *potential vorticity* play a central role in atmospheric dynamics and meteorology. They characterize the rotational state of a fluid (its tendency to "rotate" and to quantify its "amount of rotation"), and through conservation laws, analogous to those used for classical angular momentum in the mechanic of rotating bodies, they allow evaluating the particularities and evolution of geophysical fluid dynamics.

7.3.1 VORTICITY AND CIRCULATION

The vorticity $\vec{\omega}$ is defined as the curl of the velocity vector $\vec{u}(u, v, w)(u,v,w)$, given in Cartesian coordinates by

$$\vec{\omega} = \nabla \times \vec{u} = \left(\frac{\partial w}{\partial y} - \frac{\partial v}{\partial z}, \frac{\partial u}{\partial z} - \frac{\partial w}{\partial x}, \frac{\partial v}{\partial x} - \frac{\partial u}{\partial y} \right) \qquad (7.44)$$

Vorticity is a vector that measures the local rotation of a fluid (unit s^{-1}). The relevant term in (7.44) is then the vertical component of the vorticity vector ζ (perpendicular to the plane (x, y)) whose magnitude is

$$\zeta = \frac{\partial v}{\partial x} - \frac{\partial u}{\partial y} \qquad (7.45)$$

Since atmospheric motions occur in rotating frames (angular rotation speed $\vec{\Omega}$), the absolute velocity of the flow (7.7a) must be considered, and then the absolute vorticity can be derived (see Problem 7.18) given by

$$\vec{\omega}_I = \vec{\omega}_R + 2\vec{\Omega} \tag{7.46}$$

and the magnitude of the vertical component is then given by

$$\zeta + f = \left(\frac{\partial v}{\partial x} - \frac{\partial u}{\partial y} \right) + 2\Omega \sin \varphi \tag{7.47}$$

The circulation Γ_C (unit $m^2\ s^{-1}$) represents a macroscopic measure of the rotational state of a fluid. It is defined as the line integral about a closed contour of the component of the velocity vector that is locally tangent to the contour:

$$\Gamma_C = \oint_C \vec{u} \cdot d\vec{l} = \oint_C |\vec{u}| \cos \alpha\, dl \tag{7.48}$$

where

$\vec{l}(s)$ is the position vector going from the origin to the position $s(x, y, z)$ on the contour

α is the angle between $\vec{u}$ and $d\vec{l}$ that is tangent to the contour C in each point

Note that the velocity vector can be the relative or absolute (depending on the reference frame used) as related in (7.7a). According to Stokes' theorem we have

$$\Gamma_C = \oint_C \vec{u} \cdot d\vec{l} = \iint_S (\nabla \times \vec{u}) \cdot d\vec{S}, \tag{7.49}$$

where $d\vec{S}$ represents an infinitesimal surface area element enclosed by the contour. We write $d\vec{S} = dS\,\hat{n}$, where the direction of the unit normal vector $\hat{n}$ is defined by the counterclockwise rotation sense in the line integration according to the "right-hand screw rule." According to Equation 7.49, Γ_C is also defined as a "vorticity flux" or as a "vortex tube strength." From Equation 7.46, we can write Equation 7.49 in terms of the absolute and relative circulations as

$$\Gamma_{CI} = \iint_S \vec{\omega}_I \cdot d\vec{S} = \iint_S (\vec{\omega}_R + 2\vec{\Omega}) \cdot d\vec{S} = \Gamma_{CR} + 2\Omega S_n \tag{7.50}$$

where S_n is the area projected on a surface perpendicular to $\vec{\Omega}$.

As indicated above, the application of these quantities becomes relevant when they are conserved. It is possible to show that when a fluid is *barotropic* (that is, when the surfaces of constant density and constant pressure coincide) and inviscid (no frictional forces acting) on the contour C, then the absolute circulation is conserved following the motion (see, e.g., Pedlosky, 1987):

$$\frac{d\Gamma_{CI}}{dt} = 0 \qquad (7.51)$$

This is known as *Kelvin's circulation theorem*. From (7.50) we can see that in such a case, a change in the planetary term $2\Omega S_n$ (produced, e.g., by a latitude change of an atmospheric element) must be accompanied by an opposite change in the relative vortex tube strength (see Problem 7.19).

The concept of vorticity will be extensively used in Section 9.3 in its application to vortex formation in planetary atmospheres.

7.3.2 THE VORTICITY EQUATION

Equation 7.51 can be generalized to the nonconserving case by adding to its right-hand side the appropriate sources and sinks for the circulation that can be derived from the momentum equation (7.8b). However, to study the evolution of the properties of a fluid, it is more convenient to employ the vorticity field since it represents a vector and local magnitude instead of the circulation that is a scalar and integral magnitude characterizing the finite regions of the fluid. The temporal evolution of the vorticity field is derived from the momentum equation (7.8b), taking the curl and applying the vector identities. The relative vorticity evolution (using relationship [Equation 7.46]) is then given by (see, e.g., Pedlosky, 1987)

$$\frac{d\vec{\omega}_R}{dt} = \vec{\omega}_I \cdot \nabla \vec{u} - \vec{\omega}_I \nabla \cdot \vec{u} + \frac{\nabla \rho \times \nabla P}{\rho^2} + \nabla \times \frac{\vec{F}_f}{\rho} \qquad (7.52)$$

Accordingly, the relative vorticity of the fluid changes due to the action of the four terms on the right-hand side of Equation 7.52. The first two terms represent the vorticity change due to the mechanisms called "vortex tube stretching" and "tilting" (Figure 7.6). The third term is called the "baroclinic" production term of vorticity (acts when pressure and density surfaces are tilted), and the last represents the vorticity diffusion term produced by the friction force.

From (7.44), (7.45), and (7.47), neglecting the frictional term, Equation 7.52 can be rewritten in an operative form as

$$\frac{d}{dt}(\zeta + f) = -(\zeta + f)\left(\frac{\partial u}{\partial x} + \frac{\partial v}{\partial y}\right) - \left(\frac{\partial w}{\partial x}\frac{\partial v}{\partial z} - \frac{\partial w}{\partial y}\frac{\partial u}{\partial z}\right) + \frac{1}{\rho^2}\left(\frac{\partial \rho}{\partial x}\frac{\partial P}{\partial y} - \frac{\partial \rho}{\partial y}\frac{\partial P}{\partial x}\right)$$

$$(7.53)$$

The terms on the right-hand side are then the sources of absolute vorticity due to the horizontal divergence of the velocity field (first term), the tilting or twisting due to

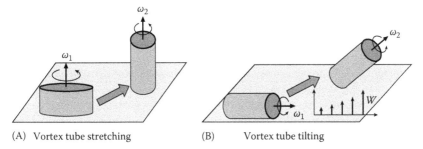

(A) Vortex tube stretching   (B)   Vortex tube tilting

FIGURE 7.6 (A) Vorticity generation of material lines enclosing a cylinder of fluid by the stretching mechanism with $\omega_2 > \omega_1$. (B) The tilting mechanism in the presence of horizontal shear of the vertical velocity.

vertical shears of the horizontal wind and horizontal shear of the vertical velocity (second term), and the solenoidal (*baroclinic*) term due to the horizontal shears of the pressure and gradient field, respectively.

Consider the case (common in some approaches) when the tilting and solenoidal terms in (7.53) are smaller than the first. Then (7.53) reduces to

$$\frac{d_h}{dt}(\zeta + f) = -(\zeta + f)\left(\frac{\partial u}{\partial x} + \frac{\partial v}{\partial y}\right) \qquad (7.54)$$

where the total derivative is $(d_h/dt) = (\partial/\partial t) + u(\partial/\partial x) + v(\partial/\partial y)$. Equation 7.54 indicates that locally, the absolute vorticity of a fluid parcel will change due to the advection and divergence of the horizontal winds.

If as a further simplification, we assume that the atmosphere (that is, a particular layer) is on the average isothermal and incompressible (constant temperature and density), the continuity equation allows Equation 7.54 to be written as

$$\frac{d_h}{dt}(\zeta + f) = (\zeta + f)\frac{\partial w}{\partial z} \qquad (7.55)$$

Now, for a flow in geostrophic balance and since the temperature is constant, it follows from the thermal wind relationship (Equation 7.30) that the wind velocity is independent of altitude. Therefore, approximating the wind speeds to their geostrophic value (u_g, v_g), and correspondingly the vorticity by its geostrophic value ζ_g, it follows from the integration of Equation 7.55 between the two levels z_1 and z_2 that bound the layer (thickness $h = z_2 - z_1$) that

$$\frac{1}{\zeta_g + f}\frac{d_h}{dt}(\zeta_g + f) = \frac{w(z_2) - w(z_1)}{h} \qquad (7.56)$$

Since $w = dz/dt$ and the thickness depends on the horizontal coordinates $h(x, y, t)$, we have that $w(z_2) - w(z_1) = d_h h/dt$ so that (7.56) can be written in a compact form as

$$\frac{d_h}{dt}\left(\frac{\zeta_g + f}{h}\right) = 0 \tag{7.57}$$

The term $(\zeta_g + f)/h$ is called the *barotropic potential vorticity* (unit $\mathrm{m^{-1}\,s^{-1}}$), so Equation 7.57 represents the conservation equation for the *barotropic potential vorticity*. This equation indicates that the evolution of the vorticity field is related to the depth of the fluid (atmospheric layer), which has been assumed to be thin relative to its horizontal length. This description of a fluid motion that has all these properties is thus called the *shallow-water layer model* (details can be found, e.g., in Pedlosky, 1987; Vallis, 2006). In essence, this equation is equivalent to the angular momentum conservation law of mechanics in which the barotropic potential vorticity plays the role of the vertical component of the angular momentum, the absolute vorticity that of the angular velocity, and the inverse depth that of the moment of inertia of the solid body.

For purely horizontal flow ($w = 0$) and more strictly when $\partial w/\partial z = 0$, Equation 7.55 simplifies to

$$\frac{d_h}{dt}(\zeta + f) = 0 \tag{7.58a}$$

or

$$\frac{\partial}{\partial t}(\zeta + f) = -\vec{V}_h \cdot \nabla(\zeta + f) \tag{7.58b}$$

which is called the *barotropic vorticity equation* and where the relationship (7.1a) has been used. It states that the absolute vorticity $(\zeta + f)$ of a parcel is conserved following the motion. A parcel moving across latitude circles must change its relative vorticity ("spin") as shown also in integral form by (7.50) and (7.51). Equation 7.58 is prognostic and can be used for a first order approximation in the study of synoptic scale nondivergent motions occurring in a rapidly rotating, inviscid, and incompressible atmosphere. This is a reasonable approach for the mid troposphere of Earth and Mars, and for the upper troposphere of giant planets.

7.3.3   POTENTIAL VORTICITY

An important extension of the above definitions of vorticity that permits a better constraint on the atmospheric motions must incorporate the thermal properties of the fluid. This is done by means of the potential temperature θ and the use of the continuity equation. We define the *potential vorticity* (PV) (also called *isentropic potential vorticity* or *Ertel's potential vorticity*, see, e.g., Pedlosky, 1987; Vallis, 2006) as

$$PV = \frac{\vec{\omega}_I}{\rho}\nabla\theta = \frac{(\vec{\omega}_R + 2\vec{\Omega})}{\rho}\nabla\theta \tag{7.59a}$$

A minus sign is usually introduced so its value is positive in the northern hemisphere. For thin fluids, using the hydrostatic approach, the *PV* can be written in terms of the vertical component of the vorticity as

$$PV = \left(\zeta_\theta + f \right) \left(\frac{1}{\rho} \frac{\partial \theta}{\partial z} \right) = \left(\zeta_\theta + f \right) \left(-g \frac{\partial \theta}{\partial P} \right) \qquad (7.59b)$$

where the relative vorticity ζ_θ is calculated on an isentropic surface (θ=constant). *PV* is conventionally expressed in units of the *potential vorticity unit* (*PVU*), which is 10^{-6} K kg^{-1} m^2 s^{-1}. Conceptually, comparing to the barotropic vorticity definition, the *PV* gives a measure of the ratio of the absolute vorticity to the effective depth defined as the distance between the potential temperature surfaces measured in pressure units $(-g\partial\theta/\partial P)^{-1}$. Its temporal evolution (following the motion) can be obtained after manipulating the equations of motion. Including all types of forces and diabatic terms, we can express it in a compact form as

$$\frac{d(PV)}{dt} = I_{PV} - S_{PV} \qquad (7.60)$$

where the two terms on the right-hand side represent the sources and sinks of potential vorticity due to diabatic heating and friction. Under the assumption that the potential temperature is conserved ($d\theta/dt=0$) and provided that the frictional force is negligible, and either the fluid is barotropic ($\nabla\rho \times \nabla P=0$, in Equation 7.52) or that θ is only a function of P and ρ, then the potential vorticity is conserved:

$$\frac{d(PV)}{dt} = 0 \rightarrow PV = \text{constant} \qquad (7.61)$$

This is called *Ertel's theorem*, equivalent to Equation 7.58 but incorporating the temperature field. For large-scale motions, planetary atmospheres behave as nearly inviscid with the processes occurring on them as approximately *isentropic*. The distribution of *PV* is uniquely determined by the wind and temperature fields. *PV* is conserved for each parcel as it moves following the flow and becomes a tracer of fluid motions, being therefore an appropriate quantity for diagnostic studies of the atmospheric dynamics (see Problem 7.20).

7.3.4 QUASI-GEOSTROPHIC POTENTIAL VORTICITY

An approximate form of the potential vorticity called *quasi-geostrophic* potential vorticity is useful in rapidly rotating atmospheres, that is, when $Ro \ll 1$ and the flow velocities (u, v) can be substituted by their geostrophic values (u_g, v_g), which, in terms of the *streamfunction* ψ (unit m^2 s^{-1}), can be written as

$$u \approx u_g = -\frac{\partial\psi}{\partial y}; \quad v \approx v_g = \frac{\partial\psi}{\partial x} \qquad (7.62)$$

Under the *Boussinesq* and *β-plane* approximations, it can be shown (e.g., Pedlosky, 1987) that the momentum and thermodynamic equations can be coupled into a single *quasi-geostrophic potential vorticity* conservation equation

$$\frac{d_g q_g}{dt} = 0 \tag{7.63}$$

where $d_g/dt = \partial/\partial t + \vec{V}_g \cdot \nabla$ and

$$q_g = f_0 + \beta y + \frac{\partial^2 \psi}{\partial x^2} + \frac{\partial^2 \psi}{\partial y^2} + \frac{1}{\rho_0}\frac{\partial}{\partial z}\left(\frac{\rho_0 f_0^2}{N_B^2}\frac{\partial \psi}{\partial z}\right) \tag{7.64}$$

is the *quasi-geostrophic potential vorticity* (QGPV). Equation 7.63 is prognostic for the streamfunction ψ and can be integrated if initial and boundary conditions are known. In terms of the single variable ψ it satisfies

$$\left(\frac{\partial}{\partial t} - \frac{\partial \psi}{\partial y}\frac{\partial}{\partial x} + \frac{\partial \psi}{\partial x}\frac{\partial}{\partial y}\right)\left[f_0 + \beta y + \frac{\partial^2 \psi}{\partial x^2} + \frac{\partial^2 \psi}{\partial y^2} + \frac{1}{\rho_0}\frac{\partial}{\partial z}\left(\frac{\rho_0 f_0^2}{N_B^2}\frac{\partial \psi}{\partial z}\right)\right] = 0. \tag{7.65}$$

The flow is determined by the structure of the geostrophic velocities and by the temperature stratification as represented by the vertical structure of the *Brunt–Väisälä frequency* $N_B(z)$. At midlatitudes of rapidly rotating bodies and assuming that friction and diabatic heating can be neglected, q_g is constant following the geostrophic flow. This is a prognostic equation for the streamfunction and can be integrated to obtain the evolution of the motion field for given initial and boundary conditions of the system.

7.4 THE PLANETARY BOUNDARY LAYER AND TURBULENT MOTIONS

So far we have neglected the effects of friction in the atmospheric motions, but frictional effects become important within the atmosphere under special circumstances and more commonly when the atmospheric flow interacts with the body surface. The layer in contact with ground that suffers most directly the frictional effects between the atmosphere and the surface is called the *planetary boundary layer* (PBL). The interaction produces vertical mixing and exchanges of momentum, heat, and moisture.

One way to express the effects of friction is through the *molecular or kinematic viscosity* (v_v) (Section 7.1.2.3), but its low value for gases $(\sim 10^{-5}$ m^2 s$^{-1})$ which means that it is not appropriate in general to represent the dominant effective friction force for large-scale motions in meteorology. This representation was used, for example, to study the friction effects on descending (falling) atmospheric particles as introduced in Section 5.3.4. Therefore, when studying the effects of friction on global atmospheric dynamics, a more convenient representation is through the introduction of an *eddy viscosity* in the frictional force, which has much larger values than the kinematic viscosity ($\sim$1–100 m^2 s^{-1}). This coefficient measures in some way the resistance of the atmospheric parcels to be deformed by stresses (associated, e.g., with shears).

The effects of friction in atmospheric motions become evident through *turbulence*. When all the elements of the atmospheric flow move along the direction of the fluid mean motion and the streamlines are nearly parallel, the flow is said to be *laminar*. If the elements move irregularly in constantly changing paths, the flow is said to be *turbulent*. In classical laboratory experiments, the classification of the flow regime, or the transition from a laminar to a turbulent situation, is controlled by the *Reynolds number*, *Re*, previously introduced in the context of planetary magnetism (Section 1.6.2) and in the microphysics study (Section 5.3). Here we redefine it in a more general form for a fluid as

$$Re = \frac{LV}{v_v} \tag{7.66}$$

where
  L is a typical length scale of the motion
  V is the velocity
  v_v is the kinematic viscosity

The transition from a laminar to a turbulent regime occurs typically at $Re \sim$ 4000–6000, but this is not universal and depends on the system being studied. Even for low values of the kinematic viscosity and using L as low as 1 m, it suffices that the flow reaches a velocity $V \sim 0.1$ m s^{-1} to become turbulent. We expect then that turbulence will be common in atmospheric flows.

One simple and practical form to express the frictional force is through the *Reynolds stresses* $\vec{\tau}_F$ that are a way of representing the action of small-scale processes on the larger scales of motion. This form comes from the description of turbulent motions that decomposes the flow magnitudes into two terms, a mean value and a fluctuation (the turbulent term). This is called the *Reynolds decomposition*. For example, the velocity is decomposed as the sum of the two terms $\vec{u} = \langle \vec{u} \rangle + \vec{u}'$, such that the mean value of the fluctuation is zero ($\langle \vec{u}' \rangle = 0$) and where $\vec{u}' = u'\hat{i} + v'\hat{j} + w'\hat{k}$. The substitution of this decomposition in the motion and continuity equations (7.2) and (7.10) results in crossed products for the wind velocities that, when averaged, lead to terms of the following type:

$$\langle uw \rangle = (\langle u \rangle + u')(\langle w \rangle + w') = \langle u \rangle \langle w \rangle + \langle u'w' \rangle \tag{7.67}$$

The average of the product of the deviation components $\langle u'w' \rangle$ is called the *covariance* and is not necessarily zero. These terms describe the *Reynolds stresses* or *eddy stresses* through the flux of momentum due to turbulent or eddy motions. For example, the *x*-component of the Reynold stress is given by

$$\tau_{Fx} = -\rho \langle u'w' \rangle \tag{7.68a}$$

(unit kg m^{-1} s^{-2}). For many atmospheric situations, in particular for the study of the planetary boundary layer, the vertical eddy fluxes are predominant (that is, those involving w'). The Reynolds stresses are then assumed to have only horizontal components $\vec{\tau}_F = \tau_{Fx}\hat{i} + \tau_{Fy}\hat{j}, (\tau_{Fz} = 0)$. One simple way to express these stresses in a manageable form is to assume that they are proportional to the vertical shear of the horizontal wind speed through a parameter called the *eddy viscosity* K_m (in m^2 s^{-1}):

$$\tau_{Fx} = \rho K_m \frac{\partial u}{\partial z}; \quad \tau_{Fy} = \rho K_m \frac{\partial v}{\partial z} \tag{7.68b}$$

This is sometimes referred to as *K*-theory. The existence of vertical wind shears is easily understood since the velocity must be zero at the surface. Then, the friction force per unit mass has the following two components:

$$F_x = -\frac{1}{\rho} \frac{\partial \tau_{Fx}}{\partial z} = -K_m \frac{\partial^2 u}{\partial z^2} \tag{7.69a}$$

$$F_y = -\frac{1}{\rho} \frac{\partial \tau_{Fy}}{\partial z} = -K_m \frac{\partial^2 v}{\partial z^2} \tag{7.69b}$$

A similar treatment can be done to other variables in the flow, most importantly, the potential temperature fluctuations θ' that give rise to an eddy heat flux $\langle w'\theta' \rangle = K_\theta(\partial\theta'/\partial z)$, where K_θ is the eddy heat diffusion coefficient, and similarly, for example, for vapor (humidity) fluxes.

7.4.1 MOTIONS CLOSE TO THE SURFACE

Consider the case in which the free planetary atmosphere (far away from the surface) is in geostrophic balance. Close to the surface, the balance is altered by the presence of friction whose intensity can intervene in the balance of forces. This is the Ekman layer.

7.4.1.1 The Ekman Layer

Using the representation (7.69) for the friction force acting on a flow that otherwise is in geostrophic balance equations (7.17), the momentum equations become

$$fu = -\frac{1}{\rho_0} \frac{\partial P}{\partial y} + K_m \frac{\partial^2 v}{\partial z^2} \tag{7.70a}$$

$$fv = \frac{1}{\rho_0}\frac{\partial P}{\partial x} - K_m\frac{\partial^2 u}{\partial z^2} \qquad (7.70b)$$

where we use the hydrostatic approximation to ignore vertical advection of mean momentum and K_m is taken as constant. From the definition of the geostrophic velocity (7.18) components

$$u_g = -\frac{1}{f\rho_0}\frac{\partial P}{\partial y}; \quad v_g = \frac{1}{f\rho_0}\frac{\partial P}{\partial x}$$

Eliminating the pressure gradient term, we can write the momentum balance as

$$K_m\frac{\partial^2 u}{\partial z^2} + f(v - v_g) = 0 \qquad (7.71a)$$

$$K_m\frac{\partial^2 v}{\partial z^2} - f(u - u_g) = 0 \qquad (7.71b)$$

and, multiplying the second equation by $i = \sqrt{-1}$ and adding to the first one, results in a single equation:

$$K_m\frac{\partial^2(u + iv)}{\partial z^2} - if(u + iv) + if(u_g + iv_g) = 0. \qquad (7.72)$$

At the surface $z = 0$ we have $u = v = 0$. Far from the surface, assuming that the geostrophic wind is purely zonal, that is, $v_g = 0$ and $u \to u_g$, a solution to (7.72) can be written as

$$u + iv = u_g\left\{1 - \exp\left[-\left(\frac{f}{2K_m}\right)^{1/2}(1+i)z\right]\right\} \qquad (7.73)$$

or in components

$$u = u_g\left[1 - \exp\left(-\gamma_E z\right)\cos\gamma_E z\right] \qquad (7.74a)$$

$$v = v_g\exp(-\gamma_E z)\sin\gamma_E z \qquad (7.74b)$$

where the parameter $\gamma_E = (f/2K_m)^{1/2}$.

These equations define the *Ekman layer*. A plot of u versus v describes a hodograph in the shape of a spiral, the *Ekman spiral* (Figure 7.7). The characteristic

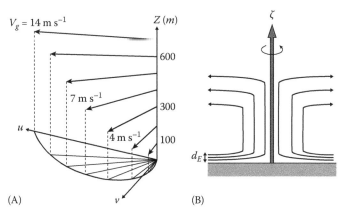

FIGURE 7.7 The Ekman layer. (A) Ekman spiral traced by the wind vertical shear. The approximate values for wind speed versus altitude correspond to the Earth case. (B) Sketch of the streamlines of the secondary circulation, forced by friction with the surface and vorticity pumping.

length scale of this layer is $d_E = \pi/\gamma_E$ ("Ekman layer thickness") such that above it (that is, when $z \gg \pi/\gamma_E$), the wind is purely zonal and in geostrophic balance. When approaching the surface, the wind deviates from geostrophic balance, reaching a deviation of 45° at the surface itself. For typical values on Earth ($f = 7 \times 10^{-5}$ s^{-1}, $K_m = 10$ m^2 s^{-1}) gives for the Ekman thickness $d_E = \pi/\gamma_E \sim 1$ km.

7.4.1.2 Ekman Pumping

The friction in the boundary layer may lead to convergent or divergent motions and from mass continuity to vertical motions. Neglecting density changes and substituting (7.74a and b) into the continuity equation (7.3), then integrating through the boundary layer (thickness d_E), we get a vertical velocity within the layer:

$$w(z = d_E) = -\int_0^d \frac{\partial u_g}{\partial y} \exp(-\gamma_E z) \sin \gamma_E z \, dz \qquad (7.75)$$

where we have used $\partial u/\partial x = 0$ and $w = 0$ at $z = 0$. Well above the layer, the wind is purely zonal and geostrophic, so the vorticity is $\zeta_g = \partial u_g/\partial y$ and (7.75) can be integrated to give

$$w(z = d_E) = \frac{1}{2} \frac{\zeta_g}{\gamma_E} \qquad (7.76)$$

For typical values found on Earth, such as those given above, one gets velocities of the order of mm s^{-1}. If the flow is convergent within the surface layer (there is inflow and then cyclonic motion), the motion produces rising air above the boundary layer. According to (7.56), the vorticity evolution equation can be written as

$$\frac{\partial \zeta_g}{\partial t} = -\frac{fw(z = d_E)}{(H - d_E)} \tag{7.77}$$

where ζ_g has been neglected when compared to f on the left-hand side of (7.56), advection has also been neglected and the motion is taken from the top of the Ekman layer to the tropopause ($z = H$, the scale height), where $w = 0$. Substituting (7.76) in (7.77) with $d_E \ll H$ gives

$$\frac{\partial \zeta_g}{\partial t} = -\frac{f}{2\gamma_E H} \zeta_g \tag{7.78}$$

Accordingly, the vorticity decreases with time (by a factor $1/e$) at a constant rate given by the *spin-down time* $\tau_E = 2\gamma_E H/f$. On Earth with $H = 10$ km and previous values, we have $\tau_E \sim 4$ days. This vortex motion in the boundary layer is a *secondary circulation* driven by friction, a mechanism known as *Ekman pumping* (Figure 7.7).

7.4.1.3 The Surface Boundary Layer and Mixing Length

In the previous section, it has been assumed that the eddy viscosity coefficient was independent of height, which is not true very close to the surface. Here, the Ekman layer model is not applicable and it is then convenient to introduce the concept of a limiting sublayer close to the surface that is called *the surface layer*. Here turbulence plays a central role and it is convenient to describe its action in terms of the *Prandtl mixing-length* model. In this context, it is assumed that an element of fluid at a level z that moves a distance ℓ (the mixing-length distance or eddy path) has a turbulent velocity proportional to the vertical shear of the zonal wind:

$$u' = -\ell \frac{\partial u}{\partial z} \tag{7.79}$$

For isotropic turbulence ($|u'| \sim |w'|$) that implies according to (7.68), a shearing stress given by

$$\tau_F = -\rho \langle u'w' \rangle = \rho \ell^2 \left(\frac{\partial u}{\partial z} \right)^2 \tag{7.80}$$

Assuming that the path distance is proportional to altitude, which occurs for statically neutral conditions, then $\ell = k_K z$, where k_K is the von Kármán's constant, we can integrate (7.80) to get the zonal wind profile structure:

$$\langle u \rangle = \frac{u_*}{k_K} \ln\left(\frac{z}{z_0} \right) \tag{7.81}$$

where
$u_* = (\tau_F/\rho)^{1/2}$ is the friction velocity
z_0 is the roughness height

This wind speed is important to understand the injection into the atmosphere of dust, sand, and other particles from the surface.

The studies of dust storms in deserts on Earth indicate that the wind speed at the surface must reach a threshold value (u_{*c}) to lift this material into the atmosphere. Its value depends on the *saltation* process that takes place when large particles lifted by surface winds fall back to the ground and dislodge small particles that can then remain suspended. Empirically, it has been found that $u_{*c} = A\sqrt{g2a(\rho_p - \rho)/\rho}$, where the parameter A depends on the particles' properties and fluid friction, a is the particle size, and ρ_p and ρ are, respectively, the particle and atmosphere densities. The threshold drag velocity on Mars, where dust is continuously lifted by different processes (convection, dust devils, and storms, see Chapter 9), is ~2–5 m s^{-1} for particles with a size of ~10 μm–1 mm and $\rho_p \sim 2.5$ g cm^{-3}, but larger for the most abundant smaller particles (1–5 μm). This indicates that *saltation* is a very important process in injecting dust into the Martian atmosphere.

A full representation of the *planetary boundary layer* is obtained by combining the logarithmic wind dependence in the surface layer with the Ekman spiral wind solution above it. Problems 7.21 through 7.23 deal with motions in the planetary boundary layer of Earth and Mars.

7.4.2 GENERALIZED TURBULENCE

Turbulent motions in the free atmosphere occur due to the development of instabilities at many different scales (Chapter 9) and their mutual interaction within the atmospheric flow. Turbulence can be generated *mechanically* (friction drag, wind shears, wakes for flow passing obstacles, etc.), *thermally* (as in free convection), and *inertially* (e.g., by vortices of different types). These turbulent motions can be envisaged as arising from the nonlinear nature of the Navier–Stokes equation (7.8b). In this equation, the total derivative can be expanded as the sum of the local derivative plus the advection terms and, in addition, the friction term is expressed from a generalization of (7.69) according to K-theory as $K_m \nabla^2 \vec{u}$. Then the Navier–Stokes equation can be written as

$$\frac{\partial \vec{u}}{\partial t} + (\vec{u}\nabla)\vec{u} = -\frac{\nabla P}{\rho} + \vec{g} - 2\vec{\Omega}\times\vec{u} - K_m\nabla^2\vec{u} \qquad (7.82)$$

where the relationship $d\vec{u}/dt = \partial\vec{u}/\partial t + (\vec{u}\nabla)\vec{u}$, similar to Equation 7.1a has been used.

Accordingly, the Reynolds number (7.66) can be expressed as the ratio between the nonlinear and friction terms:

$$Re = \frac{\text{Nonlinear}}{\text{Friction}} \approx \frac{(\vec{u}\nabla)\vec{u}}{K_m\nabla^2\vec{u}} = \frac{uL}{K_m} \qquad (7.83)$$

For flows that become turbulent according to our previous criterion that $Re > 5000$, we see from Equation 7.63 that this requires the eddy viscosity coefficient to be $K_m \sim 1$–100 m^2 s^{-1}.

Eddy motions at the different scales and turbulence can be expressed by the *Reynolds decomposition* that gives rise to the turbulent magnitudes: momentum flux $\langle u'v' \rangle$, Reynolds stresses $\rho \langle u'w' \rangle$, turbulent kinetic energy (TKE) (e.g., its zonal density $1/2\langle (u')^2 \rangle$), and heat flux $\langle u'T' \rangle$ or, as given previously, using the potential temperature, $\langle u'\theta' \rangle$. The Reynolds stress gradient terms, $\partial\langle u'v' \rangle/\partial y$, $\partial\langle u'w' \rangle/\partial z$, enter directly in the momentum equation and the gradient of the eddy heat flux $-\partial\langle v'T' \rangle/\partial y$ in the thermodynamic equation in this "zonal mean+eddy decomposition" representation.

One way to characterize or classify turbulent motions in the atmosphere is through the *turbulent kinetic energy spectrum*, that is, a representation of the turbulent kinetic energy (TKE or $(1/2)[(u')^2+(v')^2+(w')^2]$) as a function of the spatial frequency. Its form gives evidence of the different scales involved from the "macro" or "planetary-scale" to the "synoptic," "meso," and "micro" scales. For the large scales of motion, the flow can be considered two dimensional since the aspect ratio given by (7.14) $\delta = H/L \ll 1$, and then the turbulence is basically a two-dimensional process. This is mostly expected in surface planets (Earth, Mars, Venus, and Titan), but it is not evident in the case of the giant planets where the atmospheric motions can be deep (affecting a large part of the planetary radius), as will be shown later. In addition, under rapid rotation ($Ro < 1$), the Coriolis force intervenes in the development of large-scale instabilities and turbulence, and under geostrophic conditions, we refer to this case as *geostrophic* or *quasi-geostrophic turbulence*. At the smaller scales of motion, as, for example, in the previous cases close to the surface, or when rapid motions develop as during vigorous updraft convection, the turbulent motions tend to be three dimensional, which is the common case in classical fluid laboratory experiments (e.g., in intense sheared flows).

Although a full theoretical understanding of the turbulent energy spectrum in the atmosphere remains elusive, successful statistical descriptions and dimensional analysis have been proposed to explain at least for the Earth case, the measurements of the turbulent energy spectrum. Much more work remains to be done on the other planetary atmospheres.

The simplest model of atmospheric turbulence assumes that isotropic and homogeneous conditions prevail at a macroscopic level and neglects compressibility effects. Consider a finite volume V of the fluid and define $k = L^{-1}$ as the wavenumber (L is, e.g., the size of an eddy in the flow) related to the velocity fluctuation u'. Assume that the kinetic energy within this volume can be decomposed by means of a Fourier series of the wavenumber such that in the isotropic regime we define a *spectral energy density* $E(k)$ that is only a function of the modulus of k but not of its direction (the wavenumber is in fact a vector $\vec{k}$). The mean turbulent kinetic energy per unit mass can be obtained as the sum over the Fourier space:

$$E = \frac{1}{2V} \int_V \vec{u}'^2 d^3r = \frac{1}{2}\langle \vec{u}'^2 \rangle = \int_0^\infty E(k)dk \qquad (7.84)$$

$E(k)dk$ being the contribution to E from wavenumbers with modulus between k and $k + dk$. The eddies have a characteristic velocity $u'(k) = [E(k)k]^{1/2}$ and develop in timescales $\tau_k = 1/[ku'(k)]$. This could be compared, for example, with the dissipation time due to viscosity v_V given by

$$\tau_f = \frac{1}{v_V k^2} \tag{7.85}$$

This dissipation time increases with eddy size and the eddy begins to dissipate when $\tau_f < \tau_k$, or, in other words, when its wavenumber approaches k_f such that larger eddies survive for a long time. As an example, for $v_V = 10^{-5}$ m^2 s^{-1} and $u' = 0.01$ m s^{-1}, $\tau_f = \tau_k$ when the eddy size has a length $L = k^{-1} = 0.1$ m.

7.4.2.1 Kolmogorov Turbulence

Let us now introduce the quantity $\varepsilon_I = k_I [u'(k_I)]^3$ as the power per unit mass (unit J kg^{-1} s^{-1} = m^2 s^{-3}) that is fed into the system at a small wavenumber k_I (that is, at a large spatial scale k_I^{-1}). Due to the nonlinear terms in the momentum equation (7.82), the energy is transferred in a "cascade" process to higher wavenumbers (smaller spatial scales) until the dissipation wavenumber k_f is reached and dissipation dominates in the system. The wavenumber interval between k_I and k_f where neither forcing nor dissipation is important on the dynamics is called the *inertial range*. In this range, the inertial terms dominate the momentum balance.

For $k_f \gg k_I$, the *Kolmogorov theory* assumes that the energy spectrum does not depend on viscosity but only on ε_I and k and simple dimensional considerations then require that

$$E(k) = C_K \varepsilon_I^{2/3} k^{-5/3}, \tag{7.86}$$

C_K being a dimensionless constant with a value ~1.5. This is the *Kolmogorov spectrum law* with its characteristic $k^{-5/3}$ dependence. Figure 7.8 shows a schematic view of the energy spectrum according to the Kolmogorov theory (see also Problem 7.24). According to (7.84) and (7.85), the turbulent velocity is given by

$$u'(k) = C'_k \left(\frac{\varepsilon_I}{k} \right)^{1/3} \tag{7.87}$$

where C'_k is another constant of order unity, and the "turnover time" for eddies is $\tau_k = \varepsilon_I^{-1/3} k^{-2/3}$. The dissipation wavenumber, k_f, obtained from the condition $\tau_f = \tau_k$, is then given by

$$k_f \approx \left(\frac{\varepsilon_I}{v_V^3} \right)^{1/4} \tag{7.88}$$

which is also referred to as the *Kolmogorov scale* (Problem 7.25). The Kolmogorov spectrum has been observed in the turbulent boundary layer of Earth's atmosphere, and has also been claimed to occur on other planets, a subject still under study.

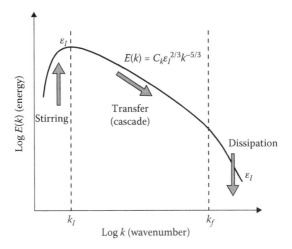

FIGURE 7.8 Schematic form of the energy spectrum as predicted by Kolmogorov's three-dimensional theory of turbulence. Energy supplied at a low wavenumber cascades to small scales where it is dissipated by viscosity.

7.4.2.2 Two-Dimensional Planetary Turbulence

Away from the surface, in the free atmosphere, the flow can be considered two dimensional in the case of most planetary atmospheres. In addition to energy, another relevant quantity for the turbulent energy spectrum in two dimensions is the vorticity (or, more explicitly, its vertical component; see, e.g., Vallis, 2006). We have already seen in Section 7.3.2 that this quantity is conserved when particular conditions are met (e.g., when viscosity is negligible). In a similar form as the definition of kinetic energy, it is convenient to introduce the concept of *enstrophy*, ξ (unit s^{-2}), defined as the mean value of the square of the vorticity ($\xi = (1/2)\langle \varsigma^2 \rangle$) (a measure of the vorticity content in the atmosphere).

Without going into the details (see Vallis, 2006 for a full description), in two dimensions, the energy spectrum of turbulence may contain two inertial subranges (see Figure 7.9):

1. *Enstrophy transfer scales*: Enstrophy supplied at some rate ε_ζ (unit s^{-3}) to the flow is transferred to smaller scales (downscale) according to a power law

$$E(k) = C_\zeta \varepsilon_\zeta^{2/3} k^{-3} \tag{7.89}$$

where the constant $C_\zeta = 5-7$. The enstrophy is transferred from large to small vortices and blocks the energy transfer toward smaller scales with an "eddy turnover time" given by $\tau_\zeta = \varepsilon_\zeta^{-1/3}$. Enstrophy dissipation by viscosity occurs at a small spatial scale $k_{f\zeta} = \left(\varepsilon_\zeta / \upsilon_V^3 \right)^{1/6}$.

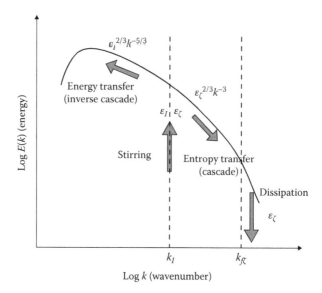

FIGURE 7.9 Schematic form of the energy spectrum for two-dimensional turbulence. Energy and enstrophy are supplied at some scale. The energy is transferred to larger scales (inverse cascade), whereas entropy is transferred to smaller scales where it dissipates by viscosity.

2. *Energy transfer*: The energy transfer follows a law similar to (7.86) but with a substantial difference. The energy injected at a given wavenumber may now flow from smaller to larger scales in a process known as an *inverse cascade*, with an isotropic spectrum given by

$$E(k) = C'_K \varepsilon_I^{2/3} k^{-5/3} \qquad (7.90)$$

where C'_K is the nondimensional Kolmogorov–Kraichnan constant. This behavior is important for large-scale atmospheric flows where large-scale structures like vortices can emerge and be organized coherently by smaller ones (escaping from the turbulent disorder).

These ideas do not perfectly apply to the atmospheres, as seen in Figure 7.10 for the Earth, which shows evidence for the existence of the $k^{-5/3}$ and k^{-3} inertial ranges in the zonal and meridional tropospheric winds but in the nonclassical places. The -3 spectrum could be associated with a forward entropy cascade, but the $-5/3$ spectrum present at the smaller scales is not yet explained.

On other planets, the measurement of the turbulent kinetic energy is difficult to perform since turbulent wind speeds are of the order of the instrumental noise limit measurements (about 1–$5\,\text{m s}^{-1}$). However, we expect in general that on rapidly rotating planets, the Coriolis force, and, more explicitly, its meridional gradient $\beta = df/dy$, becomes an efficient source of vorticity. The kinetic energy of turbulent eddies generated by this mechanism and other instabilities (see Chapter 8) lead to the *quasigeostrophic* or *QG-turbulence*. Here, the interaction of the turbulent flow and the Rossby

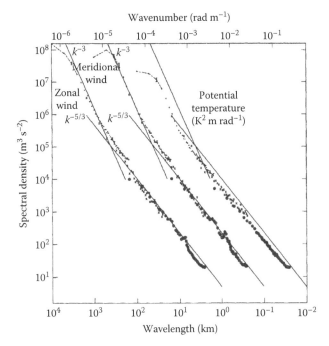

FIGURE 7.10 Spectrum of the turbulent kinetic energy (meridional and zonal wind) and potential temperature in the atmosphere of Earth. The spectra for meridional wind and temperature are shifted one and two decades to the right, respectively; lines with slopes −3 and −5/3 are entered at the same relative coordinates for each variable for comparison. (From Nastrom, G.D. and Gage, K.S., *J. Atmos. Sci.*, 42, 950, 1985. With permission.)

waves (see Chapter 8) that emerge on a rapidly rotating planet depends on the wavenumber k_β that defines the *Rhines scale*:

$$L_\beta = \frac{1}{k_\beta} = \sqrt{\frac{u_{rms}}{\beta}} \qquad (7.91)$$

It represents a cross-over scale that separates the dominance of the β-effect at large scales on the flow from that of the advection of vorticity that dominates at smaller scales. In (7.91), u_{rms} is the root-mean-square velocity at the energy-containing scales. It has been proposed that the width of the zonal jets observed in Jupiter and Saturn is directly related to this scale and that jet formation might be explained by the QG-turbulence (see Section 7.6.4 and Vallis (2006) for details on QG-turbulence).

7.5 OBSERVED LARGE-SCALE MOTIONS IN THE PLANETS

The measurements of atmospheric motions are inferred basically using different techniques: (1) direct in situ measurements using surface anemometers; (2) tracking of atmospheric balloons placed at different heights (on Earth obviously in a regular way, and on Venus on two occasions); (3) tracking of descending probe motions at localized points in space and in time (up to the present done on Venus, Mars, Jupiter, and Titan); (4) cloud tracking (mostly on Venus and the giant and icy planets that are fully cloud covered), which assumes that clouds act as passive tracers of the atmospheric flow; (5) indirectly by measuring the temperature field at different levels and then using the relevant thermal wind relationship for the corresponding planet (Section 7.2.7); (6) measurements of the Doppler shift of spectral lines (mostly in the case of the upper atmospheres of Venus, Mars, and Titan); (7) the wind speed can be constrained indirectly from determining the shape of atmospheric isobaric surfaces derived from stellar light and spacecraft radio occultation measurements. A particular case is that of Venus's mesosphere where the intense airglow emissions by oxygen (see Chapter 6) can be tracked in time and then partially used to determine atmospheric motions if interpreted as a passive air mass tracer.

7.5.1 RAPIDLY ROTATING BODIES WITH SURFACE: EARTH AND MARS

Earth and Mars show very similar large-scale circulations corresponding to a rapidly rotating planet (both planets have similar rotation periods). The main difference between them comes from the presence of the oceans on Earth that introduces important constraints on surface temperature contrasts and a water vapor condensation–evaporation cycle, and similarly on Mars, the condensation/sublimation of a large mass of CO_2 (the condensation flow) and thermal tides. Both atmospheres are, however, ultimately driven by the differential solar heat deposition on the surface and in the atmosphere.

Outside the tropical latitudes, a geostrophic regime of zonal winds dominates the circulation in both planets (Figures 7.11 and 7.12). The temperature distribution is consistent with the geostrophic thermal wind relationship (Equation 7.32). At the equator, the zonal wind is westward whereas at midlatitudes, two jet streams (one per hemisphere) flow eastward. On Earth, the jet streams are centered at ~40°–50° latitude on each hemisphere with peak speeds of ~40 m s^{-1} at an altitude $z \sim 8$ km. In the zonally and time-averaged distributions, the jets weaken above the tropopause and then increase their strength with altitude from ~25 to 70 km, reaching speeds of 60 m s^{-1} in the stratosphere-mesosphere. On Mars, due to the low atmospheric mass and lack of oceans, the jet streams are seasonally dependent. At the winter solstice in the northern hemisphere, the jet stream is centered at ~40°–50° latitude reaching a peak speed of ~40 m s^{-1} at $z \sim 5$ km (pressure level 5 mbar) but increases to ~110 m s^{-1} at $z \sim 35$ km (pressure level 0.2 mbar). At this same epoch, the jet stream in the southern hemisphere is weaker. The measurements of cloud motions and dust during local and global storms in Mars' troposphere (that occur when surface winds are >25 m s^{-1}) and in situ wind measurements by surface landers confirm this picture.

The meridional circulation is dominated on both planets by a Hadley-cell regime (Section 7.2.8). On each hemisphere on Earth, the rapid rotation breaks the meridional circulation into basically a main Hadley cell between equator and tropics (~30° latitude) and two additional cells, the Ferrel cell (circulating in a thermodynamically indirect sense) at midlatitudes and a polar cell. The deviation of the Hadley cell meridional motion by the action of the Coriolis force gives rise

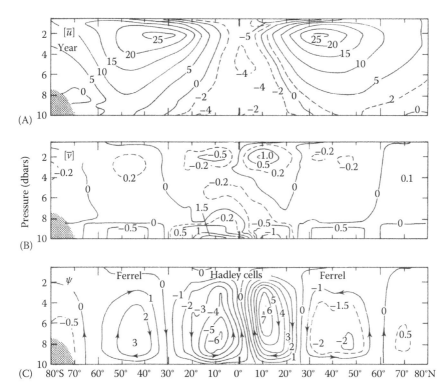

FIGURE 7.11 Zonal mean cross sections in Earth's troposphere: (A) zonal wind (m s^{-1}); (B) meridional wind component (m s^{-1}); (C) meridional atmospheric mass flow (10^{10} kg s^{-1}, meridional circulation cells indicated) for annual mean and zonally averaged conditions. The vertical coordinate is pressure P (multiply for 0.1 to get bars). (From Peixóto, J.P. and Oort, A.H., *Rev. Mod. Phys.*, 56, 365, 1984. With permission.)

to "trade winds" (in the west to east direction), with converging air in the equator that forms the Intertropical Convergence Zone (ITCZ) (see also Section 9.6.3.1). There, strong ascending moist and warm air induces the formation of dense cumulus cloud along a planetary-scale narrow cloudy belt, with corresponding dryness in the latitudes corresponding to the descending branch. Locally, the ITCZ is most active over the warm and moist continental areas (Amazon, central Africa, and Indonesia) and weaker over the cool ocean regions west of Peru and Africa, where the air tends to sink due to radiative cooling. A system of weak zonal cells called the *Walker circulation* (see also Section 9.6.3.1) develops along the equator. Long-term changes (every 3–5 years) in these trade winds and Walker cell originates pressure differences between the eastern and western Pacific that couple to temperature changes in the ocean surface and on its elevation, yielding large variations in the evaporation and latent heat release. The cell displaces eastward and air rises above the central Pacific in what is known as *El Niño* that leads to convective rain in the usually dry eastern Pacific, whereas the cold waters off the coast of Peru warm up temporarily. The pressure patterns show a simultaneous *Southern Oscillation* (see also Section 9.6.3.1) so both quasi-periodic phenomena

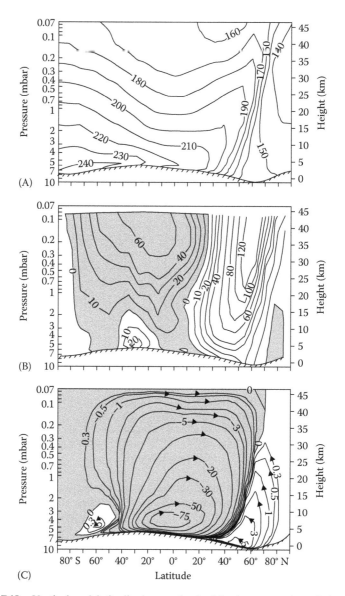

FIGURE 7.12 Vertical and latitudinal maps in the Martian atmosphere (solar longitude $L_S = 270°$) of (A) temperature (K); (B) mean zonal wind (m s^{-1}); (C) meridional streamfunction. (From Leovy, C.B., *Nature*, 412, 245, 2001. With permission.)

are known as *the El Niño Southern Oscillation* (ENSO). It has profound influences on the weather patterns on Earth. On Mars, a single Hadley cell is also present between the tropical latitudes of both hemispheres: Warm air rises in the summer hemisphere of Mars (hottest subsolar latitude ~25°–30°) and sinks in the winter hemisphere (Figure 7.12). Values for the Hadley cell circulation on Earth and Mars were given in Section 7.2.8.

However, two additional mechanisms contribute uniquely to the global motions on Mars. On the one hand, there is a seasonal flow related to the pressure gradient established from the condensation of CO_2 that generates a planet-wide circulation (a condensation flow, see Sections 6.7.2 and 7.2.9) that transports heat, momentum, and mass between the high latitudes (in the winter hemisphere where the pressure lowers) and other latitudes where pressure is higher. About 20% of the atmospheric mass is recirculated by this mechanism. The mean meridional velocities associated with the condensation flow are $\langle v \rangle \sim 0.2\text{--}0.5\,\text{m s}^{-1}$. On the other hand, due to the thin nature of the Martian atmosphere and its main CO_2 composition (that it is a good radiator), there are large temperature differences (tens of degrees Kelvin) between the illuminated and shadowed hemispheres. The temperature difference drives thermal tides (see Section 4.3.5) that propagate in the direction of motion of the Sun in the Martian skies.

In Chapter 8, we will discuss some relevant dynamical global-scale properties of the terrestrial and Martian atmospheres directly coupled to the large-scale motions as the midlatitude baroclinic wave activity, Martian global dust storms, and stationary eddies due to the large topographic contrasts.

7.5.2 SLOWLY ROTATING BODIES WITH SURFACE: VENUS AND TITAN

Venus' circulation is dominated by a strong westward zonal flow (known as *the super-rotation*) that occurs within the cloud and haze layers (altitudes 30–70 km) as inferred from the tracking of cloud motions and confirmed in detail by descending probes in particular locations. The upper atmosphere rotates in ~4 days and when compared to the 244 days rotation period of the solid body, and then it is said that it superrotates. At low latitudes (between ±50°), zonal winds are nearly constant with latitude with westward peak velocities of ~110 m s^{-1} at cloud tops (altitudes ~65–70 km) that decrease to 60–70 m s^{-1} at the cloud base (altitude ~45–50 km) (Figure 7.13). The winds then decrease to a value close to zero at the surface (Figure 7.14). Above clouds, the winds decrease with altitude until a strong solar–antisolar circulation occurs at 90–110 km as a consequence of the large temperature gradient between Venus' thermosphere and cryosphere. At high latitudes, zonal wind speeds decrease linearly with latitude and altitude, from ±50° to the pole where a vortex forms. At equatorial and midlatitudes (0°–55°), the vertical wind shear is $\partial \langle u \rangle / \partial z = 10\,\text{m s}^{-1}$ km^{-1} between altitudes $z \sim 60$ and 70 km, but is $\partial \langle u \rangle / \partial z < 1\,\text{m s}^{-1}\,\text{km}^{-1}$ between altitudes $z \sim 45$ and 60 km. In the subpolar latitudes from ~55° to the pole, the vertical wind shear within the clouds (altitude 45–65 km) is weak, $\partial \langle u \rangle / \partial z \sim 2\,\text{m s}^{-1}\,\text{km}^{-1}$ (or lower). The cloud level superrotation is approximately in cyclostrophic balance as derived from the measurements of the meridional temperature profiles at different altitudes (Equations 7.34 and 7.35). However, the balance is not perfect and some details still remain elusive to this balance interpretation.

A thermal solar tide is present at cloud tops with values of the wind speed that grow with increasing local time at a rate of the order of $2\,\text{m s}^{-1}\,\text{h}^{-1}$. Meridional winds at the cloud tops are poleward with peak speed of $10\,\text{m s}^{-1}$ at latitudes 55° but are smaller than $5\,\text{m s}^{-1}$ below the cloud tops. They have been interpreted as due to the overturning of a single hemisphere Hadley cell that goes from equator to pole at cloud tops (Section 7.2.8).

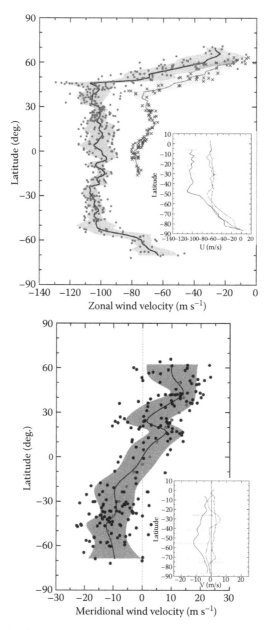

FIGURE 7.13 Latitudinal profiles of the zonal and meridional wind components derived from cloud tracking at different altitude levels and in two time periods in Venus' atmosphere. The meridional extended profiles are from measurements using Galileo images in 1990. Dots and mean continuous line (upper cloud top at $z=65–70$ km); crosses and continuous line (clouds at $z=60–65$ km). (From Peralta, J. et al., *Icarus*, 190, 469, 2007.) The inset shows the winds in the southern hemisphere using Venus Express images (period 2006–2007). Continuous line: upper cloud top at $z=65–70$ km; Dotted line: cloud at $z=60–65$ km; Dotted-dashed line: lower cloud at $z=45–50$ km. (Courtesy of R. Hueso, UPV-EHU, Spain.)

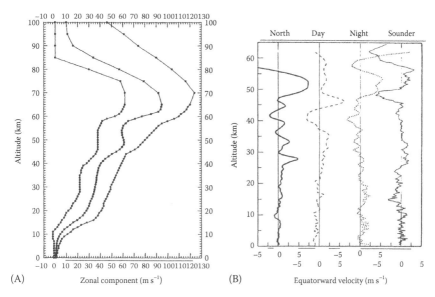

(A) Zonal component (m s^{-1})

(B) Equatorward velocity (m s^{-1})

FIGURE 7.14 (A) Vertical profiles of the zonal wind in Venus' atmosphere (VIRA model) as determined from measurements by the different entry probes (Pioneer-Venus and Venera). (B) Vertical profiles of the meridional wind speed measured by the Pioneer-Venus entry probes. (From Kerzhanovich, V.V. and Limaye, S.S., *Adv. Space Res.*, 5, 59, 1985. With permission.)

The direct measurements of the atmospheric circulation of Titan, a satellite with a rotation period of 16 days, are still scarce. The only direct wind measurements come from the tracking of the descent of the Huygens probe in January 2005 and the Doppler-shifted radio emission originated by the probe shift due to winds. The vertical profile of the wind speed was obtained in the equatorial region (Figure 7.15). Temperature measurements have been used together with the thermal wind balance for gradient winds (Equations 7.36 and 7.37) to infer global atmospheric motions (Figure 7.16). The Huygens profile is used as the reference for the application of the thermal wind gradient relationships. Accordingly, superrotating seasonally dependent winds are expected in the upper atmosphere. For the Cassini-Huygens period, a single intense eastward jet with speeds up to 190 m s^{-1} between latitudes 30°N and 50°N was deduced at the 1 mbar altitude level (about 190 km from the surface). A planetary scale Hadley circulation with overturning centered at a pressure level of 10 mbar (about 100 km altitude level) is predicted to exist from equator to pole. According to (7.38) and (7.39), the related vertical velocities are $w \sim 0.5$ mm s^{-1} with corresponding meridional velocities $v \sim 3$ cm s^{-1}.

Scarce data exist on cloud motions to constrain the wind field and these correspond to the methane–ethane discrete and sporadic clouds that form in the altitude range ~10–45 km. The haze layers that extend over large altitudes are very homogeneous in brightness and no isolated features have been detected that can be used to track motions (see Chapter 5).

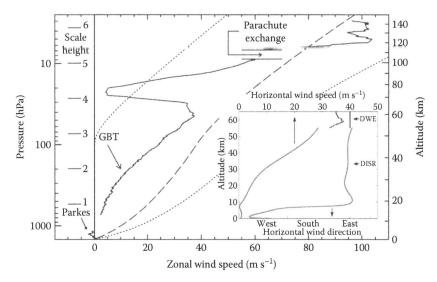

FIGURE 7.15 Vertical profiles of the zonal wind speed in Titan's equatorial atmosphere. The continuous line is the wind velocity derived from the radio tracking measurements performed during the descent of the Huygens probe in January 2005. The dashed and dotted lines are derivations from Voyager (1980 and 1981) temperature measurements and application of the cyclostrophic balance. The inset shows the wind speed and direction as a function of altitude as measured during probe descent with the image experiment (DISR). (From Bird, M.K. et al., *Nature*, 438, 800, 2005; Inset: Tomasko, M.G. et al., *Nature*, 438, 765, 2005. With permission.)

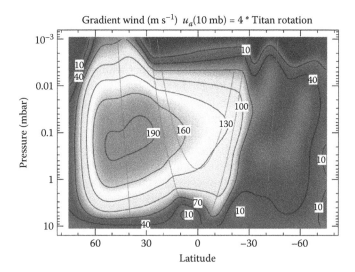

FIGURE 7.16 Vertical–meridional cross-section of the zonal wind in the upper atmosphere of Titan as derived from the application of the gradient wind balance. (From Achterberg, R.K. et al., *Icarus*, 194, 263, 2008. With permission.)

7.5.3 TENUOUS ATMOSPHERES: TRITON, PLUTO, AND IO

Only for Triton do we have indirect measurements of wind speeds. Voyager 2 images obtained in 1989 showed the following signatures of strong winds: streaks on the surface, terminator clouds confined to altitudes below 5 km (see Chapter 5), and geyser-like plumes extending up to 8 km altitude. The analysis of these motion indicators suggests that the wind direction at low altitudes (~1–3 km) was preferentially toward the northeast in the southern hemisphere with a speed of ~15 m s^{-1}. Narrow chimneys from geyser-like activity reaching 8–10 km above the surface indicate a change in the direction of winds by 180° (now directed westward) with speeds of 15 m s^{-1} at this level. These observations can to first order be explained by a combination of a condensation wind flow from the south (the sublimation region) to the north (Sections 6.7.2 and 7.2.9) coupled to a geostrophic wind at altitudes of ~8–10 km (Equations 7.17 and 7.18), and the presence of an Ekman layer in the lower ~1 km near the surface atmosphere (see Problem 7.26).

Condensation flows are also expected on Pluto and Io but we still lack any observational data on winds in these tenuous atmospheres.

7.5.4 GIANT PLANETS

Since the gaseous giant and icy planets lack a solid surface (in the sense at least of what we have in terrestrial planets and satellites, see Chapter 1), the wind speeds are measured with respect to their radio rotation period, assumed to be rooted in the massive interior representing the true rotation of the planet (see Section 1.3.4). Tracking the motion of cloud features in the upper main cloud deck is the main method used to retrieve the winds. At cloud level, all four planets show an intense zonal circulation formed by a system of jet streams that alternate their direction (East–West) with latitude and whose peak speed is typically ~20–100 m s^{-1} with maximum peak values in the velocity at the equators of Saturn and Neptune of ~450 m s^{-1}.

At the ammonia cloud level (altitude range $P \sim 0.5$–1.5 bar), Jupiter has about 8 jets per hemisphere, with maximum speed at the equator where the eastward jet reaches zonal velocities ~100–125 m s^{-1}, and at 23°N, where winds reach up to 180 m s^{-1} (Figure 7.17). Saturn has 6 jets per hemisphere with a very strong and broad eastward equatorial jet with peak speed of ~450 m s^{-1} (apparently variable), although uncertainty exists in the radio rotation period of the planet that may change this value. Eastward jets appear to dominate in intensity over westward jets (Figure 7.18).

In contrast to the two giant planets, the two icy planets, Uranus and Neptune, both show a westward equatorial jet (peak zonal velocity −100 m s^{-1} in Uranus and −400 m s^{-1} in Neptune) and two intense midlatitude eastward jets with peak speed of about 200 m s^{-1} at the altitude level of the main methane haze cloud at ~1 bar (Figures 7.19 and 7.20). At cloud level, the mean zonal winds do not show longitudinal dependence ($\partial \langle u \rangle / \partial x = 0$) in these planets, except obviously where localized disturbances (vortices, waves, or convective storms) exist (Chapters 8 and 9).

Due to the rapid rotation of these planets and the large scale of the motions and the intensity of the winds, Rossby number $Ro < 1$ and the geostrophic balance

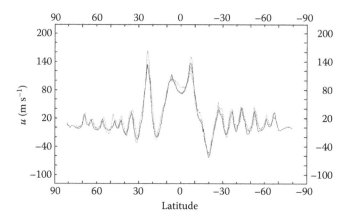

FIGURE 7.17 Meridional profile of Jupiter's zonal wind as measured from the tracking of cloud features in the upper cloud (0.5–1 bar pressure level) at three different epochs: Dotted line (Voyager, 1979: Data from Limage, S.S., *Icarus*, 65, 335, 1986.); continuous dark line (Cassini, 2000: Data from Porco, C.C. et al., *Science*, 229, 1541, 2003.); continuous gray line (Hubble Space Telescope, 1995–2000: Data from Garcia-Melendo, E., and Sánchez-Lavega, A., *Icarus*, 152, 316, 2001.).

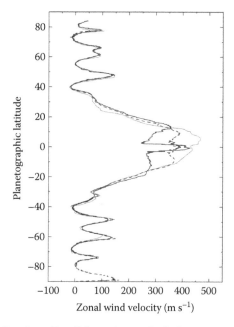

FIGURE 7.18 Meridional profile of Saturn's zonal wind as measured from the tracking of cloud features in the upper clouds (0.2–1 bar pressure level) using images obtained by Voyager 1 and 2 (1980–1981: Data from Sánchez-Lavega, A. et al., *Icarus*, 147, 405, 2000; light gray) and Cassini (2005–2008: Data from Garcia-Melendo, E. et al. (unpublished); dark continuous line—high altitude, dashed line—lower altitudes). The different profiles at the equator are due to vertical wind shears and long-term changes between the Voyager and Cassini periods.

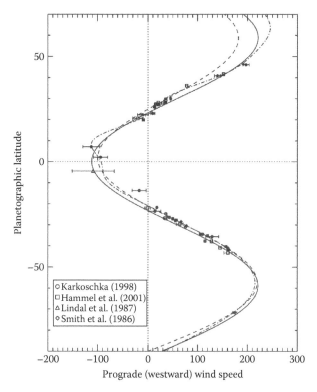

FIGURE 7.19 Uranus's zonal wind profile at cloud level between 1986 and 2004 as derived from cloud tracking measurements and a radio occultation single point, and different fitted–binned curves. (From Sromovsky, L.A. and Fry, P.M., *Icarus*, 179, 459, 2005. With permission.)

prevails in all latitudes except in the equatorial region. The measurements of the temperature field above clouds allows the use of the thermal wind relationship (7.32) and (7.33) to infer the vertical shears and derive the wind intensity (Figure 7.21). According to the current available data, the winds decrease with altitude at a typical rate of ~10–20 m s^{-1}/H. In Jupiter and Saturn, the tracking of subtle cloud structures located in the hazes above the main cloud deck confirm this picture of decreasing winds with altitude. In the case of the strong equatorial jet of Saturn, the altitude differences between cloud features are large enough to detect prominent velocity differences. For example, at 50 mbar altitude level, the wind speeds are ~250 m s^{-1} but increase to ~375 m s^{-1} at 700 mbar. The greatest speeds measured on Saturn's equator reached ~500 m s^{-1} in the Voyager epoch, indicating probably the existence of temporal changes and/or the detection of deeper motions. The vertical wind shears is ~40 m s^{-1}/H.

Below the main cloud deck, images of Jupiter and Saturn obtained in the thermal infrared spectral window at 5 μm allows the detection of clouds in the 2–4 bar altitude level by their infrared opacity to this wavelength (Figure 3.3) and the measurement of their motions. The wind system does not change significantly with respect to

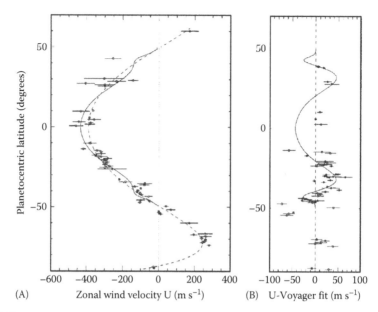

FIGURE 7.20 (A) Neptune's zonal wind profile at upper cloud level as derived from cloud motion measurements using Voyager 2 images in 1989 (dashed curve) and Hubble Space Telescope observations between 1995 and 1998 (solid curve). The Voyager results are shown as open circles with error bars. (B) Direct comparison of Voyager bin-averaged wind results with HST results. The solid line is the HST–Voyager fit difference. (Adapted from Sromovsky, L.A. et al., *Icarus*, 150, 244, 2001. With permission.)

what is measured in the upper cloud. It just tends, on the global average in both planets, to increase or decrease slightly with depth (Figure 7.22). However, in December 1995, the Galileo probe penetrated the Jovian atmosphere at a latitude of 7°N where a "hot spot" region exists (see Chapter 8) and measured the winds well below the main cloud decks, up to a depth of 24 bar (Figure 7.23). It showed that the zonal winds increase with depth up to the ~3 bar at a rate of ~30 m s^{-1}/H, and then remained nearly constant with $u \sim 180$ m s^{-1} up the depth of 24 bar. It remains to be known if this profile is a particular one for this "hot spot" region or if it is representative of the other latitudes of the planet.

The mean meridional winds have low velocities at the upper cloud compared to the zonal ones. In Jupiter and Saturn, they reach maximum values of ~5–10 m s^{-1}, but there is no evidence of a systematic meridional circulation between the jet streams. The measurements of the root mean square (rms) deviation from the mean zonal and meridional winds allows to measure the eddy velocities u', $v' \sim 3$–10 m s^{-1}. In both atmospheres, it has been found that in the jets, the eddy momentum transport $\langle u'v' \rangle$ (see Sections 7.4.2 and 7.4.3) correlates with the mean meridional shear $d\langle u \rangle/dy$ suggesting that there is a transport of momentum and energy from the eddies to the mean zonal flow, that is, in other words that the eddies can feed the jet streams. We shall go back to this idea in the next section.

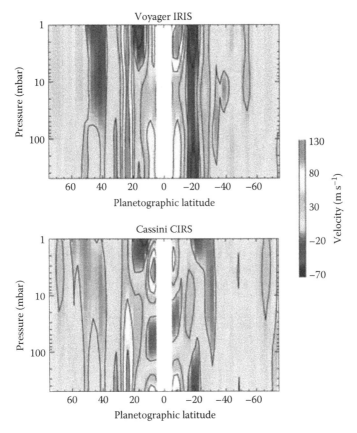

FIGURE 7.21 Jupiter's zonal thermal winds calculated from temperature maps (see Chapter 4) obtained by the Voyager IRIS (in 1979) and Cassini CIRS instruments. Contours are drawn every 30 m s⁻¹ from −60 to 120 m s⁻¹. (From Simon-Miller, A.A. et al., *Icarus*, 180, 98, 2006. With permission.)

7.6 THE GENERAL CIRCULATION OF PLANETARY ATMOSPHERES

General circulation models (GCM) are developed to explain the nature of the large-scale motions observed in planetary atmospheres and are used for the research of other atmospheric phenomena, and in the case of the Earth for weather forecasting. They are based on the physical principles described in the previous sections formulated in terms of differential equations. As already discussed, these equations are nonlinear and coupled, and no solution for the magnitudes describing the atmospheric flow (velocity, temperature, etc.), are known in terms of analytical functions. Together with the appropriate boundary conditions for the atmospheric system, and from given initial conditions, the set of equations can be solved numerically forward in time with the help of modern supercomputers. But even in this case, the nonlinear nature of the equations and uncertainties in the initial values of the variables that must be introduced in space and time to solve the system, make the results of limited

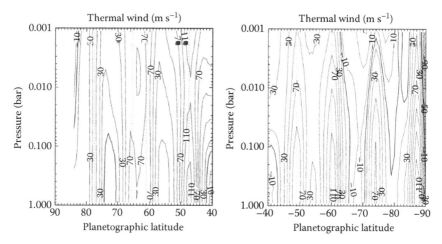

FIGURE 7.22 Saturn's zonal thermal winds calculated from temperature maps (see Figure 4.19) obtained by the Cassini CIRS instrument. (Courtesy of L.N. Fletcher, JPL-Oxford University, Oxford, U.K. and Cassini, CIRS team.)

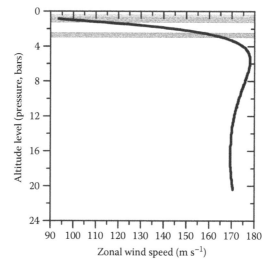

FIGURE 7.23 Vertical profile of the zonal wind velocity in Jupiter's equatorial region (a "Hot Spot area" at latitude 7°N) as measured by the Galileo entry probe during its descent in 1995. (Data from Atkinson, D.H. et al., *J. Geophys. Res.*, 103, 22911, 1998.) The parallel bands show the altitude location of detected clouds according to Ragent, B. et al., *J. Geophys. Res.*, 103, 22891, 1998.)

validity when one goes into the details. The system shows a *chaotic* behavior, even in the case of the Earth where abundant data are continuously gathered in space and time. In addition, the description of some physical processes (e.g., friction) is performed in simplified form, introducing parameterizations in the equations that limit the precision.

In the case of other planetary atmospheres, where the number of data that can be used as inputs to solve the equations is limited, these models try to reproduce the basic or the gross characteristics of the atmospheric motions. GCMs have been widely used to simulate the general circulation on Earth. They have also been used for Mars, with an emphasis on simulating the dust cycle, and in incorporating the dust and water ice cloud distribution in the thermal and dynamical description of the Martian atmosphere. The atmosphere of Venus has also been simulated using GCMs, mainly focused on the reproduction of the superrotation. On the other hand, Titan's GCM-based studies have mainly focused on the evaluation of the vertical–meridional circulation, the temperature structure and superrotation of the stratosphere, and their effects on the thick haze distribution. In the case of the giant planets, the situation is worse, since we do not know what kind of energy source (internal heat or sunlight, or both) drives the zonal jet system. Due to these complications, a variety of approaches, that is, a hierarchy of models with different degrees of complexity, have been developed to study the atmospheric phenomena.

7.6.1 Types of Dynamical Models

The atmospheric system comprises the following basic elements: the energy sources (external, internal, potential, etc.), the chemical composition (abundances and mass), the spherical geometry (and planet or satellite size) and boundary conditions (presence of surface or not, and its nature including topography), and the planetary rotation rate.

GCMs are the most complex type of model that tries to describe the evolution of the flow using the maximum number of physical and chemical processes known to occur in an atmosphere. They include different "modules" to represent the processes occurring in the system that are managed separately, but then coupled in some way. Examples of modules are (1) *dynamical core* that includes the equations of motion and thermodynamics (energy sources and their transformations); (2) *radiative transfer* module that includes the heat flow deposition and energy transfer (surface and or atmosphere), including the effects of clouds and aerosol; (3) *microphysical* module for cloud and aerosol formation; (4) *chemistry* to describe the chemical transformations in the system. Constraints from basic principles can be added to it such as those involving the angular momentum and energy budgets and their transformation as discussed later.

To make the system tractable, some simplifications are introduced into the dynamical models in a large variety of ways. The first and most obvious way is to exclude some of the modules introduced above. Then, if we consider only the dynamical core, additional simplifications can be made in a number of ways as discussed in previous sections:

1. Treating the atmosphere as an incompressible fluid or using the *Boussinesq* or *anelastic* approaches to simplify the treatment of density changes.
2. Neglecting the different terms of second- or lower-order value in the motion equations according to a scale analysis estimation (as in Section 7.2). For example, making the fluid nonviscous (*inviscid*).
3. Assuming simplified balances in some directions (e.g., typically using hydrostatic balance in the vertical), etc.

4. Parameterization or simplified representation of some forces.
5. Simplification of the geometry (from a plane parallel to a full spherical atmosphere).
6. Changing the vertical coordinate from height level representation (z) to isobaric coordinates (P, introducing the vertical velocity $\omega_P = dP/dt$), to log-pressure coordinates ($z^* = -H \ln(P/P_0)$), or to *isentropic* coordinates (θ = constant, vertical velocity $\omega\theta = d\theta/dt$).
7. Simplification of the Coriolis force effect (using constant f_0, or f-plane or β-plane approximations).
8. Boundary surface: including its physical properties (nature) and include or not the topography. The inclusion of topography is performed, for instance, using *sigma coordinate* ($\sigma(x,y,t) = P(x,y,t)/P_s(x,y,t)$) P_s being the surface pressure. This coordinate is nondimensional, so at the lower boundary, $\sigma = 1$ and the vertical velocity $d\sigma/dt = 0$.
9. Excluding seasonal variations in the sunlight input.
10. Including convective motions via a *convective adjustment* procedure (the temperature profile is adjusted to a dry adiabat under conditions of low condensable content, and to a saturated adiabatic when condensable are present) or via a *convective parameterization* scheme.
11. Inclusion or not of clouds and their effects in the radiative transfer.

Some models solve for directly measurable quantities (such as wind speeds and temperature), others work with derived quantities (such as vorticity or potential temperature). In other cases, the magnitudes are divided into the sum of two terms, its mean plus a perturbed or fluctuating value, leading to *mean variable equations* and to *perturbation equations* that can be further simplified when considering only their linearized versions.

All this means that the number of possibilities and combinations is so large that their full discussion is beyond the scope of this introductory book. Depending on the information we gather from a given atmosphere and on the degree of complexity of its physical and chemical processes, and on the phenomena we want to study, simplified models (sometimes called *toy models*) are useful to understand the basics, or in other cases, full GCM models are necessary if we want to make advanced prognostics of the atmospheric flow. Beyond the Earth, for planets with a substantial ("massive") atmosphere, the more complex models so far developed to explain the observed motions are in order of increasing simplicity: (1) Mars, (2) Venus, (3) Titan, all cases of the GCM type. For Jupiter and Saturn, Uranus and Neptune, and hot giant extrasolars, no GCMs in the pure sense have been developed due to a lack of knowledge of some basic aspects (depth reached by the motions and internal structure, role of the energy sources in driving motions), etc., as we shall present later. For dynamical studies, the most traditional models so far developed are as follows.

7.6.1.1 The Primitive Equation (PE) Model

It is complete in solving the "motion equations" in spherical geometry (7.2), (7.4), (7.8), (7.10), and (7.12) with the hydrostatic approximation for the momentum equation in the vertical direction.

7.6.1.2 Shallow Water (SW)

The atmosphere is reduced to a layer of fluid with constant and uniform density with the height of the fluid above a reference level ($z=0$) represented by the variable $h(x, y, t)$. The SW-model usually treats the fluid as inviscid.

7.6.1.3 One and a Half Model ("1½")

It consists of a "weather" layer with constant density but variable depth above a deep layer (the *abyss*), also with constant but higher density than the "weather layer." This is, for example, a familiar representation adopted for studies of giant planets. For cases (b1) and (b2), the study of the motions in the SW-layer or in the upper layer of this system needs only two equations (see, e.g., Pedlosky (1987) for their derivation):

$$\frac{\partial \vec{u}}{\partial t} + \vec{u} \cdot \nabla \vec{u} = -g\nabla h - f\hat{k} \times \vec{u}$$

$$\frac{\partial h}{\partial t} + \vec{u} \cdot \nabla h = -h\nabla \cdot \vec{u}$$

(7.92)

where
  h is the layer thickness
  $\vec{u}$ is the two-dimensional velocity

"Forcing terms" can be included by adding a friction force and an adiabatic heating term to the right-hand side of both equations, respectively. The shallow water equations contain an intrinsic length scale, the *Rossby deformation radius* given by

$$L_D = \frac{\sqrt{gh}}{f}$$

(7.93)

that results from the interaction of the planetary rotation with the gravitational relaxation of the thickness variations of the layer. Gravity and Coriolis forces tend to produce structures in the flow with a length scale close to the deformation radius.

7.6.1.4 Quasi-Geostrophic (QG)

This model represents a two-dimensional approach to describe the flow evolution when the Rossby number $R_o \ll 1$ (as in rapidly rotating planets), making use of the quasi-geostrophic barotropic vorticity equation (7.57):

$$\frac{d_h}{dt}\left(\frac{\zeta_g + f}{h}\right) = F - D$$

(7.94)

where the sources and sinks terms D and F have been included. A more complete description can be done using the quasi-geostrophic potential vorticity equation (7.65) with (7.64) defining the QGPV. The QG-models are therefore formulated for two types of flow: (1) *Barotropic* when the density depends only on the pressure ($\rho = \rho(P)$), meaning that isobaric surfaces are also surfaces of constant density. From the perfect gas law, they are also isothermal. In a barotropic atmosphere, we have $\nabla_h T = 0$ and the thermal wind equation (7.32c) reduces to $\partial \vec{V}_g / \partial \ln P = 0$, which states that the geostrophic wind is independent of height. (2) *Baroclinic* when the density depends on pressure and temperature ($\rho = \rho(T, P)$) and the geostrophic wind has a vertical shear.

In addition to the Rhines scale L_β (7.91) that has sense when f varies with latitude, the other important length scale is (7.93) called the *Rossby deformation radius* L_D that emerges as a natural scale of motions in stably stratified atmospheres and is written in the baroclinic case as

$$L_D = \frac{NH}{f}, \tag{7.95}$$

where
   $N(z)$ is the Brunt–Väisälä frequency (Section 4.5.3)
   H is as usual, the scale height

We shall use and refer to it in more detail in Chapter 8.

7.6.2 NUMERICAL SOLUTIONS

The equations that describe the atmospheric motions are partial differential equations in space and time (coupled and nonlinear), which, with the appropriate initial and boundary conditions, must be solved through numerical methods. This is a large issue that would require a whole chapter by itself or, to be treated in depth, a whole book. Here, we just outline the two basic procedures of the numerical methods usually employed.

7.6.2.1 Finite Differences

The variables are first expanded in a Taylor series. Consider, for example, an infinitesimal change in the temporal variability of the velocity that may be written as

$$u(x, t + \Delta t) = u(x, t) + \Delta t \frac{\partial u(x, t)}{\partial t} + O(\Delta t^2) \tag{7.96}$$

where the increment Δt is assumed to be small. Then the partial derivative with respect to the time ("the time step") can be calculated as

$$\frac{\partial u(x, t)}{\partial t} \approx \frac{u(x, t + \Delta t) - u(x, t)}{\Delta t} \tag{7.97}$$

with an error of the order of Δt^2 according to (7.96).

When considering the variations in space, a three-dimensional grid of lattice points is used. The infinitesimal spacing between the points is (Δx, Δy, Δz) for the eastward, northward, and vertical directions in the Cartesian system. Caution must be taken when using spherical coordinates at the poles since the grid converges there. Similar to (7.97), the "centered difference" numerical scheme allows the partial derivative with respect to space to be calculated as

$$\frac{\partial u(x,t)}{\partial x} \approx \frac{u(x+\Delta x,t)-u(x-\Delta x,t)}{2\Delta x}. \tag{7.98}$$

For the advective nonlinear terms, the "forward in time, centered in space" numerical solution is

$$u\frac{\partial \phi}{\partial x} = u(x,t)\frac{\phi(x+\Delta x,t)-\phi(x-\Delta x,t)}{2\Delta x} \tag{7.99}$$

These are some trivial examples. Other efficient numerical algorithms have been developed for a more precise calculation. To have an estimation of the iterative number of steps to be done, grid spacing typically used for a full sphere computation is $1°$ in longitude, $1°$ in latitude, and 15–30 vertical levels. The total number of horizontal grid points is given by A/Δ^2, where A is the area of the domain and Δ is the average grid distance. For bounded, smaller domains, the grid spacing in longitude and latitude can be considerably reduced (a factor of 10 or less). Examples are rectangular domains where periodic conditions are assumed in the zonal limits, and in the meridional, no-slip ($u=0$, $v=0$) and no-flux conditions can be considered. In the vertical, a rigid boundary ($w=0$) or soft frontiers with or without heat flux are used depending on the planet or satellite properties. For example, a lower rigid boundary is used, for the terrestrial planets.

7.6.2.2 Spectral Methods

In this case, the horizontal derivatives are calculated by expanding the spatial variations of the dependent variables in series of orthogonal functions whose derivatives are explicitly known. For example, assuming periodicity in the x-direction with spatial period L, the velocity $u(x, t)$ can be expanded as a Fourier series of the form

$$u(x,t) = \langle u(t)\rangle + \sum_{n=1}^{N}\left[a_n(t)\cos\left(\frac{2\pi nx}{L}\right)+b_n(t)\sin\left(\frac{2\pi nx}{L}\right)\right] \tag{7.100}$$

where N is the *truncation limit* and then the x-derivative is written as

$$\frac{\partial u(x,t)}{\partial x} = \sum_{n=1}^{N}\left[\frac{2\pi nb_n(t)}{L}\cos\left(\frac{2\pi nx}{L}\right)-\frac{2\pi na_n(t)}{L}\sin\left(\frac{2\pi nx}{L}\right)\right] \tag{7.101}$$

When using spherical coordinates, the expansion is performed in terms of the spherical harmonics $\exp(im\phi)P_n^m(\cos\varphi)$, where ϕ is longitude, φ is latitude, P_n^m is the associated Legendre polynomial with $|m|\leq n$. The maximum number of terms typically used in this type of expansion is $N\sim 200$.

These methods require the control of *numerical instabilities* that develop in the numerical calculations that are not present in the real atmosphere being modeled. For example, this requires, for finite differences, that the time step is taken such that the displacement of a material point in one time step must be smaller than the grid spacing, $u\Delta t \leq \Delta x$, or that the *Courant–Friedrichs–Lewy* (CFL) number $CFL = |u(x_0,t_0)|(\Delta t/\Delta x)$ to be ≤ 1. See further details at introductory level, for example, in Holton (2004) (Chapter 13).

7.6.3 ANGULAR MOMENTUM AND ENERGY BUDGETS

The overall balance of absolute atmospheric angular momentum (L_{atm}) and energy of the atmospheres, and their transports and conservation principles, represent an important constraint and diagnostic of the mechanisms involved in the global atmospheric motions.

7.6.3.1 Angular Momentum Balance

The angular momentum of the Earth and its atmosphere must remain constant in the absence of external torques. For a thin atmosphere (as compared with the planetary radius), its absolute angular momentum per unit mass ($m^2\ s^{-1}$) is given by sum of the terms due to Earth's rotation and atmospheric motions (taken here as zonal):

$$L_{atm} = (\Omega R_p \cos\varphi + u)R_p \cos\varphi \qquad (7.102)$$

Since the mass of the Earth is much greater than that of its atmosphere (that is, their corresponding moments of inertia) and the Earth's angular velocity is practically constant, the atmosphere's angular momentum must also be substantially constant. Therefore, the transfer of angular momentum between the Earth and atmosphere must be close to zero. The time rate of change of the absolute angular momentum of the atmosphere is due to the torques acting on it, in spherical coordinates, and according to Equation 7.10a is given by

$$\rho\frac{dL_{atm}}{dt} = -\frac{\partial P}{R_p\cos\varphi\,\partial\phi}R_p\cos\varphi + F_{f\phi}R_p\cos\varphi + \text{External} \qquad (7.103)$$

This means that the absolute angular momentum of an individual air parcel changes only by torques caused by the zonal pressure gradient force, by the friction torques (eddy stresses), and by external torques on the planetary atmosphere (*External*) due to the action of tides (gravitational or solar on thermally induced diurnal atmospheric mass variations) and other external weaker torques (e.g., those due to the solar wind). The three-dimensional nature of the winds observed in planetary atmospheres requires the existence of sources and sinks of angular momentum, and mechanisms for its vertical and horizontal redistribution.

When the zonal (East–West) atmospheric flow is more rapid on the average than that of the surface underneath (or that of the interior, as in giant planets), the atmosphere is said to be superrotating. The equatorial atmospheres of Earth, Mars, and those of Uranus and Neptune subrotate, whereas an excess of prograde angular momentum occurs in midlatitudes. On the contrary, the equatorial atmospheres of Venus, Jupiter, and Saturn superrotate, and the source for this angular momentum excess is one of the major unsolved problems in the dynamics of planetary atmospheres. Additionally, an atmosphere is said to be in "solid body rotation" when the meridional dependence of the zonal wind velocity is (see 7.102) $u = \Omega_{atm} R_p \cos\varphi$ for a constant atmospheric angular velocity (frequency) Ω_{atm}. Approximate solid body rotation seems to occur poleward of midlatitudes at the cloud tops of Venus' atmosphere.

On the terrestrial planets, frictional torques on the surface and the redistribution of angular momentum by the meridional Hadley cell, eddies of all type, and waves, are responsible for the observed zonal wind profile. In Venus, solar gravitational torques on thermally induced diurnal atmospheric mass variations could play a role. On Uranus and Neptune, the equatorial subrotation could be maintained by an axisymmetric meridional circulation. A parcel at high latitudes has low angular momentum and will rotate slowly if moved to the equator while conserving its angular momentum. Conversely, an equatorial parcel has high angular momentum and will rotate faster if it moves toward the pole. A zonally symmetric circulation cannot produce the equatorial superrotation of Venus, Jupiter, and Saturn (even though L_{atm} also decreases poleward), so transport by nonaxisymmetric features (eddies and waves) must occur, although the form in which this takes place is not known.

7.6.3.2 Energy Budget and Transformations

Energy is present in various forms in a planetary atmosphere (see Chapter 1). Motions in planetary atmospheres occur at the expense of the primary energy sources and on their subsequent transformations and exchanges. Whereas on bodies with surfaces, sunlight deposition is the primary cause for the motions we observe (for all planets and satellites), on giant planets (in the cases of Jupiter, Saturn, and Neptune), the primary source for the jets has not yet been found, either the solar radiation deposited in an upper "weather layer," or deep motions driven from the internal heat source, or both. The different forms of energy per unit mass can be grouped into the main following types: the potential energy (in terms of geopotential $\Phi_G = gz$), the internal energy ($U = C_V T$), latent heat (L_H), and kinetic energy ($K = (1/2)(u^2 + v^2 + w^2)$), so the total energy is $E = K + \Phi_G + U + L_H$.

The thermo-mechanical transformations between the different forms of energy (K, Φ_G, U, L_H) vary from one planet to another and are behind the process of jet formation in planetary atmospheres. A useful context in which to look at the global circulation is that provided by the *Lorenz energy cycle* that focuses on the transformations between the mean and eddy components of the kinetic K and potential energy terms (see Peixoto and Oort (2007)). Following a similar scheme to that presented in Section 7.4.1, it is convenient to partition both forms of energy into a zonal mean (temporarily averaged) and an eddy value (from the zonal and temporal fluctuations). In general, any magnitude X is divided into the sum of two parts, $X = \bar{X} + X'$. The approach is then to divide the total energy contained in the atmosphere into four "reservoirs": zonal mean and eddy available potential energy ($\bar{P}$, P'), and zonal mean and eddy kinetic energy ($\bar{K}$, K'). According to this scheme, the global circulation may be described in terms of the energy transport processes that link the four "reservoirs." The magnitude of the energy contained in each reservoir may be expressed as

$$\bar{K} = \frac{1}{2} \int_{Vol} \rho_0 (\bar{u}^2 + \bar{v}^2) dVol \tag{7.104a}$$

$$K' = \frac{1}{2} \int_{Vol} \rho_0 (u'^2 + v'^2) dVol \tag{7.104b}$$

$$\bar{P} = \frac{1}{2} \int_{Vol} \rho_0 \frac{R_g^{*2} \bar{T}^2}{H^2 N_B^2} dVol \tag{7.104c}$$

$$P' = \frac{1}{2} \int_{Vol} \rho_0 \frac{R_g^{*2} T'^2}{H^2 N_B^2} dVol \tag{7.104d}$$

These forms for the energy representation are valid for a quasi-geostrophic system, and the *available potential energy* terms $(\bar{P}, P')$ include both the geopotential variations and the thermal state of the atmosphere represented by the temperature (mean and eddy) and the Brunt–Väisälä frequency N_B. The substitution $\partial \Phi_G / \partial z \approx R_g^* \bar{T} / H$ (geopotential thickness) has been made (see, e.g., Holton (2004) and Peixoto and Oort (2007)).

Without going into the details, the energy transformations can be written symbolically as

$$\frac{d\bar{K}}{dt} = \{\bar{K} \cdot K'\} + \{\bar{K} \cdot \bar{P}\} + D(\bar{K}) \tag{7.105a}$$

$$\frac{d\bar{P}}{dt} = \{\bar{P} \cdot P'\} - \{\bar{K} \cdot \bar{P}\} + G(\bar{P}) \tag{7.105b}$$

$$\frac{dK'}{dt} = -\{\bar{K} \cdot K'\} + \{K' \cdot P'\} + D(K') \tag{7.105c}$$

$$\frac{dP'}{dt} = -\{\bar{P} \cdot P'\} - \{K' \cdot P'\} + G(P') \tag{7.105d}$$

The symbol $\{A \cdot B\}$, e.g., describes the energy transport from reservoir A to reservoir B. The terms $G(K,P)$ and $D(K,P)$ represent the generation and dissipation energy functions. The energy transformations are given by

$$\{\bar{P}\cdot\bar{K}\} = \int_{Vol} \frac{R_g^*}{H} \bar{w}\bar{T}\, dVol \tag{7.106a}$$

$$\{P'\cdot K'\} = \int_{Vol} \rho_0 \frac{R_g^*}{H} \langle w'T'\rangle\, dVol \tag{7.106b}$$

$$\{K'\cdot\bar{K}\} \approx \int_{Vol} \rho_0 \langle u'v'\rangle \frac{\partial \bar{u}}{\partial y}\, dVol \tag{7.106c}$$

$$\{\bar{P}\cdot P'\} \approx -\int_{Vol} \rho_0 \frac{R_g^*}{H^2 N_B^2} \langle v'T'\rangle \frac{\partial \bar{T}}{\partial y}\, dVol \tag{7.106d}$$

For $\{K'\cdot\bar{K}\}$ and $\{\bar{P}\cdot P'\}$ transformations, second-order terms have been dropped due to their low values on Earth. Figure 7.24 shows the observed energy cycle in Earth's atmosphere (this is known as an *Oort diagram*). According to this scheme, the generation of mean potential energy from differential solar heating is converted to eddy potential energy and then to eddy kinetic energy by the growing baroclinic disturbances (eddies), and from this to the mean kinetic energy of the zonal flow, which closes the cycle by transforming to mean potential energy by the meridional Hadley cell. The dissipation of kinetic energy occurs due to friction. Eddies play an important role in the energy conversions and in the thermo-mechanical cycle of the atmosphere.

The transfer of kinetic energy from eddies to the mean flow $\{K'\cdot\bar{K}\}$ has been determined on Earth, and tentatively on Jupiter and Saturn, from observations and measurements at cloud level of u' and v' (then deriving $\langle u'v'\rangle$) and from the meridional profile of the zonal wind $\bar{u}(y)$ (then deriving $d\bar{u}/dy$), and then establishing the

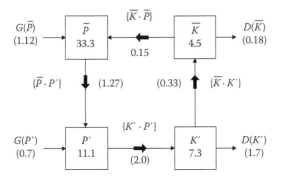

FIGURE 7.24 Schematic box diagram for the observed mean energy cycle in Earth's atmosphere. The energy amounts inside each box are given in units of 10^5 J m^{-2}, and the rates of generation, conversion, and dissipation in W m^{-2}. (Adapted from Peixoto, J.P. and Oort, A.H., *Rev. Mod. Phys.*, 56, 365, 1984.)

degree of correlation between both $\langle u'v' \rangle = a\ d\bar{u}/dy + b$ as seen from (7.106c), from linear least-square fits along the given latitude bands. This can also be understood in the frame of the *Prandtl mixing-length* concept (Equations 7.67 and 7.79), where the transfer of momentum between eddies and mean motions takes place linearly over the mixing-length distance ℓ:

$$\langle u'v' \rangle = -|u'\ell'|\frac{d\bar{u}}{dy} = K_m\frac{d\bar{u}}{dy} \tag{7.107}$$

The comparison between the energy cycles in these two different types of atmospheres is instructive within the uncertainties. The power per unit area converted from eddies to mean flow is 0.33 W m^{-2} on Earth (for a column density of 10^4 kg m^{-2}) but ~3.5 W m^{-2} on Jupiter and Saturn assuming this conversion occurs in the "weather layer" that extends down to 10 bar. These values represent ~0.1% of the radiated power by the Earth but ~15% that of Jupiter and Saturn. This indicates that the kinetic energy conversion from eddies to the mean flow is a process much more efficient on the two giant planets than on Earth.

7.6.4 THE CIRCULATION OF THE ATMOSPHERES OF THE GIANT AND ICY FLUID PLANETS

On a global context, the giant and icy planets are distinguished from the terrestrial planets and satellites in (see Chapter 1) (a) their size is ~4–10 times that of the Earth; (b) they have high angular rotation velocity (rotation periods ~10–17 h); (c) they have a significant internal energy source (with the exception of Uranus) of the order of the solar one; (d) whereas the sunlight absorption depends on the subsolar latitude (due to the planetary axis tilt), the emitted energy is independent of latitude; (e) the available thermal energy is low due to their great distance to the Sun (in Jupiter and Saturn, e.g., is a factor ~1/25 and 1/100, respectively, that received on the Earth); (f) the atmospheres of the giant planets are deep, since they occupy a significant fraction of the planetary radius (the internal structure and composition of these planets is not yet precisely known); (g) they lack a solid surface.

Despite the low available thermal energy, the observed winds at the cloud tops are at their maximum ~10 times stronger than on Earth. This is one of the major unresolved problems in the dynamics of atmospheres. At present, there is not a consensus model to explain the observed global circulations of Jupiter, Saturn, Uranus, and Neptune.

The models of the giant planets fall basically into two main categories. *Deep models* assume the internal heat source is the basic energy mechanism driving the motions that should then extend throughout the whole molecular hydrogen layer (up to pressure of ~1–3 Mbar or ~10^4 km or ~0.7–0.9 planetary radius). *Shallow layer models* have the incident solar radiation as the main driving energy source with motions extending only a few bars below the main upper cloud layer (say up to ~10 bar). Here, the thermodynamic effects, related to latent heat release from cloud condensation and to ortho-para conversion of molecular hydrogen, could have important influences on global dynamics in the upper atmosphere (the "weather layer"). Finally, *mixed models*, with solar radiation as the energy source but deeper extension (up to ~100 bar), have also been proposed. An estimation of the depth of the active "weather layer" can be obtained by an integration of the thermal wind equation using the measured wind profiles at the cloud tops (proposed as Problem 7.27).

7.6.4.1 Deep Circulation Models

Consider the vorticity equation (7.52) in the absence of friction. For $\vec{\omega}_I \ll 2\vec{\Omega}$ we have

$$\frac{d\vec{\omega}_R}{dt} = 2\vec{\Omega} \cdot \nabla \vec{u} - 2\vec{\Omega}\nabla \cdot \vec{u} + \frac{\nabla \rho \times \nabla P}{\rho^2} \qquad (7.108)$$

where

$$\frac{d\vec{\omega}_R}{dt} = \frac{\partial \vec{\omega}_R}{\partial t} + \vec{u} \cdot \nabla \vec{\omega}_R$$

Under quasi-geostrophic conditions, Equation 7.108 reduces to (see, e.g., Pedlosky (1987))

$$(2\vec{\Omega} \cdot \nabla)\vec{u} - 2\vec{\Omega}\nabla \cdot \vec{u} = -\frac{\nabla \rho \times \nabla P}{\rho^2} \qquad (7.109)$$

In the deep atmospheres of Jupiter and Saturn, density is not expected to vary along isobars due to the convective mixing driven by the internal heat source, and the fluid becomes barotropic (that is, $\nabla \rho \times \nabla P = 0$), which leads to the *Taylor–Proudman theorem* (in its compressible form):

$$(2\vec{\Omega} \cdot \nabla)\vec{u} - 2\vec{\Omega}\nabla \cdot \vec{u} = 0 \qquad (7.110)$$

In the Cartesian component system, the equations become

$$\frac{\partial u}{\partial z} = \frac{\partial v}{\partial z} = 0$$
$$\frac{\partial u}{\partial x} + \frac{\partial v}{\partial y} = 0 \qquad (7.111)$$

(z-axis parallel to the rotation axis Ω). These relationships state that the wind components in planes perpendicular to the rotation axis are independent of the coordinates parallel to it, and that the divergence of these winds in the plane perpendicular to the rotation axis is zero. The motion of the fluid takes place in cylinders parallel to the rotation axis (aligned with the rotation axis) called *Taylor columns*.

If the fluid is incompressible, then from Equation 7.3 $\nabla \cdot \vec{u} = 0$ and (7.110) reduces to the simple form

$$\left(2\vec{\Omega} \cdot \nabla\right)\vec{u} = 0 \tag{7.112}$$

The flow motion is confined to rotating columns around the rotation axis (with no vertical motions along the columns) and generates a secondary circulation in the form of counter-rotating cylinders concentric with the rotation axis (Figure 7.25). Its application to a giant planet's internal structure implies that the cylinders develop in the molecular hydrogen layer (see Chapter 1) and are assumed not to penetrate the deeper metallic hydrogen region. The columns extend along the whole molecular layer from one hemisphere to the other symmetrically. According to this model, the cylinders reaching the upper atmosphere (the curved spherical "planetary surface") produce the observed pattern of alternating jets when intersecting it. The high inertia of the cylinders should keep their motion stable, explaining the long-term stability of the giant planets' winds. Since the metallic layer should act as an impenetrable barrier for the cylinders, the jet pattern extends to the latitude intercepted by the cylinder that is tangent at the equator to the metallic region. In other words, the pattern is constrained by the width of the molecular layer.

Numerical studies that solve the full set of dynamical equations reproduce the Taylor-column motion. The standard procedure is to represent the continuity equation (7.3) (incompressible fluid), the momentum equation (7.10), and the thermodynamics

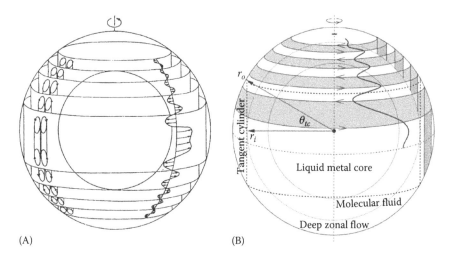

(A)   (B)

FIGURE 7.25 Deep Jovian models. Illustrations of the proposed deep circulation motions produced by convection in a spherical layer of a rapidly rotating planet. (A) Columnar convection (Taylor columns) and the generation of rotating cylinders concentric with the rotation axis and alternating pattern of zonal jets with latitude. (B) Similar scheme but showing the latitude that marks the limits of the broad equatorial jet in Jupiter and Saturn. (From Busse, F.H., *Geophys. Astrophys. Fluid Dyn.*, 23, 153, 1983. With permission.) (From Heimpel, M. et al., *Nature*, 438, 193, 2005. With permission.)

equation (7.12) (under the Boussinesq approach in most, but not all, recent models) in a nondimensional form using the nondimensional numbers (see also Section 1.5.4) and solve them in a shell (the molecular envelope) with depth D (the distance between the inner and outer radius for the shell):

$$E_K\left(\frac{\partial \vec{u}}{\partial t} - \vec{u}\cdot\nabla\vec{u}\right) - E_K\nabla^2\vec{u} + \hat{k}\times\vec{u} = -\nabla P + \left(\frac{RaE_K}{Pr}\right)\frac{g}{g_0}T\hat{r}$$

$$\nabla\vec{u} = 0 \tag{7.113}$$

$$Pr\left(\frac{\partial T}{\partial t} - \vec{u}\cdot\nabla T\right) - \nabla^2 T = \frac{1}{r}\frac{dT_0}{dr}\hat{r}\cdot\vec{u}$$

These equations have been made nondimensional using D for length, D^2/ν_V for time, ν_V/D for velocity, $\rho\nu_V\Omega$ for pressure, and ΔT for temperature (temperature difference between inner and outer radius), and the following nondimensional numbers have been used (see also Section 1.5.4):

$$\text{Prandtl:} \quad Pr = \frac{\nu_V}{D_T} \tag{7.114a}$$

$$\text{Ekman:} \quad Ek = \frac{\nu_V}{2\Omega D^2} \tag{7.114b}$$

$$\text{Rayleigh:} \quad Ra = \frac{g_0\alpha_T\Delta TD^3}{\nu_V D_T} \tag{7.114c}$$

where
  ν_V is the kinematic viscosity
  D_T is the thermal diffusivity
  α_T is the thermal expansion coefficient

A reference gravity g_0 (at the base of the layer) and temperature T_0 have been introduced. The system is typically numerically solved in a spherical layer with stress-free ($u=0$, $T=0$) or stress-rigid ($\partial u/\partial x=0$) boundary conditions. The temperature and three-dimensional velocity fields are then obtained in any point of the shell. In the molecular envelopes of the giant planets, these numbers are $E \sim 10^{-10}-10^{-15}$, $Ra \sim 10^{10}-10^{15}$, and $Pr \sim 10^{-4}-1$. However, it must be noted that these values cannot be used in existing numerical models due to computational limits, being typically employed $E \sim 10^{-6}$ and $Ra \sim 10^8-10^9$. For some particular combinations of the layer depth D, boundary conditions, and values of the numbers, a strong equatorial eastward jet and the pattern of alternating zonal jets with latitude develop (Figure 7.26). The main limitation of this model relies on the restriction for the fluid to be incompressible and the motions are essentially laminar in nature, situations that are not expected to occur in the turbulent convective layers of giant planets.

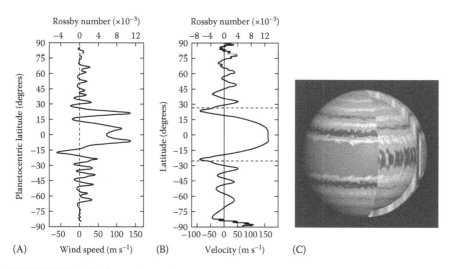

FIGURE 7.26 (A) Comparison of the observed zonal wind profile in Jupiter at cloud level and (B) that resulting from the numerical simulations of deep convective motions. (C) The structure of the azimuthal velocity field on the outer and inner spherical surface and on a meridional slice from these numerical calculations. (From Heimpel, M. et al., *Nature*, 438, 193, 2005. With permission.)

7.6.4.2 Shallow Layer Models

According to this family of models, the pattern of motions observed in giant planets is confined to the upper part of the atmospheric layer (the "troposphere," a "weather layer") where the main clouds sit (up to a depth of ~10 bars). This layer overlies a deep abyss that represents the molecular envelope. This hypothesis states that the motions in the "weather layer" are driven by eddies (vortices and waves) resulting from the temperature contrasts due to latitudinal variations in the absorbed solar radiation (e.g., *baroclinic eddies*—see Chapter 8), or due to latent heat released by water vapor condensation (moist convection), or ortho-to-para hydrogen conversion. As discussed at the end of Section 7.6.3, the measurement of eddy velocities and their cross products on Jupiter and Saturn (*Reynolds stresses* $\langle u'v' \rangle$) indicate that this could be the case. Simulations with different types of model (shallow layer, quasi-geostrophic, and primitive equation, see Section 7.6.1) shows that, under some particular conditions, the forcing of the upper shallow layer by these eddies can result in motions extending much deeper than the thickness of the layer.

The ideas developed in Section 7.4.3 on planetary turbulence apply to the shallow layer forcing, with the zonal jets resulting from an inverse energy cascade modified by the *β-effect* that introduces anisotropy and prevailing zonal jets with a typical meridional width of the order of the Rhines scale L_β given by (7.91). The stability of the layer, represented by the Rossby deformation radius L_D given by Equation 7.95, plays an important role in these models. According to one view, for small values of L_D, the *β-effect* is suppressed and the numerical calculations show a flow dominated by the presence of vortices rather than jets (Figure 7.27). Another view is that a small L_D is necessary to produce energy injection on small scales needed to feed the inverse

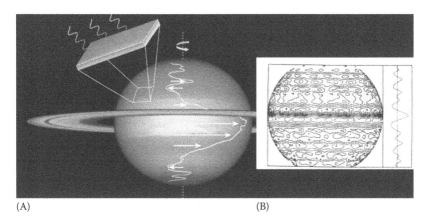

(A)   (B)

FIGURE 7.27 Shallow Jovian models. Illustration of the proposed shallow layer motions originated by solar heat deposition in the upper troposphere (in a thin shell, "the weather layer") of the giant planets. (A) Scheme of solar heat deposition in this layer for Saturn. (B) Streamlines resulting from numerical calculations and resulting zonal wind profile for Jupiter. (From Williams, G.P., *J. Atmos. Sci.*, 35, 1399, 1978. With permission.)

energy cascade over a substantial inertial range (see Section 7.4.2). The small L_D then only suppresses the β-*effect* for baroclinic motions—but not for barotropic flow; hence the jets that are formed by the anisotropic inverse cascade are mainly barotropic.

7.6.4.3 The Uranus and Neptune Case

The shape of the zonal wind profile in the troposphere (at cloud level, ~1 bar) is so similar in both planets (Figures 7.19 and 7.20) that the mechanism producing this circulation is also likely to be similar. In addition, the wind profile resembles that of the Earth, with subrotation in the equator and two superrotating jets at mid- and high latitudes. This could suggest that the solar radiation is responsible for the motions within a shallow forcing layer, as discussed in the previous section, in a dynamical regime decoupled from the deep internal motions. However, important differences in rotation axis tilt between both planets (with a strong and unusual pole–equator radiation forcing on Uranus) and internal energy source (absent in Uranus but important—compared to sunlight—on Neptune) make it complicated to see a unified vision. The winds on Uranus are mainly zonal, as on the other three fluid planets, despite the unusual deposition of the solar energy (the subsolar point changes during Uranus's year from equator to pole). Compared to the Earth, the differences are also substantial in composition, internal structure (presence of a surface), lack of internal energy source as compared to Neptune, and very weak power input per unit area (1/400 that of the Earth at Neptune). Although the energy sources are very weak on Uranus and Neptune, the winds are strong in the troposphere with $u \sim 200\text{–}400\,\text{m s}^{-1}$. This perhaps indicates that dissipative effects are very weak in these atmospheres. The emitted power in the four giant planets is independent of latitude (Figure 4.1), despite the strong latitude dependence in the deposited sunlight, suggesting that meridional heat transport must occur in the troposphere or at deeper levels, perhaps coupled to the deep internal convective motions.

7.6.4.4 Lower Stratosphere Circulation

Above the clouds, the measurements of the latitudinal temperature contrasts between the belts and zones give values ·1–5 K. Application of the thermal wind relationship for geostrophic conditions (valid in these planets since $Ro < 1$, except at equator; Equation 7.32a), rewritten in spherical coordinates as

$$f \frac{\partial \bar{u}}{d \ln P} = \frac{R_g}{r} \frac{\partial}{\partial \varphi} \left(\frac{\bar{T}}{\mu} \right) \approx \frac{R_g}{\mu r} \frac{\partial \bar{T}}{\partial \varphi} \tag{7.115}$$

indicates that the winds should globally decrease with altitude to a rest value in about 5 scale heights (see Problem 7.9). As before, the overbar denotes zonal averages (that is, along a longitude circle) of the quantities.

The thermal gradients are dynamically forced by the deeper atmospheric circulation and the corresponding decrease of the zonal winds with height is due to horizontal momentum diffusion. One way to represent the stratospheric dynamics is by means of a simple linear model in which the friction force is parameterized as a drag or *Rayleigh friction* force, such that $F_f = (u_E - u)/\tau_F$, τ_F being the Rayleigh friction damping time and u_E a prescribed zonal flow that acts as a momentum forcing. The zonal component of the momentum equation (7.10a) becomes

$$\frac{\partial u}{\partial t} - fv = \frac{u_E - u}{\tau_F} \tag{7.116}$$

where v is the northward meridional velocity. For steady conditions, this equation reduces to a balance between the Coriolis term and the friction term $fv \approx u/\tau_F$. From mass continuity (Equation 7.2), written in spherical polar coordinates as

$$\frac{1}{r \cos \varphi} \frac{\partial}{\partial \varphi} (\rho_0 v \cos \varphi) + \frac{\partial}{\partial z} (\rho_0 w) = 0 \tag{7.117}$$

a vertical motion occurs with velocity w associated with the frictionally controlled meridional velocity. The downward and upward motion will produce adiabatic heating and cooling, respectively. Because the observed meridional temperature contrasts are weak, horizontal advection of heat can be neglected. Then, the thermodynamic equation is made linear using a radiative damping scaling (*Newtonian cooling*, see Equation 7.38b) as

$$\frac{\partial T}{\partial t} + w \left(\frac{\partial T_0}{\partial z} + \frac{g}{C_p} \right) = \frac{T_e - T}{\tau_{rad}} \tag{7.118}$$

where
  $T_0(z)$ denotes the global mean temperature (averaged over latitude and longitude)
  T_e is the radiative equilibrium temperature that the atmosphere should have in the absence of motions

Note that the second term can be expressed in terms of the vertical scale height H and the Brunt–Väisälä frequency N_B since $(\partial T_0/\partial z)+(g/C_p) = HN_B^2/R_g^*$. Equation 7.118 states that temperatures decrease either as a result of advection by vertical motions or as a result of radiative cooling when T is above T_e.

Consider an eastward jet in a giant planet. This model predicts that its zonal motion is accompanied by a northward meridional velocity and an upward motion in the southern flank of the jet with corresponding descending motion in the northward side (Equations 7.116 and 7.117). From Equation 7.118, these motions yield lower temperatures south of the jet and higher temperatures north of it. The meridional temperature gradient leads, according to (7.115), to a decrease in intensity with altitude of this eastward jet (Figure 7.28). In the stratospheres of the giant and icy planets, this model gives $\tau_F \sim \tau_{rad} \sim$ years, and the meridional velocities are expected to be small and of the order of $v \approx u/(f\tau_{rad}) \sim mms^{-1}$ and even smaller for w (Problem 7.28). Changes in the stratospheric temperatures due to the seasonal insolation, as observed, for example, on Saturn, should produce the corresponding variability in this pattern.

7.6.4.5 Extrasolar Planets: Circulation in "Hot Jupiters"

Few data are yet available on this kind of planet (basically the orbital properties, the mass and radius, hence the mean density, see Table 1.9). For some of them, constraints on the upper atmosphere temperature and the detection of particular compounds have been obtained at infrared wavelengths during eclipse. There are also studies on cloud composition and altitude location based on thermochemical models. This has promoted the development of atmospheric circulation models that try to explain these few observational data and make predictions for future possible observations.

From their mean density, these bodies are basically composed of hydrogen and are strongly irradiated and heated due to their proximity to the star. These models pertain to the types listed in Section 7.6.1, parameterized to the properties of the "hot Jupiter" kind of planet. They consider the following characteristics:

1. Neglecting deep convection (the internal energy source), they fall within the category of "shallow layer" type. The driven source is the solar heat deposition in this layer (typically up to 10–100 bars). Some studies incorporate

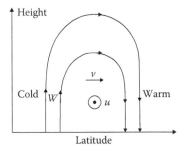

FIGURE 7.28 Schematic form of the meridional circulation associated with the decay with altitude of a zonal jet. The zonal jet is directed out of the page and the meridional velocity is directed northward. (Adapted from Conrath, B.J. and Pirraglia, J.A., *Icarus*, 53, 286, 1983. With permission.)

scale analysis of the stellar gravitational tidal heating due to the star's proximity.

2. Although the rotation rate of "hot Jupiters" is unknown, it is assumed that they have their spin synchronized to the orbital rotation period (see Chapter 1).

3. Under such a hypothesis, estimations of the Rossby number (7.13) yield $Ro \sim 1\text{--}10$, thus cyclostrophic or gradient wind balances (Section 7.2.7) may apply.

4. Simple scale analysis (see Problem 7.12) indicates that strong zonal winds must develop in these atmospheres. In the upper atmosphere, a solar-to-antisolar circulation (from the subsolar to antisolar point), similar to that observed in Venus' mesosphere, is predicted. This is confirmed by numerical shallow layer models. From cyclostrophy (Equations 7.21 and 7.25), we have

$$u \approx \sqrt{R_g^* \, \Delta T \ln \frac{P_u}{P_l}} \qquad (7.119)$$

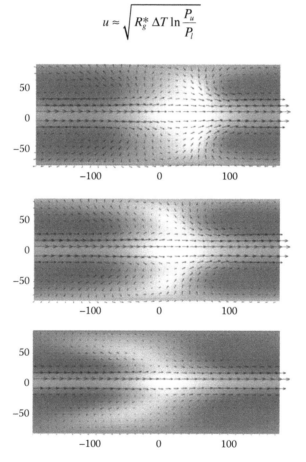

FIGURE 7.29 Numerical model calculations of the circulation in the upper atmosphere of the hot extrasolar planet HD 209458b. The maps (latitude–longitude) show the temporal evolution of the temperature contrast produced by the stellar heat deposition, and superimposed upon it, the resulting wind vectors, with the motions dominated by an intense equatorial jet. (From Showman, A.P. and Guillot, T., *Astron. Astrophys.*, 385, 166, 2002. With permission.)

where

ΔT is the subsolar to antisolar temperature difference

P_u and P_l are the upper and lower pressure levels of the "shallow layer"

The expected velocities are $u \sim 1$–3 km s^{-1}.

Figure 7.29 presents the results of the numerical calculations applied to the case of the extrasolar planet HD 209458b, one of the best studied. In this map, the substellar heat deposition occurs close to longitude 0° and latitude 0°. The day–night extreme temperature contrast generates an intense equatorial jet that reaches ~1000 m s^{-1}.

PROBLEMS

7.1 The temperature at a point on the surface of Mars decreases by 3 K per 100 km in the eastward direction. A rover moving eastward at 1 km h^{-1} measures a temperature fall of 1 K h^{-1}. (1) What is the temperature change measured on a fixed Mars Laboratory lander that the rover is passing by? Assume that the velocity field and pressure in Cartesian coordinates at the surface is given by the analytic expressions $\vec{u} = (L/t)\hat{\imath}$ and $P = Kt \exp(-x/L)$, where L is the distance and K a constant. (2) Find the material derivative for the pressure field.

Solution

(1) $\partial T/\partial t = -0.97$ K h^{-1}

(2) 0

7.2 Derive Equation 7.12b from the continuity equation (7.3), the ideal gas equation (7.4), and the *Mayer relationship* $(C_p = C_v + R_g^*)$.

7.3 Show that the zonal jet stream at midlatitudes ($\varphi = 45°$) on the Earth, Mars, and Jupiter (altitude levels $z = 10$ km (Earth), 15 km (Mars)), given by the values on the accompanying table, are in approximate geostrophic balance. Additional required data can be taken from other chapters in the book. Calculate the horizontal pressure gradient that supports them on each planet according to this balance.

Parameter	Earth	Mars	Jupiter
Zonal velocity u	50 m s^{-1}	80 m s^{-1}	50 m s^{-1}
Jet width	2,000 km	2,000 km	10,000 km
Mean pressure	100 mbar	1 mbar	1 bar
Temperature	220 K	160 K	165 K

Solution

(1) *Ro* (Earth) = 0.24, *Ro* (Mars) = 0.4, *Ro* (Jupiter) = 0.02

(2) $\partial P/\partial y$ (mbar km^{-1}) = -0.8 (Earth), -2.6×10^{-4} (Mars), -0.02 (Jupiter)

7.4 Show that the zonal superrotation at midlatitudes ($\varphi = 45°$) on Venus and Titan (altitude levels $z = 65$ km (Venus), 300 km (Titan)), given by the values in the accompanying table, are in approximate cyclostrophic and gradient wind balance, respectively. Additional required data can be taken from other chapters in the book. Calculate the horizontal pressure gradient that supports them on each planet according to these balances.

Parameter	Venus	Titan
Zonal velocity u	-100 m s^{-1}	150 m s^{-1}
Jet width	8000 km	4000 km
Mean pressure	110 mbar	0.1 mbar
Temperature	240 K	185 K

Solution

(1) Ro (Venus) = 28, Ro (Titan) = 15
(2) $\partial P/\partial y$ (mbar km^{-1}) = -4×10^{-3} (Venus), 1.9×10^{-5} (Titan)

7.5 Measurements of the geopotential field on Venus' day-side atmosphere at cloud level at equatorial latitudes has the form

$$\Phi_G(x, y) = \Phi_{G0}\left(1 + \frac{x^2 + y^2}{L^2}\right)$$

where L is a characteristic length scale. Find the zonal wind velocity.

Solution

$$u = \pm\frac{R_p}{L}\sqrt{2\Phi_{G0}}$$

7.6 Express in terms of pressure coordinates the geostrophic balance approximation, using the geopotential definition.

Solution

$$fu = -\frac{\partial \Phi_G}{\partial y}, \quad fv = \frac{\partial \Phi_G}{\partial x}$$

7.7 Assume that the global atmospheric horizontal winds are expressed as the sum of an ageostrophic component $\vec{V}_a$ and a much stronger geostrophic component $\vec{V}_g$. Find under which conditions the vertical velocity in pressure coordinates is given by $\omega = -\rho g w$. Calculate ω for Earth synoptic scale motions if $\partial P/\partial t = 0.01$ bar day^{-1}, $\vec{V}_a = 1$ms^{-1}, $\nabla P = 10^{-5}$ bar km^{-1}, $w = 1$ cm s^{-1} at the surface.

Solution

$\omega = -\rho g w \gg \partial P/\partial t$; $\vec{V}_a \cdot \nabla P = 0.001$ bar day^{-1}, $\omega = -\rho g w = 0.1$ bar day^{-1}

7.8 Derive the thermal wind relationships for geostrophic balance (Equations 7.32a and b). *Hint*: Use the geostrophic balance, hydrostatic balance, and perfect gas law, using altitude z as vertical coordinate (instead of pressure), taking the partial derivative with respect to altitude, neglecting temperature variations in the vertical but not in the horizontal direction.

7.9 Meridional temperature measurements above the upper troposphere in Jupiter and Saturn indicate that the zonal winds decrease with altitude. At midlatitude ($\varphi = 40°$), a typical eastward jet on Jupiter has a zonal velocity of 40 and $100 \, \mathrm{m \, s^{-1}}$ on Saturn, and a latitude width on both planets of $8°$. The wind measurements correspond to a pressure level of 500 mbar and the meridional temperature variation in the jet at a pressure level of 20 mbar is of order 5 K on both planets. Assuming that the layer between the two pressure levels has an average temperature of 114 K on Jupiter and 86 K on Saturn, calculate the zonal wind speed at 20 mbar. Find suitable values for the other required quantities from the tables in the book.

Solution

Jupiter $u(20 \, \mathrm{mbar}) = 7.5 \, \mathrm{m \, s^{-1}}$, Saturn $u(20 \, \mathrm{mbar}) = 67 \, \mathrm{m \, s^{-1}}$

7.10 Derive the thermal wind relationships for cyclostrophic balance (Equations 7.34). *Hint*: Use the cyclostrophic balance, hydrostatic balance, and perfect gas law, using altitude z as vertical coordinate (instead of pressure), taking the partial derivative with respect to altitude, neglecting temperature variations in the vertical but not in the horizontal direction.

7.11 Meridional temperature measurements in Venus at midlatitude ($\varphi = 50°$) at 55 km altitude (0.5 bar pressure level) show a contrast of -15 K across a latitudinal band $30°$ wide. At this level, the zonal wind velocity is $60 \, \mathrm{m \, s^{-1}}$ westward. From probes, it is known that winds increase with altitude. What is the expected wind speed at 70 km altitude (pressure level 0.035 bar)? Find values for the other required quantities from the tables in the book.

Solution

$u = 125 \, \mathrm{m \, s^{-1}}$

7.12 The extrasolar giant planet HD 209456b (a "hot Jupiter") has a mass 0.66 M_J and a radius 1.35 R_J (mass and Jovian radius) and is assumed to be composed of molecular hydrogen (adiabatic index $\gamma = 1.4$) and is in spin-orbit synchronization ($\Omega_{orbit} = 3.5$ days). If its equilibrium temperature is 1350 K, calculate (1) the sound speed (c_s) in its atmosphere; (2) the limiting Rossby number defined as $Ro(\mathrm{lim}) = c_s/\Omega R_p$ indicating the expected balance in its atmosphere; (3) the scale height; (4) the zonal wind speed for an assumed day-to-night temperature contrast of 100 and 500 K in the layer between pressure levels 0.01 and 1 bar.

Solution

(1) 2680 m s^{-1}
(2) $Ro \, (\mathrm{lim}) = 1.3$ (cyclostrophic or gradient)

(3) 640 km

(4) 1320 and 2956 m s^{-1}

7.13 Derive the thermal wind relationships for gradient wind balance (Equations 7.36 and 7.37). *Hint*: Use the gradient wind balance, hydrostatic balance, and perfect gas law, using altitude z as vertical coordinate (instead of pressure), taking the partial derivative with respect to altitude, neglecting temperature variations in the vertical but not in the horizontal direction.

7.14 Derive an expression for the thermal wind relationship in the case of geostrophic balance in the zonal wind that includes the effect of a meridional compositional gradient (mean molecular weight μ) but neglects compositional and temperature variations in the vertical.

Solution

$$\frac{\partial u}{\partial z} = -\frac{g}{fT}\frac{\partial T}{\partial y} + \frac{g}{f\mu}\frac{\partial \mu}{\partial y}$$

7.15 Meridional temperature measurements in Titan at midlatitude ($\varphi = 45°$) and at the pressure level 10 mbar show a contrast of -20 K across a latitudinal band of width $30°$. At this level, the zonal wind velocity is 40 m s^{-1}. From the Huygens probe that descended into Titan's atmosphere in 2005, it is known that winds increase with altitude. What is the expected wind speed at a pressure level of 0.1 mbar? Find values for the other required quantities from the tables in the book.

Solution

$u = 225$ m s^{-1}

7.16 On slowly rotating planets, Hadley cell motions extend across a whole hemisphere (from equator to pole). (1) In Venus' upper clouds ($z = 65$ km, $P = 100$ mbar, $T = 250$ K), measurements give mean meridional velocities $\langle v \rangle = 5$ m s^{-1}. If the radiative time constant is 9.5 days and the measured temperature lapse rate $\partial \langle T \rangle / \partial z = -4.5$ K km^{-1}, what are the related vertical velocities and the required temperature deviation from equilibrium $(T - T_e)$? In Titan's upper atmosphere ($P = 1$ mbar), the radiative time constant is 0.95 years and the temperature gradient is practically adiabatic (1.2 K km^{-1}). If the temperature deviation from equilibrium is 20 K, calculate the vertical and meridional velocities related to the Hadley motion. What is the dynamical timescale for the meridional circulation? Take additional required data from the tables in the book.

Solution

(1) Venus: $\langle w \rangle = 4.4$ mm s^{-1}, $(T - T_e) = 20$ K

(2) Titan: $\langle w \rangle = 0.5$ mm s^{-1}, $\langle v \rangle = 3$ cm s^{-1}, $\tau_{Hadley} = H/\langle w \rangle = 2.7$ years

7.17 Consider the condensation flow of CO_2 in Mars. Assume that the net sublimation mass rate is obtained from Equation 6.76 and occurs over an area $(3/4)\pi R_p^2$. Calculate the meridional speed for the flow crossing a section with an altitude of 3 km at latitude 60°. Take a mean atmospheric density of 0.01 kg m^{-3} and the rest of the data from Problem 6.15.

Solution

$\langle v \rangle = 0.5$ m s^{-1}

7.18 Derive Equations 7.46 and 7.47. *Hint:* Take the curl of (7.7a) and use the vector identity $\nabla \times (\vec{a} \times \vec{b}) = (\vec{b} \cdot \nabla)\vec{a} - (\vec{a} \cdot \nabla)\vec{b} + \vec{a}(\nabla \cdot \vec{b}) - \vec{b}(\nabla \cdot \vec{a})$.

7.19 A parcel of atmosphere has an area S_1 and is initially at latitude φ_1 being at rest relative to the rotation frame of a planet (angular speed Ω). Assuming that conditions are such that Kelvin's theorem is applicable, find the change in the circulation the parcel will suffer when moving along an isobaric surface to latitude φ_2 if the surface area changes to S_2. Apply this result to calculate the mean tangential velocity that a parcel of Neptune's atmosphere will acquire if it has a round shape (radius $r = 1000$ km) that is preserved when moving from equator to latitude 60°N.

Solution

$$\Gamma_{R2} - \Gamma_{R1} = -2\Omega(S_2 \sin\varphi_2 - S_1 \sin\varphi_1), \quad V_T = -\Omega r \sin 60° = -94 \text{ m s}^{-1}$$

7.20 A Saturnian jet has a Gaussian shape $u = u_0 \exp[-(y-y_0)^2/L^2]$, where L is its half width. Take the jet peak speed $u_0 = 146$ m s^{-1}, centered at latitude 47°N (corresponding to y_0) and $L = 5°$. Assume that the lapse rate is adiabatic such that at the 1 bar level the temperature is 135 K, and take the adiabatic coefficient to be 1.4. (1) Calculate the potential vorticity (*PV*) in the layer between the 135 K isentrope and that corresponding to pressure level 0.5 bar. (2) Assuming that *PV* is conserved in the jet, what is $\partial\theta/\partial P$ at 52°N?

Solution

(1) PV (47°N) $= 9.7 \times 10^{-7}$ K kg^{-1} s^{-1} m^{-2}
(2) $\partial\theta/\partial P = 34.8$ K bar^{-1}

7.21 A jet stream in the Martian atmosphere at 50°N has a zonal geostrophic velocity of 10 m s^{-1} and it is known that the eddy diffusivity in the Ekman layer is 70 m^2 s^{-1}. (1) Calculate the zonal wind speed 1 km above the surface. (2) A Martian lander with a sensor located 1.5 m above the surface measured a speed of 5 m s^{-1}. Taking $k_K = 0.4$ and a surface roughness $z_0 = 1$ cm, calculate the friction velocity. Dust is lifted from the Martian atmosphere when the friction velocity is above 2 m s^{-1}. (3) What would be the wind speed measured by the lander sensor?

Solution

(1) 8 m s^{-1}
(2) 0.4 m s^{-1}
(3) 25 m s^{-1}

7.22 A parabolic air stream jet above the Earth's surface, $u = u_0[1 - ((y - y_0)^2/L^2)]$, is defined by $u_0 = 5\,\mathrm{m\ s^{-1}}$ and $L = 500\,\mathrm{km}$ and is centered at latitude 40°N. Assuming that the eddy diffusivity in the Ekman layer is $15\,\mathrm{m^2\ s^{-1}}$, what is the Ekman layer scale d_E? What is the upward velocity due to convergence at the top of the Ekman layer at position $y_0 \pm (L/2)$?

Solution

$d_E = 1780\,\mathrm{m}$, $w(z = d_E) = 2.8\,\mathrm{mm\ s^{-1}}$

7.23 Find an expression for the total cross isobar flow $V_E = \displaystyle\int_0^\infty v\,dz$ and for the total mass flux per unit area $\vec{M}_E$ in the Ekman layer caused by the frictional force. Use the total velocity $U = \sqrt{u_g^2 + v_g^2}$ in (7.69a and b) instead of u_g and v_g.

Solution

(a) $V_E = \dfrac{U}{2\gamma}$

(b) $\vec{M}_E = \dfrac{U}{2\gamma}\left(-\hat{i} + \hat{j}\right)$

7.24 Using a simple scale analysis for the spectral energy density $E(k)$ (see Equation 7.84), deduce the Kolmogorov energy spectrum (7.86).

7.25 Consider the boundary layer of a planet, which is assumed to have a thickness of 100 m and velocity fluctuations of $1\,\mathrm{cm\ s^{-1}}$. Calculate the energy flux ε_l (or the power per unit mass) and the turnover timescale in the layer. Then obtain the Kolmogorov scale L_f.

Solution

$\varepsilon_l = 10^{-8}\,\mathrm{m^2\ s^{-3}}$; $\tau_k = 21.5\,\mathrm{s}$, $L_f = 0.018\,\mathrm{m}$

7.26 Using the data and results of Problem 6.17 for the vertically integrated condensation mass flow in the satellite Triton, calculate (1) the depth of the Ekman layer d_E if the eddy viscosity is $10^4\,\mathrm{m^2\ s^{-1}}$ at a mean latitude of 45°; (2) the meridional velocity of this flow if it takes place across a section with height d_E; (3) if the condensation flow takes place between the pole and equator, should it be possible to apply geostrophic conditions above the Ekman layer? (4) if so, what is the fractional change in pressure to maintain geostrophy? (5) observations of geyser plume direction and motion indicate the wind speed is $15\,\mathrm{m\ s^{-1}}$ at an altitude of 8 km. If at ground level the velocity is zero, what should be the temperature difference between pole and equator?

Solution

(1) $d_E = 1.06\,\mathrm{km}$
(2) $v = 8.5\,\mathrm{m\ s^{-1}}$
(3) $Ro = 0.23$ geostrophic conditions hold
(4) $\Delta P/P = 1.4\%$
(5) $\Delta T = 3.4\,\mathrm{K}$

7.27 Find an expression for the thickness of the zonal flow extension in the giant and icy planets, given as the ratio of the bottom to top pressure levels (P_B/P_T) from knowledge of the zonal wind meridional profile $u(\varphi)$ at the cloud tops. Integrate the thermal wind equation to find the geopotential maximum meridional contrast ($\Delta\Phi_{Gmax}$) if the vertically averaged temperature is $\bar{T}$ and its meridional contrast is $\Delta\bar{T}$. Assume the following conditions: the velocity at the bottom level is zero, the flow is purely zonal and in geostrophic balance, and the layer has constant potential temperature in the vertical but varies in the meridional direction ($\Delta\theta$). Apply to Jupiter, Saturn, and Uranus if the measurements give respectively for each planet: $|\Delta\Phi_{Gmax}/C_p| = 10, 80, 40\,K$, if for all the cases $\Delta\theta = 8\,K$.

Solution

$$\Delta\Phi_{Gmax} = -2\Omega R_p \int_0^{\varphi} u_T(\varphi)\sin\varphi\, d\varphi = R_g^* \Delta\bar{T}$$

$$\frac{P_B}{P_T} = \left[1 + \frac{\Delta\Phi_{Gmax}}{C_p\Delta\theta}\right]^{C_p/R_g^*}$$

$$\frac{P_B}{P_T} = 14.5(\mathrm{J}), 5610(\mathrm{S}), 633(\mathrm{U})$$

7.28 From the frictional damping model of an eastward zonal jet in Uranus ($u = 200\,\mathrm{m\,s^{-1}}$, $55°$ latitude) such that $\tau_F = \tau_{rad} = (5/3)$ of the orbital period, calculate the vertical and meridional velocities associated with the cell motion around the jet if the measured temperature contrasts are $5\,K$ at $100\,\mathrm{mbar}$ and the vertical temperature gradient is $1\,K\,km^{-1}$.

Solution
$v = 4.5 \times 10^{-4}\,\mathrm{m\,s^{-1}}$; $w = 1.1 \times 10^{-6}\,\mathrm{m\,s^{-1}}$.

8 Atmospheric Dynamics-I: Waves

The global atmospheric flow presented in Chapter 7 supports a variety of features in spatial and temporal scales that form and evolve as a result of disturbances generated by hydrodynamical and vertical (dry or moist convection) instabilities, and by a variety of waves resulting from different restoring force mechanisms. Atmospheric oscillations and waves phenomena manifest in many different forms, for example, as deviations relative to the mean spatial values of pressure, geopotential, temperature or wind speed, and direction or as opacity and optical depth variations if they have associated aerosol and cloud formation. Chapters 8 and 9 deal with all these features as observed in planetary atmospheres, separated into two parts: oscillations and waves and then instability phenomena.

8.1 WAVE CHARACTERISTICS

Waves are an important dynamical mechanism in atmospheres. They modify the mean atmospheric structure (temperature and composition, for example) and can intervene in the atmospheric flow motion (accelerating it, for example). Waves are characterized by their amplitude, wavelength (three-dimensional), and frequency, and when propagating by their phase and group speed. They can be classified in various ways: (1) according to the restoring mechanism (e.g., gradient pressure force for sound waves, gravity for gravity waves or "buoyancy" waves, Coriolis force for Rossby waves); (2) as *forced* when they need a continuous excitation mechanism (as thermal tides, orographic waves), or as *free modes* that do not require such mechanisms; (3) *traveling* waves whose surface of constant phase moves relative to the surface of the planet or to a given reference system (the mean atmospheric flow, for example), or *stationary* waves in the reference system (*trapped* by topographic features or by atmospheric regions with particular properties in wind shears, temperature, etc.). According to their amplitude, they can be *evanescent* when they decay with distance exponentially (along a given direction). We describe the waves in the context of the linear theory (see below).

8.1.1 WAVE PROPERTIES

A general form of a wave equation for the atmospheric field $\psi(x, y, z, t)$ (pressure, temperature, wind speed, etc.) is a differential equation that varies with space and time according to the second order differential equation

$$\frac{\partial^2 \psi}{\partial t^2} = c_w^2 \left(\frac{\partial^2 \psi}{\partial x^2} + \frac{\partial^2 \psi}{\partial y^2} + \frac{\partial^2 \psi}{\partial z^2} \right) \tag{8.1}$$

where c_w is the phase speed that depends on the wave type and the medium. By direct substitution, it can be seen that a harmonic (sinusoidal) propagating wave

$$\psi(x, y, z, t) = \psi_0 \exp[i(kx + ly + mz - \omega t)] \tag{8.2}$$

is a solution of (8.1) where ψ_0 is the amplitude, ω is the angular frequency (rad s^{-1}), and the wavenumbers in the x-, y-, and z-direction are k, l, and m, respectively (m^{-1}) corresponding to wavelengths L_x, L_y, L_z (m) such that

$$k = \frac{2\pi}{L_x}; \quad l = \frac{2\pi}{L_y}; \quad m = \frac{2\pi}{L_z} \tag{8.3}$$

The wavenumber vector marks the propagation direction of the wave and is given by $\vec{K} = k\hat{i} + l\hat{j} + m\hat{k}$ with a magnitude of $|\vec{K}| = \sqrt{k^2 + l^2 + m^2}$. Harmonic waves obey the relationship between frequency and phase speed $\omega = c_w|\vec{K}| = c_w\sqrt{k^2 + l^2 + m^2}$ (rad s^{-1}). When $\omega = \omega(|\vec{K}|)$, the phase speed depends on the wavenumber and the propagation medium is said to be *dispersive*. $\omega = \omega(|\vec{K}|)$ is then called the *dispersion relation*. In general, the atmosphere behaves as a dispersive medium.

The representation of wave disturbance by a simple sinusoidal solution as in (8.2) is an oversimplification. According to the *Fourier theory*, the wave can then be represented by a wave group that is the summation (a *Fourier series*) of harmonic terms like (8.2) with different coefficients (see, e.g., Equation 4.56). The velocity at which the center of the wave group travels is called the *group velocity* and is the velocity at which the observable disturbance and the energy propagate. It is defined as

$$\vec{c}_g = c_{gx}\hat{i} + c_{gy}\hat{j} + c_{gz}\hat{k} \tag{8.4}$$

where

$$c_{gx} = \frac{\partial \omega}{\partial k}; \quad c_{gy} = \frac{\partial \omega}{\partial l}; \quad c_{gz} = \frac{\partial \omega}{\partial m} \tag{8.5}$$

are the scalar group velocities in the respective directions (see Problem 8.1).

8.1.2   The Linear Theory

As stated above, we use the linear theory to describe the wave phenomena. It assumes that the amplitude of the wave disturbance is small enough relative to the mean state of the atmosphere so we can apply a *linear perturbation* to the dynamical magnitudes in the motion equations. It must be said that nonlinear effects are necessary in some cases to explain observations, but this is out of the scope of this introductory book. In general, the linear wave theory is a good representation for most of the observed wave phenomena in atmospheres without the need to enter into a complex mathematical treatment. Even so, the linear theory has, in some cases, formulation difficulties.

The procedure we follow consists of the following steps: (1) determine which terms in the motion and thermodynamic equations are relevant for the type of wave to be studied; (2) make linear the magnitudes intervening in the equations by means of the perturbation method; (3) derive the wave equation for a selected field magnitude; (4) introduce a harmonic solution containing the amplitude, wavenumbers, and frequency (this is again an oversimplification of the more general representation by means of a Fourier series); and (5) derive a dispersion relationship (phase speed versus wavenumber) for the wave and determine the group speed when appropriate.

In the linear perturbation method, all the field variables are divided into two parts: a basic reference state (assumed to be spatially and temporary independent) and a perturbation, which is a local deviation from the basic state. To simplify, assuming only zonal dependence, the wave field ψ is written as $\psi(x,t) = \bar{\psi} + \psi'(x,t)$. To see how this applies in the motion equation, take, for example, the term $u(\partial T/\partial x)$ that describes the advection of temperature by zonal winds. It is written in terms of this theory as

$$u\frac{\partial T}{\partial x} = \left(\bar{u} + u'\right)\frac{\partial}{\partial x}\left(\bar{T} + T'\right) = \bar{u}\frac{\partial T'}{\partial x} + u'\frac{\partial T'}{\partial x} \tag{8.6}$$

Here, two assumptions have been considered. First, the basic state variables satisfy themselves with the governing equations when the perturbations are set to zero and second, the perturbation is small enough so the products of the perturbation terms are neglected and then $\bar{u}\,(\partial T'/\partial x) \gg u'\,(\partial T'/\partial x)$. In this way, all the nonlinear terms in the equations are made linear. This does not mean that $|u'| \ll |\bar{u}|$, which would be a too restrictive condition.

8.2   ACOUSTIC WAVES

Acoustic or sound waves result from the restoration effect of the gradient pressure force acting in all type of fluids. The oscillating pressure (compression and rarefaction) can be considered adiabatic, and since sound waves are *longitudinal* (particle oscillation parallel to the propagation direction), the relevant momentum, continuity, and thermodynamic equations in Section 7.1.2 reduce to

$$\frac{du}{dt} = -\frac{1}{\rho}\frac{\partial P}{\partial x} \quad (v = w = 0) \tag{8.7a}$$

$$\frac{d\rho}{dt} + \rho \frac{\partial u}{\partial x} = 0 \tag{8.7b}$$

$$P\rho^{-\gamma} = \text{Const.} \tag{8.7c}$$

According to the perturbation method, the variables are expanded in the terms: $u(x,t) = \bar{u} + u'(x,t)$, $P(x,t) = \bar{P} + P'(x,t)$, and $\rho(x,t) = \bar{\rho} + \rho'(x,t)$ and since $|\rho'/\bar{\rho}| \ll 1$, we can use the binomial expansion for the density terms (note also that $P' \ll \bar{P}$ but not necessarily $u' < \bar{u}$). This allows for the linearization of the equations that once combined reduced to the sound wave equation

$$\left(\frac{\partial}{\partial t} + \bar{u}\frac{\partial}{\partial x}\right)^2 P' - \frac{\gamma \bar{P}}{\bar{\rho}}\frac{\partial^2 P'}{\partial x^2} = 0 \tag{8.8}$$

After substitution of the harmonic solution (the real part of it),

$$P'(x,t) = P_0 \exp\left[ik(x - c_w t)\right] \tag{8.9}$$

in Equation 8.8, with P_0 constant, we find that the wave speed is

$$c_w = \bar{u} \pm \left(\frac{\gamma \bar{P}}{\bar{\rho}}\right)^{1/2} = \bar{u} \pm \left(\gamma R_g^* T\right)^{1/2} = \bar{u} \pm c_s \tag{8.10}$$

The speed of sound relative to the background flow is therefore

$$c_s = \pm \left(\gamma R_g^* T\right)^{1/2} \tag{8.11}$$

Table 8.1 gives characteristic values for the sound speed in planetary atmospheres. The amplitude of the wave is related to its intensity (in W m^{-2}) by $I_s = (1/2)\rho\omega^2(\Delta\ell)^2 c_s = P_0^2/(2\rho c_s)$, where $\omega = 2\pi\nu$ is its angular frequency, ρ is the air density, and $\Delta\ell$ is the displacement of air molecules due to the pressure perturbation, so the pressure amplitude is $P_0 = \sqrt{2I_s\rho c_s}$ (see Problem 8.2). Sound waves are nondispersive since c_s is independent of wavenumber and then $c_s = c_g$.

8.3 GRAVITY WAVES

In an atmosphere that is stably stratified ($N_B > 0$), a parcel displaced vertically from its equilibrium position will undergo oscillations under the action of the buoyancy force and gravity (see Section 4.5.3). Gravity waves can be triggered by lower convection or by horizontal flow passing an obstacle. These include, for example, the

TABLE 8.1

Sound Speed in Planetary Atmospheres

Planet	Level	T (K)	P (Bar)	c_s (m s^{-1})
Venus	Surface	734	92	420
	$z=60$ km (clouds)	250	0.22	248
Earth	Surface	285	1	340
	$z=10$ km	217	0.1	295
Mars	Surface	240	0.007	229
	$z=80$ km	140	0.0001	185
Titan	Surface	94	1.5	194
	$z=28$ km	73	0.3	170
Jupiter	Ammonia cloud	127	0.4	815
	Water cloud	270	5	1200
Saturn	Ammonia cloud	107	0.5	763
Uranus	Methane cloud	85	1.4	655
Neptune	Methane cloud	81	1.4	640
HD 209458B	—	1350	0.5	2500

Source: Adapted from Sánchez-Lavega, A., *Phys. Teacher*, 40, 239, 2002.

lee waves that form downstream of mountains and manifest as cloud bands when condensation occurs in the crests. In the absence of boundaries, gravity waves can propagate vertically and horizontally transporting momentum and influencing the dynamics.

8.3.1 PURE GRAVITY WAVES

A simple equation to describe a gravity wave can be derived under the following assumptions: (1) consider motions in the (x, z) plane (zonal, vertical); (2) neglect rotation effects (Coriolis terms) and friction; (3) the unperturbed atmosphere is at rest; (4) use the *Boussinesq approximation* (Section 7.2.6; density is taken as constant except when coupled with gravity in the buoyancy term of the vertical momentum equation); (5) the atmosphere is adiabatic. The motion equations (momentum, continuity, and thermodynamic, for adiabatic motion, Section 7.1.2) reduce to

$$\frac{du}{dt} = -\frac{1}{\rho}\frac{\partial P}{\partial x} \tag{8.12a}$$

$$\frac{dw}{dt} = -\frac{1}{\rho}\frac{\partial P}{\partial z} - g \tag{8.12b}$$

$$\frac{\partial u}{\partial x} + \frac{\partial w}{\partial z} = 0 \tag{8.12c}$$

$$P\rho^{-\gamma} = \text{Const.,} \quad \text{or} \quad \frac{d\ln\theta}{dt} = 0 \tag{8.12d}$$

These equations are linearized using the decomposition $\rho = \rho_0 + \rho'$ (but point 4 above), $u = \bar{u} + u', P = \bar{P}(z) + P', \theta = \bar{\theta}(z) + \theta', w = w'$, where $\rho_0, \bar{u}$ are constants and the perturbed magnitude functions of (x, z, t), respectively. Introduced in Equations 8.12 and after some manipulation of the linearized versions of these equations, we arrive at the following wave equation (see, e.g., Holton (2004) for a full derivation):

$$\left(\frac{\partial}{\partial t} + \bar{u}\frac{\partial}{\partial x}\right)^2 \left(\frac{\partial^2 w'}{\partial x^2} + \frac{\partial^2 w'}{\partial z^2}\right) + N_B^2 \frac{\partial^2 w'}{\partial x^2} = 0 \tag{8.13}$$

where the Brunt–Väisälä frequency is assumed to be constant with altitude. The wave equation (8.13) has harmonic solutions for the perturbed vertical velocity of the form

$$w' = \text{Re}\left\{w_0 \exp\left[i(kx + mz - \omega t)\right]\right\} \tag{8.14}$$

where $w_0 = w_r + iw_i$ is a complex amplitude sum of real (w_r) and imaginary (w_i) terms. These oscillations occur in the (x, z) plane with the phase of the wave $(kx + mz - \omega t)$ depending on the vertical distance (assumed to vary linearly on z) on x and t. They are referred to as *internal gravity waves*. The horizontal wavenumber $k = 2\pi/L_x$ is real since the solution is sinusoidal in x, but the vertical wavenumber may be complex $m = m_r + m_i$ (the real wavenumber m_r describes the vertical oscillation distance and the imaginary part m_i describes the decay or growth of the wave in the z-direction).

For m real, $m = 2\pi/L_z$ and substituting (8.14) in (8.13) gives the dispersion relationship

$$\omega - \bar{u}k = \pm\frac{N_B k}{\left(k^2 + m^2\right)^{1/2}} \tag{8.15}$$

The positive sign is taken for phase propagation in the eastward direction relative to the mean flow and we define the *intrinsic frequency* of the gravity wave as $\hat{\omega} = \omega - \bar{u}k$, that is, the frequency relative to the mean flow. For $m < 0$ and $k > 0$, the lines of constant phase tilt eastward with respect to the vertical (Figure 8.1) since for increasing x (eastward), z must also increase to keep the phase constant $(kx + mz - \omega t)$ resulting in a downward phase propagation relative to the mean flow. The angle that the phase lines form relative to the vertical is given by

$$\cos\alpha = \frac{L_z}{\sqrt{L_x^2 + L_z^2}} = \pm\frac{k}{\sqrt{k^2 + m^2}} \tag{8.16}$$

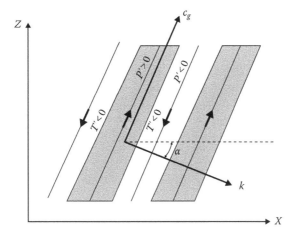

FIGURE 8.1 Vertical–longitudinal cross section showing the phases (thin sloping lines) of the pressure, temperature, and velocity perturbation for an internal gravity wave with $k>0$, $m>0$, and $c_{gz}>0$. The arrows indicate the perturbation velocity field, and the directions mark the group and phase velocities (inclined from the horizontal by the angle α). Shading regions indicate upward motion.

Thus, the gravity wave frequencies must be less than the buoyancy frequency since $\hat{\omega} = \pm N_B \cos\alpha$. The horizontal and vertical phase speeds relative to the mean flow are

$$c_x = \frac{\hat{\omega}}{k} \tag{8.17a}$$

$$c_z = \frac{\hat{\omega}}{m} \tag{8.17b}$$

and the components of the group velocity are

$$c_{gx} = \bar{u} \pm \frac{N_B m^2}{\left(k^2 + m^2\right)^{3/2}} \tag{8.18a}$$

$$c_{gz} = \pm \frac{-N_B km}{\left(k^2 + m^2\right)^{3/2}} \tag{8.18b}$$

The sign of the vertical component of the phase speed is opposite to that of the group velocity. As a consequence, group velocity propagation is perpendicular to the direction of phase propagation. Importantly, gravity waves generated when a horizontal flow passes over an obstacle (see below) or is triggered above the tops of vigorous ascending cumulus may propagate energy upward although the gravity wave oscillation remains confined to the excitation region.

8.3.2 Topographic Stationary Waves

The air flowing over topographic features in the planetary surface forces uplift and can trigger gravity waves if the atmosphere is statically stable. Assume for simplicity that the flow has a constant mean velocity $\bar{u}$ and static stability N_B, both independent of altitude. The topographic feature vertically displaces the air parcels that move with the flow from their equilibrium levels, oscillating with the buoyancy frequency N_B. Waves that are stationary relative to the surface have w' in (8.13) that is dependent on (x, z) but not with time. Then, (8.13) simplifies to

$$\left(\frac{\partial^2 w'}{\partial x^2} + \frac{\partial^2 w'}{\partial z^2}\right) + \frac{N_B^2}{\bar{u}^2} w' = 0 \tag{8.19}$$

and for harmonic solutions of the type, (8.14) yields a dispersion relationship

$$m^2 = \frac{N_B^2}{\bar{u}^2} - k^2 \tag{8.20}$$

Vertical propagation occurs for $m^2 > 0$, that is, for $|\bar{u}| < N_B/k$, which means weak wind speeds or comparatively high static stability and large zonal wavelength. Then, the solution to (8.19) is written as

$$w' = w_0 \exp\left[i\left(kx + mz\right)\right] \tag{8.21a}$$

For $m^2 < 0$ (m is imaginary, $= im_i$) that is, when $|\bar{u}| > N_B/k$, which means strong winds or comparatively low static stability, the solution to (8.19) is written as

$$w' = w_0 \exp(ikx)\exp(-m_i z) \tag{8.21b}$$

and the wave propagates horizontally but its structure varies exponentially in the vertical. These waves are referred to as *evanescent* since their energy density decreases exponentially with height.

A useful study case occurs when steady air flows over an infinite series of sinusoidal surface ridges (altitude h_0, wavenumber k) so they can be written as $h(x) = h_0 \cos kx$. Then the perturbed vertical velocity at the boundary can be calculated from the altitude change following the motion (we assume that the altitude of the ridges is much smaller than their separation distance, $h_0 k \ll 1$):

$$w'(x,0) = \left(\frac{dh}{dt}\right)_{z=0} \approx \bar{u}\left(\frac{\partial h}{\partial x}\right) = -\bar{u}kh_0\sin kx \tag{8.22}$$

and with this condition (8.21a and b) is written as

$$w'(x,z) = -\bar{u}h_0 k \sin(kx + mz) \quad (\bar{u}k < N_B) \tag{8.23a}$$

$$w'(x,z) = -\bar{u}h_0 k \exp(-\mu_H z)\sin(kx + mz) \quad (\bar{u}k > N_B) \qquad (8.23b)$$

with $\mu_H = |m_i|$ and then μ_H^{-1} is the vertical scale for the gravity wave attenuation. Gravity waves forced by orographic structures could form in Venus, Earth, Mars, and Titan and in the other satellites with tenuous atmospheres. According to (8.20), gravity waves could form near the surface of Earth and Mars even under moderate winds (~10–15 m s^{-1}) if the vertical stability is high (see Problem 8.3). On Venus, surface wind speeds measured by descending probes were very low (<5 m s^{-1}) and since the stability conditions are also low ($N_B \sim 10^{-3}$ s^{-1}), upward propagation of gravity waves will occur if they are excited with wavelengths >10 km. More details on gravity wave observations are given below.

8.3.3 INERTIO-GRAVITY WAVES

Low-frequency gravity waves are sensitive to the frequency imposed by the planet's rotation and the zonal motion equation (8.12a) must be modified to include the Coriolis term $(-fv, v$ is the meridional velocity). In addition, the meridional motion equation $dv/dt + fu = -\rho^{-1}(\partial P/\partial y)$ must be added to the system (Equation 8.12). For hydrostatic conditions and assuming a motionless basic state, the following dispersion relation is found once the linearization is performed

$$\omega^2 = f^2 + \frac{N_B^2(k^2 + l^2)}{m^2} \qquad (8.24)$$

For vertical propagation (that is, for m real), the wave frequency must satisfy $|f| < |\omega| \ll N_B$. See Problems 8.4 through 8.6.

8.3.4 GRAVITY WAVE OBSERVATIONS

Gravity waves on Earth represent a minor component of the motions in the troposphere. They are, however, frequently observed when moisture favors condensation, such as in lee waves forced by topography (Figure 8.2) or when forced by vigorous convective ascending motions (a particular case occurs at two different scales in tornados and hurricanes). A packet of arc-shaped gravity waves propagating along crest and troughs is shown in Figure 8.3. Gravity waves generated in the lower atmosphere of the Earth propagate vertically, dissipating by viscous damping in the upper atmosphere (above an altitude of 75 km) where they deposit their energy and momentum. They, therefore, play an important role in the mesosphere and thermosphere dynamics.

Gravity waves have been observed and detected in most of the dense planetary atmospheres. On Venus, Mars, and Jupiter, we have evidence of them in the cloud field by their characteristic albedo pattern (Figures 8.4 through 8.6). On Mars, they form by topographic forcing (lee waves generated by mountains and craters) as discussed above. The clouds are probably of water ice for high altitude gravity waves but of CO_2 condensing ice in the lower waves.

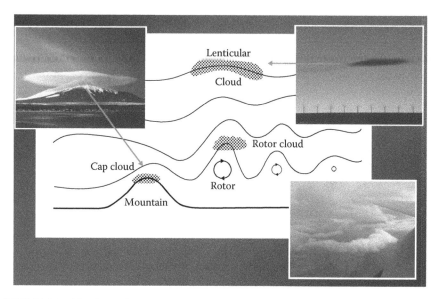

FIGURE 8.2 The formation and structure of mountain or lee waves by obstacles and associated clouds with some examples in Earth's atmosphere.

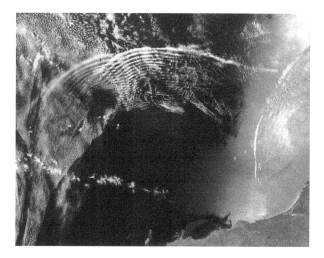

FIGURE 8.3 A packet of arc shaped gravity waves is made visible from the cloud condensation at the wave crests in Australia (November 2003). MODIS instrument onboard Terra (EOS AM) satellite. (Courtesy of NASA-GSFC, Baltimore, MD.)

On Venus and Jupiter, the gravity waves at cloud level are probably excited by convective motions in a lower (buried) unstable layer beneath the stable gravity wave-forming layer. They could also be excited by Kelvin–Helmholtz instabilities (see Section 9.4.1) produced by a strongly sheared flow when close to neutral static stability. On Venus, gravity waves are observed as alternating bands of high-low

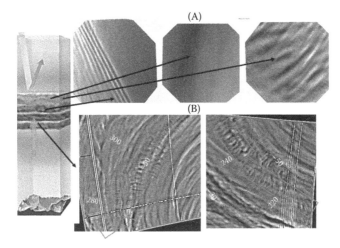

FIGURE 8.4 Gravity waves in the different cloud layers of Venus' atmosphere as observed by the ESA Venus Express spacecraft instruments in 2006–2008: (A) Upper cloud, VMC instrument; detection wavelength 980 nm (left), detection wavelength 360 nm (middle and right). (Courtesy of D. Titov, Max Planck Institute for Solar System Research, Lindau, Germany.) (B) Lower cloud, VIRTIS instrument (detection wavelength 1.27 µm). (Courtesy of J. Peralta, R. Hueso, and A. Sánchez-Lavega, UPV-EHU, Spain.)

FIGURE 8.5 Gravity waves in the Martian atmosphere traced by CO_2 ice condensation in the wave crests. From Mars Global Surveyor–MOC instrument. (Courtesy of NASA-Malin Space Science Systems (MalinSSS), San Diego, California.)

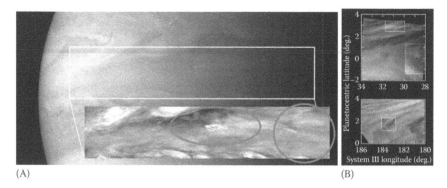

(A)   (B)

FIGURE 8.6 Gravity waves in Jupiter's equatorial clouds detected by (A) the New Horizons Spacecraft (NASA-JPL) in February 2007, and (B) the Galileo spacecraft in 1999 and 2001. (Prepared by J.F. Rojas, R. Hueso, and A. Sánchez-Lavega.)

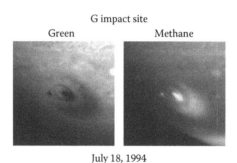

July 18, 1994

FIGURE 8.7 Ring-like gravity waves that formed following the impact of fragment G- of comet Shoemaker-Levy 9 (SL9) with Jupiter in July 1994 as observed by the Hubble Space Telescope. The wave is traced by the debris resulting from the impact. (Courtesy of HST-NASA-ESA, Baltimore, MD and Garching, Germany.)

optical depth (relative to the surroundings) within the two main cloud layers at 50 and 65 km. On Jupiter, they are probably formed by condensing ammonia ice crystals. A peculiar type of fast moving gravity wave formed in Jupiter's atmosphere following the impact of some of the fragments of comet Shoemaker-Levy 9 in 1994 (Figure 8.7) (Section 9.8). It became visible due to an enhancement in the density of the wave front produced by the dust debris resulting from the comet fragments impacts.

Gravity waves in the upper atmospheres of Venus and Mars and in the giant and icy planets (Jupiter, Uranus, and Neptune) have been detected as oscillations in the vertical profiles of the temperature field derived from radio occultation experiments and measured during the descent of probes (Venus, Mars, and Jupiter) (Figure 8.8). These oscillations are probably the manifestation of upward vertically propagating gravity waves generated close to the tropopause. In Table 8.2, we summarize the basic properties measured for gravity waves in planetary atmospheres.

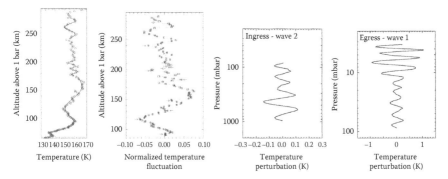

FIGURE 8.8 Vertical structure of gravity and inertio-gravity waves retrieved from temperature soundings in Jupiter (left panel) and in Neptune (right panel). The wave structure is best seen after subtraction from the mean profile. (Left: From Young, L.A. et al., *Icarus*, 173, 185, 2005. With permission; Right: From Hinson, D.P. and Magalhães, J.A., *Icarus*, 105, 142, 1993. With permission.)

8.4 ROSSBY WAVES

Rossby waves are a family of low frequency, large horizontal waves (also called *planetary waves*) that form in rotating systems. In atmospheres, they owe their existence to the change of the Coriolis force with a latitude represented by the parameter $\beta = df/dy$ (see Equation 7.16). We study their properties making use of the vorticity concept introduced in Section 7.3 for two representative situations in planetary atmospheres: the inviscid and barotropic approach and the more general case of baroclinic atmosphere.

8.4.1 BAROTROPIC ROSSBY WAVE

According to what we have seen in Section 7.3.2, in a barotropic atmosphere with constant density and for purely horizontal flow (no vertical motions), the absolute vorticity is conserved (Equation 7.58):

$$\frac{d(\zeta + f)}{dt} = 0$$

In this system, parcels displaced meridionally must change their relative vorticity since the Coriolis term changes with latitude (see Equation 7.58b). Assume that a parcel located initially in a latitude with a Coriolis parameter f_0 has $\zeta = 0$. If the parcel is displaced meridionally, a distance dy where we have f_1, the absolute vorticity conservation, indicates that $f_0 = \zeta_1 + f_1$ or $\zeta_1 = f_0 - f_1 = -\beta \, dy$ from the definition of β. This indicates that parcels displaced meridionally will be subjected to an oscillatory sinusoidal motion due to a β-effect (it acts as a restoring force proportional to distance) about the equilibrium latitude.

Let us make use of the linear small perturbation model to study the properties of this wave for a basic zonal flow. The perturbed velocity field has two components:

TABLE 8.2
Properties of Observed Gravity Waves in Planetary Atmospheres

Planet	Parameter	Altitude (z or P)	L_x (km)	L_z (km)	$c_x - \bar{u}$ (ms^{-1})
Venus	$T(z)$	>100 km	5–10	—	—
			100–600		
	$U(z)$	40 km	—	—	—
	Cloud[a]	47, 66 km	60–150	5–15	−10, +10
Earth	$U(z)$[b]	60–80 km	1000	10	—
		75–150 km	10–200	5–15	<80
	Cloud	1–10 km	0.1–1	1–10	0
Mars	$T(z)$[c]	10–70	—	6–15	—
	Cloud	<15	0.1–1	—	—
Jupiter	$T(z)$[d]	$P \sim 3$ mbar	30–40	90–290	200–430
		(430–710 km)			
	Cloud[e]	$P \sim 0.5$ bar	300	—	100
	SL9 debris[f]	$P \sim 0.2$ bar	Radial	—	450
			$r_0 \sim 4000$		180–430
Neptune	$T(z)$[g]	$P \sim 6$–80 mbar		3–8	>20

Source: Data taken from Andrews, D.G. et al., *Middle Atmospheric Dynamics*, Academic Press, San Diego, CA, 1987; Gierasch, P. et al., in *Venus II: Geology, Geophysics, Atmosphere, and Solar Wind Environment*, S.W. Bougher et al. (eds.), University of Arizona Press, Tucson, AZ, 1997; Zurek, R.W. et al., in *Mars*, B.M. Jakovsky et al. (eds.), University of Arizona Press, Tucson, AZ, 1993; Ingersoll, A.P. et al., in *Jupiter: The Planet, Satellites and Magnetosphere*, F. Bagenal et al. (eds.), Cambridge University Press, New York, 2004; Ingersoll, A.P. et al., in *Neptune and Triton*, D.P. Cruikshank (ed.), University of Arizona Press, Tucson, AZ, 1996.

[a] Wave packets with extent: $\Delta L_x = 950$ km, $\Delta L_y = 350$ km.
[b] Amplitudes 30–50 m s^{-1}.
[c] Amplitude $\Delta T = 3$–5 K.
[d] Amplitude $\Delta T = 5$, 25–40 K.
[e] Wave packets with extent: $\Delta L_x = 10{,}000$ km, $\Delta L_y = 1000$–2000 km.
[f] SL9: Comet Shoemaker-Levy 9 (see Section 9.8).
[g] Amplitude $\Delta T = 0.2$–1 K.

$u = \bar{u} + u'$ and $v = v'$ (corresponding to a perturbed vorticity field $\zeta = \bar{\zeta} + \zeta'$). Under the β-plane approximation (Equation 7.16) $f = f_0 + \beta y$, the vorticity equation (7.53) becomes

$$\left(\frac{\partial}{\partial t} + \bar{u} \frac{\partial}{\partial x} \right)\left(\frac{\partial v'}{\partial x} - \frac{\partial u'}{\partial y} \right) + \beta v' = 0 \tag{8.25}$$

The meridionally perturbed velocity field coupled with the beta term ($\beta v'$) is responsible for the Rossby wave existence (see Figure 8.9).

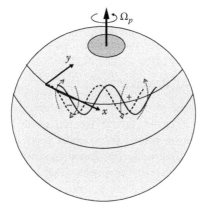

FIGURE 8.9 Rossby wave mechanism in a barotropic atmosphere. The perturbed absolute vorticity field (a conserved magnitude) induce a meridional velocity field (dotted arrows)
which moves the fluid elements to the southwest of the vorticity maximum and northwest of
the vorticity minimum, producing the oscillation about the equilibrium latitude. The heavy
wavy line represents the initial perturbation and the dashed wavy line its westward motion.

To derive a dispersion relationship for this wave, we introduce a perturbed horizontal *stream function* ψ' (see Equation 7.62) based on the fact that the flow is nondivergent such that

$$u' = -\frac{\partial \psi'}{\partial y}; \quad v' = \frac{\partial \psi'}{\partial x} \tag{8.26}$$

which satisfies the continuity equation $(\partial u/\partial x)+(\partial v/\partial y)=0$ and $\zeta'=\nabla^2\psi'$ with
$\nabla^2=(\partial^2/\partial x^2)+(\partial^2/\partial y^2)$. Substituting in (8.25) gives the following wave equation:

$$\left(\frac{\partial}{\partial t}+\bar{u}\frac{\partial}{\partial x}\right)\nabla^2\psi'+\beta\frac{\partial\psi'}{\partial x}=0 \tag{8.27}$$

Harmonic solutions of the form $\psi'=\mathrm{Re}[\psi_0\exp i(kx+ly-\omega t]$ yield the following dispersion relationship upon substitution in (8.27)

$$\omega=\bar{u}k-\frac{\beta k}{k^2+l^2} \tag{8.28}$$

We see again the decisive importance of the β-effect for the existence of the Rossby
wave (Problem 8.7). The phase speed of the wave relative to the mean zonal wind is
given by

$$c_x-\bar{u}=-\frac{\beta}{k^2+l^2} \tag{8.29}$$

which indicates that the Rossby wave drifts westward relative to the basic zonal flow since β is always >0. It can also be seen that the phase speed of the wave increases rapidly with wavelength (zonal and meridional). See Problem 8.8.

8.4.2 THREE-DIMENSIONAL ROSSBY WAVE

We consider now a more general situation when the atmosphere has a three-dimensional structure so we can infer the vertical structure of the Rossby wave. In a baroclinic atmosphere, the Rossby wave can be understood as a consequence of the conservation of the potential vorticity PV (Equations 7.59 through 7.61). However, since Rossby waves have large horizontal scales and low frequencies, the Rossby number (7.13) is low and instead of PV, it is more convenient to use the quasi-geostrophic potential vorticity $QGPV$ (q_g) that was introduced in Section 7.3.4. We remember that its conservation equation (7.60) is

$$\frac{d_g q_g}{dt} = 0$$

Consider a purely zonal base flow with components $(\bar{u},0,0)$, then according to (Equation 7.62), the geostrophic streamfunction is given by $\psi = -\bar{u}y$. In order to make the linearization, consider a streamfunction perturbation ψ' such that $\psi = -\bar{u}y + \psi'$. Then the $QGPV$ for this flow is according to (7.64)

$$q_g = f_0 + \beta y + \left[\nabla^2 + \frac{\partial}{\partial z}\left(\frac{f_0^2}{N_B^2} \frac{\partial}{\partial z} \right) \right] \psi' \tag{8.30}$$

and the linear form of (7.65) is given by

$$\left(\frac{\partial}{\partial t} + \bar{u}\frac{\partial}{\partial x} \right)\left[\nabla^2 \psi' + \frac{\partial}{\partial z}\left(\frac{f_0^2}{N_B^2} \frac{\partial}{\partial z} \right)\psi' \right] + \beta \frac{\partial \psi'}{\partial x} = 0 \tag{8.31}$$

This is the equation for the *quasi-geostrophic Rossby waves*. To further simplify it, we take N_B to be constant with altitude. Then, for plane wave solutions that include an altitude dependent amplitude (see Problem 8.6) of the form

$$\psi' = \text{Re}\left\{ \psi_0 \exp\left[\left(\frac{z}{2H} \right) + i\left(kx + ly + mz - \omega t \right) \right] \right\} \tag{8.32}$$

we find upon substituting (8.32) in (8.31) the following dispersion relationship

$$\omega = \bar{u}k - \frac{\beta k}{k^2 + l^2 + \left(f_0^2/N_B^2 \right)\left(m^2 + (1/4H^2) \right)} \tag{8.33}$$

where

m is the vertical wavenumber

H is usually taken as the atmospheric scale height

The zonal phase speed of the wave relative to the mean wind is then

$$c_x - \bar{u} = -\frac{\beta}{k^2 + l^2 + \left(f_0^2 / N_B^2\right)\left(m^2 + (1/4H^2)\right)} \tag{8.34}$$

The following considerations can be done:

- For m^2 $(1/4\ H^2)$, the vertical structure is controlled by $m^2(f_0^2/N_B^2)$, and further, if $f_0^2 \ll N_B^2$ or if the vertical wavelength $2\pi/m \rightarrow \infty$, then the three-dimensional Rossby wave dispersion relationship reduces to the barotropic case (8.28) and (8.29).
- Solving for m^2 in (8.34) gives

$$m^2 = \left(\frac{N_B^2}{f_0^2}\right)\left[\frac{\beta}{\bar{u} - c_x} - (k^2 + l^2)\right] - \frac{1}{4H^2} \tag{8.35}$$

Then, vertical propagation requires $m^2 > 0$ and (8.35) imposes the condition (Problem 8.9)

$$\bar{u} - c_x = \left[\frac{\beta}{k^2 + l^2 + \left(f_0^2 / N_B^2\right)\left(m^2 + (1/4H^2)\right)}\right] < u_c \tag{8.36a}$$

$$u_c = \frac{\beta}{k^2 + l^2 + (1/4H^2)\left(f_0^2 / N_B^2\right)} \tag{8.36b}$$

where u_c represents a critical velocity. Since $\bar{u} - c_x > 0$ (all terms are positive), this condition reduces to $0 < \bar{u} - c_x < u_c$.

- *Stationary Rossby waves* (relative to the ground or to the rotating reference System III in the fluid planets) have $c_x = 0$, and the above condition reduces to the *Charney–Drazin criterion*: $0 < \bar{u} < u_c$. It indicates that these waves propagate vertically under an eastward background flow whose velocity is $\bar{u} < u_c$. When $(k^2 + l^2) \gg (1/4H^2)\left(f_0^2 / N_B^2\right)$ in (8.36), $u_c \approx (\beta/(k^2 + l^2))$ and the above criterion indicates that long-waves (k and l small) have more possibilities to propagate vertically than short waves.
- Long Rossby waves have $(k^2 + l^2) < \left(f_0^2 / N_B^2\right)\left(m^2 + (1/4H^2)\right)$ and from (8.34) their phase speed is controlled by their vertical structure

$$c_x \approx \bar{u} - \frac{\beta}{\left(f_0^2/N_B^2\right)\left(m^2 + (1/4H^2)\right)} \tag{8.37a}$$

- Short Rossby waves have $(k^2 + l^2) \gg \left(f_0^2/N_B^2\right)\left(m^2 + (1/4H^2)\right)$ and $\beta/(k^2 + l^2)$ is small, then from (8.34) their phase speed is

$$c_x \approx \bar{u} - \frac{\beta}{k^2 + l^2} \rightarrow \bar{u} \tag{8.37b}$$

that is, they tend to move with the background flow.
- The components of the group velocity can be obtained using Equation 8.5

$$c_{gx} - \bar{u} = \beta \frac{k^2 - l^2 - \left(f_0^2/N_B^2\right)\left(m^2 + (4H^2)^{-1}\right)}{\left[k^2 + l^2 + \left(f_0^2/N_B^2\right)\left(m^2 + (4H^2)^{-1}\right)\right]^2} = \frac{\omega}{k}\left(1 + \frac{2\omega k}{\beta}\right) \tag{8.38a}$$

$$c_{gy} = \frac{2\beta kl}{\left[k^2 + l^2 + \left(f_0^2/N_B^2\right)\left(m^2 + (4H^2)^{-1}\right)\right]^2} = \frac{2l\omega^2}{\beta k} \tag{8.38b}$$

$$c_{gz} = \frac{2\beta mk\left(f_0^2/N_B^2\right)}{\left[k^2 + l^2 + \left(f_0^2/N_B^2\right)\left(m^2 + (4H^2)^{-1}\right)\right]^2} = \frac{f_0^2}{N_B^2}\frac{2m\omega^2}{\beta k} \tag{8.38c}$$

Accordingly, for $kl > 0$ and $km > 0$, the wave propagates information in the positive y and z directions and, since $\omega < 0$, it implies that the phase propagation is in the opposite direction. For example, for downward and equatorward phase propagation, the energy propagated by the Rossby wave is upward and poleward.

All these relationships hold for the case of a constant zonal wind velocity. However, if the zonal wind varies with latitude and height $\bar{u} = \bar{u}(y,z)$, we must use an *effective* β_e parameter (that is the meridional gradient of the zonal-mean vorticity) that includes the effects of the meridional and vertical wind shears instead of β (the gradient of the Coriolis parameter)

$$\beta_e = \beta - \frac{\partial^2 \bar{u}}{\partial y^2} - \frac{1}{\bar{\rho}}\frac{\partial}{\partial z}\left(\frac{f_0^2}{N_B^2}\bar{\rho}\frac{\partial \bar{u}}{\partial z}\right) \tag{8.39}$$

Note that in this equation the atmospheric static stability varies with altitude, that is, $N_B(z)$ and a simple approach to manage it is to use the vertical temperature structure $T(z)$ to calculate it as $N_B(z)^2 = g^2/C_p T(z)$. In addition, when the zonal wind varies with altitude, the wave may encounter a *critical level* defined by the condition $c_x - \bar{u} = 0$. Then, $m \rightarrow \infty$ and the wave is absorbed at this level.

8.4.3 OBSERVED NONEQUATORIAL ROSSBY WAVES IN PLANETS

Rossby waves are permanent phenomena in the Earth's mid and sub-polar latitudes. They form in the troposphere as large-scale latitudinal oscillations of the mid-latitude jet stream with a wavenumber of 4–6. Low-pressure cyclones and high-pressure anticyclones form in the crest and troughs of the wave (Figure 8.10) by a baroclinic instability (see Chapter 9), which are dominant in establishing the weather of the middle and higher latitudes.

Stationary Rossby waves propagating into the stratosphere form in our planet when the westerly flow is forced to pass over large-scale orographic structures such as the Rocky Mountains and east of the Himalayan Mountain System. They manifest as planetary waves with wavenumbers 1 and 2 and are observed in the pressure field, in the geopotential field, or in the temperature field of the stratosphere at altitudes from 15 to 45 km and at mid- and high-latitudes (45°–65°) (Problems 8.10 and 8.11). There is also evidence of free large-scale Rossby wave modes. A prominent 5 *day wave* (its oscillation period is close to 5 days) and wavenumber 1 produces at the surface a pressure fluctuation of ~0.5 mbar but its amplitude grows with altitude and at 40 km the temperature fluctuation reaches 0.5 K. Two other observed free modes have wavenumber 1 (with a 16 day period) and wavenumber 3 (with a 2 day period).

Rossby waves have also been hardly detected in the ocean surface using satellite altimetry data because they have amplitudes of 10 cm and wavelengths of 500 km. Oceanic Rossby waves are driven by large-scale winds and by the buoyancy forcing produced at the eastern boundaries of continents and oceanic interiors. Outside of the tropics, these waves are abruptly amplified by the presence of major topographic features. They play an important role in the interaction between the ocean and atmospheric circulation.

Similarly, Rossby waves form in the atmosphere surrounding the poles of Mars when the strong eastward winds at mid-latitudes blow straight into orographic

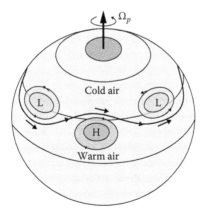

FIGURE 8.10 Development of the Rossby wave in Earth's eastward mid-latitude jet stream. The cold air occupies troughs of low pressure (cyclones) and warm areas of high pressure (anticyclones) in the crests.

TABLE 8.3

Properties of Nonequatorial Planetary Waves in Jupiter and Saturn

Planet (Wave)	Latitude	Altitude (z or P)	n (Zonal)	c_x (m s^{-1})	$c_x - \bar{u}$ (ms^{-1})
Jupiter[a]	67°S	~100 mbar	12–14	0–10	−10 to −30
Jupiter[b]	63°S	~100 mbar	11–14	9	−13
Jupiter[c]	53°S–57°S	~100 mbar	18–23	2	−30
Jupiter[d]	67°N	~100 mbar	5–12	5	−5
Saturn[e]	47°N	~500 mbar	45–60	140	0
Saturn[f]	80°N	≥500 mbar	6	0.1	0

Notes: n is the normalized zonal wavenumber ($2\pi R_p \cos\varphi / L_x$). These waves are observed at cloud-level in reflected sunlight.

[a] South Polar Wave. Conspicuous in methane band imaging (890 nm, 2.3 μm). Meridional amplitude oscillation 2° latitude. Assigned as a Rossby wave.

[b, c] South Polar Wave. Conspicuous in UV imaging (<400 nm). Assigned as a Rossby wave.

[d] North Polar Wave. Conspicuous in UV imaging (<400 nm). Assigned as a Rossby wave.

Sources of (a) through (d): Data taken from Lavega, A.S. et al., *Geophys. Res. Lett.*, 25, 4043, 1998; Barrado-Izagirre et al., *Icarus*, 194, 173, 2008.

[e] "Ribbon wave." Oscillation in meridional amplitude 2.2° latitude. Located in jet peak. Probable Rossby wave or baroclinic instability. Similar "partial waves" at 32°S and 48°S (in 1980).

[f] "Hexagon." Permanent wave (observed from 1980 to 2007). Oscillation in meridional amplitude 2° latitude. Intense embedded jet (100 m s^{-1}). Probable Rossby wave forced by a large anticyclone. No counterpart in the south pole.

Sources of (e) and (f): Data from Del Genio, A.D. et al., in *Saturn after Cassini-Huygens*, M. Dougherty et al. (eds.), Springer-Verlag, Dordrecht, the Netherlands, 2009.

structures. They are dominated by wavenumber 1 and have high amplitudes of about 20 K in the temperature field when forming in late northern fall and early northern winter. The periods for these waves range from 2.5 to 30 sols (1 sol is a Martian day). Weaker waves with amplitudes less than about 3.5 K have also been identified in the southern hemisphere. Traveling waves with wavenumbers 2 and 3 and amplitudes less than 4 K have also been observed confined near the Martian surface.

In the giant planets Jupiter and Saturn, there are a variety of wave structures that form outside the equatorial region in the upper cloud level that have been proposed to be Rossby waves (details given in Table 8.3). In Jupiter, the most conspicuous are the permanent polar waves that form as the undulating edge of the upper hazes that surround both poles, particularly prominent around the south pole (Figure 8.11). On Saturn, two prominent waves have been observed at cloud level. On one hand is the "ribbon wave" (a thin mid-latitude undulating dark band) located in the mid-latitudes of the northern hemisphere at the Voyager spacecraft times, moving with the peak speed of an eastward jet at 150 m s^{-1} (Figure 8.12). On the other hand is the "hexagon wave" that surrounds the north pole and contains an embedded eastward jet-stream with a speed of 100 m s^{-1} (Figure 8.13). These features have been

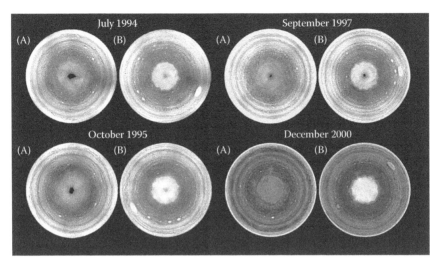

FIGURE 8.11 (A) Series of pairs of polar projection images of Jupiter's atmosphere in the ultraviolet and (B) methane absorption band at 890 nm showing the "south polar wave" as traced by the wavy boundary of a high-altitude polar haze. (Courtesy of N. Barrado, S. Pérez-Hoyos, and A. Sánchez-Lavega.)

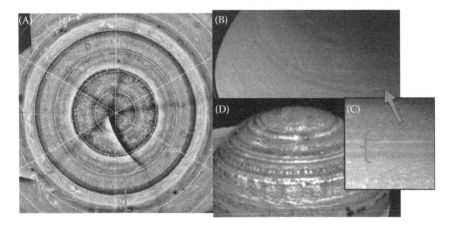

FIGURE 8.12 The "ribbon wave" band observed at two different epochs and wavelengths (marked by a bracket). (A) A polar projection of Saturn's northern hemisphere at the time of the Voyager fly-bys in 1980 and 1981. Saturn's ribbon wave stands at 47°N as a thin dark narrow wavy feature (altitude level 0.5–1 bar) (From Godfrey, D.A. and Moore, V., *Icarus*, 68, 313, 1986. With permission.); (B) Image from Cassini-ISS instrument of the same latitude on January 13, 2008 showing "no ribbon" (NASA PIA09837); (C) Details of the "ribbon area" from Cassini-ISS instrument on November 8, 2007 showing "no ribbon" (NASA PIA09796); (D) Image from Cassini-VIMS instrument of the "string of pearls" at the same latitude obtained at 5 μm that sense the lower cloud (2–5 bar) on April 27, 2006 (NASA PIA01941). (Courtesy of NASA-JPL, Los Angeles, CA.)

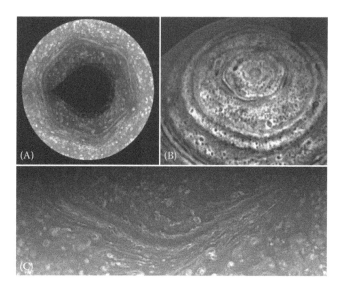

FIGURE 8.13 The hexagon wave encircling the north pole of Saturn at 80°N latitude. (A) A polar projection of Saturn's north pole region performed with images obtained by the Cassini ISS instrument on January 3, 2009 (NASA PIA 11682). (Courtesy of NASA/JPL/Space Science Institute, Los Angeles, CA.) (B) Image from Cassini-VIMS instrument of the hexagon obtained at 5 μm that senses lower clouds (2–5 bar) on October 29, 2006 (NASA PIA09188). (C) Detail of the north polar hexagon obtained by Cassini-ISS at red wavelengths on August 25, 2008 (NASA PIA10486). (Courtesy of NASA-JPL, Los Angeles, CA.)

interpreted as Rossby waves (see Problem 8.12) but they still need further analysis from the Cassini observations. Rossby waves in the giant planets could be forced by the strong circulation of large-scale vortices acting as barriers to the dominant zonal flow (see Problem 8.13).

8.5 PLANETARY-SCALE EQUATORIAL WAVES

We must first consider how the "equatorial region" is defined in a planet. For Earth and Mars, it is geographically defined as that which is bounded by the tropical latitude circles (north and south) corresponding to the planet tilt angle (the angle between the rotation axis and the line perpendicular to the orbital plane, see Chapter 1). For the fluid planets with a variety of tilts, it can be defined as the region bounded by their broad jets centered at the equatorial latitude circle: eastward jets between about ±20° for Jupiter and ±40° for Saturn or westward jets for the icy planets between about ±20° for Uranus and ±40° for Neptune (Section 7.5.4). A narrower definition for rapidly rotating planets is to consider the equatorial band as the region where pure geostrophic conditions do not hold (that is, $Ro \geq 1$). For the slowly rotating bodies (Venus and Titan) that have very different tilt angles and where $Ro \geq 1$ everywhere, the equatorial region is less precisely defined, and usually it is referred to as a broad band ~ ±20° in latitude centered about the equator.

The equatorial region is characterized by the decreasing effect of the Coriolis force ($f \to 0$) on motions. With the exception of Uranus, it is also a region where most solar radiation is deposited. The structure and types of equatorial waves can be analyzed from the linearization of the primitive equations (momentum, continuity and thermodynamics; Section 7.1.2) under the β-plane approximation (Section 7.2.1.2). In the equator, we set $\cos\varphi \sim 1$ (then $\beta = 2\Omega/R_p$), $\sin\varphi = y/R_p$, and the Coriolis parameter $f = \beta y$. Assuming no forcing, dissipation, or diabatic heating, the equations for the perturbed fields (u', v', w', Φ'_G, T') that represent the wave can be written as

$$\frac{\partial u'}{\partial t} + \bar{u}\frac{\partial u'}{\partial x} - \beta y v' + \frac{\partial \Phi'_G}{\partial x} = 0 \tag{8.40a}$$

$$\frac{\partial v'}{\partial t} + \bar{u}\frac{\partial v'}{\partial x} + \beta y u' + \frac{\partial \Phi'_G}{\partial y} = 0 \tag{8.40b}$$

$$\frac{\partial \Phi'_G}{\partial z} = \frac{R_g^* T'}{H} \tag{8.40c}$$

$$\frac{\partial u'}{\partial x} + \frac{\partial v'}{\partial y} + \frac{1}{\rho_0}\frac{\partial}{\partial z}(\rho_0 w') = 0 \tag{8.40d}$$

$$\frac{\partial T'}{\partial t} + \bar{u}\frac{\partial T'}{\partial x} + S_T w' = 0 \tag{8.40e}$$

where $\bar{u}$ is the dominant equatorial mean flow velocity and $S_T = N_B^2 T_0/g$ is the static stability of the atmosphere for which the spatial dependence has been neglected in order to maintain a simplified tractable study. We assume harmonic solutions in time and longitude (zonal) for these variables separating the (x, t) dependence (harmonic) from the (y, z) dependence. Take, for example, the meridional velocity v' (and similarly for u' and Φ'). The solution is written as

$$v'(x,y,z,t) = \hat{v}(y,z)\exp\left[ik\left(x - c_w t\right)\right] \tag{8.41}$$

Introducing it in the above system and after some manipulation, we get a single equation for $\hat{v}(y,z)$:

$$\frac{\partial^2 \hat{v}}{\partial y^2} - k^2\hat{v} - \frac{\beta}{c_w}\hat{v} + \left(H\frac{\partial}{\partial z} - 1\right)\left(\frac{\beta^2 y^2 - k^2 c_w^2}{N_B^2 H^2} H\frac{\partial \hat{v}}{\partial z}\right) = 0 \tag{8.42}$$

We now seek solutions for the distribution of $\hat{v}(y,z)$ such that the disturbance fields vanish meridionally for $|y| \to \infty$ (that is, as we separate from the equator) to make valid the equatorial condition $f = \beta y$. The solutions are then trapped equatorially. The horizontal and vertical structure equations are obtained by a separation of the variables as a series sum of modes (index-j) that give the meridional and vertical dependence

$$\hat{v}(y,z) = \sum_j V_j(y;k)G_j(z;k) \tag{8.43}$$

8.5.1 HORIZONTAL STRUCTURE

For each choice of the meridional mode number j, the horizontal structure is obtained introducing (8.43) into (8.42) with the above boundary condition, giving the equation (see Problem 8.14)

$$\frac{d^2V}{dy^2} + \left[\frac{k^2 c_w^2 - \beta^2 y^2}{gh} - \frac{\beta}{c_w} + k^2 \right] V = 0 \tag{8.44}$$

where gh is the separation constant, where g is the gravity acceleration and h is an *equivalent depth* or the mean thickness of a layer of an incompressible fluid with the same horizontal properties (this is equivalent to considering a "shallow water" layer model). This equation has the same form as the Schrödinger equation for the harmonic oscillator and the normalized solutions are written as

$$V_j = \left[\left(2^j j \ll \pi^{1/2} \right)^{-1/2} \right] e^{-\xi^2/2} H_j(\xi) \tag{8.45}$$

where a nondimensional meridional coordinate has been introduced $\xi = y \left(\beta / \sqrt{gh} \right)^{1/2}$ and where $H_j(\xi)$ are the Hermite polynomials for $j = 0, 1, 2,...$ with the first three of them having the values: $H_0(\xi) = 1$, $H_1(\xi) = 2\xi$, $H_2(\xi) = 4\xi^2 - 2$. The index j corresponds to the number of nodes in the latitude profile within the domain $|y| < \infty$. Since only decaying solutions can satisfy the boundary conditions (as $\xi \to \pm\infty$), this requires that the coefficients in the square brackets of (8.44) satisfy the relationship

$$\frac{\sqrt{gh}}{\beta} \left[\frac{k^2 c_w^2}{gh} - \frac{\beta}{c_w} - k^2 \right] = 2j + 1 \tag{8.46}$$

Finally, the horizontal solution for the meridional field (8.41) is given by

$$v'(x,y,t) = v_0 e^{-\xi^2/2} H_j(\xi) e^{ik(x - c_w t)} \tag{8.47}$$

Accordingly, the decay of the meridional solutions occurs for $y > y_d$ where

$$y_d^2 = (2j+1)\frac{\sqrt{gh}}{\beta} \tag{8.48}$$

Equation 8.46 can be rearranged into the form of a dispersion relationship $c_w(k)$ for all the equatorial wave modes j:

$$\left(\frac{c_w}{\sqrt{gh}}\right)^3 - \left(\frac{c_w}{\sqrt{gh}}\right)\left[1+\frac{(2j+1)\beta^2}{k^2\sqrt{gh}}\right] - \frac{\beta^2}{k^2\sqrt{gh}} = 0 \tag{8.49}$$

The value of the meridional mode number j describes different types of equatorial waves.

8.5.1.1 Kelvin Wave ($j=-1$)

Equation 8.49 has only one solution for $j=-1$ and corresponds to an eastward propagating wave with phase speed:

$$c_{Kelvin} = \sqrt{gh} \tag{8.50}$$

This is a nondispersive wave and resembles a shallow water gravity wave. The meridional velocity is 0 and the zonal velocity and geopotential vary in latitude as Gaussian functions centered on the equator

$$u'(x,y,t) = u_0 e^{-\xi^2/2} e^{i(kx-\omega t)} \tag{8.51}$$

with an e-folding decay width given according to (8.48) by

$$y_K = \left|\frac{2\sqrt{gh}}{\beta}\right|^{1/2} \tag{8.52}$$

Note that the solution in (8.51) initially contains the "±" sign but only the "+" sign is valid since the "−" sign will give a growing wave as $y \to \infty$.

8.5.1.2 Rossby-Gravity or Yanai Wave ($j=0$)

The only physical wave for $j=0$ has a phase speed

$$c_{Yanai} = \frac{\sqrt{gh}}{2}\left[1\pm\left(1+\frac{4\beta}{k^2\sqrt{gh}}\right)^{1/2}\right] \tag{8.53}$$

The positive choice corresponds to an eastward propagating wave moving faster than the Kelvin wave. The negative choice corresponds to a westward wave. The horizontal meridional field from (8.47) is given by

$$v'(x,y,t) = v_0 e^{-\xi^2/2} e^{i(kx-\omega t)} \tag{8.54}$$

8.5.1.3 Inertia-Gravity Waves and Equatorial Rossby ($j \geq 1$)

If $j \geq 1$, it yields *two inertia-gravity modes* (one propagating east and the other west) and one *westward Rossby mode*. The phase speed for the *inertia gravity modes* (also called Rossby–Haurwitz) can be approximately obtained from a balance between the first two terms in Equation 8.49

$$c_{IG} \simeq \pm\sqrt{gh}\left[1+(2j+1)\frac{\beta}{k^2\sqrt{gh}}\right]^{1/2} \tag{8.55}$$

whose value is always greater than that for the Kelvin wave.

The phase speed for the *equatorial Rossby mode* can be approximately obtained from a balance between the last two terms in Equation 8.49

$$c_{Rossby\,E} \simeq \frac{-\beta/k^2}{1+(2j+1)(\beta/(k^2\sqrt{gh}))} \tag{8.56}$$

that is a westward propagating wave. Note that in the short wave limit when $k \to \infty$, the phase speed $c_{Rossby\,E} = -\beta/k^2$ is independent of h. In the long wave limit $k \to 0$, the phase speed $c_{Rossby\,E} = -\sqrt{gh}/(2j+1)$, which for the gravest mode ($j=1$) gives a westward phase speed higher than that of the Kelvin wave.

8.5.2 VERTICAL STRUCTURE

From (8.42) and (8.43), the vertical wave structure equation for $G(z)$ is given by (Problem 8.14)

$$\left(H\frac{d}{dz}-1\right)H\frac{dG}{dz} = -\frac{N_B^2 H^2}{gh}G \tag{8.57}$$

This equation can be simplified using the transformations

$$\chi = e^{-z/2H}\frac{gH}{N_B^2 H^2}\frac{dG}{dz}$$

$$G = -he^{z/2H}\left(H\frac{d\chi}{dz}-\frac{1}{2}\chi\right) \tag{8.58}$$

so the vertical structure equation may be written as

$$H^2 \frac{d^2\chi}{dz^2} + \left(\frac{N_B^2 H^2}{gh} - \frac{1}{4} \right) \chi = 0 \tag{8.59}$$

whose solution is given by

$$\chi(z) = C_1 e^{imz} + C_2 e^{-imz} \tag{8.60}$$

Assuming a constant buoyancy frequency $N_B = N_0$, the vertical wavenumber m is given by $m = \left((N_0^2/gh) - (1/4H^2) \right)^{1/2}$ and the vertical wavelength is $L_z = 2\pi/m$. Vertical propagation will occur whenever $0 < h < 4N_B^2 H^2/g$.

8.5.3 REPRESENTATIVE PLANETARY-SCALE EQUATORIAL WAVES

On Earth's equatorial region, both Kelvin and Rossby-gravity modes have been identified in the wind and temperature fields of the stratosphere. The Kelvin wave has a zonal wavenumber of $n = 1–2$ and propagates eastward. The Rossby-gravity mode with $n = 4$ moving westward has been observed in the stratosphere above the equatorial Pacific. These waves and others are probably excited by the oscillations induced by the large-scale convective activity in the equatorial troposphere. Details for these waves are given in Table 8.4.

At cloud level in UV wavelengths (at altitudes of 65–70 km), Venus shows a prominent semi-permanent pattern known as the Y-, Ψ-wave, or sometimes "inverted C"-wave for its shape, centered at the equator and oriented along it (Figure 8.14). The pattern rotation period is 4 days, much shorter than the Venus surface rotation period. Its speed is slightly above that of the mean zonal westward flow, which is close to $100\,\mathrm{m\ s^{-1}}$ at this altitude level. An analysis of the zonal UV-cloud brightness distribution suggests that another 5 day wave is also present in Venus. This wave forms when the Y-mode disappears. The 4 day wave could be of a Kelvin type (Problem 8.15) or mixed Kelvin and equatorial Rossby modes.

Jupiter and Saturn exhibit a large spectrum of planetary-scale waves detected as oscillations in the temperature maps above the clouds (Figure 8.15). They move slowly relative to the radio-rotation period (System-III) implying that they drift westward relative to the mean atmospheric flow. Modeling suggests they are equatorial Rossby waves trapped in latitude by the meridional shear of the zonal wind.

At latitude 7°N there is in Jupiter's clouds a permanent pattern formed by series of elongated features that are dark in blue wavelengths and show a plume-like morphology. These dark projections received considerable interest since in 1995 the Galileo probe descended through one of them. This was the first and only available direct measurement of Jovian atmospheric properties below the tropospheric clouds. These features are called *hot spots* since they appear warmer in 5 µm infrared images relative to the rest of the planet. This wavelength senses the heat escaping from the deeper atmospheric levels (~2 bar or more, that is, below the upper ammonia clouds),

TABLE 8.4
Properties of Representative Equatorial Planetary Waves

Planet	Wave	n (Zonal)	Amplitude	c_x (m s^{-1})	$c_x - \bar{u}$ (ms^{-1})
Earth	Kelvin[a]	1–2	$u' = 8$ m s^{-1} $v' = 0$ $T' = 2$–3 K	+25	+50
Earth	Rossby-gravity[b]	4	$u' = 2$–3 m s^{-1} $v' = 2$–3 m s^{-1} $T' = 1$ K	−23	−30
Venus	Y-Ψ[c]	1	Cloud field	110–120	+5 to +10
Jupiter	Hot spots[d] Plumes	10–11	Cloud field	100–115	−40 to −70
Jupiter	NEB[e]	11	Cloud field	−5	−15
Jupiter	Temperature[f]	9–11	$T' = 0.5$ K	5–25	−70 to −90
Jupiter	Temperature[g]	1–3	$T' = 1$ K	10	−90
Jupiter	Temperature[h]	6–11	$T' = 1$–2 K	−5	−95
Saturn	Temperature[i]	1–4	$T' = 0.5$ K	0	−100

[a] Period 15 days. Meridional scale 1300–1700 km. Vertical wavelength = 6–10 km.

[b] Period 4–5 days. Meridional scale 1000–1500 km. Vertical wavelength = 4–8 km.

Sources of (a) and (b): Data from Andrews, D.G. et al., *Middle Atmospheric Dynamics*, Academic Press, San Diego, CA, 1987.

[c] Period 4 days. Altitude $z = 65$–70 km, Y-Ψ shape in UV-images (equatorially oriented). Vertical wavelength 6–11 km. Assigned as a Kelvin wave or to a mixture of Kelvin and Rossby modes.

Source of (c): Data from Schubert, G., in *Venus*, L. Colin et al. (eds.), University of Arizona Press, Tucson, AZ, 1983; Gierasch, P.J. et al., in *Venus II: Geology, Geophysics, Atmosphere, and Solar Wind Environment*, S.W. Bougher et al. (eds.), University of Arizona Press, Tucson, AZ, 1997.

[d] Latitude = 7°N. Altitude span: pressure = 0.5–2 bar (reference level). Meridional span = 2000–3000 km. Assigned as an equatorial Rossby wave.

[e] Latitude = 14°N. Altitude: pressure level = 0.2–0.5 bar. Assigned as an equatorial Rossby wave.

[f] Latitude = ±20°. Altitude: Pressure level < 0.5 bar. Assigned as an equatorial Rossby wave.

[g] Latitude = ±40°. Altitude: Pressure level 0.15–0.27 bar. Equatorial Rossby or Rossby-gravity mode.

[h] Latitude = ±20°. Altitude: Pressure level 0.02–0.25 bar. Equatorial Rossby or Rossby-gravity mode.

Sources of (d) through (h): Data taken from Ingersoll, A.P. et al., in *Jupiter: The Planet, Satellites and Magnetosphere*, F. Bagenal et al. (eds.), Cambridge University Press, New York, 2004.

[i] Latitude = 20°N–40°N. Altitude: Pressure level 0.25 bar. Equatorial Rossby mode.

Source of (i): Data taken from Del Genio, A.D. et al., in *Saturn after Cassini-Huygens*, M. Dougherty et al. (eds.), Springer-Verlag, Dordrecht, the Netherlands, 2009.

FIGURE 8.14   Venus tilted Y-wave (also named C-wave or ψ-wave) as observed in the upper cloud at ultraviolet wavelengths. Left: the open arms of the wave in an image obtained by the Galileo spacecraft in 1990. Center and right: the vertex and arms of the wave in images obtained by the Pioneer-Venus orbiter in 1979. The wave orientation is indicated. (Courtesy of NASA-JPL, Los Angeles, CA.)

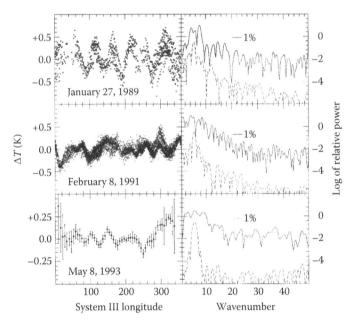

FIGURE 8.15   Jupiter's equatorial waves detected as oscillations in the temperature field (left). At right is a power spectra analysis of the wave. (From Deming, T. et al., *Icarus*, 126, 301, 1997. With permission.)

indicating that the hot spots show low opacity to the thermal radiation at this wavelength (Figure 8.16A). At visible wavelengths (0.3–1 μm), we see the features in the upper cloud layer from the reflected sunlight at a pressure level ~0.5 bar (Figure 8.16B). A comparison of their characteristics and the atmospheric properties at these two levels indicates that a Rossby wave mechanism is probably responsible for their formation (Problem 8.16). Another regularly spaced pattern of bright and dark areas has been observed at 22°N in the UV-reflected sunlight by Jupiter's tropospheric haze. Its analysis suggests that it could also be another manifestation of an equatorial Rossby wave.

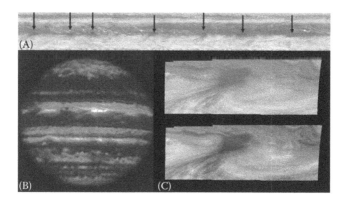

FIGURE 8.16 Images of Jupiter's dark projections (and "hot spots") at latitude 7°N. (A) Cassini-ISS image in December 2000 showing their periodic zonal structure (dark projections marked by arrows). (Courtesy of NASA-JPL, Los Angeles, CA.) (B) Jupiter image at 5 µm showing the intense emission from the hot spot where the Galileo probe penetrated in 1995 (encircled). (From G. Orton, NASA-IRTF-JPL, Los Angeles, CA.) (C) Galileo high resolution images of a dark projection (hot spot) (Courtesy of NASA-JPL, Los Angeles, CA.).

8.6 THERMAL TIDES: DYNAMICS

We have already presented in Sections 4.3.5 and 4.4 how the atmospheric temperature varies in response to the periodic heating imposed by the solar radiation when deposited in the surface or at different levels in the atmosphere. However, the day-night contrast in the solar heating and the coupling between the motion, continuity, and thermodynamic equations (see, e.g., 8.40a through e) rises in the atmosphere not only a temperature cycle but more general thermal tides, which are global-scale waves that manifest in the pressure field and in the wind components and have a dependence on local solar time. For example, thermal tides transfer mass between heated and cooled regions of the atmosphere, resulting in low atmospheric pressure where the atmosphere is warmed and high atmospheric pressure where it is cooled. They therefore play a significant role in the atmospheric dynamics, in particular in the terrestrial planets.

Migrating tides are those that follow the motion of the Sun in the sky and have a constant phase with respect to the solar heating (they are "Sun-synchronous"). The most important examples of migrating tides are the diurnal and semidiurnal modes (see Section 4.4). The classical tidal theory that considers linear disturbances (about a stationary background state) to the motion and thermodynamics equations predicts the effects of the thermal tide in the pressure and wind velocity as a function of latitude and altitude (being periodic in longitude). Details of the thermal tidal theory with application to Earth's observations can be found in Andrews et al. (1987) and in Lindzen (1990). Here, we just outline the basic concepts.

Essentially, the thermal tides in the equatorial areas are described by a set of equations like (8.40) with the appropriate boundary conditions but incorporating a specific heating term in the right-hand side of (8.40e). This heating term, as shown in Section 4.3.3, has the form $Q'/\rho C_p$ where Q' denotes the deviation of the actual heating Q

from the zonal mean net heating $\bar{Q}$ (assumed to be balanced by the infrared cooling). Using spherical coordinates, we have that $Q'(\phi+\Omega_p t, \varphi, z)$ is assumed to be periodic in longitude. This allows us to expand the heating function in Fourier harmonics

$$Q' = \text{Re} \sum_{n=1}^{\infty} Q^{(n)}(\varphi, z) \exp\left[in(\phi + \Omega_p t) \right] \tag{8.61}$$

with $n=1$ corresponding to the diurnal mode and $n=2$ corresponding to the semidiurnal mode. In tidal theory, the terms $Q^{(n)}(\varphi, z)$ are further expanded as a series of products of the *Hough functions* $\Theta_m(\varphi)$ that depend only on latitude and a heating function $Q_m(z)$ that depends only on altitude. For example, the $n=1$ diurnal forcing term $Q_m^{(1)}(\varphi, z)$ is written as

$$Q^{(1)}(\varphi, z) = \sum_m Q_m^{(1)}(z) \Theta_m^{(1)}(\sin \varphi) \tag{8.62}$$

The velocity field corresponding to the diurnal migrating wave has the form $\left\{ \left[u^{(1)}, v^{(1)}, w^{(1)} \right](\varphi, z) \right\} \exp\left[i(\phi + \Omega_p t) \right]$ where the amplitudes $u^{(1)}$ and $v^{(1)}$ can be expressed in terms of $\Phi^{(1)}$ after solving equations (8.40a through c) and then can be expressed in terms of $Q^{(1)}$ (using 8.40e but including the heating term on the right-hand side) and the vertical velocity amplitude $w^{(1)}$ is written as

$$w^{(1)}(\varphi, z) = \sum_m e^{z/2H} \chi_m^{(1)}(z) \Theta_m^{(1)}(\sin \varphi) \tag{8.63}$$

Here, $\chi_m^{(1)}(z)$ obeys a vertical structure equation similar to (8.59) but it is inhomogeneous to include the heating forcing term on the right-hand side.

Observations of solar thermal tides in planetary atmospheres show a combination of periodicities in longitude and time (solar day), so according to (8.61), it is convenient to represent them (for a given field magnitude X) as a series sum of the type

$$X(\phi, t) \approx \sum X_{n,j} \cos\left(n\Omega_p t + s\phi - \psi_{n,j} \right) \tag{8.64}$$

where
 n (=1, 2, 3,...) denotes again the subharmonic of a solar day
 j (= –3, –2, –1, 0, 1, 2, 3,...) is the zonal wavenumber
 the amplitudes $X_{n,j}$ and phases $\psi_{n,j}$ are, as indicated above, functions of altitude and latitude

The phase speed is

$$c_{n,j} = -\frac{n\Omega_p}{j} \tag{8.65}$$

n positive implies eastward propagation ($j<0$) and n negative implies westward propagation ($j>0$). It is convenient to express (8.64) in terms of the local time (LT) $t_{LT}=t+\phi/\Omega_p$ as

$$X(\phi,t) \approx \sum X_{n,j}\cos\left[n\Omega_p t_{LT} +(j-n)\phi -\psi_{n,j}\right] \qquad (8.66)$$

Migrating tides have $j=n$ so they show no longitude dependence in the solar reference frame and the phase speed from (8.65) is $-\Omega_p$, propagating westward at the same speed as the motion of the Sun in the sky in this frame. Thus, an observed s wave variation may be due to a combination of a stationary wave $X_{s,0}$ mode and a series of nonmigrating tides $X_{n,j}$ such that $j-n=\pm s$ (Problem 8.17).

There are abundant records of the Earth's surface pressure manifestation of the semidiurnal tide component and in the wind observations in the lower stratosphere up to an altitude of 30 km. Above 30 km and up to an altitude of 110 km, a number of techniques have provided wind data for the amplitudes and vertical wavelengths of the diurnal and semidiurnal modes. Typically, between the altitudes of 30 and 60 km, the amplitude (wind velocity) of the diurnal thermal tide is ~3–10 m s^{-1}. The semidiurnal tide in the Earth's mesosphere–thermosphere (80–110 km) reaches wind speed amplitudes of ~20–30 m s^{-1}. The corresponding temperature fluctuations are ~1–10 K.

The amplitude of the thermal tides in the Martian atmosphere is very large compared with the case of the terrestrial atmosphere due to the strong thermal forcing of the Mars surface and near-infrared CO_2 absorption and dusty atmosphere (Section 4.4). This results in global pressure, temperature (thus density), and wind oscillations that propagate vertically as they grow exponential, breaking and dissipating at altitudes above ~35 km (see Figures 4.12 and 4.13). The effect of the thermal tides on the mean zonal and meridional circulation is, therefore, very important in driving motions in the Martian atmosphere. In the thermosphere (see Section 6.4), the extreme UV absorption at altitudes of 110–160 km drives thermal tides that have a profound effect on the zonal variations of the atmospheric density in a level where spacecraft aerobraking occurs. Diurnal and semidiurnal modes combined with eastward Kelvin waves are involved in Martian solar tides. The diurnal modes with spatial wavenumbers 1 and 2 have long vertical wavelengths, coupling the lower and upper atmosphere. Semidiurnal tides have temperature and wind amplitudes that maximize in the winter hemisphere and are typically 10–20 K and 10–20 m s^{-1} at 50 km and a factor of 10 for winds above 150 km. The interactions between sun-fixed tides and topography and also with the albedo and thermal inertia of surface markings, local aerosol, and cloud features produce nonmigrating tides.

On Venus, the solar absorption occurs fundamentally in the cloud layers between ~45–70 km and the extreme UV radiation heats the mesosphere at an altitude of ~100–150 km (see Chapters 4 and 6; Figure 4.10). There is evidence that solar tides are involved in the deceleration of stratospheric motions and in the acceleration of the upper part of the cloud layer from morning to afternoon with speed amplitudes of ~10–15 m s^{-1}. Some studies have proposed that solar tides excited by the sunlight absorption in the cloud layer play an important role in Venus superrotation at this level (see Section 7.5.2). The joint excitation of propagating and trapped diurnal tides can lead to a horizontal distribution of momentum within the heating region. Migrating diurnal and semidiurnal tides are eastward-propagating on Venus, and thus can be expected to produce westward accelerations in the heated regions and eastward accelerations (deceleration of the mean flow) at higher levels where propagating waves dissipate.

8.7 QBO-QQO AND SAO OSCILLATIONS

Large-scale and long-term oscillations have been observed in the properties of the equatorial stratospheres of the Earth (the *quasi-biennial oscillation* [QBO]), on Mars (the *semi-annual oscillation* [SAO]), on Jupiter (the *quasi-quadrennial oscillation* [QQO]), and on Saturn (SAO).

Obviously, the better studied phenomenon is the Earth-QBO that occurs between altitudes of 16 and 50 km and is best represented by the zonal wind field intensity and direction in height-time sections at the equator (Figure 8.17). The QBO consists primarily on a downward propagation of zonally symmetric alternating easterly and westerly wind regimes with periods in the range of 24–30 months. The successive wind regimes develop above 30 km, propagating downward at 1 km month^{-1}, rapidly

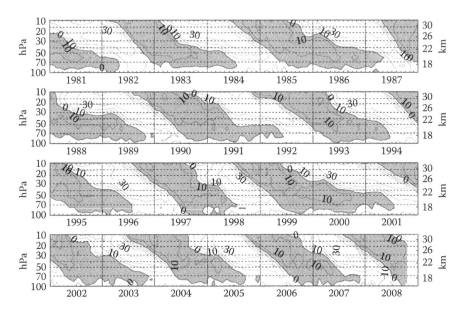

FIGURE 8.17 Time–height section of the monthly mean zonal wind component (m s^{-1}) resulting from Earth's QBO. (From Naujokat, B., in: The quasi-biennial-oscillation (QBO) data series, Institute of Meteorology, Frei Universitat, Berlin, Germany, available at: http://www.geo.fuberlin.de/en/met/ag/strat/produkte/qbo/.)

attenuating below 23 km and dissipating at the tropical tropopause. The oscillation is symmetric about the equator with a Gaussian half-width of about 12° being the amplitude of the easterly phase (about 20–30 m s^{-1}) twice as strong as that of the westerly phase (Problem 8.18).

The QBO influences the Earth's weather significantly affecting the monsoon precipitation and probably the strength of Atlantic hurricanes. It also affects the variability in the mesosphere (at 85 km) and the breakdown of the wintertime stratospheric polar vortices. It intervenes in the stratospheric flow not only at the tropics but from pole to pole by modulating the effects of extra-tropical waves. The QBO also plays a role in the distribution of chemical constituents, such as ozone, water vapor, and methane, by the circulation changes it induces.

The QBO oscillations are driven by the combined action of vertically propagating equatorially trapped Kelvin waves (that provide the westerly momentum) and Rossby-gravity waves (that provide the easterly momentum), together with intervening gravity and inertia-gravity waves. These waves are originated by the convective activity in the tropical atmosphere transferring momentum to the zonal-mean flow upward from the troposphere.

An SAO of the mean zonal wind in Earth's tropics takes place in the upper stratosphere and mesosphere. Although the Sun crosses the equator twice a year, the alternating westerly and easterly winds are driven dynamically. Easterly winds over the equator are driven by the axisymmetric cross-equatorial transport associated with the thermally induced circulation and Rossby waves, whereas the westerly winds are driven by nonaxisymmetric effects (Kelvin and gravity waves that also contribute to the easterly phase).

SAO oscillations have been recently reported in the equatorial atmosphere of Mars based on the difference between day- and night-time atmospheric temperatures as measured from the spacecraft Mars Global Surveyor. An SAO with peak values in the temperatures of 6–8 K occurs at altitudes of 25–50 km ($P \sim 0.5$–0.05 mbar). The use of a Martian GCM indicates that a zonal wind oscillation is associated with the SAO in the Mars tropics. The amplitude depends on the model properties, in particular on the amount of dust in the atmosphere, but peak values of ~20–30 m s^{-1} are expected. The simulations indicate that the thermal tides previously presented, and the quasi-stationary planetary waves induced by Martian topography, strongly contribute to accelerate these winds.

The Jovian QQO has been detected as a 4 year quasi-periodic fluctuation in the temperature maps (amplitude ±2 K) in the stratosphere between 10 and 20 mbar pressure levels and ±14° latitudes (Figure 8.18, see also Figure 4.14). It is predicted that this temperature oscillation should be accompanied by an oscillation of the mean zonal wind (see Problem 8.18).

Saturn's SAO has half Saturn's orbital period ~14.8 year and has been reported from long-term observations and from high-resolution temperature maps. The oscillating equatorial temperature (latitudes ±4° and ±15°) at stratospheric levels (0.01–10 mbar) has an amplitude variable with altitude from ±2 to ±6 K (Figure 8.19).

Both the Jovian QQO and Saturn's SAO have been proposed to be driven by the momentum injection due to vertical propagating equatorial waves, similar to the Earth QBO.

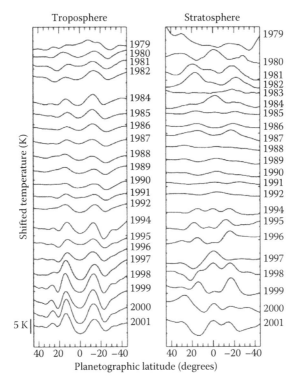

FIGURE 8.18 Oscillations in Jupiter's measured stratospheric and tropospheric temperatures as a function of latitude and time. See also Figure 4.14 where the absolute temperature values for the stratosphere are given. (From Simon-Miller, A.A. et al., *Icarus*, 180, 98, 2006. With permission.)

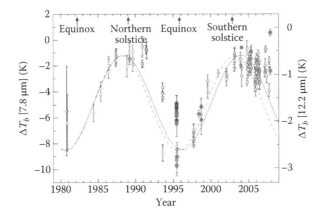

FIGURE 8.19 Saturn's Semi-Annual Oscillation (SAO) is shown from the difference of zonally averaged brightness temperature at latitudes 3.6° and 15.5° north and south. The data include emissions from methane at 7.8 μm and ethane at 12.2 μm. (From Orton, G. et al., *Nature*, 453, 196, 2008. With permission.)

PROBLEMS

8.1 Find an expression for the relationship between phase speed, group velocity (in x-, y-, z-directions), and wavenumbers. Find the relationship between phase speed and group velocity in a nondispersive medium.

Solution

$$c_{gx} = c_w \frac{k}{\left|\vec{K}\right|} + \left|\vec{K}\right| \frac{\partial c_w}{\partial k}; \quad c_{gy} = c_w \frac{l}{\left|\vec{K}\right|} + \left|\vec{K}\right| \frac{\partial c_w}{\partial l}; \quad c_{gz} = c_w \frac{m}{\left|\vec{K}\right|} + \left|\vec{K}\right| \frac{\partial c_w}{\partial m}$$

$$c_{gx} = c_w \frac{k}{\left|\vec{K}\right|}; \quad c_{gy} = c_w \frac{l}{\left|\vec{K}\right|}; \quad c_{gz} = c_w \frac{m}{\left|\vec{K}\right|}; \quad \left|\vec{c}_g\right| = c_w$$

8.2 Thunder on Earth's atmosphere produces a sound wave with an intensity of 80 dB. (a) What is the pressure amplitude considering the conditions at the surface? (b) Calculate the pressure amplitude of a sound wave on the Mars atmosphere corresponding to a sound level of 60 dB being detected by a microphone on a Martian probe.

Solution

(a) $P_0 = 2.88 \times 10^{-6}$ bar
(b) $P_0 = 2.6 \times 10^{-8}$ bar

8.3 Stationary gravity waves form on Mars when air flows with a speed of $\bar{u} = 15\,\text{m s}^{-1}$ over mountain ridges. If $N_B = 5 \times 10^{-2}\,\text{s}^{-1}$, find the limiting zonal separation of the ridges for vertical propagation of the gravity waves and the corresponding value of the perturbed vertical velocity if the ridges altitudes are 50 m.
Solution: $L_x = 1887\,\text{m}$; $w' = -2.5 \sin(kx + mz)$ (m s^{-1}).

8.4 Compare the values for the frequency of the following types of waves in an isothermal layer ($T = 160\,\text{K}$) at rest relative to System III in the atmosphere of Jupiter: (1) buoyancy (Brunt–Väisälä) frequency, (2) Coriolis frequency (latitude 45°), (3) acoustic wave frequency for a wavelength of 1 m, and (4) gravity wave frequency for zonal wavelengths of 100 km and vertical wavelengths of 10 km. (5) Calculate the zonal and vertical phase speeds for this last wave.

Solution

(1) $1.8 \times 10^{-2}\,\text{s}^{-1}$
(2) $2.5 \times 10^{-4}\,\text{s}^{-1}$
(3) $916\,\text{s}^{-1}$
(4) $\omega = 1.9 \times 10^{-3}\,\text{s}^{-1}$
(5) $c_x = 30.2\,\text{m s}^{-1}$, $c_z = 3\,\text{m s}^{-1}$

8.5 In an isothermal atmosphere, the linearized form of the system of Equations 8.12 that has wave solutions with an altitude dependent amplitude of the form $\exp(z/2H)\exp[i\,(kx+mz-\omega t)]$ gives the following *acoustic-gravity wave* dispersion relationship, see e.g., Salby (1996)

$$m^2 = k^2\left(\frac{N_B^2}{\omega^2}-1\right)+\frac{\omega^2-\omega_c^2}{c_s^2}$$

where $\omega_c=c_s/2H$ is the *acoustic cut-off frequency* that is always $\omega_c>N_B$. Consider the following two limiting cases: (a) $\omega^2 \gg N_B^2$ and (b) $\omega^2 \ll v_c^2$. Simplify the dispersion relationship for each case and then respond to (1) what kind of wave corresponds to each case? (2) Find the wave frequency for vertical propagation in both limiting cases for long-wavelengths ($k \to 0$) and for short wavelengths ($k \to \infty$).

Solution

(1) (a) High frequency acoustic waves modified by stratification and (b) high frequency gravity waves
(2) For $k \to 0$: (a) $\omega>\omega_c$, (b) $\omega \to 0$; for $k \to \infty$: (a) $\omega \to \infty$, (b) $\omega=N_B$.

8.6 Apply the results of Problem 8.5 to the vertical propagation of gravity waves in the middle cloud layer of the atmosphere of Venus ($z=56\,$km). Using the data from Table 8.2, find the vertical wavelength of upward propagating gravity waves if their zonal wavelength is $L_x=150\,$km and the Doppler shifted phase speed is $c_w - \bar{u}=10\,\text{ms}^{-1}$ assuming that $N_B=1.7\times 10^{-2}\,\text{s}^{-1}$ and $H=6\,$km. Find their frequency v and the acoustic cut-off frequency v_c and show that $|f| < |\omega| <N_B<\omega_c$ for $\langle u\rangle=100\,\text{m s}^{-1}$, latitude 30° and $T=280\,$K.

Solution

$$L_z = 3.7\text{ km}, \quad f = 3\times 10^{-7}\text{ s}^{-1}, \quad \omega=\bar{u}k+\frac{N_B k}{\sqrt{\left(k^2+m^2+(4H^2)^{-1}\right)}}=3.3\times 10^{-3}\text{ s}^{-1},$$

$$\omega_c = 2.2\times 10^{-2}\text{ s}^{-1}$$

8.7 Compare the value of the β-parameter (take it as Ω/R_p) as a diagnostic for the presence of Rossby wave formation in the atmospheres of Venus, Earth, Mars, Titan, Jupiter, Saturn, Uranus, Neptune, and the extrasolar planet HD 209458 (assuming spin-orbit synchronism). Make use of the data in the tables in Chapter 1.

Solution
Venus: -4.95×10^{-14} m^{-1} s^{-1} (rotation direction is from East to West); Earth: 1.14×10^{-11} m^{-1} s^{-1}; Mars: 2.09×10^{-11} m^{-1} s^{-1}; Titan: 1.77×10^{-12} m^{-1} s^{-1};

Jupiter: 2.46×10^{-12} m^{-1} s^{-1}; Saturn: 2.72×10^{-12} m^{-1} s^{-1}; Uranus: 3.96×10^{-12} m^{-1} s^{-1}; Neptune: 4.37×10^{-12} m^{-1} s^{-1}; and HD 209458: 2.19×10^{-13} m^{-1} s^{-1}.

8.8 Compare the zonal phase speed relative to the mean flow for barotropic Rossby waves with a wavelength of $L_x = L_y = 6000$ km in the atmospheres of Venus, Earth, and Saturn at 45° latitude. Calculate the oscillation period for this wave in the case of Earth and Saturn if the mean zonal flow is eastward in both planets and has a mean velocity of 10 m s^{-1} on Earth and 150 m s^{-1} on Saturn.

Solution

$c_x - \bar{u} = 0.035$ ms^{-1} (Venus), -8.1 m s^{-1} (Earth), -1.9 m s^{-1} (Saturn); $\tau = 6.1$ days (Earth), 1.87 h (Saturn).

8.9 Demonstrate the relationship (8.40).

8.10 A simple model of topographically forced Rossby waves assumes that the atmosphere is homogeneous but with variable depth such that the flow obeys the barotropic potential vorticity equation. The upper boundary is taken at a fixed height H and the lower boundary shows a variable height $h(x, y) \ll H$. For quasi-geostrophic conditions, using the β-plane approximation and after linearization, the flow obeys a modified version of Equation 8.29 that includes the forcing term and is given by

$$\left(\frac{\partial}{\partial t} + \bar{u} \frac{\partial}{\partial x} \right) \left(\frac{\partial v'}{\partial x} - \frac{\partial u'}{\partial y} \right) + \beta v' = -\frac{f_0}{H} \bar{u} \frac{\partial h}{\partial x} \qquad \text{(P8.10)}$$

Consider the presence of an obstacle to this flow that has a sinusoidal shape represented by the real part of the function $h(x,y) = h_0 e^{ikx} \cos ly$ and assume then that the flow is represented by the real part of the streamline function $\psi(x,y) = \psi_0 e^{ikx} \cos ly$. Find the streamline function amplitude ψ_0 for steady-state conditions. Then, consider the case for the Earth's northern mid-latitude 45° and calculate the value of ψ_0 and the peak geopotential height $\Phi_{G0} = f_0 \psi_0 / g$ for Rossby waves forced by a periodic mountain chain that has $h_0 = 2$ km, $H = 8$ km (the atmospheric scale height), a meridional wavelength of 35°, and a zonal wavelength of 180° for a flow with $\bar{u} = 20$ m s^{-1}.

Solution

$$\psi_0 = \frac{f_0 h_0}{H} \left[\frac{1}{(l^2 + k^2) - \beta / \bar{u}} \right], \quad \psi_0 = 8.6 \times 10^8 \text{ m}^2\text{s}^{-1}, \quad \Phi_{G0} = 8260 \text{ m}.$$

8.11 The solution of the previous exercise indicates that a singularity can occur in the expression of ψ_0, and for the case studied, the geopotential height is large in excess. To model a realistic situation and remove the singularity, a linear damping drag term of the form $r \, (\partial v' / \partial x - \partial u' / \partial y)$ can be added to the left-hand side of Equation P8.10, with being the inverse of the spin-down time (Section 7.4.2). Find the steady-state amplitude ψ_0 for the modified equation and its value and that of the peak geopotential height $\Phi_{G0}=f_0\psi_0/g$ for the same conditions as the previous exercise if $\tau_E=5$ days.

Solution

$$\psi_0 = \frac{f_0 h_0}{H}\left[\frac{1}{K^2 - \beta/\bar{u} - \left(irK^2/\bar{u}k\right)}\right], \quad K^2 = k^2 + l^2, \quad \psi_0 = 1.48\times10^7\,\text{m}^2\text{s}^{-1},$$

$$\Phi_{G0} = 148\ \text{m}.$$

8.12 The Saturn hexagon wave has been proposed to be a Rossby wave. The hexagonal wave is the undulation with zonal wavenumber $n=6$ of an eastward Gaussian jet-stream that can be represented by $u(y)=u_0\exp(-by^2/2u_0)$, where $y=R_p(\varphi-\varphi_0)$ with $\varphi_0=80°$ and $R_p=60,330\,\text{km}$ being Saturn's radius. Take the peak velocity $u_0=100\,\text{m s}^{-1}$ and the latitudinal curvature $b=6\times10^{-11}\,\text{m}^{-1}\,\text{s}^{-1}$, then the associated e-folding latitudinal width is $L_e=(2u_0/b)^{1/2}\sim1800\,\text{km}$. (1) Show that geostrophic conditions prevail (calculate Rossby number). (2) Neglecting the vertical wind shear of the zonal wind, compare the value of the curvature terms in Equation 8.43 averaging for the curvature of the zonal flow. (3) Then, with the wave being stationary relative to the mean flow (see Table 8.3), calculate the minimum values for this to occur in the meridional wavelength (that is, the minimum times the length L_e) and independently in the vertical wavelength (that is, the minimum value for N_B, assumed constant) if $H=38\,\text{km}$ (take $m^2\ll1/4H^2$).

Solution

(1) $Ro = u/fL_e = 0.2$
(2) $\beta = 1.1\times10^{-12}\,\text{m}^{-1}\,\text{s}^{-1}; \langle-d^2u/dy^2\rangle_e = b/e = 2.2\times10^{-11}\,\text{m}^{-1}\text{s}^{-1}$
(3) $L_y > 1.3\times2\pi L_e$, $N > 10^{-2}\,\text{s}^{-1}$

8.13 Stratospheric Rossby waves can be forced in the tropopause of the giant planets by geopotential disturbances induces by large-scale vortices. Consider the case of the Great Red Spot of Jupiter placed at 23°S latitude that induces a geopotential Gaussian disturbance with the form $\Phi_G(x,y,z=0)=\Phi_{G0}\exp[-(x^2+y^2)/a^2]$ with $a=25,000\,\text{km}$ being a representative value of its horizontal scale. Find the critical velocity for vertical propagation into the stratosphere assuming that $N=10^{-2}\,\text{s}^{-1}$ and $H=20\,\text{km}$. What do you think about the vertical propagation of the disturbance?

Solution

$$u_c \simeq \frac{\beta}{k^2+l^2} \simeq \frac{\beta}{a^{-2}} = 284\ \text{m s}^{-1}$$

Since $\bar{u} = 10\,\text{m s}^{-1}$ at the latitude of the GRS, $\bar{u} \ll u_c$ and vertical propagation can occur.

8.14 Obtain the horizontal and vertical wave structure for Equations 8.43 and 8.57 starting from Equation 8.42 and using Equation 8.43.

8.15 A Venus Y-cloud-shaped feature observed in UV has been proposed to be an equatorial Kelvin wave whose phase speed is $c_x - \bar{u} = +15\,\text{m s}^{-1}$. Find (1) the model equivalent depth for such wave speeds; (2) the meridional e-folding scale taking as a reference rotation system the atmosphere at an altitude level of 65 km that is superrotating with a period of 4.66 days; (3) the wave vertical wavelength (take $N_0 = 1.6 \times 10^{-2}\,\text{s}^{-1}$, $H = 5.2\,\text{km}$); and (4) neglecting the vertical wind shear, say if it could propagate vertically.

Solution

(1) $h = 0.03\,\text{km}$
(2) $y_K = 24{,}240\,\text{km}$
(3) $L_z = 6.44\,\text{km}$
(4) Yes since $4H^2 N_B^2/g = 3.12\ \text{km}\,(> h)$.

8.16 Jupiter's "hot spots" and plumes have been proposed to be the manifestation of an equatorial Rossby wave. Assuming for this wave (see Table 8.4) that $n = 11$, $c_x = 105\,\text{m s}^{-1}$, $\bar{u} = 140\,\text{m s}^{-1}$ (background flow velocity at 2 bar pressure level), and that $Ri \gg 1$, find: (1) the model equivalent depth for the gravest meridional j-mode; (2) the vertical wavelength (take $N_0 = 5 \times 10^{-3}\,\text{s}^{-1}$, $H = 30\,\text{km}$); (3) Could this wave propagate vertically?

Solution

(1) $h = 0.45\,\text{km}$
(2) $L_z = 4.28\,\text{km}$
(3) Yes, since $m^2 > 0$

8.17 Give the zonal and temporal thermal tidal possible diurnal and semidiurnal wavenumber components for observed modes $s = 2$ and 3 waves in the Martian atmosphere and identify their propagation direction.

Solution

For $s = 2$, we have for the diurnal wavenumber: (a) $n = 1$, $j = 3$ (westward), (b) $n = 1$, $j = -1$ (eastward) and for the semidiurnal wavenumber: (c) $n = 2$, $j = 0$, (d) $n = 2$, $j = 4$ (westward). For $s = 3$ we have for the diurnal wavenumber: (e) $n = 1$, $j = 4$ (westward), (f) $n = 1$, $j = -2$ (eastward) and for the semidiurnal wavenumber: (g) $n = 2$, $j = -1$ (eastward), (h) $n = 2$, $j = 5$ (westward).

8.18 The QBO being a symmetric oscillation about the equator causes very small mean meridional and vertical motions, so the thermal wind relationship can be applied. Using the equatorial β-plane approach, find an expression for the vertical wind shear using l'Hôpital rule for the meridional temperature gradient and the fact that equatorial symmetry implies that $\partial \overline{T}/\partial y \to 0$ when $y \to 0$. Then, from the measurements at a given QBO phase of $\partial \overline{u}/\partial z = 5 \, \text{m s}^{-1} \, \text{km}^{-1}$ and a meridional scale for the oscillation of 1200 km, calculate the temperature perturbation at the equator.

Solution

$$\frac{\partial \overline{u}}{\partial z} = -\frac{R_g^*}{H\beta} \frac{\partial^2 \overline{T}}{\partial y^2}, \quad \Delta T = \pm 3 \, \text{K}$$

9 Atmospheric Dynamics-II: Instability

Perhaps the most distinctive visual aspect of planetary atmospheres is the presence of features drawn in the cloud field. Many forms with different spatial and temporal scales of evolution are seen in images of Venus, Earth, Mars, the giant planets, and the satellite Titan. They emerge as a result of a localized instability, growing and evolving until they finally interact and mix with other features or dissipate within the atmospheric flow. Different types of measurements distinguish them as deviations of the physical or chemical atmospheric properties, relative to the mean values (e.g., pressure, temperature, geopotential, composition, and velocity). In the cloud field, they show in different ways, although in a simple scheme they can be grouped into two classes. One type has irregular textures and growths and evolves rapidly in time. The other type manifests as vortices with atmospheric parcels circulating around a central region. The first group is typically the manifestation of the convective instability with associated intense vertical motions. The second group responds to a variety of hydrodynamic instability mechanisms with dominant horizontal development. In this chapter, we first classify the main types of instabilities, and then we discuss their physical fundamentals and the properties of the different kinds of features that are observed in the planetary atmospheres.

9.1 SCALES OF MOTION AND DIMENSIONLESS NUMBERS

The two basic references for the temporal scales of disturbances and vortices evolution are the planet orbital period (the year) and the planet sidereal rotation period (or equivalently, the inverse of the Coriolis parameter $1/f$, Equation 7.15). Other time-scales important for the characterization of *meteorological phenomena* are the radiative time constant τ_{rad} (Equation 4.52 and Table 4.1), the Brunt–Väisälä frequency N_B (Equation 4.74), the diffusion scales τ_D and τ_{eddy} (Sections 6.5.1 and 6.5.2), and the friction decay scales (spin-down time τ_E, Section 7.4.1.2, and viscous dissipation τ_f, Equation 7.85). These temporal scales cover an ample spectrum of the disturbances evolution from years to seconds.

With reference to horizontal motions, the spatial scale takes the planetary radius R_p, and to the vertical motions, the spatial scale takes the atmospheric scale height H (Section 4.1.1). For example, on Earth the atmospheric disturbances are accordingly classified as *planetary* (scale ~20,000 km, few times R_p), *synoptic* (~1,000–5,000 km, sizes $\leq R_p$), and *mesoscale* (2–200 km, size ~H). We can extend this down to the *microscale* (sizes ~1 mm–200 m), to end finally at the molecular level. This terminology is usually employed in the other planets for their R_p and H. Two other important spatial scales employed to characterize atmospheric dynamics have already been introduced: the Rossby deformation radius L_D (Equation 7.95) and the Rhines scale L_β (Equation 7.91). The former is typically related to the scale of motion of baroclinic disturbances (see Section 9.3), whereas the latter is a related quasigeostrophic phenomenon that takes place in rapidly rotating planets.

It is customary in atmospheric dynamics to use dimensionless numbers to assess the relative importance of the different mechanisms that intervene in the formation of meteorological features. We have previously introduced the Rayleigh number Ra (Equations 1.131 and 7.114c) to characterize the vertical stability of the atmosphere against convective motions, the Reynolds number Re (e.g., Equations 7.66 and 7.83) indicative of the importance of laminar versus turbulent motions, the Rossby number Ro (Equation 7.13) to characterize the importance of rotation effects (Coriolis force), and the Prandtl and Ekman numbers (Equation 7.114a and b, respectively), which measure the effects of dissipation through viscosity and thermal diffusion. In laboratory experiments, both Ra and Re have been found to have critical values above which instabilities develop (see, e.g., Sections 1.5.4 and 7.4). In atmospheric dynamics, it is also customary to use the *Burger* number

$$Bu = \left(\frac{N_B H}{fL}\right)^2 = \left(\frac{L_D}{L}\right)^2 \tag{9.1}$$

that gives a measure of the importance of the vertical stratification against the planetary rotation effects for a characteristic horizontal length of the flow L. For example, on Earth and the giant planets, it is typically found that $Bu \leq 1$ at the tropopause level.

A very useful number for characterizing the dominant modes of dynamical instabilities in an atmospheric flow is the *Richardson number*:

$$Ri = \frac{N_B^2}{\left(\partial u/\partial z\right)^2 + \left(\partial v/\partial z\right)^2} = \frac{(g/\theta)\left(\partial \theta/\partial z\right)}{\left(\partial u/\partial z\right)^2 + \left(\partial v/\partial z\right)^2} \tag{9.2}$$

It gives a measure of the importance of the vertical stability of the atmosphere against the vertical shears of the wind. In other words, it also gives a measure of the intensity of the buoyancy force (or available potential energy) against the inertial force (or available mechanical energy) in the atmospheric flow. Instead of the potential temperature θ in (9.2), other definitions of Ri use the virtual potential temperature θ_v when a wet ambient is considered (see the corresponding definition for N_B, Equation 5.43, for this case). In the giant planets where the zonal motions dominate, Equation 9.2 simplifies to $Ri \approx g(\partial\theta/\partial z)/\theta(\partial u/\partial z)^2$.

According to the value of the Richardson number, the following useful classification scheme of the atmospheric instabilities can be introduced.

1. *Vertical instability* $(Ri<0)$
   The temperature decreases with height more rapidly than the adiabatic lapse rate. Then, $(\partial\theta/\partial z)<0$ and *free convection* develops (see Sections 4.5.3 and

5.2.6). The disturbances have a characteristic size about the scale height H. On the planetary boundary layer on Earth and Mars, $Ri \sim -2$ and turbulent motions occur with smaller vertical extent.

2. *Kelvin–Helmholtz instability* $(0 < Ri < 0.25)$
   This instability develops when the wind shear increases enough in a laminar flow across a density interface separating cold air below from warm air above.

3. *Inertial instability* $(0.25 < Ri < 1)$
   This perturbation is driven by the vertical eddy stresses associated with the vertical shear of thermal winds.

4. *Baroclinic instability* $(Ri > 1\text{--}1000 \text{ or more})$
   This instability grows from the horizontal temperature gradient associated with the zonal winds, converting potential energy into kinetic energy.

5. *Barotropic instability*
   This instability develops drawing energy from the kinetic energy of the mean flow by means of horizontal eddy stress. The zonal flow has no vertical shear, and so formally would correspond to the case of $Ri = \infty$.

An estimation of the Richardson number in the cloud layers of Venus, Jupiter, and Saturn is proposed in Problem 9.1.

9.2   VERTICAL INSTABILITY: CONVECTIVE MOTIONS

The stability of an atmosphere against vertical displacements, i.e., the condition that must be met by an air parcel to make it buoyant relative to its surroundings, was already presented in Sections 4.5.3 (for a dry atmosphere) and 5.2.6 (for wet conditions). A good parameter for characterizing the situation is the Brunt–Väisälä frequency N_B. An atmospheric parcel becomes unstable to convective motions when $N_B^2 < 0$, i.e., when $Ri < 0$. In this section, we focus on the dynamical behavior of the instability simplifying the situation to the one-dimensional vertical motion. Convection is a complicated issue and a detailed study will require a full three-dimensional treatment of the dynamical equations, which is beyond the scope of this book.

9.2.1   DRY CONVECTION

Consider an *air parcel* that is displaced vertically such that it adjusts, through expansion or compression, to the ambient atmospheric pressure. In addition, assume that the displacement takes place in a timescale that is short compared to the heat transfer between the parcel and its environment. The parcel can then be treated as adiabatic. From the momentum equation (7.8b), neglecting Coriolis and frictional effects, the parcel motion is given by

$$\rho' \frac{d^2 z'}{dt^2} = -\rho' g - \frac{\partial P'}{\partial z'} \qquad (9.3)$$

According to the imposed conditions, we have $P'=P$. The environment is assumed to be in hydrostatic equilibrium, i.e., $\partial P/\partial z = \rho g$, and then (9.3) is written as

$$\frac{d^2 z'}{dt^2} = \left(\frac{\rho - \rho'}{\rho'} \right) g \qquad (9.4)$$

This equation is equivalent to (4.71). As shown in Section 4.5.3, the specific buoyancy force acting on the parcel is $F_B = g(\rho - \rho')/\rho' = g(T' - T)/T$ and the parcel motion is given by (4.72), which is equivalent to rewrite the momentum equation as

$$\frac{d^2 z'}{dt^2} = \frac{g}{T}(\Gamma - \Gamma_d) z' \qquad (9.5)$$

The criteria for the parcel stability conditions discussed in Sections 4.5.3, 5.2.5, and 5.2.6 follow immediately, noting from (4.75) that the Brunt–Väisälä frequency can be rewritten as $N_B^2 = (g/T)(\Gamma_d - \Gamma)$. By defining the vertical velocity of the parcel as $w' = dz'/dt$, we also get

$$w' \frac{dw'}{dz'} = F_B \rightarrow w'^2 = w_0'^2 + 2 \int_{z_0}^{z} F_B(z') dz' \qquad (9.6)$$

Additional effects on the parcel motion previously neglected are the friction force between the parcel and the environment, and the action of nonhydrostatic pressure forces on the parcel. Frictional effects can be represented by a parameterization in terms of an *aerodynamic drag*, acting on the surface of the ascending dry parcel. A nondimensional drag coefficient C_D that takes into account the shape of the parcel (simplified usually as spherical) and that is a function of the Reynolds number is used. Under turbulent conditions (high Reynolds number), the friction force is inversely proportional to the radius r_0 of the parcel and to the square of the velocity. Per unit mass, it is given by

$$F_f = \frac{C_D w^2}{r_0} \qquad (9.7)$$

During ascension, the parcel experiments with a dynamic-pressure that tends to accelerate the parcel. It is equivalent to a force per unit mass given by

$$F_\pi = g \frac{\partial \pi_D}{\partial P} \qquad (9.8)$$

with π_D being a nonhydrostatic pressure component. Considering these two forces, the momentum equation for the parcel (9.6) is rewritten as

$$\frac{dw'}{dt} = \left(w'\frac{dw'}{dz'}\right) = g\left(\frac{T'-T}{T}\right) + g\frac{\partial \pi_D}{\partial P} - C_D\frac{w^2}{r_0} \qquad (9.9)$$

9.2.2 MOIST CONVECTION

The vertical stability of a parcel under moist conditions in the atmosphere was studied in Section 5.2.6. Wet convection occurs in the tropospheres of the planets in a great range of scales. On Earth, water moist convection produces cumulus and cumulonimbus clouds (storms) and plays a significant role in the large-scale dynamics in the tropics (see Chapters 5 and 6). Methane moist convection is thought to play an important role in Titan, and perhaps in Uranus and Neptune. In Jupiter and Saturn, there is evidence of phenomena related to moist convection that originates at the water and ammonia cloud levels. Details of the observed moist convective motions in planetary atmospheres will be given in Section 9.2.3.

The motion of a parcel in a moist environment can be described using the moisture definitions introduced in Section 5.2.1. In this case, the buoyant force per unit mass is redefined by substituting the temperatures (parcel and environment) in Equation 9.4 by their respective virtual temperatures, which account for the moisture content of air and its influence on density (5.18):

$$F_{BV} = g\left(\frac{T_V' - T_V}{T_V}\right) \qquad (9.10)$$

A parcel becomes buoyant when the difference in virtual temperature with its surroundings is positive: $\Delta T_V = T_V' - T_V > 0$. When moisture is present, a more complete description of the one-dimensional motion of the parcel requires the inclusion in (9.9) of the effects of entrainment or mixing of the ambient air with the rising parcel and the weight of the condensate that forms within the parcel as it ascends. Both effects tend to reduce the parcel buoyancy. The ambient air is generally drier and cooler than that of the parcel, and its entrainment into the parcel will tend to reduce the condensate mixing ratio and its buoyancy. Equation 9.9 is then rewritten as

$$\frac{dw'}{dt} = \left(w'\frac{dw'}{dz'}\right) = g\left(\frac{T_V' - T_V}{T_V}\right) + g\frac{\partial \pi_D}{\partial P} - C_D\frac{w^2}{r_0} - \frac{1}{m}\frac{dm}{dz'}w^2 - g\ell_C \qquad (9.11)$$

Here, ℓ_C is the condensate mixing ratio (mass of the condensed compound per unit mass of air; see Section 5.2.1). Usually, the parcel virtual temperature T_V' is calculated using the modified pseudo-adiabatic gradient (5.39) that includes the effect of the entrainment mass rate $(1/m)(dm/dz')$. A simple way to calculate the entrainment rate is parameterizing it as inversely proportional to the parcel size: $(1/m)$ $(dm/dz') = k_M/r_0$. The constant k_M is empirically determined from cumulus observations on Earth, and for approximately round parcels it can be set as 0.2. Equation 9.11 must be solved numerically to retrieve the vertical velocity of the parcel as a function of altitude $w(z)$. See Problems 9.2 and 9.3.

As stated above, moist convection forms the cumulus and cumulonimbus clouds when condensation occurs in *thermals* that are vigorous rising air parcels. On Earth, single convective cells can reach a vertical velocity of $w \sim 50\,\text{m s}^{-1}$. Once a parcel initiates its vertical motions (for example, forced by orography and surface winds), it can become saturated at a level called the *lifting condensation level* (LCL). If it continues its ascent, condensation and latent heat release will occur, and the parcel can freely accelerate upward when reaching a level called the *level of free convection* (LFC).

A useful measure of the capability of an atmosphere to develop convective storms and assess their strength is an index called the *Convective Available Potential Temperature* (CAPE). It represents a measure of the maximum possible kinetic energy that a statically unstable parcel can acquire (Figure 9.1). Neglecting the effects of condensation, mixing, and nonhydrostatic pressure, it is given according to (9.9) as

$$\text{CAPE} = \frac{w_{max}^2}{2} = \int_{z_{LFC}}^{z_{LNB}} g\left(\frac{T'-T}{T}\right) dz \tag{9.12}$$

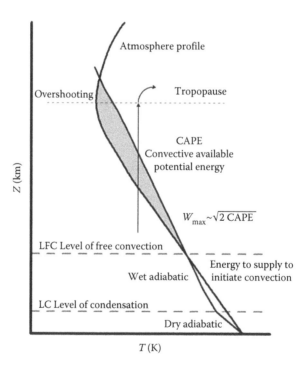

FIGURE 9.1 Scheme of the vertical stability of an atmosphere to moist convection and the different relevant altitude levels in terms of the vertical temperature profile. The approximate maximum velocity attained by an ascending parcel is given in terms of the *Convective Available Potential Temperature* (CAPE).

with the altitude LNB being the *level of neutral buoyancy* (CAPE unit is $m^2 s^{-2} = J kg^{-1}$). On Earth tropics, $\Delta T = T' - T \sim 1$–2 K along a vertical distance $\Delta z \sim 10$–12 km, thus giving from Equation 9.12 CAPE $\sim 500 m^2 s^{-2}$ and $W_{max} \sim$ 10 m s^{-1}. However, in severe mid-latitude storms, $\Delta T \sim 7$–10 K over $\Delta z \sim 10$–12 km, implying a CAPE of $\sim$2500–3000 m^2 s^{-2} (maximum vertical velocities $\sim$50 m s^{-1}).

9.2.3 CONVECTIVE PHENOMENA IN PLANETARY ATMOSPHERES

Convective phenomena are regularly observed in the cloud fields of the tropospheres of the planets with massive atmospheres (Venus, Earth, Mars [less frequently], the giants, and Titan). Here, we do not discuss the internal global heat convective transport that, for example, takes place in the deep atmospheres of the giant planets, as presented in Chapters 1 and 7. We focus on the localized, highly variable convective clouds that form in these atmospheres.

9.2.3.1 Venus

Images by the battery of spacecrafts that have orbited or penetrated Venus show evidence that cellular convection occurs in the upper cloud layer (altitudes $z \sim 60$–70 km). Ultraviolet images of the atmospheric clouds display a field of cellular features near and downwind of the subsolar region (predominantly in the afternoon and at low latitudes) with horizontal scales below 200 km (Figure 9.2). The field is composed of cells that are bright-rimmed and dark-rimmed with an individual size of 15–45 km but are organized in the large–scale clusters. The typical lifetime of a cell is roughly 2 days. The atmosphere of Venus apparently has one or two permanent layers of low static stability within the clouds (layer thickness of $\sim$5 km), and convection can develop on them under strong solar heating conditions (Problem 9.4).

9.2.3.2 Earth

On Earth, moist convection plays a major role in the circulation of the tropical atmosphere. It also represents an important fraction of the meteorological phenomena that occur at the mesoscale outside the tropics. Deep moist convection

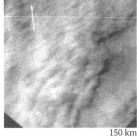

150 km

FIGURE 9.2 Convective cloud fields in Venus southern hemisphere imaged at different epochs at ultraviolet wavelengths by the VMC instrument onboard Venus Express (ESA). (Courtesy of ESA/MPIfSSR/DLR/IDA, Paris, France.)

develops in a variety of weather systems on Earth, as already presented. It occurs in summer *monsoons*, in a planetary equatorial band called the *Intertropical Convergence Zone* (ITCZ), and is locally forced by orography in the slopes and crests of the mountains, within the frontal zones of extratropical cyclones, and in hurricanes.

At the mesoscale range, *convective storms* also develop in a variety of forms, the most important being the *Mesoscale Convective Systems* (MCS) that manifest in different types of cloud features (Figure 9.3). MCSs are an organized ensemble of convective elements whose lifecycle is longer than that of the individual convective elements typically reaching sizes of ~100–500 km in length with persistence up to ~20 h. Single convective cells have horizontal sizes <2 km and temporal scales for the development of <1 h. Clusters of multi-cells form *thunderstorms* with horizontal sizes <20 km and timescales for the development of ~1 h. These systems are affected by the ambient vertical wind shear, developing rotation being split in their lower parts (few kilometers of altitude above ground) when the winds reach shears of 10–20 m s^{-1}. They are accompanied by hail formation, strong precipitation in the central parts of the system, and sometimes by *tornados* (see Section 9.2.4). *Supercell* storms organize along the *squall lines* (a particular type of MCS) reaching a linear size between 20 and 200 km. Supercells with high vertical motions reach the tropopause where the high stability above it, together with mass conservation, produces divergent motions in the cloud field that lead to the classical anvil shape storm. The growing area of the top clouds in a supercell storm can be used to infer its vertical velocity in a simple way. The horizontal divergence of the wind speed is defined in Cartesian coordinates as

$$div_h V = \nabla_h V = \frac{\partial u}{\partial x} + \frac{\partial v}{\partial y} = \frac{1}{A_H}\frac{dA_H}{dt} \tag{9.13}$$

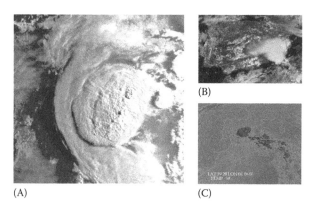

(A)   (C)

FIGURE 9.3 Satellite images of severe moist convective storms on Earth. (A) and (B) Examples of the water cumulus cloud tops; (C) temperature structure for the cloud field in (B). (B) and (C) show a typical example of the severe storms that develop over the Mediterranean Sea close to the Spanish coast during the months of September–October (this case took place on October 4, 1982). Image by Meteosat Visible channel. (Courtesy of EUMETSAT, Darmstadt, Germany.)

with A_H being the horizontal area of the cloudy region. From the continuity equation, we have $-\partial w/\partial z = \partial u/\partial x + \partial v/\partial y$. Thus, substituting and integrating the velocity along an altitude h and assuming that at the base the vertical velocity is zero, we get

$$w \leq h \frac{\ln(A_{H2}/A_{H1})}{\Delta t} \tag{9.14}$$

where A_{H1} and A_{H2} are the initial and final cloud areas measured during the temporal interval Δt (Problem 9.5).

9.2.3.3   Mars

Fields of convective clouds similar to Earth's cumulus have been observed in Mars, though they are rare. Dry convective motions are, however, active in the surface layer (see below).

9.2.3.4   Titan

On Titan, transient clouds with a convective appearance have been reported from Earth-based telescopic observations and Cassini spacecraft images in wavelength windows around 1 μm, where sunlight penetrates the dense haze layers and allows us to see deeper in the atmosphere down to the surface (Figure 5.27). These clouds evolve from near the surface to altitudes up to about 40 km and are formed by methane and ethane ice particles (Figure 9.4). Moist convection models indicate that updrafts can reach this altitude level with maximum speeds of 20 m s^{-1} (Problem 9.6).

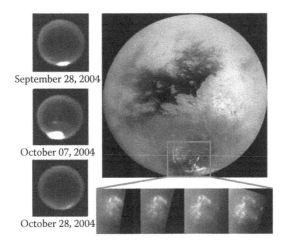

September 28, 2004

October 07, 2004

October 28, 2004

FIGURE 9.4 Convective storms in Titan's southern hemisphere assumed to be produced by methane condensation (and probably ethane). Left: Images obtained by the Keck telescope at the wavelength of 2.13 μm in September–October 2004. (Courtesy of W. M. Keck Observatory, Maunakea, HI). Right: Details of the convective clouds (decreasing phase) at higher resolution obtained by the Cassini spacecraft on October 26, 2004 (PIA06141). (Courtesy of NASA/JPL/Space Science Institute, Boulder, CO.)

9.2.3.5   Giant Planets: General

On the giant and icy planets, the development of moist convection has its peculiarities when compared with the other bodies. The saturated ascending parcels produced by the condensation of ammonia and water (in Jupiter and Saturn) and methane (in Uranus and Neptune) weigh more than the massive atmosphere that is essentially hydrogen. This is contrary to what occurs, for example, on Earth where the moist parcels (water) weigh less than the ambient air. Thus, convective motions in the giants are favored when the relative humidity of the environment is high enough to guarantee that $\Delta T_V = T_V' - T_V > 0$ (Problem 9.7). In addition, since the latent heat of condensation for ammonia is lower than that of water by a factor of 0.64 (see Table 5.2), we expect water moist convective storms in Jupiter and Saturn to be more energetic and vigorous than ammonia storms. This obviously depends on the relative mixing ratios of these two species. For a solar abundance of NH_3 and H_2O (see Table 3.6), ascending motions are predicted to be much more vigorous in the deeper water clouds than in the ammonia clouds.

9.2.3.6   Jupiter

Sudden, bright, and rapidly evolving "spots" are regularly observed in the belts of Jupiter, in particular in those close to the equator: the *South Equatorial Belt* (SEB, latitude 16°S), the *North Equatorial Belt* (NEB, latitude 10°N), and the *North Temperate Belt* (NTB, latitude 23.5°N) (see Section 9.6.4). These latitudes are fixed places of spot activity (Figure 9.5). The SEB and NEB spots emerge in latitudes where the zonal wind speed is close to zero but where it shows strong meridional shear. The NTB bright spots emerge in the peak of the strongest Jovian jet where the zonal velocity is ~165 m s^{-1}, developing a "plume-like" aspect. These bright spots grow rapidly and expand horizontally in a few days reaching sizes of ~1000–5000 km. Measurements of their growing areas during the first days give, according to (9.13) and (9.14), vertical speeds of ~1 m s^{-1} at the cloud tops. Observations at different wavelengths show that the bright spots are highly placed in the atmosphere, overshooting the tropopause, penetrating to altitude levels ~50 mbar or higher. All these properties suggest that the bright features are produced by water moist convection, and three-dimensional models support this view. Models and order of magnitude estimations using Equation 9.11 indicate that inside the single cells the velocity is ~10–100 m s^{-1}. The cloud base is at the 5–7 bar altitude level, and so the cell extends vertically more than 100 km. Smaller and less vigorous storms form due to moist convection in the ammonia clouds.

One important aspect of the bright spots in the SEB and NTB is that they trigger large-scale disturbances that encircle the planet along the latitude where they emerge. The disturbances, when they dissipate, lead to albedo changes in the banding aspect of Jupiter, and the band where they emerge changes from a "zone-like" to a "belt-like" morphology (Section 9.6.4).

9.2.3.7   Saturn

Saturn's moist convective features can be divided into two groups. Irregular bright storms with scales ~2000–4000 km form at mid-latitudes in both hemispheres at

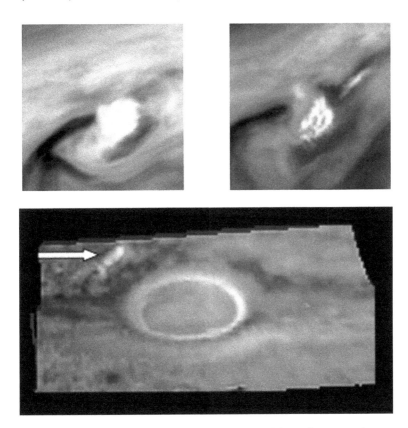

FIGURE 9.5   Examples of two types of clouds developed by moist convective storms in Jupiter. Upper part: Bright, rapidly evolving clouds generated by water convection deep in the atmosphere elevating from the 3–5 bar level to the 0.3 bar at 14°S (Galileo image, May 4, 1999). The storm has a vertical extent of at least 50 km and a length of about 4000 km and locates in the SEB convective area at 14°S. (From Gierasch, P.J. et al., *Nature*, 403, 628, 2000.) Lower part: Fresh ammonia ice clouds identified by its spectral signature in the northeast of the Great Red Spot in Galileo images of June 26, 1996 (PIA02569). (Courtesy of NASA/JPL, Los Angeles, CA.)

~35°N and 35°S. They grow rapidly in areas until they are sheared apart by the zonal winds. Their origin is probably as in Jupiter in the water and ammonia moist convection (Figure 9.6). On the other hand, there are rare but huge *Great White Spots* (GWS) that have been observed to emerge at different latitudes (most frequent at the Equator) with intervals of about 30 years (one Saturn year). The GWS bright clouds reach a horizontal size of 20,000 km and overshoot the tropopause penetrating the ~50 mbar altitude level. A GWS eruption produces a planetary-scale disturbance formed by a series of bright and dark cloud features that move away from the GWS core advected by the ambient winds. The activity prevails for 1 or 2 months, with the clouds being dispersed zonally by the meridional wind shear in opposite directions until they encircle the planet along the latitude band (Figure 9.7). For example, in the equatorial region, the GWS can suffer shears of ~200 m s^{-1} between latitude 0°

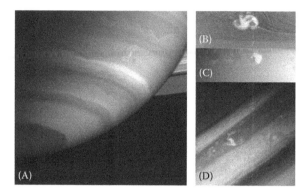

FIGURE 9.6 Examples of Saturn storms produced probably by water and ammonia moist convection. (A), (B), and (C) Cassini images of storms at latitude 38°S obtained in July 2004 (PIA06197), January 27, 2006 (PIA07789), and February 16, 2006 (PIA08142), respectively. (D) Image of a storm in the northern hemisphere (39°N) obtained by Voyager 1 on November 5, 1980. (Courtesy of NASA/ESA/ASI and NASA/JPL, Los Angeles, CA.)

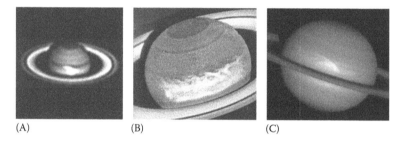

(A)   (B)   (C)

FIGURE 9.7 Saturn's "Great White Spots" (GWS), are the most intense convective storms detected in planetary atmospheres, probably produced by a mixture of ammonia and water moist instability. (A) Image of the GWS 1990 event obtained on October 2, 1990, at Pic-du-Midi Observatory (France) during the initial development phase of the storm. (Courtesy of J. Lecacheux and P. Laques.). (B) The evolution and expansion of the storm along the equator 1 month later on November 11, 1990, as imaged by the HST (NASA/ESA). (C) Image of a secondary GWS event on December 1, 1994, imaged by the HST. (Courtesy of NASA/ESA, Paris, France.)

and 20°. These spectacular storms have been proposed to be due to moist convection in the deep water clouds, probably coupled with induced, less vigorous convection in the ammonia cloud. Cloud top vertical velocities deduced from the divergence measurements give values of 0.1–$1\,\mathrm{m\,s^{-1}}$, but at the core of the ascending parcels, estimations using Equation 9.11 give velocities of $\sim$100–$150\,\mathrm{m\,s^{-1}}$.

9.2.3.8 Uranus and Neptune

Bright, irregular spots in shape have also been observed in Uranus and Neptune with cloud tops attaining the $\sim$0.1–1 bar altitude level. In Uranus, plume-like features were seen by Voyager 2 in 1986, and rapidly changing spots have been imaged by the Hubble Space Telescope and ground-based telescopes since 1994 (Figure 9.8),

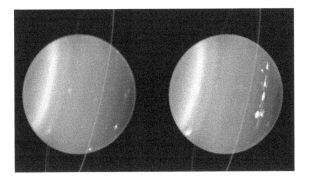

FIGURE 9.8 Uranus' moist convective storms observed in 2004 using Keck II telescope's adaptive optic system. (Courtesy of L. Sromovsky, University of Wisconsin, Madison, WI.)

both suggesting convective origin. The most prominent features by their brightness have been observed between latitudes 25°N and 45°N. The spot activity appears to be related to the pronounced seasonal cycle in Uranus (spin axis tilted ~90° relative to the orbital plane). The spots appeared in the latitudes reached by sunlight after a long period of darkness (Uranus' span 89 years).

In the case of Neptune, bright spots suggestive of convection were observed during the Voyager 2 encounter in 1989, most of the time associated with the large oval vortices. They sometimes emerged in the vortex periphery and in other cases emerged at their center (Figure 9.9). Ground-based telescopes using adaptive optic devices and

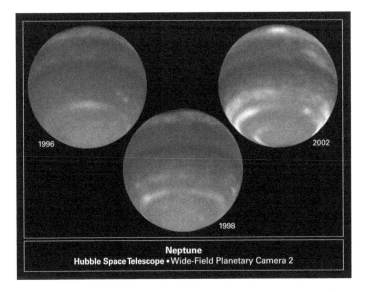

FIGURE 9.9 Neptune's storms evolution as observed in different years using the Hubble Space Telescope. (Courtesy of L. Sromovsky and P. Fry, University of Wisconsin, Madison, WI and NASA/ESA, Paris, France.)

HST imaging since 1994 recorded an enhancement of bright cloud activity (with a characteristic lifetime of 1 month) concentrated in a band around the pole at latitude 70°S. It is possible that some of these bright features are produced by methane convection at altitude levels between 1 and 3 bar. Deeper ammonia clouds (altitude levels 5–15 bar) could also be a source, but water moist convection, if occurring, is expected to be very deep (100–200 bars), thus forming clouds at unobservable levels.

9.2.4 LIGHTNING

Lightning is an electrical phenomenon that occurs mostly in the cloudy regions of planetary atmospheres. On Earth, it takes place when charges of different signs accumulated in different parts of a cloud create a difference in the electrical potential generating an electric field. The electrical charging of clouds originates through droplet and particle collision within it, followed by the gravitational separation of oppositely charged large and small particles. When this electric field becomes large enough, it ionizes the medium and an abrupt *lightning discharge* is triggered when the energy stored in the electric field is released. The lightning develops along a conducting channel from one region of a cloud to another or, at least in the case of the Earth, between the cloud and the ground. Lightning has been observed as optical flashes at visual wavelengths on the Earth and Jupiter, and has been detected at radio wavelengths on Earth, Saturn, and probably on Uranus. On Venus, the signature of lightning has been reported through magnetometer measurements in the ionosphere.

On Earth, lightning occurs predominantly within vigorous convective clouds (thunderstorms). The electrically negative charges are located in the middle of the cloud within a layer that has roughly a pancake shape (thickness < 1 km and temperatures ~ −15°C). This layer is sandwiched between positive electrical charges above and below it (Figure 9.10). Typically, the charge content in the narrow channel reaches ~10–100 C (or an equivalent density ~ few nC m^{-3}). In cumulonimbus, the transfer of charge occurs by collisions between the particles and depends on temperature and the liquid water content in the cloud, and so

 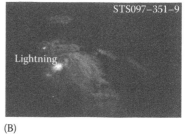

(A)   (B)

FIGURE 9.10   (A) This time-lapse photography captures multiple cloud-to-ground lightning strikes during a nighttime thunderstorm in Norman, OK (USA). (Courtesy of NOAA Photo Library, NOAA Central Library; OAR/ERL/NSSL, Tulsa, OK.) (B) This image was taken by the STS-97 Space Shuttle flying over equatorial Africa east of Lake Volta on December 11, 2000. The top of the large thunderstorm, roughly 20 km across, is illuminated by a full moon and frequent bursts of lightning. (Courtesy of NASA, Washington, DC.)

the formation of graupel and precipitation. When the electric field within the cloud reaches a critical value (~1 MV m^{-1}), the lightning discharge occurs. The temperatures in the discharge region reach ~10,000 K and the pressure in a central narrow discharge channel increases to 10–100 atm. The rapid gas expansion creates a shock wave, traveling faster than the speed of sound, generating "the thunder." The discharge process is fast (~ms) and a flow of electrons is released in a cascade (typically in steps of ~1 μs), forming the classical stepped leader lightning morphology with different associated branches. The strike moves at a speed of ~10^6 m s^{-1} following a path 1–3 km in length. This is equivalent to an electrical current of ~40,000 A. The most energetic flashes release an energy of ~10^9 J leading to an optical power per unit area of 4×10^{-7} W m^{-2}. Globally on Earth, there are about 6 flashes km^{-2} year^{-1}, mostly concentrated over tropical continental areas.

Lightning has been detected as optical flashes during nighttime observations in Jupiter by different spacecrafts (Voyager, Galileo, Cassini, and New Horizons). The lightning strikes occur in clusters with sizes of ~100 to ~1700 km with a separation of ~100 km between storms. They were predominantly detected in the belts (areas of cyclonic vorticity) within the westward jets in an ample range of latitudes between 60°S and 60°N. The New Horizons spacecraft also detected lightning in the polar area (latitudes 80°N and 75°S). The lightning flashes in Jupiter occur at an altitude level of ~0.5–1 bar but some appear to be deeper and related to the water clouds at 4–5 bar altitude level. Lightning has been observed in some of the largest Jovian storms and, as on Earth, they are related to the vigorous moist convective processes described above (Figure 9.11). The energy involved in Jupiter's lightning is ~10^9–10^{10} J, a few times larger than the largest observed on Earth. The optical power per unit area of these lightning strikes is ~3×10^{-7} W m^{-2}, about the same as that on Earth. The extrapolated global flash rates detected in Jupiter are ~4×10^{-3} flashes km^{-2} year^{-1}, smaller than on Earth, although occulted flashes can occur in deeper clouds.

Radio observations from the Voyager and Cassini spacecrafts of Saturn have detected bursts of electromagnetic waves, called *Saturn electrostatic discharges* (SEDs). The SEDs are a series of emission episodes that occur in the frequency range of 1–16 MHz. They move in the atmosphere according to the ambient zonal flow. The visible storms and SED episodes occurred at the same latitude, forming and disappearing simultaneously, confirming that they are related phenomena. Similarly as Jupiter, it is straightforward to assume that Saturn lightning also has its origin in the water clouds. Images of the right side of Saturn's mid-latitudes by the Cassini spacecraft have detected optical flashes produced by lightning. The flashes occurred once per minute over a region with a size of 200 km and were located in the NH$_4$SH or H$_2$O clouds, 125–250 km below cloud tops. The energy at visible wavelengths ranged up to 1.7×10^9 J, comparable to that on Earth and Jupiter.

Indirect magnetometer measurements from *Venus Express* report the presence of lightning activity on Venus, with electromagnetic waves of frequencies of 100 Hz, propagating from the atmosphere to the ionosphere spacecraft. The waves occur as bursts of short duration (0.25–0.5 s) and have the properties of whistler-mode signals generated by lightning discharges.

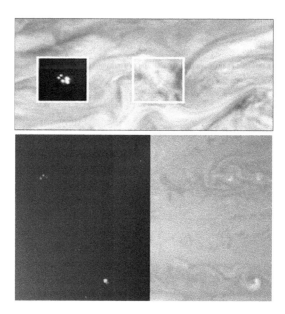

FIGURE 9.11 Jupiter's lightning. Upper: Galileo spacecraft image of a convective storm (see also Figure 9.5) and superimposed (inset) a night-side image showing the associated lightning. (Courtesy of NASA, Washington, DC). Lower: Images of Jupiter's day and night sides taken by NASA's Cassini spacecraft on January 1, 2001, show that storms visible on the day side are the sources of visible lightning when viewed on the night side. The storms occur at 34.5° and 23.5° North latitude, within one degree of the latitudes at which similar lightning features were detected by the Galileo spacecraft. (Courtesy of NASA/JPL/University of Arizona, Tucson, AZ.)

9.3 HYDRODYNAMIC INSTABILITY: VORTICES

Hydrodynamic instabilities are identified in planetary atmospheres and in laboratory fluid experiments by observing the structure and evolution of a field variable or a flow tracer (clouds in atmospheres). Not all types of features that we observe in atmospheres, resulting from an initial instability in the atmospheric flow, have been assigned to a known type of instability. The best examples are the giant planets, plentiful of vortices at a variety of scales, whose precise nature is still unknown. A flow instability grows from its mean energy (gravitational, thermal, latent heat, or kinetic) and evolves through the effects of viscous dissipation (intrinsic to the flow or related to surface friction) and planetary rotation (Coriolis force and curvature effects in the flow). Flow instabilities develop a vorticity field and, to some degree, divergent or convergent motions. Most of them fall within the different categories of what is called *a vortex*—a region of the atmosphere in rotation or spinning about its center (a rotary region that possesses vorticity). We first introduce the basic vortex properties and definitions.

9.3.1 NATURAL COORDINATES

The study of local dynamical phenomena in atmospheres is simplified if we use a *natural coordinate system* that is defined by the set of orthogonal unit vectors

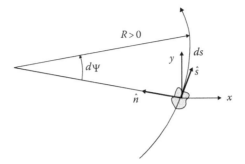

FIGURE 9.12   The natural (trajectory) coordinate system: R is the curvature radius, s is the distance along the curve with unit vector oriented parallel to the horizontal velocity, and n is the normal to it.

$(\hat{s}, \hat{n}, \hat{k})$: $\hat{s}$ oriented parallel to the horizontal velocity at each point, $\hat{n}$ oriented normal to the horizontal velocity (positive when oriented to the left of the flow direction), and $\hat{k} = \hat{s} \times \hat{n}$ oriented perpendicular to both (positive upward) (Figure 9.12). If the length of the curve followed by a parcel moving in a horizontal plane is s (x, y, t), then the scalar velocity is $V = ds/dt$ (always positive) and the velocity and acceleration are given in vector form, respectively, by

$$\vec{V} = V\hat{s} \tag{9.15}$$

$$\frac{d\vec{V}}{dt} = \frac{dV}{dt}\hat{s} + \frac{V^2}{R}\hat{n} \tag{9.16}$$

where
   d/dt is the total derivative (Section 7.1.1)
   R is the local curvature radius

The trajectory coordinates are (s, n). Since the velocity is tangent to the trajectory, we write it as V_T. The acceleration has two terms: One, tangent to the trajectory, is given by the rate change of the velocity modulus $a_T = dV/dt$. The second, perpendicular to the trajectory, is the centripetal acceleration due to the curvature $a_N = V^2/R$. When $R \to \infty$, the centripetal acceleration $\to 0$ and the motion becomes rectilinear. R is defined to be positive if the center of curvature lies in the positive $\hat{n}$ direction. For $R > 0$, the air parcel turns toward the left following the motion, and for $R < 0$, the air parcel turns toward the right, respectively. In a planetary atmosphere, the Coriolis parameter f changes sign with the hemisphere (positive in the North, negative in the South). The curvature of the flow is said to be *cyclonic* when R and f have the same sign $(fR > 0)$ and *anticyclonic* when they are of opposite signs $(fR < 0)$. A cyclonic vortex rotates counter-clockwise in the northern hemisphere and in the opposite direction in an anticyclonic vortex.

9.3.2 TRAJECTORIES AND STREAMLINES

The *trajectory* of an air parcel relative to a given reference frame is defined as the path traced by the locus of the values of its coordinates as a function of time. At any instant, the velocity field defines a family of *streamlines* that are in any position tangential to the velocity field. The streamlines represent a "snapshot" of the velocity field at any instant. If the motion field is *steady*, i.e., the local velocities do not change with time, parcel trajectories coincide with streamlines. If it is *nonsteady*, the streamlines and parcel trajectory do not coincide with the streamlines moving to a new position after a temporal interval.

In Cartesian coordinates, the horizontal trajectory of an individual element is determined by the integration over a finite time of

$$\frac{ds}{dt} = V(x, y, t) \tag{9.17}$$

remembering that d/dt represents the total derivative. The streamlines are determined by the integration with respect to x at time t_0 from the equation

$$\frac{dy}{dx} = \frac{v(x, y, t_0)}{u(x, y, t_0)} \tag{9.18}$$

9.3.3 VORTICITY AND DIVERGENCE

Let $\chi(x, y, t)$ designate the angular direction of the wind at each point on an isobaric surface. The radii of curvature of the trajectory (R_t) and streamline (R_s) are given by

$$\frac{1}{R_t} = \frac{d\chi}{ds} \tag{9.19a}$$

$$\frac{1}{R_s} = \frac{\partial\chi}{\partial s} \tag{9.19b}$$

where the first relationship gives the rate of change of the wind direction along the trajectory (positive for counter-clockwise turning), and the second relationship gives the rate of change of the wind direction along a streamline at any instant.

In natural coordinates, the vertical component of the vorticity introduced in Section 7.3 is written as

$$\zeta = -\frac{\partial V}{\partial n} + \frac{V}{R_s} \tag{9.20}$$

being the sum of the *shear* $(\partial V/\partial n)$ and *curvature* (V/R_s) vorticity terms.

The horizontal divergence of the flow (see 9.13) is defined in natural coordinates as

$$div_h V = V \frac{\partial \chi}{\partial n} + \frac{\partial V}{\partial s} \tag{9.21}$$

The first term is called the *difluence* and the second is the *stretching* term. It is useful to have the horizontal components of vorticity and divergence written in the other coordinate systems (Problem 9.8, see also Table 7.1):

- *Spherical*

$$\nabla_h \cdot \vec{u} = \frac{\partial u}{\partial x} + \frac{\partial v}{\partial y} - \frac{v \tan \varphi}{r} \tag{9.22a}$$

$$\hat{k} \cdot \nabla \times \vec{u} = \frac{\partial v}{\partial x} - \frac{\partial u}{\partial y} + \frac{u \tan \varphi}{r} \tag{9.22b}$$

- *Cartesian*

$$\nabla_h \cdot \vec{u} = \frac{\partial u}{\partial x} + \frac{\partial v}{\partial y} \tag{9.23a}$$

$$\hat{k} \cdot \nabla \times \vec{u} = \frac{\partial v}{\partial x} - \frac{\partial u}{\partial y} \tag{9.23b}$$

- *Cylindrical in the plane (polar)*

$$\nabla_h \cdot \vec{u} = \frac{\partial v_r}{\partial r} + \frac{v_r}{r} + \frac{1}{r} \frac{\partial v_\theta}{\partial \theta} \tag{9.24a}$$

$$\hat{k} \cdot \nabla \times \vec{u} = \frac{1}{r} \frac{\partial (r v_\theta)}{\partial r} - \frac{1}{r} \frac{\partial v_r}{\partial \theta} \tag{9.24b}$$

Here, (v_r, v_θ) are the *radial* and *orthoradial* components of the velocity (Problem 9.9). For the giant planets, it is important to take into account their spheroid shape (see Section 1.3.4), and (9.22a and b) are more precisely calculated as

- *Spheroidal*

$$\nabla_h \cdot \vec{u} = \frac{1}{r_M} \frac{\partial v}{\partial \varphi_c} - \frac{v}{r_Z} \sin \varphi_c + \frac{1}{r_Z} \frac{\partial u}{\partial \phi} \tag{9.25a}$$

$$\hat{k} \cdot \nabla \times \vec{u} = \frac{-1}{r_M} \frac{\partial u}{\partial \varphi_c} + \frac{u}{r_Z} \sin \varphi_c + \frac{1}{r_Z} \frac{\partial v}{\partial \phi} \qquad (9.25b)$$

where r_Z and r_M are the radii of curvature in the spheroid (in Equation 1.64a and b, they were given in terms of the planetographic latitude). See Problem 9.10 to derive a formula that is useful for the giant planets for the vorticity of an elliptical (oval-shaped) vortex and Problem 9.11 for its application to Jupiter's Great Red Spot (GRS).

9.3.4 MEAN VORTICITY AND DIVERGENCE

The *mean values* of the vorticity and divergence can be defined from the vector integral relationships. The application of the Stoke's theorem (7.49) to a vortex with tangential velocity V_T (see Problem 9.10) is written as $\oint_C \vec{V}_T \cdot d\vec{\ell} = \iint_S \left(\nabla \times \vec{V}_T \right) \cdot d\vec{S}$. We define the mean vorticity of this vortex as

$$\langle \zeta \rangle = \frac{\oint_C \vec{V}_T \cdot d\vec{\ell}}{S} \qquad (9.26)$$

with S being the area enclosed by the vortex and the circulation is obtained from the tangential component of the velocity along the periphery.

Similarly, from Gauss's theorem applied to a purely two-dimensional flow, with the radial (outward) component of the velocity across the curve, V_R is given by

$$\oint V_R \, d\ell = \iint_S \left(\nabla_h \cdot \vec{V} \right) dS \qquad (9.27a)$$

and we define a mean vortex divergence as

$$\langle divV \rangle = \frac{\oint V_R \, d\ell}{S} \qquad (9.27b)$$

An application to these definitions can be found in Problem 9.12.

9.3.5 BALANCED FLOW

We have previously presented in Section 7.2 the force balance approaches to the momentum equation for mean zonal and meridional flows. Here, we particularize the balanced flow to vortices in terms of the natural coordinates introduced in Section 9.3.1. First note that the Coriolis and pressure gradient forces expressed in terms of the geopotential in natural coordinates are given by

$$-f\vec{k} \times \vec{V} = -fV\hat{n} \tag{9.28a}$$

$$-\nabla_p \Phi_G = -\left(\hat{s} \frac{\partial \Phi_G}{\partial s} + \hat{n} \frac{\partial \Phi_G}{\partial n} \right) \tag{9.28b}$$

Using (9.16) and considering only these two forces in the momentum balance (see also Equation 7.8), we get the following two relationships:

$$\frac{dV}{dt} = -\frac{\partial \Phi_G}{\partial s} \tag{9.29a}$$

$$\frac{V^2}{R} + fV = -\frac{\partial \Phi_G}{\partial n} \tag{9.29b}$$

These equations give the force balances in directions parallel to and normal to the wind speed V (here d/dt is the total derivate, Equation 7.1a). The flow speed is constant if the motion is parallel to the geopotential height contour ($\partial \Phi_G/\partial s = 0$) and if, in addition, $\partial \Phi_G/\partial n = $ constant, then the radius of curvature of the trajectory R also remains constant. For such conditions, the following force balances are permitted leading to the following vortex types (Figures 9.13 and 9.14).

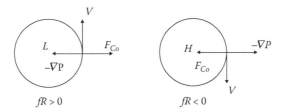

Geostrophic flow

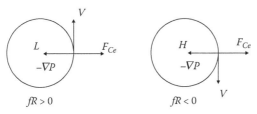

Cyclostrophic flow

FIGURE 9.13 Force balances in the Northern Hemisphere of a planet for geostrophic flow and for cyclostrophic flow. The represented forces (per unit mass) are $-\Delta P$ (gradient pressure: $-\partial \Phi_G/\partial n$), F_{Co} (Coriolis: fV), and F_{Ce} (centrifugal: V^2/R). L and H represent low and high pressure centers.

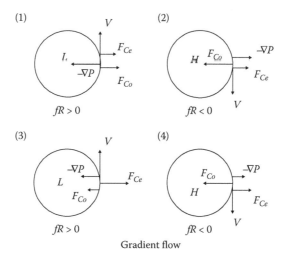

Gradient flow

FIGURE 9.14 Force balances for gradient wind flow in the Northern Hemisphere: (1) regular low (cyclonic flow); (2) regular high (anticyclonic flow); (3) anomalous low (antibaric flow); (4) anomalous high (anticyclonic flow). Symbols as in Figure 9.13.

9.3.5.1 Geostrophic Vortices

A geostrophic vortex occurs when a balance holds between the Coriolis force and the pressure gradient force:

$$fV_g = -\frac{\partial \Phi_G}{\partial n} \tag{9.30}$$

where we have written $V = V_g$ (for geostrophy) and where n gives the direction normal to the isobars (or geopotential height contours) and points toward increasing values. In a geostrophic vortex, the wind has constant speed and blows parallel to the isobars. The flow is cyclonic (in either hemisphere) for motion about a low pressure center and is anticyclonic about high pressure centers (Figure 9.13). Geostrophy is a reasonable balance approximation for most synoptic scale and mid-latitude planetary vortices in rapidly rotating bodies, in particular in Earth's extratropical cyclones (Problem 9.13).

9.3.5.2 Inertial Motion

When the geopotential field is uniform, a balance occurs between the Coriolis and centrifugal forces. Then, from (Equation 9.29b) with $\partial \Phi_G / \partial n = 0$, we get

$$\frac{V^2}{R} + fV = 0 \tag{9.31}$$

The radius of curvature is given by

$$R = -\frac{V}{f} \tag{9.32}$$

In the f-plane approximation (Equation 7.15), we have $f=f_0$ so that the curvature radius is constant and the streamlines draw circular anticyclone paths (clockwise rotation in the northern hemisphere and counter-clockwise in the south). The rotation period of a parcel is $\tau=|2\pi R/V|=2\pi/|f|$. This motion has been observed in terrestrial oceans, but is in general unimportant in atmospheric motions.

9.3.5.3 Cyclostrophic Vortices

When the Coriolis term is small in Equation 9.29b, a balance occurs between the centrifugal and pressure gradient forces

$$\frac{V^2}{R} = -\frac{\partial \Phi_G}{\partial n}$$ (9.33)

and the vortex rotation velocity is given by

$$V = \sqrt{-R\frac{\partial \Phi_G}{\partial n}}$$ (9.34)

For this expression to make sense, the radius of the curvature and the geopotential gradient must have opposite signs. A cyclostrophic vortex may be either cyclonic (R and $-\partial \Phi_G/\partial n$ both positive) or anticyclonic (R and $-\partial \Phi_G/\partial n$ both negative), but will always be moving about a low pressure center (the pressure gradient force is directed toward the center of curvature and the centrifugal force is directed away from the center of curvature) (Figure 9.13). For this vortex to form, the Coriolis force term must be low, which occurs in slowly rotating bodies, such as Venus or Titan, or in rapidly rotating bodies in places close to the equator. It also occurs when the flow curvature is strong (R small) as in small-scale rapidly rotating vortices like Earth's *tornados* (Problems 9.13 through 9.15).

9.3.5.4 Gradient Wind Vortices

If we retain the three terms in the balance equation (9.29b), and with $dV/dt=0$ in Equation 9.29a, the resulting force balance describes a gradient wind vortex:

$$\frac{V^2}{R} + fV = -\frac{\partial \Phi_G}{\partial n}$$ (9.35a)

It follows the gradient wind speed

$$V = -\frac{fR}{2} \pm \left(\frac{f^2 R^2}{4} - R\frac{\partial \Phi_G}{\partial n} \right)^{1/2}$$ (9.35b)

The physically possible solutions are limited by the fact that the root must be real (($f^2R^2/4$)$>R(\partial \Phi_G/\partial n)$) and V must be positive. In addition, the centrifugal term

V^2/R must point outward to the vortex center. In the northern hemisphere ($f>0$), there are four possible solutions corresponding to the two possibilities $R>0$ and $R<0$. For $R>0$ (anticlockwise rotation), we have $V>0$ only if $-\partial\Phi_G/\partial n>0$ (i.e., $\partial P/\partial n<0$) indicating a low pressure at the center of curvature (this is called a *regular low*, cyclonic vortex). For $R<0$ (clockwise rotation), there are three possible solutions: (a) $-\partial\Phi_G/\partial n<0$ (i.e., $\partial P/\partial n>0$) corresponding to a low pressure center (called *anomalous low*, antibaric vortex); (b) two high pressure anticyclonic vortices ($-\partial\Phi_G/\partial n>0$, i.e., $\partial P/\partial n<0$). One is the *anomalous high* (corresponding to the positive root) with $V>-fR/2$, and the other is the *regular high* (negative root) with $V<-fR/2$ (Figure 9.14). See Problems 9.16 through 9.18.

9.4   TYPES OF HYDRODYNAMIC INSTABILITIES

In what follows, we describe the different types of hydrodynamic instabilities according to the classification scheme based on the Richardson number, as presented in Section 9.1.

9.4.1   Kelvin–Helmholtz Instability

This instability develops when the vertical shear of the horizontal wind becomes strong enough and $0<Ri<0.25$. Consider, for example, the case of a laminar flow that contains two layers separated by a density interface with cold air below and warm air above. As the shear flow between the layers increases, it can become unstable with waves forming across the interface. In the atmosphere, the waves grow in amplitude and curl over themselves becoming visible when the moisture on them forms clouds (Figure 9.15). The crests make the Kelvin–Helmholtz (KH) instability visible by the *billow clouds* oriented perpendicularly to the wind shear vector ($\partial\vec{V}/\partial z$). Such instability also accounts for the formation of waves at the surface of the sea.

To quantify the KH instability, consider the simplest case where the overlaying layer has density ρ_2 and velocity u_2, whereas the underlying layer has density ρ_1 and velocity u_1. Disturbances of the form e^{ikx} become unstable and grow when the wavenumber k obeys the relationship

$$k > g\frac{\rho_1^2 - \rho_2^2}{\rho_1\rho_2\left(u_1 - u_2\right)^2} \tag{9.36}$$

Perturbations with small wavelengths (large k) become unstable if the velocity difference is small enough. Highly sheared layers under low stability conditions can occur in the cloud layers of Venus and in the upper clouds in the equatorial tropospheres of Jupiter and Saturn (see data in Problem 9.1). Although the Ri range of values for the KH instability is narrow, it is possible that this instability develops in thin cloudy layers of atmospheres producing turbulent motions there (Problem 9.19).

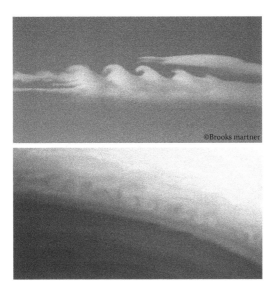

FIGURE 9.15 Wave patterns in the cloud field produced by the Kelvin–Helmholtz instability. Upper: Earth example with wave forming and breaking (altitude is in the vertical) over Laramie (Wyoming, USA). Photo by Brooks Martner. Lower: Saturn case. Image taken by the Cassini spacecraft on October 9, 2004. The turbulent boundary between two latitudinal bands with wind shear in Saturn's atmosphere curls repeatedly probably due to the Kelvin–Helmholtz instability. Image PIA06502. (Courtesy of NASA/JPL/Space Science Institute, Boulder, CO.)

9.4.2 INERTIAL INSTABILITY

Consider a purely zonal flow $[\bar{u}(y, z), 0, 0]$ on a β-plane. If the atmosphere is statically stable according to the criteria presented in Sections 4.5.3 and 5.2.6, i.e., for Ri positive, an instability develops in the meridional direction when the planetary and ambient (shear) vorticity obey the relationship

$$Ri < \left(1 - \frac{1}{f}\frac{\partial u}{\partial y}\right)^{-1} \tag{9.37}$$

The instability leads to axially symmetric motions independent of the longitude. It transports heat down the horizontal temperature gradient and upward, tending to vertically stabilize the atmosphere. According to (9.37), the instability is favored in regions of anticyclonic vorticity and, therefore, it has been proposed as a mechanism to explain the banded (belt-zone) cloud distribution in the giant planets and the clouds that form in the giant anticyclones like the GRS of Jupiter.

Assuming no pressure perturbations and no vertical wind shear ($\partial u/\partial z = 0$), and using the geostrophic balance to substitute the pressure gradient force in terms of the mean zonal flow $\bar{u}$, the zonal and meridional momentum balance equations are written as

$$\frac{du}{dt} = fv \qquad (9.38a)$$

$$\frac{dv}{dt} = f(\bar{u} - u) \qquad (9.38b)$$

A parcel displaced in the meridional direction at a distance y will have a meridional velocity $v = dy/dt$ and (9.38a) is rewritten as

$$\frac{du}{dt} = f\frac{dy}{dt} \qquad (9.39)$$

If the parcel moves from an initial position y_0 to $y_0 + y'$, the integration of (9.39) gives $u(y_0 + y') - \bar{u}(y_0) = fy'$, and then substituting for $u(y_0 + y')$ in (9.38b), we get

$$\frac{d^2 y'}{dt^2} = f\left(\frac{\partial \bar{u}}{\partial y} - f\right) y' \qquad (9.40)$$

When $\partial \bar{u}/\partial y = 0$, Equation 9.40 reduces to a simple inertial oscillation. In the presence of a meridional shear, Equation 9.40 admits oscillatory or exponentially decaying or growing solutions. Instabilities will grow when the absolute vorticity of the mean flow $(f - \partial \bar{u}/\partial y)$ has a sign opposite to the planetary vorticity. For example, in the northern hemisphere $f > 0$, and the instability (exponentially growing disturbances) will occur whereby $(f - \partial \bar{u}/\partial y) < 0$ (see Problem 9.20).

9.4.3 BAROTROPIC INSTABILITY

In a *barotropic* atmosphere, there are no temperature gradients on constant pressure surfaces. Then, according to the thermal wind relationship, the vertical shear in the zonal flow is zero ($\partial \bar{u}/\partial z = 0$) and the flow has only meridional dependence [$u(y)$] and $Ri = \infty$. Instability is initiated when the horizontal curvature of flow becomes strong. To see how this occurs, we start assuming that the flow conserves the absolute vorticity (i.e., it obeys the *barotropic vorticity* equation 7.58b). Following the same perturbation treatment that we did for barotropic Rossby waves in Section 8.4.1, but allowing now for the latitude dependence of the zonal wind, the vorticity can be decomposed as $\zeta = \bar{\zeta} + \zeta'$ and Equation 8.25 becomes

$$\left(\frac{\partial}{\partial t} + \bar{u}\frac{\partial}{\partial x}\right)\zeta' + v'\frac{\partial(\bar{\zeta} + f)}{\partial y} = 0 \qquad (9.41)$$

with $\zeta' = (\partial v'/\partial x - \partial u'/\partial y)$. Introducing the stream function definition (8.26), we get

$$\left(\frac{\partial}{\partial t} + \bar{u}\frac{\partial}{\partial x}\right)\nabla^2 \psi' + \frac{\partial \psi'}{\partial x}\frac{\partial(\bar{\zeta} + f)}{\partial y} = 0 \qquad (9.42a)$$

Since $\partial \bar{\zeta}/\partial y = -\partial^2 \bar{u}/\partial y^2$ and $\partial f/\partial y = \beta$, it is convenient to rewrite this equation as

$$\left(\frac{\partial}{\partial t} + \bar{u}\frac{\partial}{\partial x}\right)\nabla^2 \psi' + \frac{\partial \psi'}{\partial x}\left(\beta - \frac{\partial^2 \bar{u}}{\partial y^2}\right) = 0 \qquad (9.42b)$$

Consider now wave solutions of the form $\psi'(x, y) = Re\{\psi_0(y)\exp[ik(x-ct)]\}$, where it must be noted that the phase velocity c may be complex $(c = c_r + ic_i)$. When $c_i > 0$, the wave will grow exponentially with time like $\exp(kc_i t)$ and the flow becomes unstable. The quantity kc_i is called the *growth rate* and represents the e-folding time for the instability. Upon substitution of this solution in (9.42b), we obtain a second-order differential equation:

$$(\bar{u} - c)\left(\frac{d^2 \psi_0}{dy^2} - k^2 \psi_0\right) + \left(\beta - \frac{\partial^2 \bar{u}}{\partial y^2}\right)\psi_0 = 0 \qquad (9.43)$$

To solve this equation, we must impose the boundary conditions. We assume that the growing wave is confined to a channel of width L so that $\psi(y) = 0$ at $y = 0$ and at $y = L$. There are solutions for particular values of c, but we are interested in the exponentially growing case $(c_i > 0)$ and, noting that the stream function amplitude is complex, we write it as $\psi_0 = \psi_r + i\psi_i$. We now handle (9.43) multiplying it by the complex conjugate ψ_0^* of ψ_0, dividing by $(\bar{u} - c)$, and integrating it over y, then equating the real and imaginary parts. For the imaginary part, this gives

$$\int_0^L \left(\psi_i \frac{d^2 \psi_r}{dy^2} - \psi_r \frac{d^2 \psi_i}{dy^2}\right)dy = c_i \int_0^L \left(\beta - \frac{\partial^2 \bar{u}}{\partial y^2}\right)\frac{|\psi_0|^2}{|\bar{u} - c|^2}dy \qquad (9.44)$$

The integrand on the left-hand side can also be written as $(d/dy)(\psi_i(d\psi_r/dy) - \psi_r(d\psi_i/dy))$, and from the boundary conditions at $y = 0, L$, we have $\psi_i = \psi_r = 0$ and the integral is zero. Then, (9.44) represents a way of establishing a criterion for the growing waves to develop, known as the *Kuo theorem*:

$$c_i \int_0^L \left(\beta - \frac{\partial^2 \bar{u}}{\partial y^2}\right)\frac{|\psi_0|^2}{|\bar{u} - c|^2}dy = 0 \qquad (9.45)$$

For exponentially growing solutions $c_i > 0$ (i.e., c_i being nonzero), the integral term in (9.45) must vanish. Since $|\psi_0|^2/|\bar{u} - c|^2$ is positive, the integral is zero when $(\beta - \partial^2 \bar{u}/\partial y^2)$ changes signs within the channel $0 < y < L$, or equivalently when the meridional gradient of the potential vorticity changes signs. This is a necessary condition for the barotropic instability to occur and is known as the *Rayleigh–Kuo criterion*, i.e.,

$$\beta - \frac{d^2\bar{u}}{dy^2} = 0 \quad \left(\text{somewhere within the latitude band}\right) \tag{9.46}$$

Since $\beta > 0$, the instability is favored where $d^2\bar{u}/dy^2$ reaches a high value. In general, a large β tends to stabilize the zonal flow, and broad currents with weak horizontal shear tend to be stable against the barotropic instability (Problem 9.21). The instability draws its energy from the kinetic energy of the mean flow by means of horizontal eddy stress. The most unstable modes are three-dimensional, and the half-wavelength scale in the downstream direction of the most unstable mode is $\sim\pi L_\beta$ (see Equation 7.91) in the form of waves and vortices. The characteristic growing timescale for the instability is $\sim 10/\sqrt{u\beta}$.

The barotropic instability is believed to be responsible for large-scale disturbances in the Earth's tropical region such as the depressions in the ITCZ waves in the African easterly jet, the disturbances in the equatorial Pacific, and eddies in the stratosphere. It could also occur in Venus mid-latitude winds and in a very strong circumpolar jet that develops in the Martian winter with zonal wavenumbers 1–2 and periods of 1–3 days.

According to this criterion, the jets in the giant planets Jupiter and Saturn show signs of instability, in particular in the westward jets where $d^2\bar{u}/dy^2 = \pm 2\beta$ (Problem 9.22). The instability could also develop at particular latitudes in the broad jet systems of Uranus and Neptune (Problem 9.23). It is, however, unknown how the vertical structure of the jets influences their stability against the barotropic growing disturbances.

9.4.4 Baroclinic Instability

9.4.4.1 Fundamentals

A *baroclinic* atmosphere shows temperature gradients on constant pressure surfaces, in particular $\partial T/\partial y \neq 0$, and thus it follows from the thermal wind equation (Section 7.2.7) that $\partial\bar{u}/\partial z \neq 0$. If the flow is statically stable $(N_B^2 > 0)$, then $Ri > 1$ and a type of disturbance called the *baroclinic instability* can develop. According to the classification scheme of Section 9.1, this occurs for an ample range of values of the Richardson number $(Ri \sim 1\text{–}1000)$.

Assume for simplicity that we have a pure zonal flow with velocity $(\bar{u}(z), 0, 0)$. Under the Boussinesq and f-plane approximations (Sections 7.2.6 and 7.2.1) and for geostrophic and hydrostatic conditions, the thermal-wind shear relationship (7.32a) can be written in terms of the density as

$$f_0 \frac{d\bar{u}}{dz} = \frac{g}{\rho_0} \frac{\partial\rho}{\partial y} \tag{9.47}$$

Here, ρ_0 is a reference density value, but the density function can be understood to be the sum of a mean value and a perturbation $\rho(y,z) = \bar{\rho}(z) + \rho'$. Consider the case of a flow in the northern hemisphere $(f_0 > 0)$ whose wind speed increases with altitude

$(d\bar{u}/dz>0)$. The density then increases with the northward distance $(\partial\rho/\partial y>0)$. In the meridional-vertical (y, z) plane, the density surfaces tilt with a slope α given according to partial differentiation rules by

$$\tan\alpha = \left(\frac{\partial z}{\partial y}\right)_\rho = -\frac{\partial\rho/\partial y}{\partial\rho/\partial z} \tag{9.48}$$

Since the flow is statically stable in the vertical $(\partial\rho/\partial z<0)$, it follows that $\tan\alpha>0$ and the density surfaces slope poleward and upward. A parcel moving from a low density surface to a higher one along a tilted northward-upward path in the (y, z) plane will change its gravitational potential energy (per unit volume) by an amount $\Delta E_p\sim -g\Delta\rho\ \Delta z$, releasing potential energy despite the stable density stratification $(\Delta\rho>0, \Delta z>0)$ (Problem 9.24). Add to this the internal energy change due to the temperature change modifying the above estimation by an amount $\sim (C_p/R_g^*)$. The resulting tilted motion is also called *sloping convection* with the northward moving parcels moving upward and the southward moving parcels moving downward. The released potential energy converted to kinetic energy (the reverse is valid when changing the slope sign) is used to feed a baroclinic disturbance that can be regarded as an eddy where the air parcels are interchanged along a slant path with slope α. It must be noted that only a small fraction of the total potential energy of the atmosphere is available for kinetic energy conversion, as discussed in Section 7.6.3.2.

9.4.4.2   Eady Model

The simplest formulation of the baroclinic instability mechanism is provided by the *Eady model* that relies on the following simplifications for the atmosphere: (1) the vertical shear of the zonal wind is constant $\bar{u}=\Lambda_0 z$ ($\Lambda_0=$constant), thus the streamfunction is given by $\psi=-\Lambda_0 yz$; (2) assume f-plane geometry ($\beta=0$); (3) rigid horizontal boundaries at $z=0$ and $z=H$ (no vertical motion across them); and (4) statically stable atmosphere, $N_B=$constant (Boussinesq approximation; the background density is a linear function of the meridional distance, y). Under such simplifications, from the linear form of the QGPV presented in Equation 8.31, the disturbed streamfunction ψ' obeys

$$\left(\frac{\partial}{\partial t}+\bar{u}\frac{\partial}{\partial x}\right)\left[\nabla^2\psi'+\frac{f_0^2}{N_B^2}\frac{\partial^2\psi'}{\partial z^2}\right]=0 \tag{9.49}$$

The boundary condition $w=0$ at $z=0$, and H implies that $(d_g/dt)(\partial\psi/\partial z)=0$ at these boundaries (see Equations 7.62 and 7.63). Its linearization about the background flow with $\partial\bar{u}/\partial z=\Lambda_0$ gives

$$\left(\frac{\partial}{\partial t}+\bar{u}\frac{\partial}{\partial x}\right)\frac{\partial\psi'}{\partial z}-\Lambda_0\frac{\partial\psi'}{\partial x}=0 \quad \text{at } z=0, H \tag{9.50}$$

We look for the normal mode solutions of Equations 9.49 and 9.50 of the form

$$\psi'(x, y, z) = Re\left\{\psi_0(z)\cos(ly)\exp\left[ik(x - ct)\right]\right\} \tag{9.51}$$

where $k>0$ and the phase speed may be complex $(c=c_r+ic_i)$, allowing for growing disturbances. Note that this form of solution implicitly presumes that the disturbances are trapped meridionally with rigid walls at $y=\pm\pi/2l$ and with the amplitude dependent only on the vertical. The substitution of (9.51) into Equation 9.49 leads to

$$ik(\bar{u} - c)\left[\frac{f_0^2}{N_B^2}\frac{\partial^2\psi_0}{\partial z^2} - (k^2 + l^2)\psi_0\right] = 0 \tag{9.52}$$

Hence, if $\bar{u}(z)\neq c$, it must be $d^2\psi_0/dz^2 - \alpha_B^2\psi_0 = 0$ where $\alpha_B^2 = \left(N_B/f_0\right)^2\left(k^2 + l^2\right)$. Equation 9.52 has solutions of the form

$$\psi_0(z) = A\sinh(\alpha_B z) + B\cosh(\alpha_B z) \tag{9.53}$$

where A and B are complex constants. Substitution of (9.51) into the boundary condition equation (9.50) leads to

$$\left(z - \frac{c}{\Lambda_0}\right)\frac{\partial\psi_0}{\partial z} - \psi_0 = 0 \quad \text{at } z = 0, H \tag{9.54}$$

The next step is the substitution of (9.53) into (9.54) that leads to a homogeneous system of two equations for the coefficients A and B. Consistency between the solutions requires that (Problem 9.25)

$$\left(c - \frac{\Lambda_0 H}{2}\right)^2 = \Lambda_0^2 H^2\left[\frac{1}{4} - \frac{\coth(\alpha_B H)}{\alpha_B H} + \frac{1}{\alpha_B^2 H^2}\right] \tag{9.55}$$

Since $c=c_r+ic_i$, we note first that

$$c_r = \frac{\Lambda_0 H}{2} \tag{9.56}$$

so the baroclinic wave is advected eastward with a phase speed that is equal to the mean flow at the middle of the layer (this is called the *steering* level). Note that the unstable baroclinic wave is not a Rossby wave since, as shown by this model, it exists for $\beta=0$. Second, the condition for the wave to grow exponentially $(c_i\neq0)$ requires that

$$1 - \frac{4}{\alpha_B H}\coth(\alpha_B H) + \frac{4}{\alpha_B^2 H^2} < 0 \tag{9.57}$$

The flow is *neutrally stable* for this quantity equal to zero and a critical value for $\alpha_B (= \alpha_{Bc})$ occurs when $\alpha_{Bc}H/2 = \coth(\alpha_{Bc}H/2)$, which implies that $\alpha_{Bc}H \sim 2.4$ (Problem 9.26). The instability development requires $\alpha_B < \alpha_{Bc}$, i.e.,

$$\left(k^2 + l^2\right) < \left(\frac{\alpha_{Bc}^2 f_0^2}{N_B^2}\right) \approx \frac{5.76}{L_D^2} \tag{9.58}$$

where $L_D = N_B H/f_0$ is the Rossby deformation radius scale (Equation 7.95). For squared waves with equal zonal and meridional wavenumbers $(k = l)$, the above condition gives the wavelength corresponding to the maximum growth rate:

$$L_{Bcl} = \frac{2\pi\sqrt{2}L_D}{(H\alpha_{Bm})} \approx 5.5 L_D \tag{9.59}$$

where $H\alpha_{Bm} = 1.61$ is the value for which kc_i is maximum when $l = 0$. The maximum growth rate is $kc_i \sim 0.31\Lambda_0 f_0/N_0$, which is proportional to the wind shear coefficient Λ_0 and also, according to (9.47), to the meridional temperature gradient. For simplicity, we have taken a constant Brunt–Väisälä frequency of N_0.

A complete stability analysis similar to that developed for the barotropic case in the previous section (Equation 9.46) indicates that the zonal flow becomes unstable to the baroclinic wave disturbance when, in a latitude band, the quantity

$$\beta - \frac{d^2\bar{u}}{dy^2} - \frac{1}{\rho_0}\frac{\partial}{\partial z}\left(\frac{\rho_0 f_0^2}{N_B(z)^2}\frac{\partial\bar{u}}{\partial z}\right) \tag{9.60}$$

changes sign—a condition known as *Charney–Stern* criterion. The third term in (9.60) is the baroclinic contribution to Equation 9.46 and contains two main factors: the vertical shear of the zonal wind and the vertical stratification given by $N_B(z)$, and as before, ρ_0 is the reference density. See Problems 9.27 and 9.28 for their application to Earth, Mars, and Saturn where baroclinic formations have been observed as presented in the following section.

9.4.4.3 Baroclinic Instability in Planets

Baroclinic waves form continuously in the Earth's mid-latitude jetstream (Northern and Southern hemispheres), being responsible for the classical eastward propagating extratropical anticyclone—the cyclone pattern in weather maps (Figure 9.16). The vortices with high- and low-pressure centers form in the crests and thoughts of the wave. In Earth's mid-latitudes, the Rossby deformation radius is $L_D \sim 1000\,km$ and then the wavelength of the maximum growth rate is

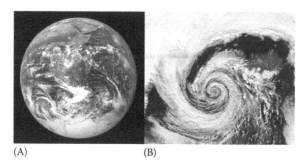

(A)   (B)

FIGURE 9.16 Earth's extratropical cyclones. (A) Global view of Earth's Southern Hemisphere with a series of cyclone vortices with anticyclonic areas between them, along the polar circle surrounding the Antarctic continent. Apollo 17 image; (B) A deep low pressure cyclone vortex over the southeastern coast of Iceland in September 4, 2003, obtained by the satellite Aqua. (Courtesy of NASA, Washington, DC.)

TABLE 9.1
Observed Baroclinic Features

Planet	Latitude	Altitude (z or P)	n (Zonal)	c (m s⁻¹)	V_T (m s⁻¹)	L (km)
Earth[a]	30°–60°	0–10 km	5–6	5–10	10–30	5000
Mars[b]	60°–80°	6–7 km	2–4	5–15	25–30	100–500
Saturn[c]	47°N	~0.5 bar	45–60	140	—	4000–5000

[a] Extratropical cyclones. Anticyclone–cyclone pattern (high- and low-pressure centers).

[b] Spirals, probably extratropical cyclones.

[c] "Ribbon wave," accompanied by anticyclone–cyclone pairs in crests and thoughts. It probably represents a manifestation of a Rossby wave or a baroclinic instability (see also Table 8.3).

~5500 km. At 45° latitude, this leads to about 5–6 such alternating anticyclone–cyclone structures (Problem 9.29). See Table 9.1 for baroclinic wave properties in planetary atmospheres.

The cyclones (also called extratropical storms) are characterized by their spiral shape in satellite images as drawn by *fronts* of cloudy bands. The fronts are narrow regions of intense temperature gradients that emerge as the baroclinic eddy develops. The process associated with the front generation is called *frontogenesis* and its discussion is beyond the scope of this text. A simple model takes the front as a sloping surface across which the temperature and the along-frontal wind are discontinuous, but with pressure and cross-front wind being continuous (Figure 9.17). Obviously, the warm air overlies the cold air, and the front is called warm when the warm air follows cold air and vice versa for the cold front. The warm air forced to ascend along the front condenses and forms cloud bands with associated rain. If the temperatures and along-frontal wind speeds are (T_1, v_1) for the cold air mass and (T_2, v_2) for the warm air mass, application of the hydrostatic and geostrophic wind relationships across the front gives its slope relative to the surface:

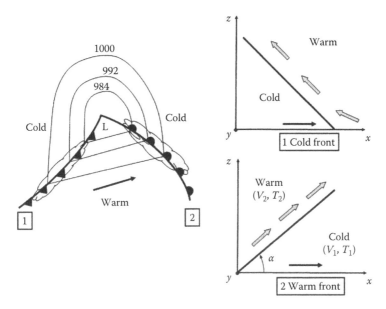

FIGURE 9.17 Earth's extratropical cyclone structure resulting from baroclinic instability as appears on surface weather maps. The central depression (L, low) is drawn by the isobars given in mbar with the fronts (1 cold, 2 warm) indicated. On the right are the cross-sections through the two front types (see text for the other symbols).

$$\tan \chi_F = \frac{f\left(v_1 T_2 - v_2 T_1\right)}{g\left(T_2 - T_1\right)} \tag{9.61}$$

At mid-latitudes 45°–50°, the typical values are $T_2 - T_1 = 3\,\mathrm{K}$ and $v_2 - v_1 = 10\,\mathrm{m\ s^{-1}}$, and for mean conditions with $T = 280\,\mathrm{K}$ and $v = 10\,\mathrm{m\ s^{-1}}$, the slope of the fronts is 1/50–1/300 or $\chi_F \sim 1.1° - 0.1°$.

Transient spiral cloud systems with cyclonic vorticity have been observed to evolve in the high-latitude jetstreams (60° in both hemispheres) of the Martian atmosphere (Figure 9.18). This is a season-dependent phenomenon. From the morphology ("front-like" cloud structure and a central eye in the storm), they are probably baroclinic cyclones. Models for mid-winter conditions in Mars indicate that a number of ~2 baroclinic waves could grow during the fall and spring seasons, increasing to ~3–4 in summer and winter due to the increase in the ambient static stability. The smaller wavenumber in Mars compared with Earth is due to its smaller planet radius since the Rossby deformation length is similar on both planets (~1000 km). The calculated growth rates for the instability are ~1 Martian day and the observations suggest that the cyclones are vertically shallow (at least in their vertical cloud extent). The spiral storms, in a gradient wind balance, propagate zonally with speeds ~5–15 m s⁻¹ (see Table 9.1 and Problem 9.17).

The applicability of the previous barotropic and baroclinic instability models to the giant planet (i.e., their weather layer in the altitude pressure range ~0.1–10 bar) is

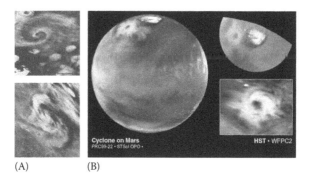

FIGURE 9.18 Images of spiral cloud bands in the atmosphere of Mars evolving around the pole. (A) Viking 1 orbiter images; (B) Hubble Space Telescope images. (Courtesy of J. Bell, Cornell University, Ithara, NY; S. Lee, University of Colorado, Boulder, CO; M. Wolff, SSI, NASA and NASA/ESA, Paris, France.)

probably conditioned by the nature of their deep adiabatic fluid structure. The lack of a lower rigid surface modifies the stability criteria (9.60). The baroclinic stability of deep zonal flows and mode development depends on three parameters: the value of the Rossby number, the ratio of the Rossby deformation radius to the flow width (L_D/L), and the ratio $\beta L_D^2 / u$. For purely geostrophic conditions, westward jets are more unstable than eastward ones. The baroclinic instability mechanism has been proposed as an explanation of Saturn's "ribbon" wave and its related pattern of anticyclone–cyclone eddies seen in the crests and thoughts of the wave (see Section 8.5.3, Figure 8.12 and Table 8.3). Images by the Voyagers 1 and 2 spacecrafts showed that these vortices rotate in an opposite sense depending on the position North or South of the wave. In addition, a significant meridional temperature gradient was measured across the wave. Both aspects are suggestive of a baroclinic instability mechanism. The estimated Rossby deformation radius is ~1100 km, also consistent with the length of the eddies.

Baroclinic waves are not expected to play a major role in Venus and Titan because of the low rotation rate of these bodies (Coriolis parameter $f_0 \sim 4 \times 10^{-7}$ s^{-1} at 45° latitude in Venus). The deformation radius for L_D reaches ~150,000 km, for N_B the deformation radius reaches ~10^{-2} s^{-1}, and for H the deformation radius reaches ~6 km typically for Venus, a value that exceeds the length of the parallel circle at this latitude.

9.5   VORTICES IN THE GIANT PLANETS

Large-scale vortices, say a size above 0.1 the planetary radius, are abundant in Jupiter and Saturn, scarce in Neptune, and practically nonexistent in Uranus. The differences are probably a consequence of the different number of alternating jets, their meridional width, and their peak-to-peak intensity among these planets (Chapter 7). Vortices can be defined as coherent structures in their cloud distribution, with closed parcel trajectories and small divergent or convergent motions. They detach by their albedo contrast relative to the surroundings. Some of Jupiter's vortices are long-lived, such as the famous GRS with a lifetime of 100 years and the former three White Oval Spots (WOS) that survived 50 years before merging into a single vortex

TABLE 9.2
Main Properties of Representative Vortices in the Giant Planets

Planet	Vortex	Latitude	$L_x - L_y$ (km)	V_T (m s⁻¹)	u_v (m s⁻¹)	ζ (max) (s⁻¹)	du/dy (s⁻¹)	f_0 (s⁻¹)
Jupiter	GRS[a]	22.5°S	24,000–11,000	100–160	−3	-6.1×10^{-5} A	-2×10^{-5}	-1.3×10^{-5}
	WOS[b]	33°S	10,000–6,000	100–120	+2	-6.4×10^{-5} A	-1.5×10^{-5}	-1.9×10^{-5}
	Barge[c]	15°N	12,000–3,000	55	−2.5	-4×10^{-5} C	-10^{-5}	$+9 \times 10^{-5}$
	A[d]	40°S	4,000–2,000	60	7	-8×10^{-5} A	-10^{-5}	-2.2×10^{-4}
Saturn	NPS[e]	75°N	10,000–5,000	—	0.2	−A	$+2.5 \times 10^{-5}$	$+3.1 \times 10^{-4}$
	BS[f]	42°N	6,000–4,000	55	+5	$+4 \times 10^{-5}$ A	$+3.3 \times 10^{-5}$	$+2.1 \times 10^{-5}$
	CS[g]	45.5°S	3,700–1,720	21	+68	-5.8×10^{-5}	-2.5×10^{-5}	-1.6×10^{-4}
Uranus	UDS[h]	28°N	2,700–1,300	—	43	—	1.5×10^{-6}	-9.5×10^{-5}
Neptune	GDS[i]	20°S	15,000–6,000	—	−350	−A	-1.6×10^{-5}	-7.4×10^{-5}
	DS2[j]	52.5°S	5,200–2,600	—	0	−A	-2.9×10^{-5}	-1.7×10^{-4}

Notes: A = Anticyclone [$(\partial\bar{u}/\partial y)/f_0 > 0$], C = Cyclone [$(\partial\bar{u}/\partial y)/f_0 < 0$].

[a] GRS (Great Red Spot). Its lifetime is >130 years (probably >300 years). It is a single vortex in its latitude band, although transient smaller anticyclones form from time to time within its band. The GRS size shrunk from $L_x \sim 39,000$ km in 1880 to ~24,000 km in 2008. The vortex oscillates in latitude with a period of 90 days and amplitude 1° longitude.

[b] WOS (White Oval Spots). These three ovals formed in 1939–1940 and were called BC, DE, and FA. They merged sequentially into a single oval, first BC and DE (in 1998) to form BE, and then BE and FA (2000) to form the present BA. They shrunk in their zonal length from 90,000 to 10,000 km.

[c] Barges are cyclones typically in a number of 3–5 in the latitude band and lifetimes ~1–2 years.

[d] White anticyclones that form in a number of 9–12 in this latitude band. Similar ovals form at many other anticyclone latitudes. Their characteristic lifetime is ~1 year and most disappear following mutual mergers.

[e] NPS (North Polar Spot) is a large single anticyclone observed during the Voyagers' encounters and later with telescopes, reaching a lifetime above 15 years.

[f] BS (Brown Spots) are typical anticyclones of Saturn's mid-latitudes where they are abundant (2–5 per latitude circle). Their characteristic lifetime is ~1 year and most disappear following mutual mergers. Similar ovals form at many other anticyclone latitudes, but their colors are variable.

[g] CS (cyclone spot) was a long-lived cyclone (surviving more than 4 years, 2004–2008) observed with the Cassini spacecraft ISS instrument.

[h] UDS (Uranus Dark Spot) was an isolated dark spot—the only one so far observed, tracked for 4 months in 2006. It had associated bright companion clouds.

[i] GDS (Great Dark Spot) was an isolated vortex and was the largest vortex in Neptune at the time of the Voyager 2 encounter (in 1989). The GDS migrated in latitude from 27°S to 17°S in 8 months, disappearing when close to the equator. It showed oscillations in longitude position (period 36 days, amplitude 9°) and in shape (L_x from 12,000 to 18,000 km, L_y from 5,200 to 7,400 km with a period of 8 days).

Similar spots were observed using HST and ground-based facilities at 32°N (from 1994 to 2000) and at 15°N (1996–1998) with the peculiarity when compared to the GDS that they did not migrate in latitude.

[j] DS2 (Dark Spot 2) was isolated and showed an active nucleus with bright clouds. It showed oscillations in longitude (period 36 days, amplitude 46°) and in latitude (amplitude 2.5°).

called BA in 1998–2000. Details of the measured properties of the giant planets' synoptic vortices are given in Tables 9.2 and 9.3.

9.5.1 DATA FROM OBSERVATIONS

On the giant planets, the synoptic scale vortices are embedded in between the eastward-westward opposed jets, i.e., in a meridionally sheared zonal wind. Their

TABLE 9.3

Compared Properties of Hurricanes and Jupiter's Great Red Spot

Altitude level	Property	Hurricane	GRS
Upper troposphere	Pressure	100 mbar	150–300 mbar
	$T_{center} - T_{outside}$	−2 K	−10 K[b]
	V_T^a	−10 m s^{-1}	−50 m s^{-1}
Troposphere	Pressure	200–900 mbar	500 mbar–1 bar
	$T_{center} - T_{outside}$	+15 K	+5 K
	V_T	+90 m s^{-1}	−100 m s^{-1}

[a] From the thermal wind relationship.
[b] The temperature difference (vortex–surroundings) decays with altitude to 0 K at altitude levels $P < 50$ mbar.

vorticity agrees in sign with the wind shear (du/dy) being a factor 2–10 of it but smaller than the planetary vorticity (f_0) at their latitude. Within the vortex, the vorticity is usually concentrated in a peripheric ring with a width of ~0.1 the length scale of the vortex. In Jupiter, the latitudinal extent of the vortices is that of the distance between the contiguous opposed jets. In Saturn, the vortices are smaller in width than this distance. In Uranus and Neptune, the jets are broad and the vortices have a much smaller meridional scale than the jets' peak-to-peak separation. There are no measurements of the vorticity of Neptune vortices but their cloud morphology indicates that it is the same as that of the ambient wind shear. Anticyclones are by far more abundant than cyclones in a ratio of approximately 9:1 for Jupiter and Saturn. In general, anticyclones are more stable (long-lived) than cyclones.

Anticyclones are oval in shape, with a 1.5–2 zonal to meridional length ratio although some of them show significant short-term changes in their shape. The Jovian GRS and WOS shrunk by a factor of 2 in their size along their lifetime. Jupiter and Saturn cyclones also show an oval shape and coherent structure, but in general cyclonic areas they are more irregular in their cloud morphology distribution than anticyclones. Their properties (size, vorticity, and motions) are determined from the patterns observed in the upper clouds (~0.5–1 bar), but they also manifest in the temperature field above clouds. For the largest vortices, it is also possible to delineate their periphery tracing their streamlines. Figures 9.19 through 9.22 show images of typical Jovian vortices.

Jupiter and Saturn anticyclones cloud tops are above surroundings (they are elevated structures). Their upper clouds are also thicker, more efficiently blocking the internal 5-μm infrared emission when compared with their surroundings, with colder temperatures at the tropopause level (see Figure 9.20 and Table 9.3). The contrary occurs with cyclones that have deeper and thinner upper clouds. In Jupiter, the cyclones have less organized clouds and show more chaotic internal distribution with filamentary and folded morphology. On Saturn, the anticyclones show internal laminar patterns with scarce large-scale structures (Figure 9.23). Only one dark spot, tentatively an anticyclone, has been detected in Uranus (Figure 9.24), although a bright, long-lived feature

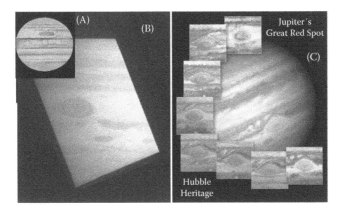

FIGURE 9.19 Views of Jupiter's Great Red Spot at different epochs: (A) Upper left, in 1880; (B) Left, by Pioneer 11 spacecraft during its fly-by in 1974 (Courtesy of NASA, Washington, DC); (C) A series of images taken during the 1990s by the Hubble Space Telescope. (Courtesy of NASA/ESA, Paris, France.)

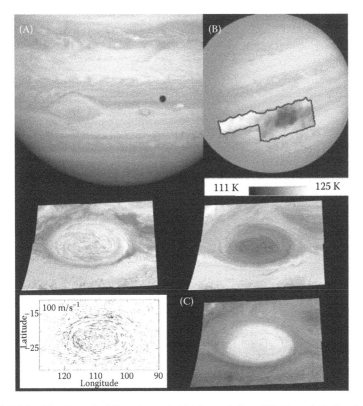

FIGURE 9.20 The Great Red Spot imaged at high resolution: (A) Upper left, Cassini spacecraft in 2000; (B) Upper right, temperature map above clouds (Galileo spacecraft, 1996); (C) Lower mosaic of four images, Galileo images in red, blue, methane, and wind field (top left to right, bottom right to left). (Courtesy of NASA/JPL, Los Angeles, CA.)

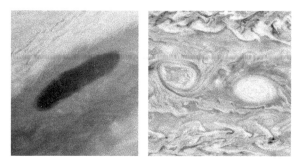

FIGURE 9.21   Images of cyclones in Jupiter: Left a "barge" at 16°N (Voyager 2); right, a cyclone–anticyclone pair at mid-latitudes (Galileo). (Courtesy of NASA/JPL, Los Angeles, CA.)

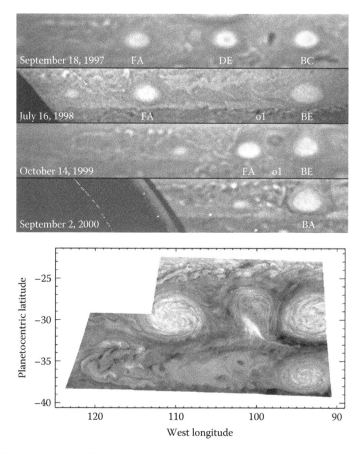

FIGURE 9.22   Jupiter's White Ovals evolution and mutual mergers between 1997 and 2000 (upper mosaic). (Courtesy of GCP-UPV-EHU, Spain.) In the lower image the aspect of White Ovals BC and DE in 1996 by Galileo spacecraft with a pear shape cyclone between both. (Courtesy of NASA/JPL, Los Angeles, CA.)

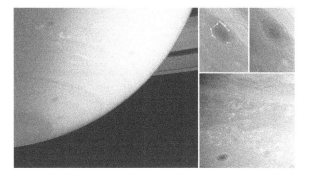

FIGURE 9.23 Families of Saturn anticyclones seen as dark ovals in red light in Cassini ISS images (left and bottom at right, located in the middle and high latitudes in the southern hemisphere), and in Voyager 2 images (upper right frames, located at 45°N). (Courtesy of NASA/JPL, Los Angeles, CA.)

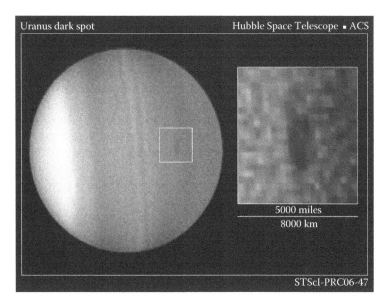

FIGURE 9.24 The "dark spot" of Uranus, a probably anticyclone vortex observed with the Hubble Space Telescope in 2006. (Courtesy of NASA, ESA, and L. Sromovsky, University of Wisconsin, Madison, WI.)

observed at latitude 34°S from 1986 to 2004 is also probably associated with a vortex. Neptune's vortices are featureless at the synoptic scale and in this aspect resemble Saturn, with the peculiarity that some of them develop bright clouds in their centers suggestive of probable ascending motions (Figure 9.25). Others, as Neptune GDS, show dynamically evolving "bright clouds" along their northern and southern edges.

In general, Jupiter and Saturn vortices are nearly stationary or move zonally at low speeds, relative to the background sheared flow. Most of them sit close to the latitude where the flow has zero velocity, i.e., in between the opposed jets, although there are

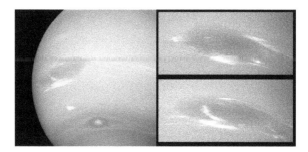

FIGURE 9.25 The Great Dark Spot of Neptune images by Voyager 2 cameras in 1989. (Courtesy of NASA, Washington, DC.)

exceptions such as the anticyclones that form in the intense eastward jet of Jupiter at 23°N. However, vortices are not directly entrained by the zonal flow (relative to System III), but move at ~0.1–0.3 times the mean flow speed. The Neptune GDS anticyclone has been observed migrating toward the equator in latitude until its disappearance. Latitude migration is not observed in Jupiter and Saturn vortices. The vortices of both planets remain almost fixed in latitude, with any possible meridional movement constrained by the jets that flow in opposite directions along their north and south flanks. Oscillations in the longitude and latitude position have been observed in some of the largest vortices in Jupiter and Neptune (details are given in the footnote of Table 9.2).

The formation process of the large-scale vortices has been observed in few cases. For example, the Jovian WOS (Figure 9.22) was formed from the fragmentation in three white areas of a whole latitude band, probably due to the growth of recirculation barriers within the band, in a process similar to that observed in the south tropical zone disturbance (STrD) (see Section 9.6.4). In Saturn, there are observations showing vortex formation following the development of a convective storm. A mechanism that has sometimes been invoked in the giant planets is that storms generated by moist convection develop intense upward motions that evolve to eddies and, by mixing, evolve to larger vortices.

Interactions between close vortices occur when the "impact parameter" defined as the latitude distance between their centers is smaller than their half meridional size. Observations show the following type of interactions: (a) mergers that follow a pair orbiting; (b) close encounters, with disruption of the minor vortex by the major one, entraining part of the smaller; and (c) close passage of one vortex next to another without any appreciable change in their structure or with slight changes in their motions.

Noninteracting large-scale single vortices (size > 2000 km) that do not merge with other vortices tend to survive for years. The slow decay in their vorticity probably reflects the low dissipation, as compared to what occurs on Earth or Mars, due to the lack of a rigid surface that acts as an effective dissipating boundary for the eddy.

9.5.2 MODELS

One of the major unknowns in understanding vortex nature in the giant planets is their vertical extent below the upper cloud at 0.5–1 bar. The temperature maps reveal that large vortices rapidly decay in their circulation above the tropopause (located at the ~100 mbar altitude level in all these planets), but their deep structure is a mystery. A simple scaling argument between the vertical-horizontal extents and the relevant oscillation frequencies in the vertical and horizontal directions (N_B and f_0, respectively) gives a rough depth scale ~$(f_0/N_B)L$. At 45°N, we have $f_0 \sim 2.5 \times 10^{-4}$ s^{-1} (Jupiter and Saturn) and $f_0 \sim 1.5 \times 10^{-4}$ s^{-1} (Uranus and Neptune). For $N_B \sim 10^{-2}$ s^{-1} (typical tropopause values) and vortex size $L \sim 5,000$–$10,000$ km, the resulting depth is ~100–200 km. This is similar to the vertical extent of the "weather" layer in the Jovian planets (the distance between the tropopause and the water cloud base for a solar abundance). If this is the case, the largest vortices such as the GRS and the WOS would be "pancake-like" with horizontal to vertical depth ratios of ~50. The horizontal scale of the largest vortices seems to be well above the Rossby deformation radius, which, at the tropopause level, is $L_D \sim 1500$–2000 km.

The behavior and basic properties of Jupiter and Saturn vortices (stability, geostrophic balance as a first approach [Problem 9.17], the oval shape, and merger when they interact) is reproduced by a variety of models and laboratory experiments. The smaller Jupiter and Saturn eddies (sizes < 500–1000 km) could have their origin in moist convection storms when the vertical ascending motions transform due to divergent motions into a close vortex. It is possible that the largest vortices survive against dissipation by absorbing smaller ones produced by moist convection, barotropic instability, or baroclinic instability.

Some time ago it was proposed that the GRS could be a gigantic hurricane-like feature. Some similarities between both phenomena are in fact observed (see Table 9.3). However, the lack of ocean surface as well as other important differences rule out such a hypothesis. It has also been suggested that the large vortices are "free modes" that grow in the wind sheared ambient under small dissipation. A simple model that reproduces some of the observations of giant planet vortices is the "Kida vortex." In a two-dimensional inviscid and nondivergent flow with uniform vorticity ($\partial u/\partial y$), a steady and stable elliptical vortex with vorticity $\langle \zeta \rangle$ can exist when the aspect ratio $\varepsilon_V = L_y/L_x$ (north-south to east-west length ratios) follows the relationship

$$\frac{1-\varepsilon_V}{\varepsilon_V(1+\varepsilon_V)} = \frac{|\partial u/\partial y|}{|\langle \zeta \rangle| - |\partial u/\partial y|} \tag{9.62}$$

The Kida vortex model is not a vertical structure, and does not need for its existence a gradient in the planetary vorticity (β-effect). Although it is a very simple representation of a vortex, it can suffer stable oscillations in its aspect ratio (east–west length to north–south width) that are similar to those observed in Neptune vortices. However, it is not evident that the relationship (9.62) holds for the vortices in other planets (Problem 9.30).

9.6 OTHER DYNAMICAL PHENOMENA

In this section, we present a summary of other dynamical phenomena that occur in planetary atmospheres and pertain to other types of instabilities not presented above or to unclassified dynamical phenomena. Here is a selection of some remarkable disturbances, not to be exhaustively covered.

9.6.1 MESOSCALE VORTICES

9.6.1.1 Dust Devils

Dust devils are short lived (typically minutes), rotating columns or tubes of rising convective (warm) air in spiraling flows that become visible when sand and dust are laden from the surface (Figure 9.26). They form both on Earth and Mars under strong insolation conditions around noon local time when a superadiabatic lapse rate and convective motions develop. Dust devils usually occur in arid regions of the Earth (southwestern USA, Africa, and Australia are typical) and in many areas of Mars following a period of maximum insolation. They are visually characterized by their dusty core and high rotational wind speeds that loft sand and dust into the core of the dust devil via the *saltation* mechanism. The particles have diameters of $1-100\,\mu m$ and are lifted into the air by winds, suspended temporarily but quickly falling by sedimentation. For the saltation to work, the surface speed must be above a certain critical value $u_* > u_{*c} \sim 5\,m\,s^{-1}$ and $30\,m\,s^{-1}$ for Earth and Mars, respectively (see Equation 7.81). The visible core represents only a small part of the complete vortex that typically extends horizontally $1-100\,m$ with heights up to several hundred meters or more in the case of the Earth. On Mars, whose boundary layer is much thicker, the dust devils can reach several hundred meters in diameter and up to $7\,km$ in height (Figure 9.27). The morphology of dust devils is highly variable but in general they range from tilted columns to inverted cones.

FIGURE 9.26   Image of a dust devil on Earth. (Courtesy of NASA, Washington, DC.)

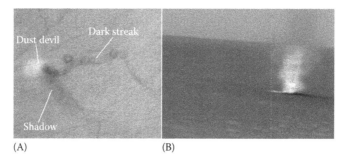

(A)   (B)

FIGURE 9.27   Images of dust devils on Mars. (A) Tracks produced on the dusty surface; (B) Image taken from the Spirit rover in March 15, 2005. (Courtesy of NASA/JPL, Los Angeles, CA.)

Dust devils have no tendency toward a sense of rotation since they are small and not influenced by Coriolis forces. The tangential wind speed increases toward the core of the vortex, reaching a maximum at the core boundary. The tangential wind velocity then drops linearly to zero at the core center. Dust devils can be represented as a *Rankine vortex* type with a core in solid body rotation (constant angular velocity ω) plus an outside irrotational vortex (Problem 9.31) and where the tangential velocity is null (the radial and vertical velocity are zero)

$$V_r = V_z = 0$$

$$V_T(r) = \frac{V_0 r}{r_0} = \omega r \quad r \leq r_0 \tag{9.63}$$

$$V_T(r) = \frac{V_0 r_0}{r} = \frac{\omega r_0^2}{r} \quad r \leq r_0$$

with V_0 being the maximum velocity at the core edge radius r_0. Typically, V_0 is 5–10 m s^{-1} but peak velocities of ~25 m s^{-1} can be reached. The Rossby number for Earth's dust devils is in the range of $Ro \sim 500$–2000 (for $V_0 = 20$–10 m s^{-1} and $r_0 = 100$–10 m, respectively) and $Ro \sim 100$–400 for Mars (for $V_0 = 100$–30 m s^{-1} and $r_0 = 2$ km–100 m, respectively). Therefore, they can be considered to be dynamically in cyclostrophic balance (Equation 9.33). Accordingly, the core pressure drops by ~2–5 mbar and 10–50 μbar in the cases of the Earth and Mars, respectively, with a corresponding core temperature rise of 1–10 K in both cases.

Dust devils were first discovered on the Martian surface from Viking orbiter images and have been observed and measured *in situ* by rovers and by the tracks they leave on the surface. They are particularly important phenomena on Mars because they might provide a mechanism for the injection of micron-sized particles into the atmosphere, significantly contributing to the Martian dust cycle.

9.6.1.2 Earth's Tornado

A tornado is an intensely rotating narrow column of air that extends downward to the ground from cumuliform stormy clouds. Tornados form when a warm and moist layer is topped by a dry layer with a thermal inversion occurring between both. The tornado becomes visible from a funnel-shaped cloud pendant from the base of the cumuliform storm and/or from the dust and debris that spiral upward from the ground (Figure 9.28). When they form over the sea, they are called *waterspouts*.

Tornados can reach maximum velocities of 110–125 m s^{-1} at the periphery of a column with a radius of 100–200 m (the core), although their radius of influence can reach a few km. The highest velocities are concentrated within a radius of ~50 m and in altitudes from the ground to 200 m. The vertical velocity can reach up to 80 m s^{-1} with radial inflows in some cases, concentrated in narrow bands, reaching velocities of ~50 m s^{-1}. The radial dependence of the velocity field in a tornado is well represented by the Rankine vortex model equation (9.63). Due to the intense rotation, the Rossby number is so large that the cyclostrophic balance (Equation 9.33) applies, and then in terms of the pressure gradient, we have $V^2/r = \rho^{-1} (\partial P/\partial r)$. As a consequence,

FIGURE 9.28   Tornado images over Earth's ocean and land. The upper frame shows series of *waterspouts* (tornados over the sea) on the Albanian coast. (Courtesy of Robert Giudici) Lower image is from World Press.

the pressure drops at its center by 10–100 mbar relative to its surroundings (see Problems 9.15 and 9.32).

The rotating air column increases its vorticity exponentially in time due to column stretching, approximately following the law

$$\zeta(t) = \zeta_0 \exp\left(\frac{t}{\tau}\right) \tag{9.64}$$

where τ is the e-folding time for tornado amplification, which is typically of the order of hours or less (see Problem 9.33). Tornados produce much damage and are abundant in the south and center of the United States during the months of April, May, and June, e.g., in the Great Plains there is an area known as "Tornado Alley" where about 900 tornados occur per year.

9.6.1.3   Von Kármán Streets

In a classical laboratory experiment, a solid body placed in a low viscosity laminar gas or a liquid flow can, under certain conditions, give rise to a periodic vortex shedding in its wake, leaving behind a trail of eddies in the flowing medium. Under low flow velocities, two stationary counter-rotating vortices form behind the obstacle, remaining anchored to it. As the flow speed increases, one of the vortices detaches and begins to travel downstream on one side of the wake. Next, the vortex on the other side does the same while a second vortex forms in the place of the first, and so on sequentially—a *Von Kármán vortex street* forms. In the atmosphere, the repeated pattern of swirling vortices becomes visible when clouds are present. In the case of the Earth, this occurs in the lee of isolated islands that act as the solid obstacle to the flow (Figure 9.29). The other known case of a probable atmospheric vortex street has been observed in Saturn, where the barrier to the zonal flow was produced by a

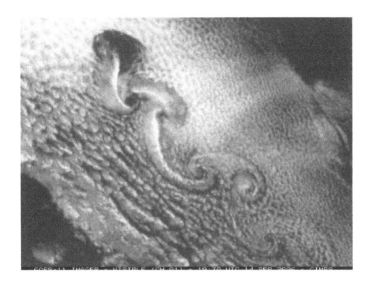

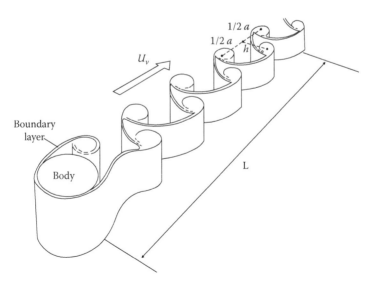

FIGURE 9.29 Von Kármán street pattern of vortices formed on the trail of the Guadalupe island (west of Baja California) formed in the field of marine stratocumulus (from GOESS 11 satellite), and the parameters defining this kind of instability. (Courtesy of NOAA/NASA-GSFC, Washington, DC.)

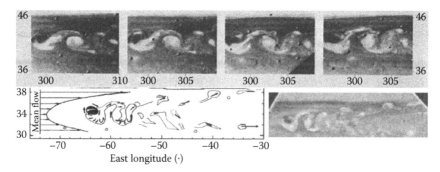

FIGURE 9.30 A Von Kármán street of eddies formed at mid latitudes in the atmosphere of Saturn as observed in 1981 by the Voyager 2 spacecraft. (Courtesy of NASA, Washington, DC.)

compact anticyclone vortex (Figure 9.30). The pattern was observed at mid-latitudes around 40°N—a region particularly active in forming vortices and storms.

Figure 9.29 also shows a scheme of the Von Kármán vortex street geometry and the main parameters intervening in its formation: the horizontal and vertical distance between eddies (a, h), the pattern velocity relative to the obstacle (U_V), and the pattern length (L). For a uniform flow, laboratory studies show that the only arrangement for the pattern to become stable occurs when $h/a = ar\cosh\left(\sqrt{2}\right)/\pi = 0.281$. The rate at which vortex pairs are shed from the obstacle is given by

$$N = \frac{U_V}{a} \qquad (9.65)$$

This number is related to the undisturbed flow velocity U_0 and size of the obstacle (for a circular section of diameter D) through the *Strouhal number*

$$St = \frac{ND}{U_0} \qquad (9.66)$$

which is empirically determined from laboratory data. For cylinders, $St=0.2$. The velocity V induced by one row of vortices in the other is $V=(\Gamma_c\pi/a)\tanh(\pi h/a)$ and we have approximately that $U_V=U_0-V$. Here, $\Gamma_c=V_T r$ is the circulation of the induced vortices and r is their radius (Problem 9.34). In Table 9.4, we give the values for the parameters involved in *Von Kármán streets* as observed on the Earth and Saturn.

9.6.2 MARTIAN DUST STORMS

In the atmosphere of Mars, there is a permanent layer of suspended dust. The dust particles have a typical radius of ~1–2 μm and are made in a 60% proportion by SiO_2. The optical depth of this dust cover varies cyclically ranging from $\tau\sim0.3$–0.5 in the clearest seasons (northern summer) to $\tau\sim2$–5 during the development of large-scale

TABLE 9.4

Parameters for Von Kármán Streets Observed on Earth and Saturn

Parameter	Symbol	Saturn (Mid-latitudes)	Earth (Canary I.)
Obstacle diameter	D	1700 km	45 km
Bandwidth of street	h	800 km	55 km
Distance between eddies	a	5000 km	150 km
Zonal flow velocity	U_0	2–10 m s^{-1}	7.5–10 m s^{-1}
Vortices zonal velocity	U_V	2–6 m s^{-1}	5 m s^{-1}
Length of vortex street	L	9000 km	800 km
Von Kármán ratio	h/a	0.16	0.35–0.4
Vortex production rate	N	~10^{-6} s^{-1}	4×10^{-5} s^{-1}
Vortex lifetime	t_L	~17–95 days	45 h

dust storms. The dust cycle intervenes in the radiative-dynamical state of the atmosphere since the dust injected by the winds modifies the solar radiation heating rates, the temperature and pressure fields, and the dynamics and the winds from the surface up to ~50 km altitude. In the altitude in the first 20 km, the heating rates vary from 20 K per Martian day (a sol) during the clearest periods to 100–150 K sol^{-1} during the dusty seasons.

Dust storms are identified by the massive injection of dust particles in the atmosphere up to altitudes of 45–50 km (Section 7.4.1) (Figure 9.31). Historically, they were first described, from telescopic visual observations, as "yellow clouds" to distinguish them from other types of "bright" and "white" condensate clouds. Storms take place at three different scales: local (extension < 2000 km), regional (extension > 2000 km, not encircling the planet), and planet-wide scales, which are called *major dust storms*. In this last case, various centers of activity occur and the ambient winds expand the dust along the whole planet. The major storms develop during the southern spring and summer seasons when Mars is within a solar longitude $L_s = \pm 90°$ from the perihelion. The southern subtropics are a usual trigger place. Two famous planet-wide storms (probably the only true "global" cases) occurred in August 1956 (at $L_s = 249°$) and in September 1971 (at $L_s = 260°$). A recent global storm took place in 2001 (Figure 9.31). There is some seasonal recurrence in the storm production with one or two planet-encircling storms occurring every Martian year, although there are also years without a major storm occurring. All this suggests an apparent interannual variability.

There have been a variety of proposals to explain the onset of dust storms, all of them relying on positive feedback between the injected dust and enhanced winds. The trigger occurs when a disturbance close to the surface generates winds with enough intensity to be able to lift massively small particles, probably through the saltation mechanism (Section 7.4.1) or by dust devils. The Viking landers measured wind speeds ~15–30 m s^{-1} at an altitude of 1.5 m from the soil during local storms. However, the mechanism by which a localized dust storm evolves to a planet-encircling storm remains obscure. One possibility is that winds increase in some places by a positive

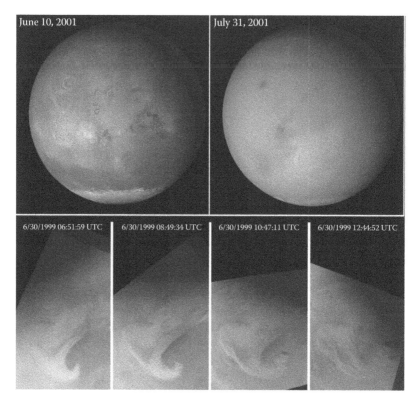

FIGURE 9.31   Martian dust storms from images obtained by the MOC camera onboard the Mars Global Surveyor. Upper image: The global dust storm that occurred in 2001 is evident in this image pair comparison with a one and a half month separation. Lower image: Evolution of the frontal-like dust storm in 1999 (dates indicated). (Courtesy of NASA/JPL/Malin Space Science Systems, Paris, France.)

coupling between the seasonally variable Hadley cell, tides, and planetary-scale topographic winds. The superposition of these effects occurring during the summer heating maximum, enhanced near perihelion, could produce the required winds.

The dust storm decays rapidly, i.e., the dust is settled into the ground, in an exponential form in around 2 months. Other slower removal processes (sedimentation by water and CO_2 condensation on the grains) act on a longer timescale.

9.6.3   EARTH EQUATORIAL–TROPICAL PHENOMENA

We now describe phenomena that are unique to the Earth since they relay in the existence of the oceans (tremendous water reservoir unique to our planet) and to their potential energy put in action by the intense radiation heating in the equatorial latitudes. The first difference between mid-scale phenomena occurring at low latitudes (e.g., between latitudes $\pm 30°$) and mid-latitudes resides in the weakness of the Coriolis force. The Coriolis parameter at tropics is $f \le 10^{-5}$ s^{-1}, and thus the Rossby number is >1. The second difference resides in their energy source.

Assume first that no precipitation occurs so the synoptic-scale temperature fluctuations are produced by diabatic heating due primarily to the emission of long-wave radiation (Section 4.3.3), which cools the troposphere at a rate of ~1 K day^{-1}. The temperature fluctuations are small (~0.3 K day^{-1}) and the radiative cooling is approximately balanced by adiabatic warming due to subsidence. A scale analysis of the large-scale tropical motions shows that the divergence of the horizontal wind is very low ~10^{-7} s^{-1}. Synoptic-scale disturbances (no precipitating) with vertical extent comparable to the scale height are barotropic. The absolute vorticity is conserved following the nondivergent wind and no conversion of potential energy to kinetic energy occurs. They must be driven by a barotropic conversion of the mean-flow kinetic energy or by a meridional coupling to mid-latitude synoptic systems.

When precipitation occurs, the latent-heat release by moist convection becomes the main energy source (see Sections 9.2.2 and 9.2.3). Massive moist convection at the equatorial latitude produces the large-scale cumulus cloud clusters that dominate the visual aspect in satellite images. Typical precipitation rates in the equatorial area are $m_w \sim 2$ cm day^{-1}, implying a condensation of 20 kg of water in an atmospheric column with a cross-section of 1 m^2. Since the latent heat of condensation of water is $L_H \sim 2.5 \times 10^6$ J kg^{-1}, the energy added to the atmosphere is $\sim m_w L_H \sim 5 \times 10^7$ J m^{-2} day^{-1}. Distributing uniformly over the atmospheric column of mass ($P_s/g \sim 10^4$ kg m^{-2}) gives a heating rate per unit mass of $L_H m_w / C_p (P_s/g) \sim 5$ K day^{-1}. The condensation heating in deep cumulus towers occurs in fact within a layer between 300 and 400 mbar leading to heating rates as high as 10 K day^{-1}. This produces strong updrafts and divergence in the layer, and the synoptic motions are not barotropic. The condensation heating is, therefore, a fundamental dynamical ingredient in the equatorial region, and the way these "localized" deep cumulus towers couple to the synoptic scale structures is a complex problem. Different techniques have been proposed to explain coupling, usually within a *parameterization scheme* of the cumulus convective heating, but their description is beyond the scope of this introductory text. We comment in the following text on the main phenomena that takes place in the equatorial-tropical tropospheric dynamics.

9.6.3.1 The Intertropical Convergence Zone and Equatorial Disturbances

The ITCZ (see Section 7.5.1) is a large undulating band of cumulus clouds that forms close to the equator (between latitudes ±10°) as a result of the convergence of winds in the lower troposphere due to the thermally direct Hadley circulation (the equatorward motion of the *trade winds*) discussed in Section 7.2.8 (Figure 9.32). Continuity requires the air to rise following, e.g., a pseudoadiabatic ascent (Equation 5.42), condensing and forming clusters of cumulonimbus clouds.

The ITCZ clouds show a zonal variance in their organization that is caused by weak westward-moving *equatorial wave disturbances* driven by the release of latent heat in the convective precipitation regions. The wave organizes a pattern of cloudy areas separated by ~3000–4000 km, moving westward with a speed of ~10 m s^{-1} and periods of ~4–5 days. The land-sea distribution in the equator strongly influences the ITZC pattern, as occurs with the *African wave disturbance* that forms under the strong insolation conditions over this continent.

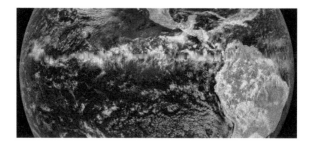

FIGURE 9.32   The Intertropical Convergence Zone (ITCZ) as imaged by the GOES-11 satellite. (Courtesy of NOAA/NASA-GSFC, Washington, DC.)

An important seasonal reversing circulation system is the *monsoon* that results from the contrast in the thermal properties of land and sea. The most notorious case occurs in the tropical region of Asia dominating the climate in the whole Indian area. The low heat capacity of the land compared with that of the sea produces enhanced cumulus convection and latent heat release with warm temperatures throughout the troposphere. Convergence (and low pressures) over the land are accompanied by divergent motions in the upper troposphere due to the updrafts of warm air. These are accompanied by divergence (and high pressures) over the sea surface with low pressures occurring in the upper troposphere that produce the descent of cold air over the sea. These circulations generate cyclonic and anticyclonic vorticity centers that are driven directly by the latent heat release and by radiative heating and cooling. This is the "moist phase" of the Asian monsoon. During the winter time, the land-sea thermal contrast reverses and the circulation is just the opposite of that described, with continents becoming cool and dry and precipitation occurring over the warm ocean (Figure 9.33). The monsoon seasons are regulated by the north and south displacements of the ITCZ across the equator: southward for northeastern monsoons and northward for southeastern monsoons.

Other longitudinal variations in the ITCZ structure are due to the sea surface temperature distribution that induces zonally symmetric circulations with east-west overturning cells, the most prominent occurring over the equatorial Pacific and in the Indic, known as the *Walker circulation* (Figure 9.34, see Section 7.5.1). This cell circulation has east-west associated pressure gradients that suffer inter annual variations. The global pressure pattern and its associated changes in wind, temperature, sea surface elevation, and precipitation is known as the *southern oscillation* (SO) (see Section 9.5.1) and has periods of around 2–5 years. To characterize this oscillation, several indexes are employed: the sea surface temperature anomalies (SST) over the eastern Pacific (the deviations from the mean value; peak ~ ±2.5°C); the anomalies in the sea surface pressure (deviations from the mean; peaks ±2.5 mbar) measured as the difference between the pressure in Tahiti and Darwin locations; and sea surface elevation deviation (up to ~60 cm). The SO manifests in the circulation of the ocean and produces the warming of the coastal waters of Peru and Ecuador during Christmas time. Over the following months, it spreads westward along the equator inducing a large anomaly in the sea surface temperatures and elevation, a phenomenon known as *El Niño* (Figure 9.34). During this "warm" phase,

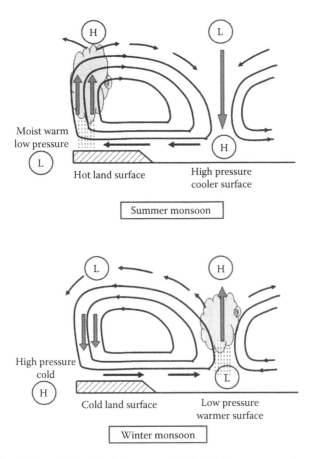

Moist warm
low pressure

Hot land surface

High pressure
cooler surface

Summer monsoon

High pressure
cold

Cold land surface

Low pressure
warmer surface

Winter monsoon

FIGURE 9.33 Scheme of the circulations associated with the monsoon phenomena for the summer and winter seasons.

the trade winds weaken, the pressure anomaly is low, and the wind-driven oceanic eastward motion reduces the upwelling over the equatorial Pacific (warm waters occurring over Peru) with enhanced convection there. In the opposite "cold" phase known as *La Niña*, the anomaly results in a high in the pressure field with the trade winds increasing their intensity (flowing westward) and with cool waters emerging over the equatorial Pacific coast. These complex phenomena are collectively known as El Niño-southern oscillation (ENSO).

A key feature of the ocean-atmosphere interaction is the positive feedback between the intensity of the trade winds and the SST contrasts (referred to sometimes as the *Bjerknes feedback*). A simple description of the ENSO is provided by the "delayed-oscillator" model in which the sea surface temperature anomaly T in the eastern Pacific is described by a differential-delay equation

$$\frac{dT}{dt} = cT(t) - bT(t - \tau) \tag{9.67}$$

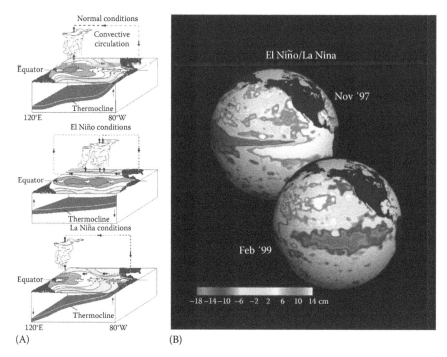

FIGURE 9.34   (A) Scheme of the Walker circulation and "El Niño–La Niña" phenomenon. (Courtesy of NOAA); (B) Sea surface elevation variability cycle due to "El Niño–La Niña" as measured by the Topex-Poseidon experiment between 1997 and 1999. (Courtesy of NASA/CNES, Paris, France.)

where
  c and b are the positive constants
  τ is the timescale associated with the dynamical adjustment of the ocean

The first term represents the positive feedback associated with the pressure anomaly (as proposed by Bjerknes), whereas the second term provides negative feedback due to the adjustment to the ocean temperature caused by propagating equatorial waves in the ocean (Kelvin and Rossby types). The waves carry the energy and momentum they receive from the wind stress at the ocean surface and provide the "memory" for the year-to-year variability and ENSO.

9.6.3.2   Tropical Cyclones

Tropical cyclones are intense vortices that develop close to the equatorial latitudes following the period of strong radiative heating of the oceans (Figure 9.35). They are called *hurricanes* when forming over the Atlantic and eastern North Pacific, and *typhoons* when forming over the western North Pacific. Their primary energy source is the latent heat released during massive moist convection developing over the warm oceans (surface temperature $T_s \geq 26^\circ C$) (Section 9.2.2). The initial rotation is initially provided by the weak Coriolis force, requiring the storm formation at latitudes above

FIGURE 9.35 Series of images of Hurricane Andrew that developed and evolved over the Caribbean Sea between August 16 and 28, 1992. (Courtesy of GSFC/NASA, Washington, DC.)

~5°–10°. The hurricane develops when a saturated lapse rate gradient (Section 5.2.5) is achieved near the center of rotation of the growing storm. Low vertical wind shear, especially in the upper level of the atmosphere, and high relative humidity values from the surface to the mid levels of the atmosphere also favor the storm formation.

A hurricane can be viewed as a thermodynamic machine that converts the potential energy to kinetic energy obtained from the fluxes of water, latent heat, and sensible heat acquired at the air-sea interface. Energy dissipation occurs, due to friction, at the boundary layer. The presence of a lower boundary layer (the ocean surface) provides convergence and moisture, and at the top, an upper boundary layer, the tropopause, causes the air to diverge and dry (Figure 9.36). In a crude approximation, the hurricane can be viewed as a *Carnot* cycle heat engine in which the heat is extracted from a warm focus (the ocean) at a temperature $T_s = 300\,\mathrm{K}$ and deposited into a cool focus by radiative heating to space at a temperature of $T_0 = 200\,\mathrm{K}$ (the upper troposphere). This *Carnot heat machine* has a high efficiency $\sim 1 - (T_0/T_s) \sim 0.33$.

A particular feature of the hurricane is the presence of a central *eye*, a region with a radius of about 5–50 km free of clouds. Descending dry air and clear skies occur there. The eye is enclosed by an *eyewall* (radius ~ 100 km) where the most massive cumulus clouds and strong convection develops and where the winds reach velocities up to 70–90 m s^{-1} in a cyclonic rotation path. Within the eye, the wind speeds are low and vertically its structure is dominated by a warm and moist boundary layer 2–3 km thick, overlaid by a hot dry air cap at temperatures >30°C but with humidity <50%. The vortex global structure extends horizontally ~3000 km formed by cloud bands (ascending cumulus towers) spiralling toward the eye. Some properties of hurricanes are given in Table 9.3.

A hurricane can be represented by an axisymmetric vortex in gradient wind balance since $Ro = V/fL \sim 10$; for typical conditions $V = 50\,\mathrm{m\ s^{-1}}$, $f = 5 \times 10^{-5}\,\mathrm{s^{-1}}$ (latitude 20°), and $L \sim 100\,\mathrm{km}$ (eyewall radius). Using the radial coordinate (the normal direction positive outward) and pressure instead of geopotential with $\partial\Phi/\partial r = \rho^{-1}(\partial P/\partial r)$, Equation 9.29b is written as

FIGURE 9.36 A scheme of the structure model of Hurricane Andrew. The highest winds reached 285 km h^{-1}, and the lower pressure at its center was 922 mbar. (Courtesy of GFDL/NOAA, Washington, DC.)

$$\frac{V^2}{r} + fV = \frac{1}{\rho}\frac{\partial P}{\partial r} \qquad (9.68)$$

The cyclonic rotation implies low pressures at the hurricane center ~950 mbar (the record was established at 870 mbar by typhoon Tip in 1979). Using the thermal wind relationship for this gradient balance (assuming hydrostatic conditions in log-pressure coordinates, i.e., $\partial\Phi/\partial z^* = R_g^* T/H$), after taking the vertical derivative of (9.68), we get

$$\frac{\partial V}{\partial z^*}\left(\frac{2V}{r} + f\right) = \frac{R_g^*}{H}\frac{\partial T}{\partial r} \qquad (9.69)$$

The cyclonic flow in a hurricane is maximum above the boundary layer (see Table 9.3) and decreases with altitude ($\partial V/\partial z^* < 0$), which from (9.69) implies $\partial T/\partial r < 0$. At the lower level, the hurricane has, therefore, a warm core with enhanced temperatures by ~10°C (Problem 9.35). Above the boundary layer, the absolute angular momentum (defined as the momentum per unit mass $L_{hur} = Vr + fr^2/2$) is approximately conserved following the motion, decreasing with altitude ($\partial L_{hur}/\partial z^* < 0$) (Problem 9.36). Thus, surfaces of constant angular momentum tilt radially outward with increasing height. If no azimuthal forces are present, air parcels ascending by convection conserve L_{hur} above the boundary layer and spiral outward following the L_{hur} surfaces.

Tropical cyclones move relative to the Earth's surface with a speed of ~5–10 m s^{-1}, first westward (advected by the mean flow) and then poleward by altering the vorticity conditions of their environment. When reaching higher latitudes, the cyclone enters in contact with lower sea temperatures that are unable to sustain it, or moves onto land, losing the moisture supply. The hurricane decays in its azimuthal circulation within the eyewall in few days, spinning down due to the friction force, becoming a regular storm system until its dissipation.

9.6.4 BELTS AND ZONES AND PLANETARY-SCALE DISTURBANCES IN JUPITER AND SATURN

The visual aspect of Jupiter and Saturn is dominated by an alternating pattern of low and high albedo bands with latitude, called *belts* and *zones*. The dynamics and cloud and haze microphysics behind their nature is unknown. Above the tropopause (pressure altitude range ~10–100 mbar), the UV images (wavelengths ~250–350 nm) and methane absorption band images (wavelengths from 890 nm to 2.3 μm) show the correlated banding distribution. The aerosols make the high altitude hazes bright in the methane bands (relative to surroundings) by sunlight reflection at high altitudes and dark by their absorption in UV (see Figures 5.22 through 5.25). Below the tropospheric hazes (pressure range ~1–4 bar), the opacity of the NH_3–NH_4SH clouds to the escaping thermal radiation at a wavelength at 5 μm shows that the banding distribution is present. The pattern is linked to the system of zonal jets (Figure 9.37),

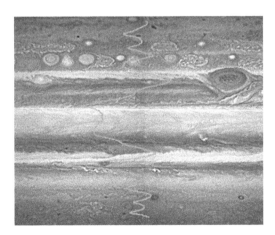

FIGURE 9.37 Map projection of Jupiter showing the band and zone pattern as observed at visual wavelengths with a superimposed zonal wind profile. South is up and East to the right (inverted astronomical view). (Courtesy of E. García-Melendo.)

but the mechanism beyond this relationship is not obvious, and apparently differs in Jupiter and Saturn. In Jupiter, the belts are confined between the peak of the two contiguous opposed jets (one westward and the other eastward). The belts are regions of cyclonic vorticity and the zones are regions of anticyclonic vorticity. But in Saturn, there is not a systematic relation between the position of the belts and zones and the peaks of the jets. In addition, the meridional extent of the bands and their reflectivity varies significantly in time without a correlated change in the jet system that remains globally stable. When mapping the temperature field measured above clouds (Chapter 4, Figures 4.16 and 4.17), and comparing with the aerosol/haze optical depth and wind distribution ($P < 0.5$–1 bar), the picture that emerges is that the belts and zones are related to meridional cell distribution. Air rises in the zones (where high and thick clouds form) toward the equator of an eastward jet peak and descends in the belts (where clouds are lower and thin) toward the pole of an eastward jet peak (Figure 7.28).

Some belts and zones of Jupiter and Saturn suffer major changes in their reflectivity, called disturbances. In Jupiter, "outbreaks" of activity in the belts occur when one or more bright spots "erupt" as "plumes" (Section 9.2.3), followed by a planetary-scale disturbance that leads to turbulent and turmoil features, finally changing the reflectivity of the band upon mixing and homogenization (Figure 9.38). The two best known examples are the south equatorial belt disturbance (SEBD) at 16°S and the north equatorial belt disturbance (NTBD) at 23.5°N. Another form of albedo change occurs in the STrD at 24°S. It consists in the formation of a curved dark column

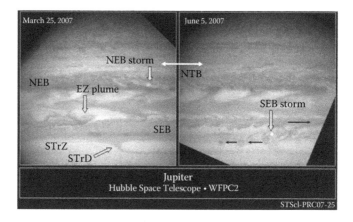

FIGURE 9.38 Images of Jupiter in 2007 by the HST showing the group of the most important large-scale disturbances. The bands and zones that suffer from them are identified by their acronym and disturbances are signaled by the broad arrow: North Equatorial Belt (NEB) and convective storm at 9°N; South Equatorial Belt (SEB) with a convective storm at 16°S and disturbances propagating in the direction marked by the thin arrow (relative to the source) at 9°S and 21°S; South Tropical Zone (STrZ) and its disturbance with columnar shape (STrD) at 23°S; Equatorial Zone large plume at 6°S. Note also the change in the albedo of the North Temperate Belt (NTB) that followed a large-scale disturbance between both dates. (Courtesy of NASA, ESA, A. Sanchez-Lavega, University of the Basque Country and A. Simon-Miller, NASA Goddard Space Flight Center, Greenbelt, MD.)

that grows between two opposed jets and acts as a barrier to the flow, giving rise to a recirculation pattern. In Saturn, the major albedo change in the bands occurs following the rare outbreaks of bright clouds known as the GWS (Figures 9.6 and 9.7). The large storm generates periodically spaced bright and dark cloud features that propagate along the band until encircling the whole latitude circle in scales ~ month. As indicated above, it is striking to see that although the bands change their reflectivity and morphology following these disturbances, the zonal wind jet system keeps its structure both in velocity and location.

9.6.4.1 The South Equatorial Belt Disturbance in Jupiter

This disturbance occurs in the latitude band from 10°S to 20°S, usually occupied by a low reflectivity "belt" (at visible wavelengths) named the South Equatorial Belt (SEB). However, the SEB mutates cyclically with a not well-defined temporal interval (of the order of years) from a "zone"-like state (as a "bright" band) to its regular "belt" aspect, and so on. The major SEBDs occur when the band is in the "zone"-like state, also known also as a "fade phase." The disturbance initiates with the apparition of one, two, or even three very bright "white" spots that emerge within ~1 month at the latitude 16°S (always at this fixed latitude, where the wind speed is close to zero). The spot locates between two opposed jets, one eastward at 9°S with a peak speed of 100 m s^{-1} and the other westward with peak velocity of -60 m s^{-1} at 20°S. The spot grows rapidly in size up to ~6000 km (Figure 9.39), being dragged by the ambient zonal flow and simultaneously sheared apart, with its poleward edge moving westward and the equatorward edge moving eastward, acquiring an "S"-like shape. The bright initial spot is most probably originated by moist convection in the water clouds at 4–5 bar depth (Sections 9.2.2 and 9.2.3).

When the spot impinges violently on the zonal flow and SEB-faded band, it disturbs the band, producing a series of dark spots with a scale of ~2000 km separated in longitude by a similar distance, moving in opposite directions according to the ambient shear of the zonal flow (Figure 9.39). The turbulent patches generated in the SEBD, moving in opposite directions at these two latitudes, reach the same meridian in a timescale $\Delta t = 2\pi R_p \cos \varphi/(u_1 - u_2)$ at ~35 days with u_1 and u_2 being the velocity (with their signs) of the opposed jets at 9°S and 20°S. Simultaneously, the central part of the cyclonic belt enters in a state of turmoil with alternating bright and dark spots moving away from the initial bright spot source. When the spots reach the position of the GRS, they are obliged to flow around its northern periphery following its anticyclonic rotation sense. A pattern of turmoil clouds is generated along the GRS wake. In about 2 months, the mixing and merging of the spots in the "chaotic" ambient ends when the band becomes more or less homogeneous in its albedo.

9.6.4.2 The North Temperate Belt Disturbance in Jupiter

In its origin, the NTBD resembles the SEBD since it also starts with the eruption of one or two bright spots (called "plumes" by their elongated morphology) but with a substantial difference. The outbreak occurs in the peak of the most intense jet of Jupiter at cloud level with velocities of ~170 m s^{-1} at latitude 23.5°N (Figure 9.40). The initial spot grows rapidly in 2–3 days in its horizontal extent to a size of

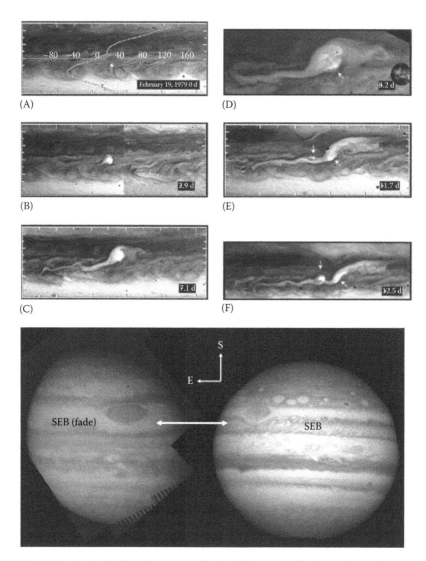

FIGURE 9.39 Jupiter's SEB disturbance. Series (A) through (F) shows the eruption, expansion, and shearing of the initial bright convective storm, source of the disturbance (Voyager 1 images in 1979). The wind profile is indicated in A. The lower pair of images shows how the SEB albedo changes from a "fade" or zone-like state (left, Pioneer 10 image in 1973) to a belt-like phase that follows a major disturbance (right, HST image). Note also the reduction in size of the Great Red Spot. (Courtesy of NASA and NASA/ESA, Paris, France.)

~5000–7000 km. As for the SEBD, the plumes originate most probably by moist convection at the water cloud, moving at the speed of the jet peak when it emerges and becomes visible. To the west of the plumes, a pattern of dark patches is generated sequentially (longitudinal size ~10,000 km, meridional width ~3,000 km). They are regularly separated by distances between 12,000 and 20,000 km (equivalent to

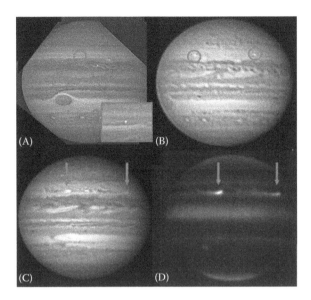

FIGURE 9.40 Jupiter's NTB disturbance in 2007 (latitude 23°N). (A) Outbreak of the activity with the emergence of a growing bright spot (magnified in the inset); (B) the two bright sources 2 days later; (C) the two bright sources and dark spots that follow them and conform the disturbance 11 days after the eruptions; (D) the two bright sources seen as "plumes" in the 2.3 μm methane absorption band. (From Sánchez-Lavega, A. et al., *Nature*, 451, 437, 2009.)

wavenumbers $k = 34$–20), moving slower than the plumes with speeds in the range of 100–150 m s^{-1}. This is a westward motion with speeds of −40 to −60 m s^{-1} relative to the plume. The behavior suggests that the dark patches could be the result of a wave, and modeling for the known belt conditions indicates that a forced Rossby wave is the best candidate. The dark patches are dispersed zonally by the shear and their mergers give rise to a generalized turmoil state of the NTB. The pattern finally encircles the planet, encountering the plumes in their opposite voyage, destroying them. In a matter of months, the spots originated from the disturbed patches mix and homogenize in their zonal distribution, forming a new dark belt between latitudes ~22°N and 30°N that has a predominant brown-red color (see Figure 9.38). During this process, the NTB jet profile suffers few changes similarly to what is observed in the SEBD. The jets keep robust against these huge disturbances and models trying to reproduce them suggesting that this requires the winds to extend in depth with low vertical shear up to at least the 7 bar pressure level.

The NTBD and the SEBD are two examples of planetary-scale shear instabilities and wave formation triggered by a localized moist convective event.

9.6.4.3 The South Tropical Zone Disturbance (STrD) in Jupiter

The South Tropical Zone (STrZ) is a "bright" band located between latitudes 20°S and 30°S. The GRS sits in its interior at latitude 23.5°S. The zone becomes disturbed when growing "undulations" at the border with the SEB at 20°S expand southward

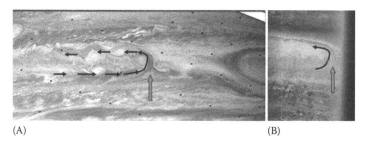

(A)   (B)

FIGURE 9.41 Jupiter's STrD and its curved shape as marked by the broad arrow (latitude 23°S). (A) The curved column is fully developed and the bright spots arriving to the column are forced to recirculate as indicated by the arrows (Voyager 2 image in 1979); (B) Details of the curved STrD feature and its related turbulence as observed by the New Horizons spacecraft in 2007. (Courtesy of NASA/JPL, Los Angeles, CA.)

and penetrate "inside" the STrZ (Figure 9.38). Typically, this occurs when the SEB is in the state of activity that follows a SEBD or when a series of bright anticyclonic spots form actively in its southern edge (Figure 9.38). The disturbed state is reached when one or two of these undulations (dark since they came from the SEB) penetrate as "curved columns" in the STrZ, being deflected eastward by the ambient flow at 26°S (Figure 9.41). The column acts like a barrier for the spots that move westward along latitude ~21°S. When they reach the STrD, they follow the curved trajectory marked by the column, being deflected eastward along latitude 26°S. These features have been historically identified as "recirculating" spots due to their change in latitude and in their direction of motion.

The STrD has been observed many times, particularly eastward of the GRS, during the encounters of Voyagers 1 and 2 with Jupiter in 1979 (Figure 9.41). The "center of the column" located at 24°S–25°S remains nearly stationary relative to the flow. The GRS moves typically westward with speeds of $-3\,m\,s^{-1}$ encountering the STrD on its west side. The column deflects southward, passing along the GRS southern periphery and apparently reforming when emerging on its east side. This behavior was interpreted some time ago as "a lack of interaction" between the GRS and the STrD, so these features were proposed to be a manifestation of a "solitary wave." However, the observations suggest that the interaction in fact occurs, so this hypothesis seems unlikely. The STrD probably results from a nonlinear amplification of a wave disturbance at the peak of a westward jet.

9.7   POLAR DYNAMICS

The poles are singular regions of planetary atmosphere. The spherical or spheroid geometry makes the Coriolis parameter $f \to 2\Omega$ and, depending on the orientation of the spin of the planet relative to the orbital plane, particular insolation conditions can occur there (Chapter 1). All planets show peculiarities in their polar dynamics. A common element is the formation of polar vortex, occurring both in the terrestrial planets and Titan, and in the giants and icy planets.

Polar vortices are usually represented by contours of Ertel's potential vorticity PV (Equation 7.59) on isentropic surfaces that is a conserved magnitude for adiabatic frictionless flow. However, *PV* increases exponentially with height, since the potential temperature $\theta \propto (1/P)_\kappa$ makes the *PV* horizontal gradients hard to determine. Then a modified scaled version of *PV* taken relative to a potential temperature reference is usually employed: $PV' = PV \, (\theta/\theta_{ref})^n$, where the index $n = 9/2$ for an isothermal atmosphere. Polar vortices can also be represented by means of a tracer distribution (O_3 for the Earth and hydrocarbons for Saturn), by temperature maps, or by wind maps.

9.7.1 EARTH POLAR VORTEX

Both poles on the Earth (Arctic and Antarctic) develop vortices in their stratospheres. The interest for this phenomenon increased when the Antarctic ozone hole was discovered in the 1980s. The polar vortex is defined as the region in the stratosphere surrounding the pole at a latitude of ~60° that is enclosed by an undulating circumpolar westerly circulation (a cyclonic vortex with speed ~60 m s⁻¹ at 100 mbar level) (Figure 9.42). The vortex develops at an altitude of ~9–25 km or equivalently between the 200 and 30 mbar pressure levels. The vortex lies in the sinking branch of the *Brewer–Dobson circulation* (a stratospheric equator-to-pole residual circulation) where cold air in the upper stratosphere descends to the tropopause. Its edge is marked by a sharp gradient in the *PV*, giving a measure of the vortex strength and of its nonpermeability to isentropic mixing.

The polar vortex develops and weakens in area, temperature structure, and lifetime following the seasonal insolation cycle, separating the polar and mid-latitude

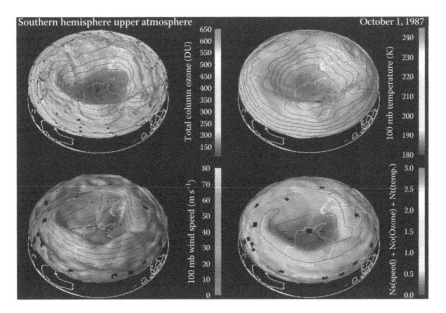

FIGURE 9.42 Earth's Antarctic polar vortex is shown in different fields. Upper: total ozone abundance and 100 mbar temperature. Lower: Wind speed at 100 mbar and a combination of speed-ozone and temperature. (Courtesy of IBM Thomas J. Watson Research Center, New York.)

stratospheres from each other at its maximum strength. It grows under the low temperatures present in wintertime (known as the *polar-night*). For example, over the Antarctic continent, the vortex forms in June–September when temperatures descend to ~180 K at the 20 mbar level and weakens and disappears by December when temperatures increase to ~240 K at the 20 mbar level. The dissipation of the vortex is taken to occur when the maximum wind speed, averaged around the vortex edge, drops below 15 m s^{-1}.

The polar vortex is controlled by the interaction between different dynamical processes related on the one hand to planetary and gravity wave propagation, and on the other hand to radiation that acts to restore the stratosphere to a radiative equilibrium (Figure 9.43). The radiative cooling of the polar air (8 K day^{-1}) is counteracted by adiabatic warming caused by planetary and gravity wave dissipation in the middle atmosphere. Because of differences in topography, the upward planetary wave flux from the troposphere is stronger in the northern hemisphere than in the southern hemisphere. This results in a vortex asymmetry between the two hemispheres with the Antarctic vortex being less disturbed, colder, and persisting for a longer time than the Arctic one. The absence of solar heating leads to low temperatures that induce polar stratospheric cloud formation over the Antarctic when the temperatures drop to −80°C. This and the intense circulation of the vortex that enhances compositional differences with mid-latitudes favors the ozone destruction over the colder Antarctic (see Sections 2.5.7 and 6.2.2).

Earth's polar vortex has other important collateral effects. The weakening-strengthening oscillation of the polar vortex jet (known as Earth's stratospheric annular modes) has been shown to be coupled to, and possibly even be, a driving mechanism for the tropospheric *Arctic oscillation* (AO)–*North Atlantic oscillation* (NAO). This is a key player phenomenon in northern mid-latitude weather patterns

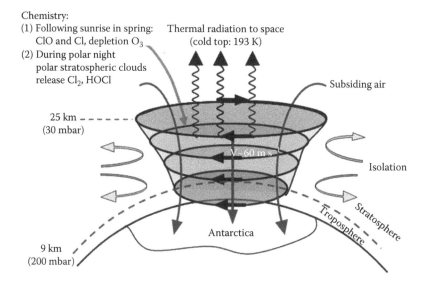

FIGURE 9.43   Scheme of the Earth's polar vortex over Antarctica.

and also appears to be correlated with the global warming atmospheric trends. In addition, the terrestrial polar vortex exhibits a dramatic phenomenon known as a "sudden stratospheric warming" characterized by an increase of 40–60 K in the average polar temperatures within a short 1 week period.

9.7.2 Polar Dynamics in the Planets

9.7.2.1 Mars

Temperature measurements at different altitude levels obtained by spacecrafts over the north pole of Mars during the wintertime have been used to calculate the geopotential height and winds. This reveals the formation of a polar vortex in maps of the Ertel potential vorticity on isentropic surfaces (Figure 9.44). The Martian polar vortex is annular in shape, covering a band from latitudes 60° to 80°, which contrast with the terrestrial polar vortex where potential vorticity normally increases monotonically toward the pole. Over the south pole, the polar vortex is less prominent than over the north pole.

9.7.2.2 Venus North and South Dipole

Polar maps of ultraviolet images (with wavelengths of 350–400 nm) that show the upper clouds of Venus (with altitudes of ~65 km) in reflected sunlight (Chapter 5) reveal the existence of a spiral pattern organization of the clouds around the poles, suggestive of the presence of a polar vortex (Figure 9.45a). The vortex is also present in the lower clouds (altitudes of ~45–50 km) as observed in nighttime images (wavelengths 1.74 and 2.3 μm) (Figure 9.45b). Infrared observations at longer wavelengths, sensitive to temperature and cloud opacity, show the detailed vortex structure at cloud levels (altitudes of ~65–70 km) in both hemispheres. The vortex manifests as

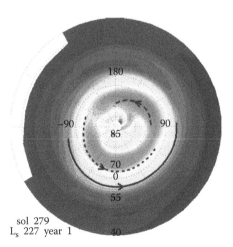

FIGURE 9.44 Structure of Mars polar vortex as drawn by Ertel's potential vorticity on isentropic surfaces as calculated from winds derived from temperature measurements with Mars Global Surveyor TES instrument. (Courtesy of T.H. McConnochie and NASA/JPL/ASI, Los Angeles, CA, October 19, 1999.)

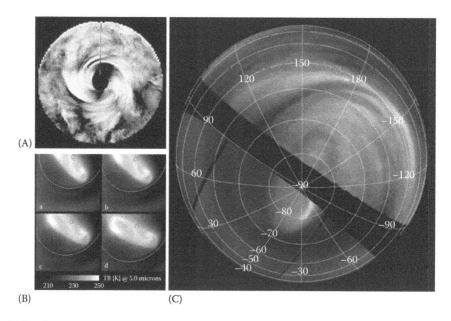

FIGURE 9.45 Different views of Venus polar vortices. (A) Image of the north polar vortex formed in the upper cloud observed with the Mariner-10 spacecraft in ultraviolet light. (Courtesy of NASA, Washington, DC.) (B) The south polar dipole observed at nighttime at 5 μm (thermal emission) with the VIRTIS instrument onboard Venus Express (VEX). (C) Combined images of the day and night sides of the south pole of Venus showing the arc-like feature (a vortex arm) in reflected light (dark) and in thermal emission (bright) observed with the VMC and VIRTIS instruments onboard VEX. (Courtesy of ESA/MPS/INAF-IASF/Obs. de Paris-LESIA, Paris, France.)

a "bright" rotating "dipole" where the temperatures are high ~240 K, surrounded by a "dark" cold collar with temperatures of ~210–220 K (Figure 9.45c). The dipole has two rotation centers and has a filamentary morphology with an inverted "S"-shape but displays a high variability. The filaments twist in a matter of a month as the dipole rotates in a retrograde sense (as the whole atmosphere does), with a period between 2.8 and 3.2 days (northern dipole) and 2.5 days (southern dipole) (Figure 9.45d).

It is not known how the dipole forms, but a combination of the super-rotation of the decreasing zonal winds toward the pole with the meridional Hadley cell circulation is probably responsible for its formation.

9.7.2.3 Titan Polar Vortex

Temperature measurements obtained by the Cassini's CIRS instrument during Titan's midwinter season (years 2006 and 2007) have provided strong evidence of a polar vortex at a high altitude (~300 km) in the northern hemisphere. A circumpolar jet with winds of up to 190 m s^{-1} blows at 30°N–50°N. These are the fastest winds anywhere on Titan (Figures 4.15d and 7.16). Gradients in the spatial distribution of different chemical compounds used as atmospheric tracers (HCN, HC$_3$N, C$_2$H$_2$, C$_3$H$_4$, and C$_4$H$_2$) provide strong evidence that the vortex acts as a barrier for mixing at latitude 60°N, isolating the vortex interior compounds from the mid-latitude atmosphere.

9.7.2.4 Jupiter Poles

There is no evidence yet for a polar vortex on Jupiter. As presented in Section 8.4.3 (Figure 8.11 and Table 8.3), the most prominent polar features in Jupiter are the system of waves that stand at sub-polar latitudes (~60°). In addition, images in ultraviolet light (250 nm) obtained by HST in 1997 and during the Cassini spacecraft fly-by at the end of 2000 revealed the formation of a transient large dark oval with a lifetime ~ months, with a shape and size similar to the GRS in the upper northern polar haze layer. Its origin is probably dynamically linked to auroral processes (Figure 9.46).

9.7.2.5 Saturn Polar Vortices

Thermal infrared images of the southern pole showed a small warm cap with edges at latitude 88.5°S. The feature is in fact a hole in the upper cloud layers surrounded by two walls of clouds that elevate about 70 and 40 km relative to the surrounding background cloud level. They form a cyclonic vortex enclosed by a jet stream at 88°S with an eastward peak velocity of 160 m s⁻¹. Rotating, spiral in shape bright features are observed within the hole (Figure 9.47). The mechanisms that form this cyclonic vortex and the spiral internal structures still remain a mystery.

As stated in Section 8.4.3 (Figure 8.13 and Table 8.3), a persistent hexagonal wave that contains a strong eastward jet with a speed of 100 m s⁻¹ surrounds the northern pole at latitude 80°N. The hexagon is visible in reflected sunlight images

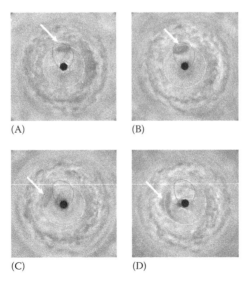

(A)   (B)

(C)   (D)

FIGURE 9.46 Jupiter's large dark auroral oval in rotation (marked by an arrow) as observed on November 13, 2000, in the northern Jovian stratosphere at ultraviolet wavelengths (258 nm). The 60°N latitude circle is shown. The location of the main auroral oval is plotted as a dashed curve. The sequence A through D shows the dark oval rotation (anticlockwise direction).(Courtesy of NASA/JPL, Los Angeles, CA.)

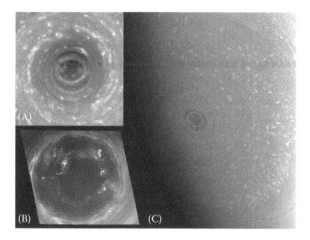

FIGURE 9.47 Saturn's south polar vortex as imaged by the Cassini ISS instrument. The global view for latitude above 60°S is shown in (C), in an image taken on August 27, 2008, at red wavelengths (PIA10490). Details of the cloudy walls and features interior to the hole (upward moving and spiralling) can be seen in (A) in an image taken on October 11, 2006 (PIA08332), and in (B) in an image taken on July 14, 2008 (PIA11104). (Courtesy of NASA/ JPL/Space Science Institute, Boulder, CO.)

and in the $5\,\mu m$ images sensitive to the opacity to thermal radiation by the lower NH_4SH cloud, extending vertically from the altitude level of 50 mbar to the 3 bar level or deeper.

9.7.2.6 Uranus and Neptune Poles

There is no evidence for the existence of a polar vortex at the cloud level in these planets. The lack of resolution in the available images might be the reason. Uranus' northern pole has never been observed, and the southern pole appears featureless in visible images except for a dark cap that could indicate the existence of a vortex. Neptune's southern polar region showed enhanced stratospheric temperatures (by about 3 K over surroundings), and high resolution images from Voyager 2 and ground-based telescopes indicate the presence of a small bright spot at the pole. The Voyager 2 wind data also revealed the existence of a broad eastward jet at latitude 70°S with a peak velocity of $\sim\!250\,m\,s^{-1}$ rapidly decreasing to $-10\,m\,s^{-1}$ at latitude 87°S. It is possible that this is an additional signature of a polar vortex.

9.8 ATMOSPHERIC EFFECTS OF THE IMPACT OF COMET SHOEMAKER-LEVY 9 WITH JUPITER

A unique event in the history of the observations of the solar system occurred between July 16 and 22, 1994, when 16 fragments of comet Shoemaker-Levy 9 (abbreviated as SL9) impacted sequentially with the planet. The comet was previously captured and fragmented by Jupiter's gravity. As a result, a series of atmospheric phenomena took place during the impacts and later, allowing the exploration of the properties of the upper atmosphere and its response to the energetic impacts. The impacts occurred behind the dawn limb of Jupiter as observed from Earth (just by 3.4°–8.8° in longitude), and only the Galileo spacecraft that was in route to Jupiter observed them directly. A large battery of ground-based and Earth-orbiting telescopes (in particular the Hubble space telescope) observed the events in the whole spectrum: the bright plumes that emerged behind the disk and the debris left in the atmosphere that became visible as the planet rotated toward visibility from Earth. Here, we briefly describe the basic physics and response of the atmosphere to impacts.

The fragments were named alphabetically in impact chronology order, A the first one and so on, although some letters were omitted since they were catalogued as fragments but they did not produce impact signs and other fragments disappeared before impact. The most energetic impacts corresponded to fragments with a size of ~0.5–1 km (the total comet size was 1.5–2 km and the density was ~0.5 g cm^{-3}). They were the L, G, and K fragments. The impacts occurred with a velocity of 60 km s^{-1} relative to Jupiter at an entry angle of 43° relative to the vertical and azimuth angle 16° from the south. The fragments impacted in a narrow band between latitudes 43°S and 45°S and the typical distance separating each event was ~40° in longitude. A typical impact produced the following observed event phases.

9.8.1 IMPACT

Each comet fragment disrupted when falling in the upper atmosphere, vaporizing and depositing most of its kinetic energy at a terminal atmospheric depth. This occurred in about 5 s during a "meteoroid" entry phase along a narrow channel producing a temperature increase in the heated atmospheric layer of 40,000 K that resulted in a brief "light flash." The most energetic fragments deposited ~1.5–2 × 10^{21} J and probably reached the 1 bar pressure level atmospheric depth.

9.8.2 "FIREBALL" AND "BLOWOUT"

During this phase, the atmosphere responded to the energy, momentum, and material deposited by the fragment producing a shockwave that propagated in all directions. The superheated comet material and entrained Jovian air at the high pressure and temperatures reached high in the atmosphere along a cone-like shape and was directed back up at speeds of 10 km s^{-1}. This produced a fireball and a high-altitude plume initially detected only by the Galileo spacecraft that had direct vision of the night Jovian hemisphere during its route to the planet. The altitude reached by the fireball cone was estimated to be ~3000 km above the impact point, influencing Jupiter's ionosphere and magnetosphere. The study of this phase involved complex shockwave physics and nonhydrostatic hydrodynamics, which are not the purposes of this book.

9.8.3 PLUME FLIGHT

The plume material injected became visible high in the atmosphere (altitudes 2300–3100 km) by the sunlight reflected in the debris (Figure 9.48) and by the thermal

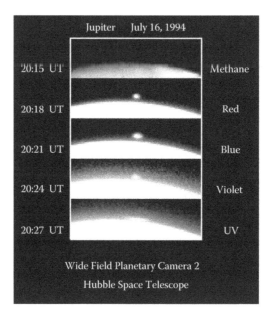

FIGURE 9.48   The plume formed by one of the impacts of the fragmented comet SL9 when emerging above the limb as imaged at different wavelengths by the HST in July 1994. (Courtesy of NASA/ESA, Paris, France.)

emission produced by the high temperatures (~8000 K), as detected by the Galileo spacecraft instruments. The ground-based telescopes observed the bright plume emerging behind the limb in the 2.3 μm methane absorption band where the planet is dark (Figure 9.49). The material (gas and dust from the comet) traveled in ballistic trajectories, cooling in a matter of 20 min to just 10 K above the normal Jupiter temperature. It is probably at this time that the ring-shape waves were created (see below).

9.8.4 PLUME SPLASH AND AEROSOL CLOUD FORMATION

The grain debris resulting from the ablated comet material and those created during the impact, transported by the gas along with the plume up to 3000 km, fell back under gravity and was deposited above the tropopause. Dark "aerosol clouds" were observed as soon as the impact points rotated into visibility. The structure of the aerosol clouds is shown in Figure 9.50. The largest impacts consisted of a dark core (like a streak centered at the impact point, presumably centered at the impact site) surrounded by two expanding rings (waves) and an outward arc-like crescent area resulting from the splash of the plume. This outer arc-ring had an inner size of 6,000 km and an outer ring size of 14,000 km. The two rings propagated at velocities of 454 m s^{-1} (outer one) and 180–350 m s^{-1} (inner one) and have been interpreted as gravity waves (see Figure 8.7). In addition to this, large rings with radii of ~20,000 km were seen rapidly expanding (velocities ~1.5 km s^{-1}) in the impact events *G* and *K*. Their nature is still unknown.

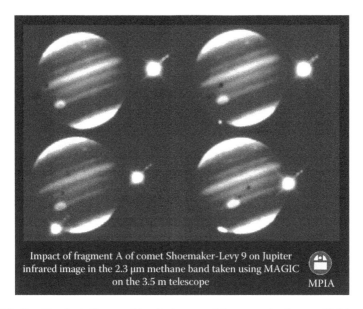

Impact of fragment A of comet Shoemaker-Levy 9 on Jupiter
infrared image in the 2.3 μm methane band taken using MAGIC
on the 3.5 m telescope   MPIA

FIGURE 9.49 The fireball created by the impact of fragment A of comet SL9 is seen as a bright evolving spot in the lower left part of the mosaic. Images taken in the 2.3 μm absorption band produced by methane. (Courtesy of CAHA/MPIA; Calar Alto Observatory, Spain, July 16, 1994.)

The aerosol clouds were formed by thin particles with a size of 0.15–0.3 μm and had a dark appearance in ultraviolet and visible light, but were bright in the 2.3 μm methane absorption band, indicating they were spread at altitudes from ~1 to 200 mbar. In some areas of the debris, their optical depth reached values of τ (210 nm) ~10 and τ (890 nm) ~3.

A rich chemistry accompanied the impacts. New species were detected in the stratosphere, above ~10 mbar, and others showed an enhancement in their abundances within the plume and later in the latitude band of the impacts. Among them S_2, CS_2, CS, OCS, NH_3, HCN, C_2H_4, H_2O, CO, and possibly H_2S were detected. Some molecules were generated during the impact by the strong shock heating at high pressures, but others came from Jupiter lower levels and from the comet material left in the atmosphere. Some species were found to be transient (enduring days to months), but others had long-term residence times (such as CO and CS_2). The highest abundances were observed for the following compounds: CO (7.5×10^{14} g), NH_3 (10^{14} g), H_2O (5×10^{13} g), COS (1.5×10^{13} g), and HCN (10^{13} g). Dust grains (silicates, Al_2O_3, and organics) were also detected.

9.8.5 CLOUD AND ENVIRONMENT EVOLUTION

The panorama of Jupiter aerosol clouds following the impact week is shown in Figures 9.51 and 9.52. The impact clouds were quasi-regularly spaced and were

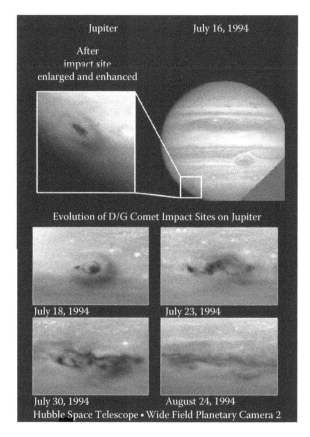

FIGURE 9.50 HST images of the aerosol debris left in the atmosphere of Jupiter following the SL9 impacts A (upper image pairs) and D/G (lower four images). The four lower frames show the complex patterns produced by the dispersion of the aerosols produced by the expansion, wind shears and interactions of the debris with Jupiter's dynamical features. (Courtesy of NASA/ESA, Paris, France.)

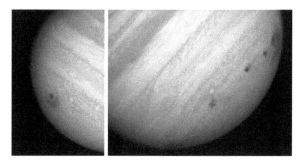

FIGURE 9.51 Details of the visual aspect of the SL9 aerosol debris left by the impact of features G (left) and a series of smaller impact spots (right) obtained with the HST. (Courtesy of NASA/ESA, Paris, France.)

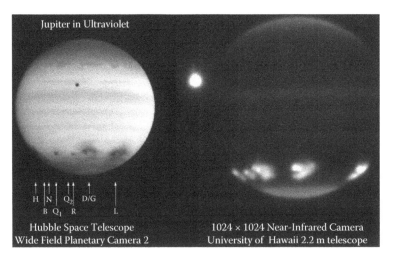

FIGURE 9.52   The SL9 impact aerosol debris produced by different impacts observed at two different wavelengths sensitive to high altitudes above Jupiter's regular cloud deck: ultraviolet (left) and in the 2.3 μm methane absorption band (right).

subjected to the advection by the ambient zonal winds $u(y,z)$ (see Figure 7.17), and to the interaction with vortices and other regular Jupiter features. The meridional and vertical shears spread and expanded zonally to the aerosol clouds, producing their mutual mixing and finally, in a matter of a month, produced a dark belt of debris along the latitude band from 42°S to 47°S. On average, the expanding velocities of the debris were lower than those measured in the Jupiter cloud deck (by about 30%). Accordingly, the zonal winds should decrease with altitude as expected from the measured meridional temperature gradient and the thermal wind balance. At the same time, they spread in latitude with velocities of 0.4 m s^{-1}, mainly in the equatorward direction, until reaching latitude 20°S. The aerosol optical depth gradually decreased by sedimentation but their signature was still present ~3 years later.

9.9   THE JULY 19th 2009 IMPACT

A second, single large impact on Jupiter occurred on July 19, 2009. The impact took place again in the nonilluminated, nonvisible hemisphere from Earth at planetocentric latitude 55°S, about 12° poleward of the SL9 impact. It was recorded by the aerosol debris left in the atmosphere by the impactor disintegration (Figure 9.53), but no fireball was detected (it occurred in Jupiter's night side hemisphere. The dark bruise was first noticed on CCD images obtained by amateur astronomer Antony Wesley on July 19th, just emerging in Jupiter's west limb. The impact nature of the spot was confirmed by images obtained in the methane band filters, where it detached as the brighter feature in Jupiter. The debris consisted of a main spot with an east-to-west size of 4800 km, a north-to-south size of 2500 km,

FIGURE 9.53   The visual aspect of the aerosol debris left by the impact of a single body with Jupiter on July 19, 2009. The feature imaged by the HST WF3 camera shows the debris dispersion and expansion 4 days after the impact, on July 23. (Courtesy of NASA, ESA, H. Hammel, Space Science Institute and the Jupiter Impact Team, Boulder, CO.)

and a halo-arc shape feature nord east of the main spot. The entry trajectory of the impactor was in a direction opposite to that of the SL9 fragments, i.e., from NW to SE, with a lower incidence angle relative to vertical. From the debris left by the impact on Jupiter, it was concluded that it was an icy or low density body with intermediate size between the largest and smaller SL9 fragments (probably ~0.5 km in size). However, spectroscopic analysis of the impact site suggests that the body could have been of asteroidal nature.

At the central latitude of the impact, the Jovian flow has nearly zero speed, but anticyclonic vorticity bounded by jets at $-51.5°$ (directed westward with a velocity of $-10\,\mathrm{m\ s^{-1}}$) and at $-57.5°$ (directed eastward with a velocity of $25\,\mathrm{m\ s^{-1}}$) (Figure 7.17). Accordingly, the aerosol debris were advected by the zonal winds and expanded in longitude at different speeds depending on latitude due to the shear. Since the debris extended in altitude from ~1 mbar to the ammonia cloud top at ~500 mbar, the tracking of their motion allowed retrieval of the winds at high altitudes, indicating that they increased with altitude by about 1–$2\,\mathrm{m\ s^{-1}}$ per scale height. Enhanced thermal emission was observed in the impact area from hydrocarbons and ammonia gas in the lower stratosphere, coincident with the location of high-altitude particulate debris. A temperature increase up to 8 K and ammonia mixing ratio enhancement was measured between altitude levels of 10 and 100 mbar.

Statistically, such impacts must not be too uncommon on Jupiter and could amount to one every 10 years. Detection of such events with dedicated telescopes surveying the planet in the methane bands ($0.89\,\mu m$ in the CCD optical range and 2.12–$2.3\,\mu m$ in the near infrared, see Figure 9.49) will allow for determination of the impact statistics frequency and the study of the properties of Jupiter's upper atmosphere and its response at different latitudes to different energetic impacts.

PROBLEMS

9.1 Calculate and compare the value of the Richardson number and the instability conditions in the equatorial clouds of the planets Venus, Jupiter, and Saturn for the measured data given in the table. Take other required data from previous chapters in the book.

Planet	Level (P or z)	Wind Speed (m s⁻¹)	Mean Temperature (K)	Mean Static Stability (K km⁻¹)
Venus layer 1	66 km	110	248	5
Venus layer 1	61 km	70		
Venus layer 2	61 km	70	300	−1
Venus layer 2	51 km	50		
Jupiter	1 bar	100	205	0.2
Jupiter	3.4 bar	170		
Saturn	50 mbar	265	100	0.66
Saturn	700 mbar	365		

Solution

Ri (Venus layer 1) $=2.8$; Ri (Venus layer 2) $=-7$; Ri (Jupiter) $=7$; Ri (Saturn) $=114$. Venus middle layers (layer 2) are convectively unstable.

9.2 Make a scale analysis of the different acceleration and deceleration terms in Equation 9.11 for water moist convective clouds in the Earth and Jupiter for the following conditions and values of the parameters: $C_D=0.2$, $k_M=0.2$, $w=10$ m s⁻¹, $\ell_C=5$ g of water per kg of air, nonhydrostatic pressure 0.26 mbar along a layer from 0.5 to 1 bar level, parcel virtual temperature 303 K, ambient virtual temperature 300 K. Take the other required data from other sections in the book.

Solution

Earth: 0.1 m s⁻² (buoyancy term), −0.02 m s⁻² (drag term), −0.02 m s⁻² (entrainment term), 0.0052 m s⁻² (nonhydrostatic term), −0.05 m s⁻² (condensate weight term); Jupiter: 0.23 m s⁻² (buoyancy term), −0.02 m s⁻² (drag term), −0.02 m s⁻² (entrainment term), 0.012 m s⁻² (nonhydrostatic term), −0.116 m s⁻² (condensate weight term).

9.3 Find an expression for the vertical velocity acquired by a buoyant ascending parcel as a function of altitude z assuming the terms in Equation 9.11 remain constant along a short path starting at level z_0 where the velocity is w_0. Assume the effect of the entrainment rate on the virtual temperature of the parcel is negligible.

Solution

$$w^2 = \frac{a}{b} - \left[\frac{a}{b} - w_0^2\right]\exp\left[-2b(z-z_0)\right]; \quad a = g\left(\frac{T_V'-T_V}{T_V}\right) + g\frac{\partial \pi_D}{\partial P} - g\ell_C$$

$$b = \frac{(C_D+k_M)}{r_0}$$

9.4 Using the formula from Section 1.5.4.3, derive a modified version of the Rayleigh number in terms of the volumetric heating rate and of the convective vertical velocity in terms of the mixing-length scale. Then calculate the Rayleigh number for convection in Venus cloud layers and the convective vertical velocity ($z=60$ km). Then calculate the radiative time constant and compare it with the convective timescale. Use the following data (other necessary data can be taken from different chapters): $D_T=\upsilon=155\,\mathrm{m^2\,s^{-1}}$, $P=0.3$ bar, $T=268$ K, $\rho=0.429\,\mathrm{kg\,m^{-3}}$, heating rate $H_Q=0.0106\,\mathrm{W\,m^{-3}}$, solar flux at this level $F=129\,\mathrm{W\,m^{-2}}$, mixing-length scale $d=7$ km (the thickness of the unstable layer where convection develops).

Solution

$$Ra_Q = \frac{H_Q g d^5}{T\rho C_p D_T^2 \upsilon} = 6.8\times 10^6; \quad w = \left(\frac{Fgd}{\rho C_p T}\right)^{1/3} = 4.3\ \mathrm{m\,s^{-1}};$$

$$\tau_{rad} = 47.6\ \mathrm{days} \gg \tau_{conv} = 0.45\ \mathrm{h}$$

9.5 The area of cumulus clouds formed by convective storms is observed to increase in time on satellite images. Assuming that this increase represents the average divergence within an altitude range h at whose base the vertical velocity is zero, calculate the vertical velocity at the top of the layer for the following two cases: (1) an Earth storm whose area increases by 30% in 10 min within an altitude range of 3 km; (2) for Saturn's GWS, whose elliptical shape increases from a size of 5,000 km times 3,000 km to 20,000 km times 10,000 km in 7 days between the base of the water clouds (10 bar) and the tropopause (100 mbar). Take the scale height of 40 km.

Solution

(1) Earth: $w=1.3\,\mathrm{m\ s^{-1}}$
(2) Saturn: $w=0.75\,\mathrm{m\ s^{-1}}$

9.6 Calculate the maximum upward velocity and CAPE for methane storms in Titan's atmosphere if the parcel virtual temperature is 78 K and the environment virtual temperature is 76.5 K if the parcel ascends convectively between the altitude levels from the surface by 10 and 30 km.

Solution
$w_{max}=32.3\,\mathrm{m\ s^{-1}}$, CAPE$=523\,\mathrm{m^2\ s^{-2}}$.

9.7 Compare water moist convection in Jupiter and Earth. Consider a mole water fraction in both planets of $3\,\mathrm{g\ kg^{-1}}$. (1) Calculate the parcel and environment virtual temperatures in Jupiter and Earth for parcel and environment temperatures 277 and 275 K, respectively. (2) Calculate the maximum buoyant updraft velocity for H_2O cumulus in Jupiter if moving from the pressure level of 5 bar up to 0.1 bar maintaining the virtual temperature difference and similarly for Earth when moving from 1 to 0.1 bar. Take the scale height of 20 km for Jupiter and 7 km for Earth.

Solution

(1) $T_V'(J) = 276.17$ K, $T_V'(E) = 277.5$ K; $T_V(J) = 274.17$ K, $T_V(E) = 275.5$ K

(2) $w_{max}(J) = 162.2$ m s^{-1}, $w_{max}(E) = 47.7$ m s^{-1}

9.8 A typical measure of the horizontal shear flow velocity on Earth (E) and Jupiter (J) gives the following values: $\partial u/\partial y = 1.2 \times 10^{-5}$ s^{-1} (E), $\partial u/\partial y = 2 \times 10^{-5}$ s^{-1} (J), $\partial v/\partial x = 0$ (E and J); $\partial u/\partial x = 3.5 \times 10^{-7}$ s^{-1} (E), $\partial u/\partial x = 1.6 \times 10^{-8}$ s^{-1} (J), $\partial v/\partial y = 0$ (E and J). Compare and comment on these values with the contribution of the tangent terms to the divergence and vorticity in both planets for $u = 50$ m s^{-1} and $v = 10$ m s^{-1} at latitudes 10° and 80° of the northern hemisphere.

Solution

Earth: $-v \tan \varphi/R_{Earth}$ (s^{-1}) $= -2.64 \times 10^{-7}$ ($\varphi = 10°$), -8.5×10^{-6} ($\varphi = 80°$)
Jupiter: $-v \tan \varphi/R_{Jup}$ (s^{-1}) $= -2.46 \times 10^{-8}$ ($\varphi = 10°$), -7.9×10^{-7} ($\varphi = 80°$)
Earth: $u \tan \varphi/R_{Earth}$ (s^{-1}) $= 1.3 \times 10^{-6}$ ($\varphi = 10°$), 4.25×10^{-5} ($\varphi = 80°$)
Jupiter: $u \tan \varphi/R_{Jup}$ (s^{-1}) $= 1.23 \times 10^{-7}$ ($\varphi = 10°$), 3.95×10^{-6} ($\varphi = 80°$)

The tangent term is important for the divergence calculations in both planets and latitude. For vorticity calculations, it is only important for polar flow on Earth.

9.9 Calculate the vorticity and horizontal divergence in cylindrical coordinates for (1) a Rankine vortex ($v_r = 0$, $v_\theta = C/r$, $C = $ const.) and (2) for a vortex in solid body rotation (angular velocity ω_0).

Solution

(1) $\nabla_h \cdot \vec{V} = 0$; $\hat{k} \cdot \nabla \times \vec{V} = 0$

(2) $\nabla_h \cdot \vec{V} = 0$; $\hat{k} \cdot \nabla \times \vec{V} = 2\omega_0$

9.10 A model of (oval) vortices of the giant planet proposes that the streamlines are concentric ellipses with the geometry shown in the figure. The ellipse is centered at (0, 0) in the Cartesian axis system (x, y) and has major and minor semi-axes (a, b). The position angles are related by $\tan \chi = (a/b)^2 \tan \theta$ with $\tan \theta = y/x$. Let (u, v) be the zonal and meridional velocity components and V_T and V_R be the tangential and radial components (both orthogonal, relative to angle χ). (1) Find the relationship between the velocity components and (2) assuming that $V_R = 0$ and V_T (a), find an expression for the vorticity from its Cartesian equation (9.23b).

From Mitchell, J.L. et al., *J. Geophys. Res.*, 86, 8751, 1981.

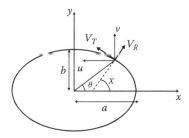

Solution

(1) $u = V_R\cos\chi - V_T\sin\chi$

      $v = V_R\sin\chi + V_T\cos\chi$

(2) $\zeta(a,\theta) = \dfrac{a}{b^2\eta^3}V_T + \eta\dfrac{dV_T}{da}$

   $\eta = \left[\dfrac{\cos^2\theta + (a/b)^4\sin^2\theta}{\cos^2\theta + (a/b)^2\sin^2\theta}\right]^{1/2}$

9.11 Apply the above results to the case of Jupiter's Great Red Spot whose tangential velocity can be written as a function of $V_T(a) = C_1a + C_2a^2 + C_3a^3 + C_4a^4$, where a is the semi-major axis (unit 10^6 m) and the coefficients have the following values (unit appropriate to get V_T in m s^{-1}): $C_1 = 9.66$, $C_2 = -7.96$, $C_3 = 1.66$, and $C_4 = -0.086$ (measurements performed during the Voyager fly-bys in 1979). Find the local vorticity at a position $a = 10.2 \times 10^6$ m and $\theta = 0°$ (GRS size: $a = 12.08 \times 10^6$ m, $b = 5.58 \times 10^6$ m).

Solution
$\zeta = 3.9 \times 10^{-5}$ s^{-1}.

9.12 Obtain an expression for the mean vorticity and horizontal divergence for an elliptical vortex with semi-axes (a, b) and mean tangential and radial velocities $\langle V_T\rangle$ and $\langle V_R\rangle$. Use for the length of the ellipse the approach $L_e = 2\pi a[1 - (e^2/4)]$ with $e = \sqrt{1 - (b^2/a^2)}$. Apply these expressions to calculate the mean vorticity, divergence, and Rossby number to a Saturn's Brown Spot vortex that has $a = 2000$ km, $b = 1500$ km, $\langle V_T\rangle = 50$ m s^{-1}, $\langle V_R\rangle = 5$ m s^{-1} (outward), and latitude $42°$N.

Solution
$\langle\zeta\rangle = \langle V_T\rangle X$; $\langle div_h V\rangle = \langle V_R\rangle X$ with $X = b^{-1}\left(1 - (1/4)\sqrt{1 - (b/a)^2}\right)$. $\langle\zeta\rangle = 5.5 \times 10^{-6}$ s^{-1}; $\langle div_h V\rangle = 5.5 \times 10^{-7}$ s^{-1}; $Ro = 0.025$.

9.13 Consider two vortices on Earth (assumed for simplicity to be circular): an extratropical cyclone at $45°$N latitude (tangential velocity $V = 10$ m s^{-1}, curvature radius $R = 1000$ km) and a tropical cyclone at $10°$N latitude (tangential velocity $V=100\,\text{m s}^{-1}$, curvature radius $R=100\,\text{km}$). Find the Rossby number for each case and their first-order balance. Determine the edge to center pressure difference if the surface conditions outside the vortices are $P=1$ bar, $T=27°\text{C}$.

Solution

(1) Extratropical cyclone: $Ro=0.2$ (geostrophic), $\Delta P=11.6$ mbar
(2) Tropical cyclone: $Ro=40$ (cyclostrophic), $\Delta P=116$ mbar

9.14 A mesoscale circular eddy in Mars at latitude $15°$ has a tangential velocity profile of $V_T=V_0(r/r_0)^2$ with $V_0=10\,\text{m s}^{-1}$ at radius $r_0=100\,\text{m}$. (1) Determine its dynamical balance and Rossby number; (2) find the vorticity distribution in the vortex and its value at r_0; (3) find the radial pressure gradient and estimate the center to edge pressure difference if the surface pressure is $P=6$ mbar and the temperature $T=-30°\text{C}$.

Solution

(1) $Ro=2725$ (cyclostrophic balance)
(2) $\zeta(r)=\left(3V_0/r_0^3\right)r$; $\zeta(r_0)=0.3\,\text{s}^{-1}$
(3) $\partial P/\partial r = \rho\left(V_0^2/r_0^4\right)r^3$, $\Delta P=0.013$ mbar

9.15 A tornado on Earth is placed at latitude $45°\text{N}$ and has a radius $r_0=100\,\text{m}$ where the velocity is $V_0=10\,\text{m s}^{-1}$ and is in solid body rotation. Find an expression for the pressure at the center of the tornado as a function of the temperature and angular velocity ω_0 if the surface pressure at the tornado edge is P_0. Find the central pressure if $P_0=1$ bar and the temperature is $30°\text{C}$.

Solution

$$P_c = P_0\exp\left(-\frac{\omega_0^2 r_0^2}{2R_g^*T}\right); \quad P_c = 990 \text{ mbar}.$$

9.16 Write the momentum equations (7.10a and b) in cylindrical coordinates (v_r, v_θ) and then solve for $dv_r/dt=0$ to find the expression for the gradient wind equation.

Solution

$$\frac{dv_r}{dt} = fv_\theta - \frac{1}{\rho}\frac{\partial P}{\partial r} + \frac{v_\theta^2}{r}; \quad v_\theta = -\frac{fr}{2} \pm \frac{r}{2}\sqrt{f^2 + \frac{4}{\rho r}\frac{\partial P}{\partial r}}$$

$$\frac{dv_\theta}{dt} = -fv_r - \frac{v_r v_\theta}{r}$$

9.17 Find an expression for the ratio between the geostrophic wind and the gradient wind (V_g/V) velocities and calculate it for the following vortices: (1) an Earth anticyclone and a cyclone at 45°N (for both $R=1000$ km and $V=10$ m s^{-1}); (2) a Martian cyclone at 66°N ($R=300$ km, $V-25$ m s^{-1}), (3) Jupiter's Great Red Spot anticyclone at 23°S ($R=10,000$ km, $V=110$ m s^{-1}); (4) Saturn's North Polar Spot, an anticyclone at 75°N ($R=8000$ km, $V=100$ m s^{-1}).

Solution
$(V_g/V)=1+(V/fR)$; (1) $V_g/V=0.9$ (anticyclone), 1.1 (cyclone); (2) $V_g/V=1.64$; (3) $V_g/V=0.92$; (4) $V_g/V=0.96$.

9.18 Calculate the value of the acceleration terms in the gradient wind balance equation and the vorticity of a Martian extratropical cyclone (anticlockwise rotation) close to the pole (latitude 81°N) that has a radius of 100 km and tangential wind speed of 32 m s^{-1}. What is the pressure gradient?

Solution
$fV=0.0045$ m s^{-2}, $V^2/R=0.01$ m s^{-2}, $(-1/\rho)$ $(\partial P/\partial r)=0.0145$ ms^{-2}; $\langle \zeta \rangle=6.4\times10^{-4}$ s^{-1}; $(\partial P/\partial r)=-2.13\times10^{-3}$ mbar km^{-1}.

9.19 An isothermal and hydrostatic layer forms close to the surface on Earth ($T=290$ K) and in Venus' lower cloud ($T=384$ K). Assume that a Kelvin–Helmholtz instability develops in both planets within this layer when the wind shear is strong so the Richardson number is $Ri=0.15$. (a) Find the required wind shear. (b) What is the thickness of the layer if the pressures in the lower and upper parts are 1 and 0.965 bar (Earth), and 2 and 1.975 bar (Venus clouds). (c) Calculate the zonal wavelength of the instability in both planets.

Solution

(a) $\partial u/\partial z=0.048$ s^{-1} (Earth), $\partial u/\partial z=0.04$ s^{-1} (Venus)
(b) $\Delta z=300$ m (Earth), $\Delta z=100$ m (Venus)
(c) $\lambda=1852$ m (Earth), $\lambda=460$ m (Venus)

9.20 Study the stability of a westward zonal flow with a velocity profile of $\bar{u}(y)=-u_0 \cos(2\varphi)$ against the inertial instability if $u_0=0.55\Omega R_p$ with Ω and R_p being the planetary rotation rate and radius.

Solution
Unstable for $\varphi<24.6°$.

9.21 Study the stability of a westward jet with profile $\bar{u}(y)=-u_0 \sin^2[L(y-y_0)]$ against barotropic instability. Find the position y_c where a change of sign in the vorticity gradient occurs and the maximum value allowed is L.

Solution
Unstable for $2u_0L^2>\beta$

$$y_c = y_0 + \left(\frac{1}{2L}\right)\cos^{-1}\left[-\frac{\beta}{2u_0L^2}\right]; \quad L(\max) = \sqrt{\frac{2u_0}{\beta}}.$$

9.22 Study the conditions for barotropic instability to develop in a Jovian westward jet with a parabolic profile $\bar{u}(y) = -u_0 + \beta_0(y - y_0)^2$. Find the position y_c where a change of sign in the vorticity gradient occurs. If the jet has its peak at latitude 30°, the peak speed is 30 m s^{-1} and its half-width is 3°. Find β_0.

Solution
Unstable for $\beta(y) = 2\beta_0$; $y_c = (\pi R_J/180) \cos^{-1}(\beta_0 R_J/\Omega)$ in meters and symbols as usual; $\beta_0 = 2.14 \times 10^{-12}$ m^{-1} s^{-1}.

9.23 An analytical approach to Uranus' zonal wind profile at the cloud level is provided by the expression $u(\varphi) = 200 (0.4 \cos \varphi - \cos 3\varphi)$ in m s^{-1}, with φ being the latitude. Give the latitude band within 1° thickness where this profile can develop the barotropic instability. Take the required data from the book.

Solution
Between latitudes 29°–30°.

9.24 Calculate the change in gravitational potential energy per unit volume ($\Delta E_P/Vol$) of a parcel that moves 10 km in the vertical from a level at (T_1, P_1) to (T_2, P_2) in the following planets and cases: (1) Earth ($T_1 = 288$ K, $P_1 = 1$ bar; $T_2 = 220$ K, $P_2 = 0.24$ bar); (2) Mars ($T_1 = 214$ K, $P_1 = 0.007$ bar; $T_2 = 169$ K, $P_2 = 0.035$ bar); (3) Jupiter ($T_1 = 165$ K, $P_1 = 1$ bar; $T_2 = 145$ K, $P_2 = 0.6$ bar). Then assume that all this energy is converted to kinetic energy. What would be the wind speed at the upper level for each case? Comment on this result.

Solution

(1) $\Delta E_P/Vol = 81{,}242$ J m^{-3} (E), 234 J m^{-3} (M), 11,436 J m^{-3} (J)
(2) $V = 248$ m s-1 (E), $V = 2.3$ m s^{-1} (M), $V = 50$ m s^{-1} (J)

The high values for the Earth indicate that only a small fraction of the potential energy is available for baroclinic eddy development (see Section 7.6.3.2).

9.25 Derive expressions (9.54) and (9.55) from the determinant formed by coefficients A, B.

9.26 Demonstrate for (9.57) that $\alpha_{Bc}H/2 = \coth(\alpha_{Bc}H/2)$ and then find graphically that $\alpha_{Bc}H \sim 2.4$.
Hint: Use the identity $\tanh x = (2 \tanh(x/2))/(1 + \tanh^2(x/2))$.

9.27 Compare the three terms in the baroclinic vorticity gradient equation (9.60) in the case of an eastward jet stream with profile $\bar{u}(y,z) = u_0 \cdot (z/H) \sin^2 [L(y - y_0)]$ that extends across a layer with thickness H (the scale height) where the Brunt–Väisälä decreases with altitude according to $N_B(z) = N_0/(z/H)$ in the latitude of the peak of the jet and at altitude $z = H$, with u_0 and N_0 being constant.

Solution

$$\beta - 2u_0 L^2 \frac{z}{H} + 2 \frac{u_0 f_0^2}{N_0 H^2}$$

9.28 Apply the result of Problem 9.27 to compare the values of the three terms in (9.60) for the following planets and appropriate conditions of the jet (take latitude 45°N and $N_0 = 10^{-2}$ s^{-1} in all the cases): (1) Earth ($u_0 = 30$ m s^{-1}; $\pi/L = 2000$ km; $H = 8$ km); (2) Mars ($u_0 = 30$ m s^{-1}; $\pi/L = 1000$ km; $H = 11$ km); (3) Saturn ($u_0 = 150$ m s^{-1}; $\pi/L = 5000$ km; $H = 50$ km).

Solution: (unit m^{-1} s^{-1})

(1) Earth: $\beta = 1.6 \times 10^{-13}$; $d^2\bar{u}/dy^2 = -1.48 \times 10^{-10}$; $2u_0 f_0^2/(N_0 H^2) = 7.5 \times 10^{-9}$
(2) Mars: $\beta = 2.9 \times 10^{-13}$; $d^2\bar{u}/dy^2 = -5.9 \times 10^{-10}$; $2u_0 f_0^2/(N_0 H^2) = 5.4 \times 10^{-9}$
(3) Saturn: $\beta = 3.8 \times 10^{-14}$; $d^2\bar{u}/dy^2 = -1.2 \times 10^{-10}$; $2u_0 f_0^2/(N_0 H^2) = 3.17 \times 10^{-8}$

9.29 Earth's mid-latitude zonal flow is observed to be centered at 50°N and has a wind shear velocity of 25 m s^{-1} over 8 km altitude, with the Brunt–Väisälä frequency being 10^{-2} s^{-1} and the vertically averaged temperature of the layer being 275 K. Calculate the Richardson number and Rossby deformation radius. Using the Eady model, find the maximum zonal length for baroclinic disturbances, their growth rate timescale, the number of total anticyclones and cyclones along the latitude band, and their propagation speed. What is the meridional temperature drop across the wave if its meridional length is 1/3 its zonal size?

Solution
$Ri = 2.6$; $L_D = 715$ km; $L_{Bcl} = 3934$ km; $n = 6$–7; $c_r = 12.5$ m s^{-1}; $\Delta T = 12.8°$.

9.30 Calculate the two terms in the vortex Kida relationship (9.62) for the following vortices in Table 9.2: GRS, WOS, Barge, A, and BS.

Solution
For all cases, $(1 - \varepsilon_V)/[\varepsilon_V(1 + \varepsilon_V)]$ and $|\partial u/\partial y|/[|\langle \zeta \rangle| - |\partial u/\partial y|]$ are given. GRS (0.81–0.49), WOS (0.42–0.31), Barge (2.4–0.3), A (0.67–0.14), BS (0.31–4.7).

9.31 Find the vorticity and tangential velocity in terms of the vortex strength or circulation for the Rankine vortex described by Equation 9.63.

Solution

(1) (a) $0 \le r \le r_0$, $\zeta = 2V_0/r_0$; (b) $r > r_0$, $\zeta = 0$
(2) (a) $0 \le r \le r_0$, $V_T = \Gamma_C r/2\pi r_0^2$; (b) $r > r_0$, $V_T = \Gamma_C/2\pi r$.

9.32 A model of a tornado uses the Rankine vortex described by Equation 9.63 with the following parameters: maximum speed 100 m s^{-1}, core radius 50 m, air density (assumed independent of pressure) 1.25 kg m^{-3}. (1) Calculate the pressure deficit relative to the ambient air pressure using the appropriate dynamical balance.

Solution

$$\Delta P = \rho V_0^2 = 125 \text{ mbar.}$$

9.33 Generation of a tornado starts from a surface rotating and vertically upwelling supercell air mass. Assume that initially the air mass vorticity is 10^{-2} s^{-1} and that its ascent velocity w increases from zero to $5\,\text{m s}^{-1}$ at an altitude of $1\,\text{km}$. The rotating air column increases vorticity by stretching in the vertical direction. From the definition of the average vorticity, find a relationship between the rate of vorticity change and that of area change. Then, using the continuity equation (9.13), define a characteristic timescale for a tornado to grow. What would be the time required to form a tornado with a Rankine vortex structure that has a maximum velocity of $100\,\text{m s}^{-1}$ and a radius of $250\,\text{m}$?

Solution

$$\frac{1}{\zeta}\frac{d\zeta}{dt} = -\frac{1}{A}\frac{dA}{dt}; \quad \tau = \left(\frac{\partial w}{\partial z}\right)^{-1}; \quad \tau = 200 \text{ s, then } t = 14 \text{ min.}$$

9.34 Calculate the tangential velocity at the periphery of the eddies generated in a Von Kármán street in Saturn's atmosphere (parameters: $h=800\,\text{km}$ and $a=5000\,\text{km}$) produced by an anticyclone that has a diameter of $1700\,\text{km}$ if the zonal flow has a velocity of $10\,\text{m s}^{-1}$ and the vortex street pattern moves with a velocity of $5\,\text{m s}^{-1}$ in the same direction. Take the radius of the eddy $r=D/4$ (see Figures 9.29 and 9.30).

Solution
$V_T = 40\,\text{m s}^{-1}$.

9.35 Find the pressure drop and temperature rise at the center of a hurricane at an altitude level close to the surface, if the temperature and pressure outside the hurricane are $T_s = 300\,\text{K}$ and $P_s = 1$ bar. The hurricane data are latitude $20°$, tangential velocity $V = 50\,\text{m s}^{-1}$, vertical extent $H = 8\,\text{km}$, and eyewall radius of $100\,\text{km}$.

Solution
$\Delta P_s = -32$ mbar, $\Delta T_s = 9\,\text{K}$.

9.36 Using the definition of absolute angular momentum for a hurricane in cyclostrophic balance and the formulation for the vertical shear of the geopotential field (Section 9.6.3), find an expression for the radial temperature gradient in a hurricane above the boundary layer in terms of the vertical shear of the angular momentum surfaces.

Solution

$$\frac{\partial L_{hur}^2}{\partial z^*} = r^3\frac{R_g^*}{H}\frac{\partial T}{\partial r}$$

Appendix: Methods to Study Planetary Atmospheres

The study of planetary atmospheres is a large multidisciplinary field that involves remote sensing studies such as the use of astronomical techniques as telescopic observations (and their data processing methods) and spacecraft vehicles covering from x-rays to radio wavelengths, and in situ measurements by probes, balloons, and landers, laboratory studies (fluid dynamics and chemistry), and advanced modeling (computational numerical simulations). In this appendix, we give a synthetic overview of the methods employed and their application to the study of planetary atmospheres.

A.1 REMOTE SENSING MEASUREMENTS

Remote sensing is the most common method used to explore atmospheres (except obviously in the case of the Earth). Because of the distance that separates the Earth from the planets, the use of astronomical instrumentation is necessary. However, due to the limitations imposed by the Earth's atmospheric transparency produced by the opacity of gases and clouds and to the image blurring produced by refractivity changes and motions (called the *seeing*), space telescopes on Earth orbit are a very important tool. They allow observing the atmospheres at wavelengths not accessible from the ground and, at the same time, they give in general higher quality images in particular at optical/IR wavelengths. All this information comes as electromagnetic radiation (photons) that are emitted, absorbed, and reflected by atmospheres.

With the advent of astronautics during the twentieth century, the exploration of the solar system by spacecrafts gave a major impulse to the knowledge of the planets, their satellites, and atmospheres. The first spacecrafts in the 1960s performed fly-bys to the closest planets (Venus and Mars) with orbiting and landing on them occurring in the 1970s. The exploration of the outer planets, Jupiter and Saturn, started in 1973–1974 but was rapidly followed in 1979 with a grand tour to them, with the Voyager 2 reaching Uranus and Neptune in the second half of the 1980s. Since then, orbiting and probing the atmospheres of the two giants, and landing on distant bodies, was achieved (Galileo orbiter and probe of Jupiter in 1995, Cassini orbiter of Saturn, and Huyghens landing on Titan in 2005). Fly-by and orbiting involve remote sensing methodologies, but descending probes, balloons, and landers and mobile rovers perform in-situ measurements. Additionally, the spacecrafts perform indirect atmospheric studies through stellar occultation experiments and radio-occultation that use the attenuation of the radio signal sent to the Earth by the spacecraft. In the case of the major planets, another relevant study is the sounding of their interiors (i.e., their deep atmospheres) by measurements of the structure of the external gravitational field.

A.1.1   GROUND TELESCOPES AND RADIOTELESCOPES

Telescopes are basically characterized by the diameter of the objective D (also called aperture) and its focal length f (the distance from the objective to its focus, i.e., the point where the collected rays converge). A large diameter means a large collecting area, making possible to study very faint sources (or equivalently allowing a reduction of the exposure time of the source). It also allows increasing the spatial resolution to the limit imposed by diffraction that for a circular objective is given by

$$\sin\theta \approx \theta = \frac{1.22\lambda}{D} \quad \left[\text{rad}\right] \tag{A.1}$$

where
   θ is the angular size of the radius of the diffraction spot and
   λ is the wavelength

The focal length determines field of view of the telescope and thus the scale of the image that forms in the focal plane of the telescope and hence magnification. For an object or feature with angular size α, the linear scale s is given by

$$s = f\tan\alpha \approx f\alpha \tag{A.2}$$

As stated above, the telescopic observations through the atmosphere are limited by two major aspects: (1) The seeing σ or scintillation of the object imposed by the changing refraction index along the ray path (diminishing significantly the spatial resolution). (2) The absorption of this radiation at some wavelengths by the Earth's atmosphere, that reduces the observations to the so-called atmospheric windows of the electromagnetic spectrum.

   The *seeing* σ is measured in arc-seconds (is the "mean size" of the scintillating spot resulting from a point source) and it typically ranges under the best conditions from 0.1 arcsec to the worst case when it reaches 2–3 arcsec or more. As a reference, the disk of Uranus and Neptune as seen from the Earth has a size of ~4 and 2 arcsec, respectively. Generally for optical and near infrared observations we have $\sigma > \theta$ and thus the seeing effectively limits the spatial resolution of a telescope. At greater wavelengths (submillimeter and radio) $\theta < \sigma$ and the seeing is not a problem. The seeing can be improved significantly using methods than compensate for the atmospheric scintillation, called adaptive optic (AO). It acts on the telescope focus image before the detection and permits to reach the diffraction limit of the telescope. Although the method is limited to a small field of view of the telescope (~30 arcsec), it has been used very successfully in the wavelength range from ~1 to 2.5 μm. For planetary atmospheres, it has produced significant results in the study of Titan, Uranus, and Neptune. In the visible range, a method usually employed to improve image resolution is the capture of sequences of target frames in times below or close to those due to the seeing scintillation frequency ("lucky imaging"). The resulting image series are selected by their quality, and their final composition (summation) is done by aligning using the appropriate software. This method has allowed well equipped amateur astronomers to use "webcams" and "stacking" software to get diffraction limited images of the planets.

The spatial resolution is also significantly improved by the use of interferometric techniques that combine the light from various telescopes and radiotelescopes appropriately. For two telescopes separated by a distance L, the effective angular resolution in the direction parallel to the line connecting the telescopes is λ/L and the angular resolution in the perpendicular to this direction is λ/D. Table A.1 lists major operative ground-based telescopes and radiotelescopes.

A.1.2 The Atmospheric Window

The absorption of the terrestrial atmospheric gases varies with altitude and so do the strength of some of the atmospheric windows. The shortest wavelengths (gamma and x-rays) are absorbed by oxygen and nitrogen molecules and other free atoms (in particular oxygen) and planetary observations must be done from space. Ultraviolet wavelengths below ~300 nm are absorbed by ozone. Above the wavelength of 1.3 μm, the water absorption bands become important absorbers. The optical window extends from ~300 nm to 1 μm and in the near infrared from wavelengths ~1–2.5 μm. At higher wavelengths, the altitude dependent absorption produced by water vapor limits the observations to high-altitude observatories in dry environments (wavelength variable window from 3 to 20 μm, known as the thermal infrared). For wavelengths between 20 μm and 1 mm, the absorption by water and carbon dioxide molecules fully impedes the ground observations and space observatories are necessary. The radio window finally extends from about ~1 mm to 20 m, with the millimeter range observations (~1–10 mm) still sensitive to altitude.

Astronomical observations from the ground are also sensitive to the extinction of radiation produced by the scattering of light due to gases and particles (aerosols and clouds). This reduces the observations to nighttime and clear skies for the optical and near infrared spectral ranges (submillimeter and radio observations can be done in day time and cloudy skies).

A.1.3 Remote Sensing Telescopes and Spacecrafts

A.1.3.1 Earth Orbiting Space Telescopes

Outside the atmospheric windows, remote sensing observations are performed from space (x-ray, UV, and IR ranges) using space telescopes in orbit around Earth (see Table A.2). Complementary airplanes, such as the Kuiper Airborne Observatory and coming SOFIA flying at altitudes above the tropopause (~14 km), i.e., above the major water vapor absorption bands, can make observations in the infrared to submillimeter ranges (1–500 μm). In the optical and near-infrared ranges, telescopes are used to improve the spatial resolution till the diffraction limit. The best example is the Hubble Space Telescope [HST] (and planned future James Webb Telescope).

Obviously the better resolution and ample wavelength coverage by remote sensing can be obtained using telescopes and instruments onboard spacecrafts that approach the planets (see Table A.3).

TABLE A.1
Ground-Based Major Telescopes (Operative in 2009)

1. Optical–IR Telescopes (Aperture > 8 m)

Name of the Telescope	Location	Aperture (m) (Type)	Date of Operation
Gemini North	Mauna Kea, Hawaii	8.0 (meniscus)[b]	1998
Gemini South	Cerro Pachon, Chile	8.0 (meniscus)	2000
Very Large Telescope (VLT) UT1 Antu[a]	Cerro Paranal, Chile	8.2 (meniscus)	1998
Very Large Telescope (VLT) UT2 Kueyen	Cerro Paranal, Chile	8.2 (meniscus)	1999
Very Large Telescope (VLT) UT3 Melipal	Cerro Paranal, Chile	8.2 (meniscus)	2000
Very Large Telescope (VLT) UT4 Yepun	Cerro Paranal, Chile	8.2 (meniscus)	2000
Subaru	Mauna Kea, Hawaii	8.2 (meniscus)	1999
South African Large Telescope (SALT)	Sutherland, South Africa	9.2 (segmented)[c]	2005
Hobby-Eberly Telescope	Mt. Fowles, Texas	9.2 (segmented)	1997
Keck I	Mauna Kea, Hawaii	10.0 (segmented)	1993
Keck II	Mauna Kea, Hawaii	10.0 (segmented)	1996
Gran Telescopio Canarias (Grantecan, GTC)	La Palma, Canary Islands (Spain)	10.2 (segmented)	2008
Large Binocular Telescope (LBT)	Mt. Graham, Arizona	11.8 (Honneycomb)	2006

[a] VLT operates individually or in combined mode equivalent to a 16 m aperture telescope. Eventually it can be used in an interferometer mode together with three 1.8 m auxiliary telescopes to form the VLTI interferometer.

[b] Meniscus Telescopes have flexible mirrors and use the *active optics* control of the primary objective (i.e., its shape) and move the secondary mirror to improve image quality.

[c] Segmented Telescopes are composed of individual hexagonal elements that operate together as a single, high precision mirror.

2. Other ground-based Observatories with dedication to planetary (Solar System) observations:
   - European Southern Observatory (ESO): Has two observatories in the Atacama's desert in La Silla (Chile) with different telescopes that have apertures ranging from 0.6 to 3.6 m and includes the SEST 15 m submillimeter antenna. The Paranal observatory is the site of the VLT telescopes.
   - Mauna Kea Observatories. In addition to the telescopes listed above, this observatory homes the NASA Infrared Telescope Facility (IRTF) 3 m in aperture (used extensively for monitoring the planets in the IR), the Canada-France-Hawaii (CFHT) 3.6 m aperture, the 3.8 m UKIRT and the 15 m JCMT sub-millimeter antenna.
   - Calar Alto Observatory (Centro Astronómico Hispano Alemán). Located in Almeria (Spain) homes 1.23, 1.5, 2.2, and 3.5 m aperture telescopes.
   - Pic-du-Midi Observatory (France). Located in French Pyrennes operates 1 and 2 m telescopes. The 1 m telescope is dedicated to Solar System and has made a number of important contributions to planetary atmospheres studies.

TABLE A.1 (continued)
Ground-Based Major Telescopes (Operative in 2009)

- Canary Island Observatories. There are two observatories one is located at the Teide volcano in Tenerife that is mainly dedicated to solar studies, and the other at La Palma that homes the GTC, the Isaac Newton Group of telescopes (apertures 4.2 and 2.5 m), the Telescopio Nazionale Galileo TNG (3.6 m) and the Nordic Optical Telescope NOT (2.5 m) among others.
- There are numerous other Observatories that participate in the search and characterization of extrasolar planets (see list in http://exoplanet.eu/searches.php). Among them, the major observatories are Anglo-Australian Observatory (AAO, Australia); Kitt Peak National Observatory (Arizona); Palomar Observatory (California); Las Campanas Observatory (Chile).
- An archive of planetary images (from the early photographs in the nineteenth century up to ~1980) resides at Meudon Observatory (France) and at Lowell Observatory (Arizona).
- Not to forget is the contribution to planetary imaging that is routinely performed by a large battery of amateur telescopes around the word. Contributions to the observations of Venus, Mars, Jupiter, and Saturn have been important in surveying the long-term evolution of cloudy features. Their quality has significantly increased in recent times with the advent of the CCD detectors and particularly with the use of "webcams" and image selection and summation methods that have allowed reaching diffraction limited images of these planets. This effort is coordinated by different groups, for example their scientific use for the outer planets is performed by the International Outer Planet Watch at http://www-ssc.igpp.ucla.edu/IJW/
- Atmospheric node at http://www.ehu.es/iopw/
- Image archive at http://www.pvol.ehu.es/

3. Airborne Observatories:
- Kuiper Airborne Observatory (KBO) is an airplane flying up to 14 km. The telescope diameter has 0.91 m and operated in the wavelength range $\lambda = 1 - 500\,\mu m$ (currently decommissioned).
- SOFIA (Stratospheric Observatory and Infrared Astronomy) is a NASA partnership with the German Space Agency (DLR) to develop a Boeing 747SP airliner fitted with a 2.5 m reflecting telescope. SOFIA will be the largest airborne observatory in the world, and will begin flight testing during 2009.

4. Radio Telescopes (microwave range)
   Microwave windows of interest for atmospheres are at: 0.85, 1.3, 2, 3, and 7 mm.

Facility	Location	Configuration
Institut de Radio Astronomie	Pico Veleta, Spain	30 m antenna[a]
Millimétrique (IRAM)	Plateau Bure, France	6×15 m antennas[b]
Very Large Array (VLA)	New Mexico	27×25 m antennas (Y-shape)[c]
Very Large Baseline Array (VLBA)	Different sites in USA	10×25 m antennas[d]
Berkeley Illinois Maryland Association (BIMA)	Hat Creek, California	10×6.1 m antennas
Owens Valley Radio Observatory (OVRO)	Bishop, CA	6×10.4 m antennas
Nobeyama Millimeter Array (NMA)	Minamisaku, Japan	6×10 m antennas (transportable) 45 m antenna

[a] At a wavelength of 1.3 mm the resolution is 10 arcsec.
[b] Movable on rail tracks up to a distance of 408 m. Best resolution achieved is 0.5 arcsec at 1.3 mm.
[c] Y-shaped system with 9 antennas per arm forming 351 individual interferometer pairs. Each arm is 21 km in length. The collecting area is equivalent to a 130 m diameter dish. The minimum detectable wavelength is 7 mm (43 GHz) where the achieved resolution is 0.05 arcsec.
[d] Minimum wavelength 7 mm (43 GHz) with a highest resolution of 0.001 arcsec.

TABLE A.2
Major Space Telescopes Used for Planetary Atmospheres Studies (Up to 2009)

Space Telescope	Launch Date	Notes
International Ultraviolet Explorer (IUE) (USA/ESA)	January 26, 1978	Earth orbit Ultraviolet observatory[a]
Infrared Astronomical Observatory (IRAS) (NASA, NIRV, SERC)	January 25, 1983	Earth orbit Infrared Observatory[b]
Infrared Space Observatory (ISO) (ESA)	November 7, 1995	Earth orbit Infrared Observatory[c]
Hubble Space Telescope (HST) (USA/ESA)	April 15, 1990	Earth orbit Observatory UV-Visible-Near IR[d]
Spitzer Space Telescope (SST) (USA)	August 25, 2003	Earth orbit Infrared Observatory[e]

Notes: Satellites in Earth orbit contributing to the search and characterization of extrasolar planets by photometry are MOST (launched June 30, 2003) and COROT (launched December 27, 2006).

[a] IUE: UV-spectroscopy ($\lambda = 115-320$ nm).

[b] IRAS: 57 cm telescope ($\lambda = 12-100$ μm).

[c] ISO: 60 cm telescope ($\lambda = 2.5-240$ μm).

[d] HST: 2.2 m telescope ($\lambda = 115$ nm-2.5 μm).

[e] SST: 85 cm telescope ($\lambda = 3.6-106$ μm).

A.1.3.2 Planet Fly-by

In this case, the spacecraft passes the planet at a given distance for a planed mission exploration or when it is in route to another planet. This comprises the cases where a vehicle makes many passages around a planet to get a gravitational acceleration. The effective resolution reached by onboard instruments depends on the distance to the target and on the own capabilities of the sensors and telescopic cameras. The time span of the mission is limited to the approach and closest fly-by time (typically 1–3 months).

A particular fly-by configuration is that of an orbiting spacecraft around a planet that makes planned fly-bys of the natural satellites, as has been the case of the Galileo in Jupiter and Cassini in Saturn.

A.1.3.3 Planet Orbiting

In this case, the spacecraft enters in orbit around the planet and becomes its artificial satellite. This is the case of Earth observing and application satellites. The spacecraft orbit can change in time in a natural way due to the gravitational interactions with natural satellites or with the Sun, or to friction with the upper atmosphere of the body, or in an artificial and planned way using its own propulsion system. This allows seeing the planet under different geometries and solar illumination angles. The orbits are basically of two types: equatorial (the spacecraft lies in approximately the equatorial plane) and polar (the orbital plane passes through the poles), although other configurations are also in use to get different perspectives. The eccentricity of

TABLE A.3
Successful Missions of Interest for Planetary Atmosphere Studies
(Up to 2009)

Spacecraft	Launch Date	Target	Notes
Venera 4 (USSR)	June 12, 1967	Venus	Atmospheric probe: October 18, 1967
Venera 5 (USSR)	January 5, 1669	Venus	Atmospheric entry probe: May 16, 1969
Venera 6 (USSR)	January 10, 1969	Venus	Atmospheric entry probe: May 17, 1969
Venera 7 (USSR)	August 17, 1970	Venus	Lander: December 15, 1970
Mariner 9 (USA)	May 30, 1971	Mars	Orbiter: November 13, 1971
Pioneer 10 (USA)	March 3, 1972	Jupiter	Fly-by: December 3, 1973
Venera 8 (USSR)	March 27, 1972	Venus	Lander: July 22, 1972
Pioneer 11 (USA)	April 6, 1973	Jupiter	Fly-by: December 4, 1974
		Saturn	Fly-by: September 1, 1979
Mariner 10 (USA)	November 3, 1973	Venus	Fly-by: February 5, 1974
		Mercury	Fly-bys: 1974
Venera 9 (USSR)	June 8, 1975	Venus	Lander: October 22, 1975
			Orbiter: October 22, 1975
Venera 10 (USSR)	June 14, 1975	Venus	Lander: October 25, 1975
			Orbiter: October 25, 1975
Viking 1 (USA)	August 20, 1975	Mars	Orbiter: June 19, 1976
			Lander: July 20, 1976
Viking 2 (USA)	September 9, 1975	Mars	Orbiter: August 7, 1976
			Lander: September 3, 1976
Voyager 1 (USA)	September 5, 1977	Jupiter	Fly-by: March 5, 1979
		Saturn	Fly-by: November 12, 1980
Voyager 2 (USA)	August 20, 1977	Jupiter	Fly-by: July 9, 1979
		Saturn	Fly-by: August 26, 1981
		Uranus	Fly-by: January 24, 1986
		Neptune	Fly-by: August 25, 1989
Pioneer-Venus 1 (USA)	May 20, 1978	Venus	Orbiter: December 8, 1978
Pioneer-Venus 2 (USA)	August 8, 1978	Venus	Large probe: December 8, 1978 (day time)
			North probe: December 8, 1978 (night time)
			Day probe: December 8, 1978 (day time)
			Night probe: December 8, 1978 (night time)
Venera 11 (USSR)	September 9, 1978	Venus	Lander: December 25, 1978
Venera 12 (USSR)	September 14, 1978	Venus	Lander: December 21, 1978
Venera 13 (USSR)	October 14, 1981	Venus	Lander: February 27, 1982
Venera 14 (USSR)	November 4, 1981	Venus	Lander: March 5, 1982
Venera 15 (USSR)	June 2, 1983	Venus	Orbiter (radar): October 10, 1983
Venera 16 (USSR)	June 7, 1983	Venus	Orbiter (radar): October 14, 1983
Vega 1 (USSR)	December 15, 1984	Venus	Lander: June 11, 1985
			Balloon in Venus Atmosphere

(*continued*)

TABLE A.3 (continued)
Successful Missions of Interest for Planetary Atmosphere Studies
(Up to 2009)

Spacecraft	Launch Date	Target	Notes
Vega 2 (USSR)	December 21, 1984	Venus	Lander: June 15, 1985
			Balloon in Venus Atmosphere
Phobos 2 (USSR)	July 12, 1988	Mars	Orbiter: January 29, 1989 (failed prior to land)
Magellan (USA)	May 5, 1989	Venus	Orbiter (radar): August 10, 1990
Galileo (USA)	October 18, 1989	Venus	Fly-by: February 10, 1990
		Jupiter	Orbiter: December 7, 1995
		Jupiter	Atmospheric probe: December 7, 1995
Mars Global Surveyor (USA)	November 7, 1996	Mars	Orbiter: September 12, 1997
Mars Pathfinder (USA)	December 2, 1996	Mars	Lander: July 4, 1997
			Rover: July 4, 1997
Cassini (USA)	October 15, 1997	Jupiter	Fly-by: December 30, 2000
Huyghens (ESA)		Saturn	Orbiter: July 1, 2004
		Titan	Lander: December 15, 2005
Mars Odissey (USA)	April 7, 2001	Mars	Orbiter: October 23, 2001
Mars Express (ESA)	June 2, 2003	Mars	Orbiter: December 25, 2003
Spirit (USA)	June 10, 2003	Mars	Rover: January 3, 2004
Opportunity (USA)	July 7, 2003	Mars	Rover: January 24, 2004
Messenger (USA)	August 3, 2004	Venus	Fly-by: June 5, 2007
		Mercury	*Fly-bys-Orbiter*: (March 18, 2011)
Mars Reconnaissance Orbiter (MRO)	August 12, 2005	Mars	Orbiter: March 10, 2006
Venus Express (ESA)	November 9, 2005	Venus	Orbiter: April 11, 2006
New Horizons (USA)	January 19, 2006	Jupiter	Fly-by: February 28, 2007
		Pluto	*Fly-by*: (July 2015)
Phoenix (USA)	August 4, 2007	Mars	Lander: May 25, 2008

the orbit can also be selected to have (a) uniform visibility and resolution (circular orbits); (b) change from high resolution imaging (perihelion observations, sampling part of the atmosphere) to lower resolution (aphelion observations, whole planet viewing), elliptical orbits.

Viewing geometries from orbit include: (a) nadir view, the spacecraft observes the planet (atmosphere) in "front of it," vertically downwards toward the center, sampling short atmospheric path lengths; (b) limb view in which the instrument views the atmosphere tangentially, in direction of the limb, sampling long-path lengths; (c) limb occultation that uses a natural source (Sun, stars, other satellites) or the radio signals sent from the spacecraft as detected from Earth to vertically sample the atmosphere (see Section A.1.4).

A.1.4 Remote Sensing Techniques

Remote sensing studies using electromagnetic radiation as a source of information can be classified in two broad groups: imaging at selected spectral intervals and spectroscopy. Complementary to them are the already cited occultation techniques, and the measurements of the gravitational field that sound the deep atmospheres of the giant planets and in general the planetary interiors. Finally, the measurements of the magnetic field structure are necessary to interpret the electromagnetic phenomena occurring in the upper and tenuous atmosphere, and establish or constrain the rotation period of the giant planets.

We summarize in Table A.4 the main techniques employed in planetary atmospheres studies and the basic scientific results they produce. The intensity and wavelength units employed in spectroscopy and for image calibration vary with the spectral range, and are given as a footnote in this table. The intensity (absolute reflectivity and radiance) measurements are the basis for radiative transfer modeling. The most important retrievals are (vertical and meridional directions): (a) Temperature structure and thermal winds; (b) chemical composition (spatial and temporal distributions); and (c) optical properties of aerosols and cloud particles and their spatial distribution and temporal variability; (d) wind speed and direction from the tracking of cloud motions.

A.1.4.1 Imaging

Different types of instruments are used to obtain images of the planets depending on the spectral range. In x-rays, telescopes use the grazing reflection principle to prevent photon absorption and imaging is performed through different types of detectors with limited resolution. In the UV-optical-near IR ranges, imaging is performed with cameras (coupled to telescopes) that consist of an optical system (that appropriately transmits the required wavelength range) and a digital detector that converts photons directly to digitized electrical signals. The cameras onboard planetary spacecrafts act as small telescopes, and are characterized by their field of view (FOV) given in arc-seconds or in radian units. The detection is performed by charge-coupled devices (CCD) of different types. CCD detectors marked a revolution in astronomical studies at the beginning of the 1980s, changing the classical photographic plates and emulsions for imaging, to a direct digital format. The archives of planetary images before 1980s are in photographic format (although a large part has been digitalized), but more recently they are in digital format as received from spacecrafts. The first cameras used vidicon systems (TV-like) that had important problems in geometrical and intensity calibration. As an example this was the case for the large data set of images gathered by Voyagers 1 and 2 in their fly-bys of Jupiter, Saturn, Uranus, and Neptune. Imaging of the planets at thermal wavelengths (above $5\,\mu m$) is performed with infrared detector arrays and bolometers (detectors sensitive to temperature changes). In the microwave range the currently used detectors are bolometers and heterodyne receivers that convert the microwave signal to a lower frequency and mix it nonlinearly using a local oscillator before amplification and conventional electronic detection.

The detectors are characterized and compared according to their figures of merit. Most important are the spectral and time response, the sources and levels of noise,

TABLE A.4
Observation Techniques and Main Objectives in Wavelength Dependent Remote Sensing Studies of Planetary Atmospheres

Wavelength Range[a,b,c]	Instrument	Technique	Studies
X-rays (0.01–1 nm)	X-ray Space Telescope	Spectroscopy Imaging	Aurora[d]
Ultraviolet (UV)	UV-Space Telescope	Spectroscopy	Chemical composition upper atmosphere
EUV (50–100 nm)	Optical Space Telescope (NUV) (HST)	Photometry	Aurora, airglow, non-aurora processes[d]
FUV (100–200 nm)		Imaging	Rayleigh and Raman scattering
NUV (200–350 nm)	Planetary spacecraft (UV-Onboard Instrument)		Aerosol and cloud properties (FUV-NUV) (reflected sunlight, upper atmosphere)
Visible (Optical) (350 nm–1 μm)	Earth Telescopes Optical Space Telescope Planetary spacecraft	Spectroscopy Imaging	Chemical composition Aerosol and cloud properties (reflected sunlight) Dynamics (motions upper cloud)
Near Infrared (1–5 μm)	Earth Telescopes Optical Space Telescope Planetary spacecraft	Spectroscopy Imaging	Chemical composition Aerosol and cloud properties (reflected sunlight and opacity thermal emission) Dynamics (motions upper and deeper clouds) Aurora[d] Temperature profile
Thermal Infrared (7–20 μm)	1. Earth Telescopes (high altitude) Airborne Telescope	Spectroscopy Imaging	Chemical composition Aerosol and cloud properties (opacity to thermal emission)
Sub-mm (20 μm to 1 mm)	IR Space Telescope Planetary spacecraft		Dynamics (thermal waves, global oscillations)
	2. Airborne Telescope IR Space Telescope Planetary spacecraft		Temperature sounding (dynamics: thermal winds)
Radio	1. Earth Antennas (microwave)	Spectroscopy	Chemistry (deep atmosphere)

TABLE A.4 (continued)
Observation Techniques and Main Objectives in Wavelength Dependent Remote Sensing Studies of Planetary Atmospheres

Wavelength Range[a,b,c]	Instrument	Technique	Studies
Thermal (≤ 1 cm)	2. Earth Antennas Radio Interferometry	Imaging (time integration)	Temperature sounding Temperature/molecular weight vertical profiles
Nonthermal (1 cm to 20 m)	Planetary spacecraft	Radio occultation Radio emission	Rotation period (fluid planets)

[a] Intensity units employed in each spectral range:

For all spectral ranges: Relative units referred to a value at a given wavelength.

X-rays: (keV s sr cm^2)$^{-1}$, cts ks^{-1} keV^{-1} arc min^{-2}.

UV: Rayleigh Å^{-1}, geometric albedo (%).

Visible: geometric albedo (%), reflectivity (I/F).

Infrared: W cm^{-2} sr^{-1} cm^{-1}, brightness temperature (K), W cm^{-2}.

Radio: Jansky, W m^{-2} sr^{-1} Hz^{-1}, brightness temperature (K).

[b] Wavelength units employed in each spectral range:

X-rays: Photon energy (keV), Angstrom (Å).

UV: Angstrom (Å), nanometer (nm).

Visible: Angstrom (Å), nanometer (nm).

Infrared: micrometer (μm), wavenumber (cm^{-1}).

Radio: millimeter (mm), centimeter (cm); frequency (GHz, MHz).

[c] Wavelength unit conversions:

(a) From energy: $\lambda = hc/E$

(b) From frequency: $\lambda = c/f$

(c) From wavenumber: $\lambda(\mu m) = 10^4 / \bar{v}(cm^{-1})$

with:

Planck constant: $h = 6.6256 \times 10^{-24}$ J s

Light speed: $c = 2,9979 \times 10^8$ m s^{-1}

E is the energy (in J, 1 J = 6.242 × 10^{18} eV).

[d] Aurora observing wavelengths: see Tables 6.7 and 6.8.

their size (or area), the number of image elements (pixels) and pixel size. The spectral response is usually given in terms of a magnitude that depends of the type of detector used. Typically the quantum efficiency QE(λ) (which represents the percentage of photons that produce a detection event at a given wavelength) and the responsivity $R(\lambda)$ (electric current produced per incident watt, unit mA W^{-1}) are employed. The pixel size is matched to the diffraction resolving power of the optical system, and the camera size together with the FOV, determine the fraction of the planet that is observed. Finally, to reduce noise, sensible detectors are cooled, specially at infrared wavelengths.

Most employed planetary imaging operates in two broad wavelength ranges: (1) Optical (near ultraviolet and visible wavelengths, from ~200 nm to 1 μm) that employs CCD detectors and (2) near-infrared (from 1 to 5 μm) that employs infrared detector arrays. The wavelength selection for the image is performed usually with filters that have different pass-bands transmission wavelengths to match aerosol (continuum) and gaseous absorption bands. These are basically of two types: narrow band interference filters (pass-bands ~ 0.1–10 nm) and broad band filters (pass-bands ~ 20–50 nm). In the infrared it is also customary to use circular variable filters (CVF) for narrow spectral IR imaging (~0.2–0.3 μm). These are multilayer thin film interference filters mounted on a filter wheel where its substrate is a sector of a ring. The thickness of each layer is a function of the angle on the sector and acts as a narrow band pass filter. The central wavelength and its band-pass are functions of the angle on the sector.

Recently, spectral imaging devices have been introduced that combine spectral resolution and image formation by scanning the spectral dispersion grating along the field of view, performing a "three-dimensional" image detection known as an "image cube." It contains the 2D images formed by the scan in the grating dispersion direction at different wavelengths that constitutes the third dimension. A low-resolution spectrum can be retrieved at each image pixel. Examples of such instrument are NIMS onboard Galileo, VIMS onboard Cassini, and MARSIS and VIRTIS onboard Mars Express and Venus Express. These instruments cover usually the two channels (optical and near infrared) containing also a higher spectral resolution channel.

Planetary images used for atmospheric studies require (1) to be calibrated in intensity (i.e., the absolute radiance coming from a point in the image must be known as a function of wavelength) and, (2) to be calibrated in geometry and "navigated," i.e., each point in the image (x, y) must be converted to its planetary coordinates (latitude and longitude) and the solar phase angle retrieved to determine the scattering angles of incoming radiation. A good spacecraft pointing is required and the presence of the planet limb helps to make an accurate positioning. The intensity calibration requires a flat field correction necessary to leave the image free of the nonuniform response of the pixels (is obtained by dividing pixel to pixel the work image by a uniformly illuminated image). For cloud morphology studies and cloud motions, image processing methods that do not conserve the absolute intensity are necessary to enhance subtle cloud details and improve the contrast between features. A variety of digital filters are available in the software packages. If a point-like reference source is available (a star or a small satellite), a deconvolution of the image can be performed using the point spread function (PSF) of the source that allows determining the degree of blurring of each pixel element of the image and mathematically correct it.

Intensity calibration of the spectrum is performed using calibrated lamps in the laboratory for spacecraft instrumentation or standard stars in the sky. This step allow to convert the "digital number" (DN) of each pixel element to absolute radiance units. Some spacecrafts incorporate a photopolarimeter to make measurements of the intensity of polarized light as a function of the wavelength and viewing geometry (scattering angles), important to characterize cloud and aerosol particles properties.

A.1.4.2 Spectroscopy

Spectrometers are characterized by their resolving power

$$R = \frac{\lambda_0}{\Delta\lambda} \qquad (A.3)$$

i.e., the observed wavelength divided by the minimum resolved wavelengths $\Delta\lambda = \lambda - \lambda'$, or in other words, the capability to separate two contiguous wavelengths. Typical values are: $R \sim 100$ (low resolution spectroscopy), $\sim 10^3$ (mid-resolution) and $\sim 10^4$ (high-resolution). The spectrum should have ideally the appropriate spatial resolution on the planet disk (length of the slit) to isolate the properties of atmospheric structures. The most commonly used are the grating spectrometers that employ a diffraction grating as spectral dispersion element. The grating is characterized by the number of fringes per unit length (e.g., 10,000 cm^{-1}) with a total number of fringes N. The grating is "blazed" in order to maximize the throughput at the central wavelength of the spectrum. From the diffraction theory we known that the dispersion angle θ_s (relative to the direction perpendicular to the grating), the diffraction order ($m = 0, \pm 1, \pm 2, \ldots$) and the separation of the fringes (a) are related according to $a \sin\theta_s = m\lambda$, then the grating dispersion D_s is given by

$$D_s = \frac{\mathrm{d}\theta_s}{\mathrm{d}\lambda} = \frac{m}{a\cos\theta_s} \qquad (A.4)$$

with unit for dispersion in nm (of wavelength) per mm in the spectrum. A high-resolution dispersion is for example ~ 0.1 nm mm^{-1}. For this case, the resolved power is given by

$$R = \frac{\lambda_0}{\Delta\lambda} = Nm \qquad (A.5)$$

whereas the dispersion increases with diffraction order, the intensity diminishes as m increases, so a compromise must be taken to have the necessary high dispersion but at the same time enough signal in the detector for the spectrum to be recorded.

When high spectral resolution is required, the use of interferometer techniques is necessary, particularly in the mid- to far-infrared wavelengths. The most employed instruments are the Fabry–Pérot spectrometer that uses multiple beam interferences in etalons formed by two highly reflecting surfaces on two transmission layers separated by a variable distance, and the Michelson interferometer that uses mirrors (one of them movable) to form the interference patterns. The technique is commonly called Fourier-Transform spectroscopy (FTS) since the interferogram of the incident light can be theoretically and perfectly reconstructed using the inverse Fourier transform. These spectrometers provide resolutions $R \sim 10^4$–10^5.

Intensity calibration of the spectrum is performed in a similar way as for imaging. Wavelength calibration is performed through the comparison with a reference spectrum obtained from a well-known source.

A.1.4.3   Remote Sounding from the Ground

Remote sounding from the ground of atmospheres is restricted to our planet but has been applied in some planetary missions that used landers and descending probes. The techniques can be divided in two broad groups: (1) active systems—radar and lidar; (2) passive systems—radiometers and spectrometers.

The active systems use a device to measure the backscatter pulses of radio waves (radar) and pulses of light ("lidars," from "light detection and ranging" or "laser imaging and ranging") sent to the atmosphere from the ground. The radars operating at wavelengths of the order of centimeters are used for weather forecasting and for the estimation of precipitation rates in cloudy areas and their motions (pulse radar technique). The backscattered radiowave signal depends on the droplet and ice crystals sizes, and their concentration (number per unit volume). The radar operating at wavelengths of the order of meters is able to sense variability in the atmospheric refractive index of the lower and middle atmosphere. These are produced by fluctuations in the temperature and water vapor concentration due to turbulence whose profiles can then be reconstructed from the signal.

A lidar device was mounted onboard the Mars polar mission Phoenix to study dust and ice particles suspended in the atmosphere. It consisted in a Neodymium-YAG laser working at two wavelengths 1064 and 532 nm that emitted pulses of width 10 ns at a frequency of 100 Hz. The light green and near-IR wavelengths scattered in the atmosphere are then captured by two detectors.

The passive systems are radiometers and spectrometers that use a bright source in the sky (the Sun or the Moon) to measure the radiance at ground, which depends on the source position in the sky. The column abundance of a given compound along this path length can be retrieved. The radiometers use filters to match the molecular absorption bands operating usually at visible and infrared wavelengths. The most common application on Earth is the measurement of the water vapor, CO_2, and ozone (Dobson spectrometer) column abundances. They are also used to study the atmospheric extinction by aerosols along the optical path.

A.1.4.4 Occultation Techniques

The occultation technique is used to sample the atmosphere when a point-like source passes behind the planet limb. A usual source is a star (stellar occultation) and the observations can be done from Earth (ground-based and orbiting telescopes) or from a planetary spacecraft in orbit around the planet. The Sun can be also used as a source. The planetary spacecraft itself can be used as the source using the radio signals it sends to the Earth when passing behind the planet limb in its orbit around the planet. If a battery of spacecrafts orbit a planet (a minimum of two is required) one can be used as the source for the radio signal occultation and the others as receivers.

The occultation involves two effects that can be use to sample vertically the atmosphere. One is the differential refraction that reduces the apparent brightness of the source. When the source is the spacecraft, the radiowaves they emit suffer a frequency Doppler phase shift due to the spacecraft motion and to atmospheric refraction, $\Delta f = (f_0/c)v_s\theta$, where f_0 is the frequency at the spacecraft, $\theta(t)$ the refraction angle, and v_s the component of the spacecraft velocity perpendicular to the Earth's direction.

From the relative brightness occultation light curve of the starlight $I(t)/I_0$ (which can be measured in immersion and in emersion), it is possible to get the refraction angle variability $\theta(t)$. This allows the refractivity profiles ($n-1$ versus altitude) to be retrieved, being n the refraction index, and by inversion under appropriate considerations, the vertical profile of the temperature to molecular weight ratio $T/\mu(z)$ of the upper atmosphere and the presence of discontinuities due for example to high altitude hazes.

A.1.4.5   Gravitational Field Measurements

The measurement of the gravitational field of a fluid planet (Jupiter, Saturn, Uranus, and Neptune) provides information about its internal (deep atmospheric) structure and of its motions, which can manifest in the upper, visible cloud layers as zonal jets (see Chapters 1 and 7). A spacecraft can sample the deviations in the gravitational acceleration produced by a planet from those produced by a homogeneous spherical internal mass distribution. There is a planned mission *Juno* to Jupiter to study in detail its gravitational field from a polar orbit and infer or constraint internal motions.

The gravitational field of a planet is also determined from measurements of the motions of natural satellites and space vehicles. Precise measurements of outer satellites and asteroid orbits and the use of the perturbation theory for orbit determination have been employed to constraint the internal structure of Jupiter. Similarly this has been done for Saturn, Uranus, and Neptune using their satellites and rings. For a spacecraft performing a fly-by or orbiting a planet, the tracking of the trajectory using the radio link permits to measure its velocity from the Doppler shift and deduce the gravitational field. This can be understood easily by using the energy conservation theorem

$$\frac{1}{2}v_s^2 = E + V(r,\varphi) \tag{A.6}$$

where

v_s is the spacecraft velocity
E its total energy (calculated at large distances from the planet)
$V(r,\varphi)$ the gravitational potential (1.37)

The classical Doppler frequency measured along the line of sight is given by

$$f' = f_0 \frac{1 - v_E/c}{1 - v_s/c} \tag{A.7}$$

which is v_E is the Earth's orbital velocity.

A.2   IN SITU ATMOSPHERIC MEASUREMENTS

In situ measurements are performed when the instruments onboard the spacecraft (a descending probe, floating balloons, and landers and rovers) enter in direct contact with the atmosphere. This is the most common situation in the Earth's atmosphere and has also been used in Venus, Mars, Jupiter, and Titan (see Table A.3).

A.2.1   Descending Probes

They perform local measurements of different atmospheric parameters along the vertical descent path. If the body has a surface the probe can land and deploy a battery of instruments for in situ ground measurements. Probes, some ending as landers, have been used on Venus (Veneras 4–14, Pioneer-Venus 4 probes, Vega 1–2), Mars (e.g., Viking 1–2, Pathfinder, Phoenix), Jupiter (Galileo), and the satellite Titan of Saturn (Huygens). The most common experiments onboard planetary probes are (1) atmospheric instruments that determine temperature, pressure, density, and molecular weight as a function of altitude. Typically the probe incorporates temperature and pressure sensors and accelerometers to measure the probe trajectory and the deceleration to determine the density as a function of altitude; (2) Mass spectrometer for chemical and isotopic composition as a function of altitude; (3) Nephelometer and aerosol collectors to detect and characterize aerosol and cloud particles (that measure cloud density through light reflection); (4) Radiometer (spectrophotometer) with optical–infrared channels to determine the ambient thermal and solar energy as a function of altitude; (5) Cameras to image the ambient clouds during descent and the surface if landing is possible; (6) Lightning sensor and radio antenna to search for lightning phenomena in the clouds, which can be complemented with electrical permittivity and electromagnetic wave analyzer for measurements of charged particles concentration, and microphones for detection of acoustical events. In addition, measurements of the Doppler shift of the radio signals sent by the probe to an orbital vehicle or to Earth are used to determine the wind speeds and turbulent velocities as a function of altitude.

A.2.2 BALLOONS

Balloons have so far only been employed in Venus by the Soviet missions Vega 1 and 2. In June 1985, they deployed two surface landers and meteorological balloons at near equatorial latitudes (7°). They were floating at altitudes 50–55 km and translated each one about half the parallel circle of Venus. They had onboard thermometers, a spinner-driven wind meter or "anemometer," a light-level meter, a pressure sensor, and a "nephelometer." Wind speed measurements from Doppler shift tracking were also performed.

Possible future applications of robot balloons ("aerobots") for the exploration of the atmospheres of Venus, Mars, and Titan are under study.

A.2.3 LANDERS AND ROVERS

They performed in situ measurements of the local meteorological properties close to the surface usually during large periods of time (as done for example in Mars by Viking 1 and 2 in 1976), or in areas nearby to the landing place using rovers (e.g., Mars Pathfinder, Sojourney, Spirit, and Opportunity). Successful landing has also been performed on Mars polar region (Phoenix) in different places and local times in Venus (seven Venera landers, numbers 8 through 14) and in Titan (Huygens probe) were the transmitted meteorological information for ~1 h.

The atmospheric instrumental package on landers include (1) meteorological sensors (pressure, temperature, wind speed and direction); (2) gas chromatographer and mass spectrometer; (3) cameras to measure sunlight diffusion and extinction in the atmosphere to determine aerosol properties over the lander skies, and image clouds and surface dust motions and frost deposition. A lidar was incorporated to the Martian Phoenix mission. In Mars, the success of the Viking and Pathfinder missions has permitted to record meteorological information on a highly sampled temporal interval (years for climate studies, daily and hourly).

A.3 LABORATORY STUDIES AND NUMERICAL MODELING

To support the research in planetary atmospheres, different studies and experiences are performed in the laboratory as well as with computer numerical modeling. We present an overview of the main topics typically treated.

A.3.1 FLUID DYNAMIC LABORATORIES

Fluid dynamic experiments are usually carried out in rotating cylindrical annular tanks to simulate the global atmospheric motions and the formation of turbulent patterns, waves, and vortices. The tank is filled with a liquid such as water (that mimics the atmosphere) and, placed on a turntable, is put in rotation with constant angular velocity to simulate planet rotation. Different experiences can be carried out with the tank: (1) Establishing a temperature difference between the inner and outer walls of the cylinder to simulate planet meridional temperature gradients, and generate a Hadley-like circulation or baroclinic instabilities. (2) Injecting (pumping) and extracting water into the tank through holes at its base in a number that can be controlled, or spraying with a rotating arm denser fluid (e.g., salty water) in the upper free surface of the tank. These experiments try to simulate upward and downward (convective) motions. (3) Changing the geometry of the bottom of the annulus surface, for example making it parabolic or just planar but tilted.

In the case of the giant fluid planets, to better simulate the geometry of deep convection, other experiments use rotating spherical shells with controlled thickness and rotation rate, being differentially heated in their outer and inner walls.

Different techniques are used to track the motions in the surface of the rotating liquid. Particles injected into the fluid become suspended and are made visible by illuminating them with a flat light or with a laser across the outer wall of the tank. Then their motions are tracked by short interval photographs using "streak cameras" mounted on the rotating apparatus. A method called correlation image velocimetry (CIV) is used in the analysis of the images providing good temporal and high spatial resolutions in the derived velocities.

Although the details of the flow properties differ from that of the atmosphere (for example the incompressible nature of the fluid and its viscosity), the theoretical models developed to explain these motions can be compared to the observations to test the similarities and differences. The basic intervening parameters are: the rotation rate, the fluid density, the impressed contrasts in temperature or density, the kinematic viscosity and the vertical depth and radial size of the annulus.

Different dynamical fluid laboratories work on these aspects. A major facility is the European Coriolis rotating platform located in Grenoble (France) that has a cylindrical annulus with an arm 13 m in radius. Fluid dynamical experiments with interest for planetary atmospheres have also been realized in the Skylab and in the

International Space Station in microgravity conditions by different astronaut teams that allows approaching the spherical geometry of planetary atmospheres.

A.3.2   CHEMISTRY

Theoretical and experimental work is performed on chemical species of interest for planetary atmospheres, fundamentally in relation with molecular and atomic spectroscopy. These are some of the research areas: (1) Theoretical calculations of spectroscopic parameters (absorption coefficients) of a variety of species (and interacting molecules) under the temperatures and pressures found in planets. (2) Laboratory measurements of the absorption coefficient for molecular bands as a function of pressure and temperature. Spectrometers like the intracavity laser spectrometer (ILS) are of current use in chemical laboratories. (3) Laboratory studies on stimulated desorption and surface ionization products on thin films in high vacuum to simulate the formation and conditions of tenuous planetary atmospheres; (4) Chemistry (reactions, rates and products) of different compounds.

A.3.3   AEROSOL PROPERTIES

Theoretical and experimental work is performed on particle optical properties and their ensemble scattering processes with interest for condensates and suspended dust in atmospheres. These are some research areas: (1) Laboratory spectroscopy (optical properties: imaginary and real refraction indexes) of pure ices or contaminated with chromophore agents of interest for the giant planets and Titan. (2) Laboratory measurements of light scattering by different types of particles (changing the size, nonspherical geometry and optical properties) and their comparison with theoretical models. (3) Laboratory studies of the formation and growth (adsorption, nucleation, condensation) of cloud droplets, ice crystals, organic particles, with instrumentation designed to reproduce the thermodynamic conditions in planetary atmospheres for laminar and turbulent flows.

A.3.4   COMPUTER SIMULATIONS: PREDICTIONS

Computer models that simulate the general circulation and the dynamical processes are a fundamental tool to understand the physical and chemical mechanisms operating in atmospheres. In the case of the Earth, complex models (at least for the troposphere and lower stratosphere) are developed and improved continuously since they are the base of the numerical weather prediction (NWP), the forecast of weather maps for few days to 2 weeks ahead, and to make predictions of the development and evolution of severe damaging storm systems (such as the tropical cyclones and deep extratropical cyclones). Predictability of local phenomena (not just their occurrence, but their intensity) is much more complex on the long term due to the sensitivity to the parameterizations and initial conditions. Among these are subgrid scale processes, inclusion of local conditions (e.g., orography), and the initial measurement errors in some variables, as well as the inevitable inhomogeneous spatial distribution of the data. The atmospheric system forms part of the variety of *chaotic processes* that occur in nature, being highly sensitive in their evolution to the initial conditions. Accordingly, infinitesimal differences in the values of the magnitudes at the starting points, can lead to very different evolutionary paths of the system. This defines the predictability horizon as the time required for substantial divergence in model prediction that occurs when starting from two close initial conditions. Long-term climatic forecast (for years to decades) also suffers from these drawbacks, being a good example the different predictions of the greenhouse effect evolution.

The NWP models are being used increasingly for data assimilation, a procedure by which the observed data are inserted into the model gradually, over discrete periods of time. By means of an iterative procedure analysis (the analysis correction scheme), the model adjusts the observations being modified to move in small increments (within the error bars) toward the observations.

Weather prediction models have been run for the case of the Martian atmosphere using GCM so far developed and data assimilation from different spacecraft missions. Depending on the season (in particular the presence of suspended dust), predictability can be done reasonably within 3–8 Martian sols. For the slowly rotating Venus and Titan, the gross aspect of the circulations can be simulated but the mechanisms that impulse the superrotations are not well known. However, in Venus, the stability of the circulation, the temperature and cloud global properties (for example) make "weather predictions" easier than on Earth. A meteorologist in Venus troposphere has simple daily forecast with permanent winds (vertical and meridional shears known), total cloud coverage, and known temperature profiles. On the contrary, for Titan, the present coarse circulation models do not yet allow to make predictions on cloud formation and precipitation rates, or how the liquid lakes are resupplied. For the giant planets, the origin of the winds is unknown, but as in Venus, long-term stability seems to occur at least on the best studied planet, Jupiter. Its wind jet system shows few changes (in space and time and in intensity), however we are still not able to predict when a convective storm will occur, or when a large vortex will form. Obviously we still lack data to understand the physics operating below upper clouds. But the force of numerical modeling is large and the first GCM models have already been developed for giant extrasolar planets, although we do not yet see these planets. These models are able to make predictions for example on heat transport and winds (from the illuminated to nonilluminated hemispheres), predictions that can be tested using as inputs the first available measurements of temperature contrasts on these planets.

A.4 FUTURE OBSERVING FACILITIES

A number of large projects directly involved in the observation and study of the planets or indirectly, employing huge telescopes of general purpose with advanced instrumentation are under development. The closest in time projects are listed in Table A.5 with details on their status available on Internet. The research of planetary atmospheres will continue to progress from these explorations and since the field is highly multidisciplinary, from other planned investigations not indicated in the list. Evidently, the growth in the knowledge of planetary atmospheres is highly marked

TABLE A.5
Facilities and Missions under Development to Investigate the Planets

1. Ground-Based Optical–IR Telescopes
   Interferometer configurations combining telescopes: VLT Interferometer (VLTI), Keck
   Interferometer, LBT Interferometer (LBTI).
   Extremely Large Telescopes (ELT): Aperture ~ 30 m (various under consideration).
   OverWhelming Large Telescope (OWL): Aperture ~ 100 m.
2. Radio–mm Telescopes
   Atacama Large Millimeter Array (ALMA) in Chile: 64×12 m dishes ($\lambda = 0.3$–10 mm).
   Maximum baseline of 10 km.
   CARMA: interferometer combination for OVRO and BIMA.
3. Space Telescopes
   Herschel Space Observatory (ESA). Infrared telescope (3.5 m aperture) operating between
   $\lambda = 60$–670 μm. Orbit about the second Lagrangian point (L2) of the Earth-Sun system.
   Successful launch on May 14, 2009.
   James Webb Space Telescope (JWST, NASA/ESA): Optical-IR telescope (6 m aperture) operating
   between $\lambda = 0.6$–28 μm. Orbit about the second Lagrangian point (L2) of the Earth-Sun system.
   Kepler (NASA/ESA): Space-borne telescope (0.95 m aperture) designed to search for transits of
   extrasolar planets by the photometric technique. Successful launch on March 6, 2009.
   Gaia (ESA): A dual telescope concept (1.45×0.5 m each) with 106s CCD. Detect and
   characterize exoplanets by astrometry and transits (lauch in 2011). Orbit about the second
   Lagrangian point (L2) of the Earth-Sun system.
   A list of projects for the search of extrasolar planets is given at: http://exoplanet.eu/searches.php
4. Planetary Missions
   Venus Climate Orbiter (Japan): May 2010. Venus orbiter mission.
   Bepi-Colombo (ESA): 2013. Mercury two orbiters (MMO and MPO) mission.
   Mars Science Laboratory (USA): 2011. Mars Lander and Rover.
   Juno (USA): August 2011. Jupiter polar orbiter.
   ExoMars (ESA): January 2013. Orbiter and Lander plus a Rover.

by the advances produced in the knowledge of our own atmosphere (fundamentally in the theoretical area). But at the same time, the comparative view of the atmospheres with a variety of compositions and physical properties, and their response to the variable boundary conditions (energy, rotation rate, etc.) that we find in the solar system and in the increasingly "planetodiversity" in other systems, will significantly improve the knowledge of Earth's atmosphere.

Bibliography

List of the books and textbooks used as references or consulted in the elaboration of this textbook.

Planets (University of Arizona Press Space Science Series)

S. K. Atreya, J. B. Pollack, and M. S. Matthews (eds.), *Origin and Evolution of Planetary and Satellite Atmospheres*, The Arizona University Press, Tucson, AZ, 1989.

J. T. Bergstralh, E. D. Miner, and M. S. Matthews (eds.), *Uranus*, The Arizona University Press, Tucson, AZ, 1991.

S. W. Bougher, D. M. Hunten, and R. J. Phillips (eds.), *Venus II*, The Arizona University Press, Tucson, AZ, 1997.

J. A. Burns (ed.), *Planetary Satellites*, The Arizona University Press, Tucson, AZ, 1977.

D. P. Cruikshank (ed.), *Neptune*, The Arizona University Press, Tucson, AZ, 1995.

T. Gehrels (ed.), *Jupiter*, The Arizona University Press, Tucson, AZ, 1976.

T. Gehrels and M. S. Matthews (eds.), *Saturn*, The Arizona University Press, Tucson, AZ, 1984.

D. M. Hunten, L. Colin, T. M. Donahue, and V. I. Moroz (eds.), *Venus*, The Arizona University Press, Tucson, AZ, 1983.

H. H. Kieffer, B. M. Jakosky, C. W. Snyder, and M. S. Matthews (eds.), *Mars*, The Arizona University Press, Tucson, AZ, 1992.

D. Morrison (ed.), *Satellites of Jupiter*, The Arizona University Press, Tucson, AZ, 1982.

F. Vilas, C. R. Chapman, and M. S. Matthews (eds.), *Mercury*, The Arizona University Press, Tucson, AZ, 1988.

Planets and the Solar System

F. Bagenal, T. Dowling, and W. McKinnon (eds.), *Jupiter: The Planet, Satellites and Magnetosphere*, Cambridge University Press, Cambridge, U.K., 2004.

B. Bertotti, P. Farinella, and D. Vokrouhlicky, *Physics of the Solar System, Astrophysics and Space Science Library*, Vol. 293, Kluwer Academic Publishers, New York, 2003.

R. H. Brown, J. P. Lebreton, and J. Hunter Waite (eds.), *Titan from Cassini-Huygens*, Springer, New York, 2009.

J. W. Chamberlain and D. M. Hunten, *Theory of Planetary Atmospheres*, Academic Press, New York, 1987.

G. H. A. Cole and M. M. Woolfson, *Planetary Science*, Institute of Physics, Bristol, U.K., 2002.

I. de Pater and J. J. Lissauer, *Planetary Sciences*, Cambridge University Press, Cambridge, U.K., 2001.

M. Dougherty, L. Esposito, and S. M. Krimigis (eds.), *Saturn from Cassini-Huygens*, Springer, Dordrecht, the Netherlands, 2009.

T. Encrenaz, J. P. Bibring, M. Blanc, M. A. Barucci, F. Roques, and Ph. Zarka, *The Solar System*, 3rd edn., Springer-Verlag, Heidelberg, Germany, 2004.

G. Faure and T. M. Mensing, *Introduction to Planetary Science*, Springer, New York, 2007.

R. A. Hanel, B. J. Conrath, D. E. Jennings, and R. E. Samuelson, *Exploration of the Solar System by Infrared Remote Sensing*, Cambridge University Press, Cambridge, U.K., 2003.

P. G. J. Irwin, *Giant Planets of Our Solar System: An Introduction*, Springer, Berlin Heidelberg, Germany; Praxis, Chichester, U.K., 2009.

J. Kelly Beatty, C. Collins Petersen, and A. Chaikin, *The New Solar System*, The Sky Publishing Corporation, Cambridge University Press, Cambridge, U.K., 1999.

J. S. Lewis, *Physics and Chemistry of the Solar System*, Academic Press, San Diego, CA, 1997.

J. R. Lewis and R. G. Prinn, *Planets and Their Atmospheres: Origins and Evolution*, Elsevier, Amsterdam, the Netherlands, 1983.

N. McBride and I. Gilmour, *An Introduction to the Solar System*, The Open University, Cambridge University Press, Cambridge, U.K., 2003.

P. L. Read and S. R. Lewis, *The Martian Climate Revisited: Atmosphere and Environment of a Desert Planet*, Springer, Berlin Heidelberg, Germany; Praxis, Chichester, U.K., 2004.

J. H. Rogers, *The Giant Planet Jupiter*, Cambridge University Press, New York, 1995.

J. H. Shirley and R. W. Fairbridge, *Encyclopedia of Planetary Sciences*, Chapman & Hall, London, U.K., 1997.

P. R. Weissman, L. A. McFadden, and T. V. Johnson, *Encyclopedia of the Solar System*, Academic Press, San Diego, CA, 2006.

Earth Atmosphere (General)

D. G. Andrews, *An Introduction to Atmospheric Physics*, Cambridge University Press, Cambridge, U.K., 2000.

J. Holton, J. Pyle, and J. Curry (eds.), *Encyclopedia of Atmospheric Science*, Vols. 1–6, Academic Press and Elsevier, London, U.K., 2002.

J. Houghton, *The Physics of Atmospheres*, Cambridge University Press, Cambridge, U.K., 2002.

M. L. Salby, *Fundamentals of Atmospheric Physics*, Elsevier, San Diego, CA, 1996.

J. M. Wallace and P. V. Hobbs, *Atmospheric Science: An Introductory Survey*, Academic Press, San Diego, CA, 2006.

Upper Atmosphere

S. J. Bauer and H. Lammer, *Planetary Aeronomy: Atmosphere Environments in Planetary Systems*, Springer, Berlin, Germany, 2004.

J. K. Hargreaves, *The Solar-Terrestrial Environment: An Introduction to Geospace—The Science of the Terrestrial Upper Atmosphere, Ionosphere, and Magnetosphere*, Cambridge University Press, Cambridge, U.K., 1992.

M. G. Kivelson and C. T. Rusell, *Introduction to Space Physics*, Cambridge University Press, Cambridge, U.K., 1995.

M. H. Rees, *Physics and Chemistry of the Upper Atmosphere*, Cambridge University Press, New York, 1989.

Atmospheric Dynamics and Meteorology

D. G. Andrews, C. B. Leovy, and J. R. Holton, *Middle Atmospheric Dynamics*, Academic Press, San Diego, CA, 1987.

W. R. Cotton and R. A. Anthes, *Storm and Cloud Dynamics*, Academic Press, San Diego, CA, 1989.

J. A. Dutton, *The Ceaseless Wind: An Introduction to the Theory of Atmospheric Motion*, Dover Publications, Mineola, NY, 1986.

A. E. Gill, *Atmosphere-Ocean Dynamics*, Academic Press, San Diego, CA, 1982.

J. R. Holton, *An Introduction to Dynamical Meteorology*, Elsevier, San Diego, CA, 2004.

M. Z. Jacobson, *Fundamentals of Atmospheric Modelling*, Cambridge University Press, Cambridge, U.K., 2005.

E. Kessler, *Thunderstorms Morphology and Dynamics*, University of Oklahoma Press, Norman, OK, 1992.

R. A. Lindzen, *Dynamics in Atmospheric Physics*, Cambridge University Press, Cambridge, U.K., 1990.

J. Pedlosky, *Geophysical Fluid Dynamics*, Springer-Verlag, New York, 1987.

J. P. Peixoto and A. H. Oort, *Physics of Climate*, Springer, New York, 2007.

R. B. Stull, *An Introduction to Boundary Layer Meteorology*, Springer, Dordrecht, the Netherlands, 1987.

G. K. Vallis, *Atmospheric and Oceanic Fluid Dynamics*, Cambridge University Press, Cambridge, U.K., 2006.

Atmospheric Radiation and Clouds

C. F. Bohren and E. E. Clothiaux, *Fundamentals of Atmospheric Radiation*, John Wiley & Sons, New York, 2006.

S. Chandrasekhar, *Radiative Transfer*, Dover, New York, 1960.

R. M. Goody and Y. L. Yung, *Atmospheric Radiation: Theoretical Basis*, Oxford University Press, New York, 1995.

R. A. Houze, *Cloud Dynamics*, Academic Press, San Diego, CA, 1993.

K. N. Liou, *Radiation and Cloud Processes in the Atmosphere: Theory, Observation and Modeling*, Oxford University Press, New York, 1992.

K. N. Liou, *An Introduction to Atmospheric Radiation*, Elsevier, New York, 2002.

G. E. Thomas, K. Stammes, and A. J. Dessler, *Radiative Transfer in the Atmosphere and Ocean*, Cambridge University Press, Cambridge, U.K., 2002.

H. C. van de Hulst, *Light Scattering by Small Particles*, Dover, New York, 1981.

M. K. You and R. R. Rogers, *Short Course in Cloud Physics*, Elsevier, New York, 1989.

W. Zdunkowski, T. Trautmann, and A. Bott, *Radiation in the Atmosphere: A Course in Theoretical Meteorology*, Cambridge University Press, Cambridge, U.K., 2007.

Fluids Mechanics

P. K. Kundu and I. M. Cohen, *Fluid Mechanics*, Academic Press, New York, 2008.

L. D. Landau and E. M. Lifshotz, *Fluid Mechanics*, Vol. 6, Butterworth-Heinemann, Oxford, U.K., 1987.

Index